Brute Souls,
Happy Beasts,
and Evolution

Brute Souls,
Happy Beasts,
and Evolution:

The Historical Status
of Animals

Rod Preece

UBC Press • Vancouver • Toronto

15 14 13 12 11 10 09 08 07 06 05   5 4 3 2 1

Printed in Canada on acid-free paper that is 100% post-consumer recycled, processed chlorine-free, and printed with vegetable-based, low-VOC inks.

Library and Archives Canada Cataloguing in Publication

Preece, Rod, 1939-
   Brute souls, happy beasts, and evolution : the historical status of animals / Rod Preece.

   Includes bibliographical references and index.
   ISBN-13: 978-0-7748-1156-9 (bound) ; 978-0-7748-1157-6 (pbk.)
   ISBN-10: 0-7748-1156-0 (bound) ; 0-7748-1157-9 (pbk.)

   1. Animal welfare – History. 2. Human-animal relationships. 3. Animal welfare – Moral and ethical aspects – History. 4. Animal welfare – Philosophy – History. I. Title.

QL85 P732005   179' .3   C2005-903169-7

Canadä

UBC Press gratefully acknowledges the financial support for our publishing program of the Government of Canada through the Book Publishing Industry Development Program (BPIDP), and of the Canada Council for the Arts, and the British Columbia Arts Council.

This book has been published with the help of a grant from the Canadian Federation for the Humanities and Social Sciences, through the Aid to Scholarly Publications Programme, using funds provided by the Social Sciences and Humanities Research Council of Canada.

Printed and bound in Canada by Friesens
Set in Adobe Garamond by Brenda and Neil West, BN Typographics West
Copy editor: Robert Lewis
Proofreader: Deborah Kerr

UBC Press
The University of British Columbia
2029 West Mall
Vancouver, BC V6T 1Z2
604-822-5959 / Fax: 604-822-6083
www.ubcpress.ca

For my Mother and Father,
who made everything possible.

Pythagoras and Empedocles ... say that we have a fellowship not only with one another and the gods but also with the irrational animals. For there is a single spirit which pervades the whole world as a sort of soul and which unites us with them. That is why, if we kill them and eat their flesh, we commit injustice and impiety, inasmuch as we are killing our kin. Hence these philosophers urged us to abstain from meat.

– Sextus Empiricus, *Against the Mathematicians* (c. third century)

... the thing that moved me most, and changed my way of thinking, was that Reason ruled and cared for the beasts, except only for man and his mate; for many a time they wandered ungoverned by Reason.

– William Langland, *Piers the Ploughman* (c. 1360)

The pig does not appear to have been formed upon an original, special, and perfect plan, since it is a compound of other animals; it has evidently useless parts, or rather parts of which it cannot make any use – toes, all the bones of which are perfectly formed, and which, nevertheless, are of no service to it. Nature is far from subjecting herself to final causes in the formation of her creatures.

– Comte de Buffon, *Histoire naturelle,* vol. 5 (1755)

The deepest minds of all ages have had pity for animals because they suffer from life and have not the power to turn the sting of their suffering against themselves, and understand their being metaphysically.

– Friedrich Nietzsche, *Thoughts Out of Season* (1873)

We need another and a wiser and perhaps a more mystical concept of animals ... they are not brethren, they are not underlings; they are other nations, caught with ourselves in the net of life and time, fellow prisoners of the splendor and travail of the earth ... the animal shall not be measured by man.

– Henry Beston, *The Outermost House* (1928)

Contents

Preface

I apologize at the very outset for using in this book certain arguments, information, and quotations that I have previously used elsewhere. I have done so where I found it apposite in the interest of explaining myself as well as I am able to my readers and as an introduction to evidence, argument, and theses that I have not previously developed. It would be both arrogant and entirely unwarranted to imagine that all readers will have read my previous books and, even if a few have, that they will have remembered all that I wrote therein. The primary respect in which I have used material that I have employed before is in reciting a number of the quotations presented in my *Awe for the Tiger, Love for the Lamb: A Chronicle of Sensibility to Animals* (2002). By and large, the passages selected for that book were allowed to stand on their own, the commentary being restricted mainly to giving the passages a historical context. Here some of them are interwoven into the story to show their implications for understanding the human-animal relationship. Indeed, I have used that book as I primarily intended it to be used: as a source book for scholars investigating particular issues in the development of the history of ideas as it pertains to animal ethics.

I have also adapted some of the material on evolutionary ideas in my Introduction to the Mellen Animal Rights Library edition (2003) of William Youatt's *The Obligation and Extent of Humanity to Brutes* (1839) and in my article on "Darwinism, Christianity and the Great Vivisection Debate" in the *Journal of the History of Ideas* (2003) for a new context. The reader will also find that some of the material on examples of Christian belief in animal immortality has been developed from, but is far more extensive than, the evidence offered in Rod Preece and David Fraser, "The Status of Animals in Biblical and Christian Thought: A Study in Colliding

Values" (2000). One or two of my comments on primitivism are borrowed and reoriented from my Introduction (1999) to George Nicholson's
On the Primeval Diet of Man (1801). Finally, I have reprised some of the
examples of Aboriginal and Oriental cultural attitudes that I employed in
Animals and Nature: Cultural Myths, Cultural Realities (1999). The bor-
rowings are, in fact, considerably less than this listing might intimate, but
they should nonetheless be mentioned. Even those who may have read
and can recall my previous work will find the substance of this book quite
different from anything of mine they may have read before.

   One further apology is due, concerning the reiteration of gender exclusive language, a problem common to all historical work since gender in-
clusivity did not become frequent until the 1990s, even in books written
by women. It is instructive today to recall that insightful and sensitive
feminist writers, Mary Wollstonecraft in the late eighteenth century, for
example, would continue to employ "man" as the abstraction for people
of both genders and "woman" as the abstraction for females alone, an
inconsistency fraught with problems. Traditional gender inconsistencies
and sexism create certain reporting difficulties for the historian of ideas,
which I have attempted to overcome where the context permitted, but I
fear that on occasion my doing so has produced a certain clumsiness. On
other occasions, in order not to be unduly pedantic, I have used "man"
and "mankind" where the historical context would have made any other
usage appear contrived. I will illustrate the problem by quoting from the
gender-induced plight of the British magazine *Punch* in 1893: "Nearly all
our best men are dead! Carlyle, Tennyson, Browning, George Eliot! – I'm
not feeling very well myself."

Acknowledgments

Above all, I must record my indebtedness to the gargantuan endeavours of George Boas. Many of my classical sources in the second part of Chapter 4 are derived from the chapter on "The Superiority of the Animals" in Arthur O. Lovejoy and George Boas, *Primitivism and Related Ideas in Antiquity* (1935) – which I assume to be primarily the work of Boas. In Chapter 5 I have relied for most of my French and Italian Renaissance sources on George Boas's *The Happy Beast in French Thought of the Seventeenth Century* (1933). While my commentary and interpretation of the evidence differ substantially from that of Boas, especially on the competing emphases in Montaigne's work – I do not find Montaigne's theriophily a great deal more restrained than that of those he followed – and while I have used much material ignored by Boas, especially outside France, my analysis would not have been possible without Boas's scholarly unearthing of the most important classical and French material or without his translation of the Greek sources into English, of which I have on a few occasions availed myself unashamedly. In *The Happy Beast* Boas was a great deal less generous, for there his quotations, which are many and sometimes long, are left in the original Latin, Renaissance French, even Italian. It is for this reason, I believe, that *The Happy Beast* has been alluded to frequently in recent literature yet almost never quoted. More important is that, despite the allusions, the lessons of Boas's studies are almost entirely ignored. I have, consequently, spent a portion of Chapters 4 and 5 of this book elucidating the findings of Boas, and, at least on occasion, repeating his message, and sometimes extending it, partly by providing an English translation of some of the quotations he left in the original. Of course, I do not wish to imply that the conclusions I draw are essentially the same as those of Boas – indeed, were he alive, I am sure he would deny it with a degree of enthusiasm – but, without the previous research undertaken by Boas, I would not have possessed the evidence to draw some of the conclusions to which I have been led.

The work of Hester Hastings has also been of considerable value. Perhaps because she was writing in 1936 and much of her work on the Enlightenment has been superseded by subsequent scholarship, I find myself less indebted to her than I am to Boas; even though his work came earlier, less frequently was his area the subject of later scholarly reinterpretation. Nonetheless, she too has provided an abundance of original material. Yet she is not as frequently cited in the literature as she might be. This is perhaps because, like Boas in *The Happy Beast,* she too has presumed linguistic accomplishments among her readers that, unfortunately, are increasingly rare. She has left almost all her quotations in the original French. Again, I have availed myself of them occasionally and provided an English translation.

I am also obligated to my Wilfrid Laurier University colleagues Steve Brown and Ute Lischke for offering sound advice, and generous encouragement, on the Introduction and Chapter 7 respectively. Bernard Rollin, University Distinguished Professor of Philosophy at Colorado State University, very kindly sent me an early version of a conference paper he was preparing, entitled "Reasonable Partiality and Animal Ethics," which proved very useful in helping me to clarify and hone the draft I had already developed for my final chapter. Prior to reading his paper I had failed to recognize the degree to which the work of Aristotle and Hume was relevant to the issue of "reasonable partiality." Moreover, the ideas expressed in that chapter with regard to the relationship between species needs and animal ethics correspond fairly closely with the conclusions of the discussions I have had over the years with David Fraser, National Sciences and Engineering Research Council Research Professor in Animal Welfare at the University of British Columbia. And I fear that any originality may come more from David's mind than from my own. Certainly, some of the animal-welfare science evidence I offer in that chapter had been first brought to my attention by Professor Fraser. The evidence provided in Chapter 8 was bolstered by the ever helpful editor of *Veterinary History,* John Clewlow, who sent me a copy of an important, but to me unknown, article from a 1989 issue of the journal. Likewise, my valued associate and coauthor Chien-hui Li, Fellow of Wolfson College, Cambridge University, discovered, and sent to me, the 1839/40 articles in *The Veterinarian* by Karkeek, Spooner, and Manthorp concerning what Karkeek called "The Future Existence of the Brute Creation." The cooperation of colleagues and friends lightens the burdens and magnifies the joys of scholarship.

It is a pleasure to record my gratitude to Professors Angus Taylor and Bernard Rollin for their appraisals of the manuscript. Their erudition and goodwill were extremely valuable in the evaluation process, and they agreed to the release of their names so that I could make appropriate recognition here of their invaluable comments, which both persuaded the publishers that the manuscript was worthy of publication and helped me to refine some of the points I make. Much as I have benefitted from their insights, any inadequacies that remain are entirely my own.

I need also to thank my editor, Randy Schmidt. His advice to me has ever been wise, sound, and considerate. He makes the task of authorship a much more bearable weight. Finally, as always, my indebtedness to my wife, Lorna Chamberlain, is enormous. On this occasion, the debt owed is even more than customary, for without her endeavours in a time of medical crisis, I doubt I would have been here to complete the task.

Brute Souls,
Happy Beasts,
and Evolution

Introduction

I

SOME MAY READ THE TITLE OF THIS book as disparaging of our fellow creatures since "brute" derives from the Latin "brutus," meaning dull or threatening, and has come to mean coarse and cruel as well as devoid of reason and refinement; and "beast" has even been used to identify the Antichrist. However, the "brutus" I have uppermost in mind is Lucius Junius Brutus (fl. 510 BC), the founder of the Roman republic, who *feigned* stupidity, but who *in fact* possessed greater abilities and wisdom than were recognized by his compatriots, as indeed do animals far exceed the abilities many ascribe to them. Nonetheless, if the insulting interpretation must prevail, the reader is invited to think of the Cartesian brutes who refused to acknowledge the sentient souls of our fellow creatures.

"Beast" is derived from the Latin "bestia" and from the Old French "beste," where it possessed in its early usage no negative connotation. By the seventeenth century the usage had changed sufficiently that one could find the descriptive and derogatory senses in a single sentence – as with Molière's cogent quip in the prologue to *Amphitryon* (1688): "Les bêtes ne sont pas si bêtes que l'on pense" (The beasts aren't as beastly as one may think). Such was also the case in English, where the term, although essentially a synonym for "animal," was reserved for those animals of a more specific nature, the category often, but by no means always, denoting wild or feral animals. Only later did it acquire the pejorative sense. And even today it is occasionally used without any attitudinal implications. Of course, the historical reality is that animals were often termed "brutes" and "beasts," and my use of "brutes" and "beasts" reflects the fact that I am concerned with the historical understanding of the status of animals. Even the term "brute" was not necessarily disparaging. For example, Mary Shelley, whose daemon in *Frankenstein* (1818) was an animal-respecting, compassionate

1

vegetarian, still referred in *The Last Man* (1826) to "the charity girl's" play-fully boisterous and protective "large Newfoundland dog," duly named "Lion," as the girl's "brute companion."[1] The term was, in this instance, one of endearment. Moreover, in 1903, in repeating a newspaper story of a sensitive mastiff, the proponent of animal immortality Elijah Buckner reported him as "that noble brute."[2] The term was, in this instance, one of admiration. Easy generalizations about apparently pejorative terms are not always warranted.

II

This book traces historical development relative to four customary claims that, in my view, show the conclusions of current scholarship to be, by and large, seriously impaired: (1) that *the* Christian doctrine, typically presented as an unchanging monolith, has denied immortality to animals, with the corresponding implication that they were thereby denied ethical consider-ation; (2) that there was a near universal belief that animals were intended for human use, with the corresponding implication that they were not ends in themselves and thus not entitled to ethical consideration; (3) that Charles Darwin's theory of evolution had a profoundly positive impact on the way in which nonhuman animals were regarded and treated; and (4) that the idea of the "happy beast" was merely a trope to condemn humans for their hubris and was not at all a sincere attempt to raise the status of animals.

The theme of this study is that there is no orthodoxy in the history of animal ethics. These claims engender far from simple, one-sided issues. It is here argued that many Christian thinkers did ascribe immortal souls to animals; that many who did not do so nevertheless acknowledged our obli-gations toward them; that some who did not do so were of the view that an animal's lack of an immortal soul required a greater obligation on our part to give it due moral consideration during its lifetime; that, paradoxi-cally, the notion of animal "irrationality" did not deny animals the capacity for reason; that denigrations of nature – by, for example, Francis Bacon, Leonardo da Vinci, and Thomas Hardy – were often not at the expense of animals but born out of exasperation at the tribulations animals endure; that the Cartesian conception of animals as insentient machines was very largely rejected, both in the seventeenth century and later; that many of those who regarded animals as intended for human use were also of the opinion that they were entitled to considerate treatment; and that many gifted scholars, such as Henry More, John Ray, Leibniz, and Goethe, even Descartes, questioned the view that animals were so intended, arguing explicitly in some instances that they were, at least in part, ends in themselves. It is here argued further that the almost universal belief that Charles Darwin's discovery of evolution by natural selection brought about a metamorphosis in our attitudes to animals is entirely untenable; it did not produce any significant change in attitudes, at least none that was more favourable to animal interests.

"Theriophily" will also receive considerable attention. Unlike the other three claims, expressions of theriophily are rarely treated at all seriously in the literature of recent decades. George Boas coined the term in the first half of the twentieth century, combining the Greek words *therion* (beast; wild or feral animal) with *philos* (loving; dear).[3] Theriophily – "admiration for animals" might be the most appropriate rendering, although "animal superiority" is sometimes closer – is associated by Boas with the idea of "the happy beast," the notion that the life of a nonhuman animal is in some manner preferable to, more fulfilling than, that of a human being, an idea notoriously expressed by Gryllus ("grunter") in Plutarch's essay "On the Use of Reason by 'Irrational' Animals."[4] Animals, so Gryllus opines, possess "morality, intelligence, courage and all the other virtues," and the faculty of reason they possess "is both practically better and more impressive than human intelligence." Among the classics, we can find similar arguments for animal superiority in the writings of Democritus, Ovid, and Pliny. It is customarily assumed that those who engage in such attitudinizing are not being serious. It is often suggested that such authors as Plutarch, Montaigne, and Byron engage in this form of argument less to elevate animals than to oust arrogant humans from their usurped towers. And no doubt there is some truth in the claim – and much justification in the cause. Nonetheless, there is often also an underlying notion that there are respects in which animal life is really better. Animals are thought to fulfill their needs more readily; they possess more untrammelled honesty and compassion; and they are less subject to suffering from unfulfilled expectations.

In his mid-eighteenth-century *Discourse on Inequality*, Rousseau tells us "we observe every day the repugnance of horses to tread a living body under foot, and an animal does not, without some uneasiness, pass close by a dead member of its own species ... the sorrowful lowing of cattle entering a slaughterhouse bespeaks the impression of the horrible spectacle which confronts them."[5] In Rousseau's view, "compassion is a natural sentiment"[6] among both humans and animals, but as humans become civilized and formally educated they lose their natural attributes. "I venture to affirm that the state of reflection is contrary to nature and that the man who meditates is a depraved animal."[7] The benevolent passions become the exclusive prerogative of primitive humans and nonhuman animals:

By the activity of the passions, our reason is perfected; we seek knowledge only because we desire enjoyment, and it is impossible to conceive why a person who has neither desires nor fears would take the trouble to reason. The passions, in their turn, find their origin in our needs, and owe their progress to our knowledge, for things can be desired or feared only on the basis of ideas that we can form about them or through the simple impulsion of nature, and savage man, bereft of every sort of enlightenment, experiences only passions of this last kind; his desires do not exceed his physical needs. The only goods he knows in the universe are food, a female, and sleep; the only evils he fears are pain and hunger; I say pain, and not death, for an animal will never know what it is to die, and knowledge of death and its terrors is one of the first acquisitions that man made in leaving the animal state.[8]

To be sure, for Rousseau, humankind possesses the capacity for self-perfection and choice, which he denies to animals. Nonetheless, Rousseau believes that reason is valuable, in both humans and animals, only to the extent that it is instrumental to the fulfillment of the ends determined by the passions and the needs determined by the species' constitution. Animals and "savages" possess such reason. Civilized humans have long lost it, replacing it with the self-alienating activity of intellectual reflection. The reflections encouraged by the refinements and luxuries of Enlightenment civilization produce a far greater likelihood of self-estrangement than of satisfaction. And nonhuman animals do not experience such estrangement from their natural selves. For Rousseau, animals, like primitive humans, live at least in some respects the superior life. Indeed, having no knowledge of death and hence no fear of it, unlike even the most primitive of humans, animals live in at least this one respect a better life. The idea of a Golden Age in which humans were able to achieve their ends more fruitfully by living lives that conformed more closely to those of the animals has played a significant role in human thought. Rousseau is, as we shall see later, no rare exception.

III

A premise of this book is that, despite some works of great merit, much recent writing on the development of the status of animals is seriously misleading. We have created a discipline whose intellectual integrity has often been subordinated to politically correct goals concerning the value of animals. I share many of these goals, but consider it neither ethical nor scholarly to allow such goals to drive research or determine conclusions. Many scholars set out to prove the importance of animal wellbeing – in my view, a matter beyond legitimate dispute – and then pretend that they are in the vanguard of thought in suggesting such a novel proposition, when, in fact, the selfsame view has been proclaimed throughout human history as one side of a continuous debate about the relative status of humans and animals.

Some of the more influential studies in the discipline have been written in a spirit of ideological antipathy to Western norms – without taking more than a momentary glance at their historical reality, and a prejudicial glance at that. There is in play a certain elitist arrogance in putting down the past in order to raise oneself above it. And it involves a human hubris that sees an author, in seeking to raise the animals, at least in equal part demeaning our forbears and proclaiming in preening glory the abundant wisdom of the new authorities above and beyond history's traditional errors. The grating self-righteousness of those who take pride in sneering at those they regard as their moral inferiors of the past should encourage us to recognize the value of a healthy dose of humility and a closer inspection of the details of history.

At the turn of the nineteenth century, the proto-anarchist William Godwin noted that cruelty to animals by "gentlemen" increased proportionately in opposition to the more compassionate views of dissenters. Thus, during the reign of Charles II, when animal sports were opposed adamantly by the Puritans (and their public display was generally outlawed during the Commonwealth), conspicuous delight was taken by the royalists both in bear- and bull-baiting and cockfighting; likewise by the French nobility and their associates in fox-hunting at the time of the French Revolution.[9] The intolerant certainty of the Puritans and Revolutionaries persuaded those who remained unconvinced of the truth of these ethical novelties to be even more vigorous in their adversity. An ethical vegetarian such as Godwin – who opposed what he called in *Fleetwood* "the sports of the field" as much as the baiting of animals – found it difficult to determine whether to be more appalled at the animal cruelty of the squirearchy or at the sectarian prudery of their fanatical enemies. At least sometimes, judicious contemplation is the intellectually and politically wiser as well as the more virtuous course. Moral outrage may be less effective than intelligent compromise. And historical analysis written in the spirit of compromise may uncover more truth than that written with a sense of moral outrage, however justifiable that outrage may be on occasion.

What we sometimes get in animal-rights literature is predicated more on vitriol than careful reasoning; it is more polemic than analysis, more advocacy than scholarship. The less than surprising consequence is that the authors have managed to paint in broad strokes an undeviating history of venomous tyranny. Much of the rhetoric serves as self-indulgent bombast. If there is a particularly merciless individual, brutal event, or less than respectful cultural practice, it is treated as the norm, the cultural standard, rather than in the context of a multiplicity of competing values and characteristics. If there is more than one plausible interpretation of a particular phenomenon, the analyst will start from the assumption that the least favourable is the most probable. Instead, the task should be to strike a balance that corresponds with historical reality. For example, Dr. Samuel Parr of Hatton, an Anglican parson regarded in the late eighteenth century as one of England's preeminent philosophers – now almost entirely, and justifiably, forgotten – was subject to fits of rage. He relieved his aggressive feelings by personally butchering the cattle at the slaughterhouse with an axe. If we are outraged at his behaviour – and it is nigh impossible not to be so – we would provide an unbalanced perspective of his age, his church, and his compatriots if we related his story without also noting that his actions contravened socially acceptable norms, that his bishop required that he cease forthwith, and that he was subject to general calumniation, specifically and loudly by the members of the Godwin circle.[10] The telling of such stories usually emphasizes the atrocity and omits the wider context without which the atrocity cannot be fully understood.

Writers' ideological premises will be found on occasion to inform their conclusions unduly. Indeed, the author will have set out to prove a case rather than to investigate a hypothesis. As Charles Kingsley explained clearly in his Chartist novel *Yeast* (1851): "The prejudice and feelings of their heart give them some idea or theory, and then they find facts at their leisure to prove their theory true."[11] The research is conclusion driven. The wish is father to the thought. In many instances, the books are written to pursue a political end – to be sure, the laudable end of promoting the consideration of animal interests – instead of to further understanding; and the data are selected and interpreted in a manner consistent with this purpose. There would appear to be a greater striving for originality and impact than for historical accuracy. And what good is originality merely for the sake of nonconformity? There would appear to be a penchant for criticism for the sake of being a critic. It is difficult to be confident of the context of the history of animal ethics if what is written about it is unduly influenced by the writer's predilections rather than being the product of impartial investigation. Unfortunately, even a number of studies devoted to discovering a truth, rather than to propounding a prejudice, have been influenced by the damning tones of those who have set the malevolent stage. Even the admirable Mary Midgley has succumbed. She states that "It became the tradition ... as a result of a deliberate and sustained campaign by Christian thinkers, using some very strange material gathered from Rationalist philosophers, to crush a natural respect for animals, and for nature generally, which they saw as superstition."[12] This is pervasively accepted. Yet it is less than half the truth, as we shall see. The result is that one can have very little confidence that the conclusions one encounters in a number of the relevant books ever found an evidential match in the real world. It is my hope that this book will add a different, and convincing, dimension to those customarily encountered.

The disconcerting reality is that the academic field of animals and civilization is strewn with holier than history attitudes whereby all was dark and dismal until our modern-day heroes strode onto the field of honour and announced for the first time the rights of animals and our obligatory respect for them. To be sure, the authors offer an occasional isolated historical knight in shining armour – Jeremy Bentham as often as not, occasionally a Shelley, sometimes Mary, sometimes Percy, frequently a Charles Darwin – who are depicted larger than life in stirring contrast to their contemporaries and forerunners, when in reality they are often little different from them and sometimes rather less dedicated to the cause of animal respect than others of their time. Proclaiming heroes and villains becomes the quest rather than understanding the complexities of past realities.

The evidence suggests that in many instances – although, of course, by no means all – modern standards of scholarship lag behind those of a half-century, a century, or even longer ago. There is nothing of late on the history of vegetarianism to match the work of Howard Williams's *The Ethics of Diet* (1883)[13] – other than a fine study restricted almost entirely to the classical Greek tradition. And only one book on pre-nineteenth-century British thought comes close to Dix Harwood's *Love for Animals and How It Developed in Great Britain* (1928),[14] with this one book depending heavily for its contribution to the history of animal ethics on Harwood's evidence and findings. Many authors seek a striking distinctiveness – which others then follow in oxymoronic imitation of nonconformity. But in the desire to be authentic and different, it is more often the wisdoms than the errors of historical scholarship that are discarded.

The theme of this book is certainly not that Western attitudes to animals have been healthy, generous, or generally commendable – far from it – although, as I have argued elsewhere, they have been, on the whole, no worse than those of their Aboriginal and Oriental counterparts.[15] My intention is most assuredly not to offer Western cultural history as a moral paradigm but to attempt to provide a balanced perspective: the aquiline nose

along with the double chin, the curls along with the paunch, the contemplative frown instead of the unremitting scowl. In the worlds of theology
and art appreciation, the twentieth-century Franco-Russian surrealist Marc
Chagall is reasonably well known to have proclaimed himself "something
of a Christian" of the kind of Saint Francis of Assisi "with an uncondi-
tional love for other beings."[16] I have never encountered an acknowledgment of this statement – or the many others like it from a multiplicity
of sources – in the secular animal-rights literature. The *Panarion* by the
Christian-despising Epiphanius and the *Recognitions* and *Homilies* by
pseudo-Clement point to rigorous ethical vegetarian sects in early Christianity.[17] The apocryphal second-century *Acts of Andrew,* widely employed
by Gnostic and heretical groups, claimed a connection between demons
and meat. As Stephen H. Webb has argued, apocryphal or not, "it demonstrates that the Christian revulsion against animal sacrifices" – the customary occasions of flesh consumption – "was widespread."[18] If the apostle
Paul emphasized the importance of meat (Romans 14-15) – and whether
he did is a debated issue[19] – it is only a reflection of the fact that many
early Christians were adamant vegetarians, whom Paul opposed not for
eschewing flesh but for putting too stringent restrictions on Christianity,
thereby hindering its appeal. Indeed, for several centuries, there were fears
that the "rigorist position," both in questions of diet and sexual morality,
would deter ordinary mortals from the faith. St. Augustine claimed Christian vegetarians to exist in the fifth century "without number," much to his
chagrin.[20] Moreover, it is clear that Augustine believed that they were vegetarian because they considered it unethical to kill animals for food. Indeed,
the finest of the commentators on the Desert Fathers, Benedicta Ward, has
remarked that the "aim of the monks' lives was not asceticism, but God,
and the way to God was charity. The gentle charity of the desert was the
pivot of all their work and the test of their way of life. Charity was to be
total and complete."[21]

    The following instructive story, reflective of distinctions between the
Church hierarchy and the groundlings, is related by Theophilus, who was
Archbishop of Alexandria at the same time that Augustine was Bishop of
Hippo: "As the fathers were eating with [the archbishop], they were brought
some veal for food and they ate it without realising what it was. The bishop,
taking a piece of meat, offered it to the old man beside him, saying, 'Here
is a nice piece of meat, abba, eat it.' But he replied, 'Till this moment, we
believed we were eating vegetables, but if it is meat, we do not eat it.' None
of them tasted any more of the meat which was brought."[22] In the long
run, the meat of the bishops outweighed the vegetables of the abbas in the
development of Christian attitudes. But the primitive values underscoring

the consciousness of the abbas never disappeared entirely from Western mores.

The monastic communitarian Essenes of the second century BC to the late first century AD, strict Jewish followers of the laws of Moses, as they understood themselves to be, who *may* have fused with Jewish Christianity – a final separation between Judaism and Christianity occurred only in about AD 85 – and *probably* influenced some early Christians, were vegetarians, according to Porphyry (*De abstinentia* 4.3) and Jerome (*Against Jovinianus* 2.14); and the Jewish historiographer and military leader Flavius Josephus (*Antiquities* 15.10.4) stated unequivocally, if not completely convincingly, at least if the traditional view of Pythagoras is the correct one, that the Essene and Pythagorean principles and practices were identical. The Essenes, in fact, permitted themselves grasshoppers and fish, but no other flesh. Philo of Alexandria affirmed that the Essenes eschewed animal sacrifice, as did Jesus, who avowed "Mercy is what pleases me, not sacrifice" (Matthew 9:13) and who proclaimed, according to the *Panarion,* "I have come to abolish the sacrifices," a view of Jesus repeated by pseudo-Clement – who reported of another Jewish sect, the Therapeutae, that they restricted their diet to bread, hyssop, salt, and water.[23] Many Gnostic groups – second-century heretical Christians who proclaimed an esoteric knowledge (gnosis) of the supreme divine being – were vegetarian.

The orthodox were no less notable. Thus St. Athanasius wrote of the desert-dwelling St. Anthony that his "food was bread and salt, his drink only water. Of meat and wine it is needless to speak, for nothing of this sort was to be found among the other monks either." Nor were the grounds for vegetarianism solely ascetic, for "the wild beasts kept peace with him," as enjoined by Job 5:23. When there was a drought, he released his camel that it might seek for itself and survive. "My book is nature," he is reported to have said, "and whenever I will I can read the words of God."[24] Among others of the Desert Fathers, St. Macarius was renowned for being gentler with a blind hyena-kitten than with his own kind.[25] Simeon Stylites lived for forty years on a fifty-foot column outside Antioch. Legend has it that he converted and cured a blind dragon. Such monks, Benedicta Ward indicates, "chose to live at the limits of human nature, close to the animals, the angels, and the demons."[26] St. Jerome told the story of Theon, one of the vegetarian monks of Egypt, whose "food was cooked on no fire. They said of him that at night he would go out to the desert, and for company a great troop of the beasts of the desert would go with him. And he would draw water from his well and offer them cups of it in return for their kindness in attending him. One evidence of this was plain to see, for the tracks of gazelle and goat and the wild ass were thick about his cell."[27]

Apocryphal? Perhaps. Exaggerated? Probably. But even if so – for such stories of whatever tradition are prone to reflect more of the desirable than the real – the tale imparts what was thought an epitomizing attribute of a saintly person. Those who have studied the exploits of T.E. Lawrence (Lawrence of Arabia) will know the lengths to which the inhabitants of the desert would go to deprive all but close kin of the use of their wells. To share one's well with the animals was a mark of great respect and an intimation of closely felt kinship.

In classical Greece, vegetarianism was hotly debated, and many renounced flesh. In Rome vegetarianism was viewed as part of the cultural heritage. To be sure, we encounter in Roman history some of the grossest cases of the public display of carnivorous gluttony. But on the other hand, Nero outlawed the selling of meat in public eating places, while permitting its consumption in private at home, a reflection of the belief that there was something less than wholesome in meat consumption. In general, the Romans admired those who denied themselves a flesh diet, both for ascetic reasons and because, despite the Roman games, there was widespread recognition that humans and animals alike shared something of a common identity. Had there not been so, the Stoics would not have thought it necessary to oppose the idea. Somewhat earlier than Nero, the Stoic Cicero, in general no great friend to the animal cause, described a boring day at the games to his correspondent Marcus Marius, in which he questioned "what pleasure a cultivated man can get out of seeing a weak human being torn apart by a powerful animal or a splendid animal transfixed by a hunting spear?" Moreover, he added, the public showed no enjoyment at the treatment of the animals. Instead, there "was even an impulse of compassion, a feeling that the [elephants] had something human about them."[28] Even some Stoics were tempted to recognize the kinship of humans and animals.

Many will rejoice in the Jain principle of *ahimsa* (nonharm, noninjury); some will even recognize the importance of the Judaic dictum of *Bal Taschit* (do not destroy). But where in the animal-rights literature will one ever encounter the fifth of the seven sayings of the early Christian abbot Moses: "A man ought to do no harm to any"[29] (as derived from Isaiah 11:9: "They shall not hurt nor destroy in all my holy mountain")? The three traditions are here identical – expressed in all, and lived up to in none. The renowned patristic scholar and translator Helen Waddell wrote of the early-seventh-century monk John Moschus, author of the *Pratum Spirituale,* that he "was a lover alike of man and beasts, and never weary of stories about the goodness and guilelessness of lions, and the wisdom of the little dog of the abbot Subena Syrorum."[30] He told tales also of Abba John the Eunuch, who "had more compassion than anybody we ever saw, not only for men but also for animals." In her delightful book on *Beasts and Saints* (1934), which consists of some forty-odd translations of medieval Latin works on what she calls "the mutual charities" between them, Waddell depicts a world alive with interspecies respect among "The Desert Fathers," "The Saints of the West," and "The Saints of Ireland." She also has important things to say about what she is compelled to exclude:

> By restricting these sources to Latin, I have had to forego the race of the ver-
> nacular; Robert Mannyng's bear who came to keep a lonely hermit in com-
> pany, and as he came up to the gate louted and made fair cheer – "Feyre
> chere as a bere myght": the four Irish scholars who went to sea for the love
> of God and took nothing with them, only that the youngest said, "I think
> I will take the little cat": the grasshopper who came to Portiuncula through
> a winter's night to sing the midnight office for St. Francis, and left his small
> tracks in the snow, to the compunction of the slothful brethren who had
> lain warm in their beds. The truth is that the Middle Ages are so rich in this
> kind of story that without some arbitrary principle of selection one is ham-
> pered by too much liberty.[31]

The much admired anti-vivisectionist Anna Bonus Kingsford noted in her 1880 University of Paris medical doctoral dissertation that in her own day: "S. Benedict's rule prohibits the flesh of quadrupeds to all except the feeble and sick. The rule of S. Francis of Paola is severely vegetarian, for-bidding even eggs and milk. The Trappist monks, the religious of S. Dom-inic's order (friar preachers), and of S. Basil's order, are all vegetarian; and among the orders of women, the rule of life of the Poor Clares is similar."[32]

Of course, one must not exaggerate the role of such ideas, groups, and individuals in the early, or even later, Christian tradition – or, as is com-monplace, in any other tradition. Certainly, one should take note of the dismissal of serious animal consideration in Saint Paul's First Letter to the Corinthians (9:9), where we are told that the Hebrew Bible's concern for oxen must be interpreted to imply a preeminent concern for humans. In *The Oxford History of Christianity* Henry Chadwick tells us that such utter-ances arose because the early Christian communities were "urban, and only slowly penetrated rural societies." And, when they did, we "hear of farm-ers alarmed to learn" of such pronouncements. They were aware of our obligations to other species.[33] Despite the obvious failure of Christian doc-trine in too many instances to enshrine animal interests, surely one must reject out of hand the conclusion of James Turner in one of the books to set the contemporary standard of interpretation – *Reckoning with the Beast* of 1980 – that in the premodern era: "Pity was reserved for people."[34] It is an error that has plagued our understanding ever since.

Until animal-respecting references are commonplace, the study of the history of animal ethics will suffer. Why, one must ask, does one not encounter in this literature the poem "The Righteous Man" by the Victorian bard Samuel Butler? It was a popular satire that reflected much of the Victorian antipathy to those who disregarded the rights of animals:

The righteous man will rob none but the defenceless,
Whatsoever can reckon with him he will neither plunder nor kill;
He will steal an egg from a hen or a lamb from an ewe.
For his sheep and his hens cannot reckon with him hereafter –
They live not in any odour of defencelessness:
Therefore right is with the righteous man, and he taketh advantage
    righteously,
Praising God and plundering.

The righteous man will enslave his horse and his dog,
Making them serve him for their bare keep and nothing further,
Shooting them, selling them for vivisection when they can no longer
    profit him,
Backbiting them and beating them if they fail to please him;
For his horse and his dog can bring no action for damages,
Wherefore, then, should he not enslave them, shoot them, sell them for
    vivisection?

But the righteous man will not plunder the defenceful –
Not if he be alone and unarmed – for his conscience will smite him;
He will not rob a she-bear of her cubs, nor an eagle of her eaglets – Unless
    he have a rifle to purge him from the fear of sin:
Then may he shoot rejoicing in innocency – from ambush or a safe
    distance;
Or he will beguile them, lay poison for them, keep no faith with them;
For what faith is there with that which cannot reckon hereafter,
Neither by itself, nor by another, nor by any residuum of ill
    consequences?
Surely, where weakness is utter, honour ceaseth.

Nay, I will do what is right in the eyes of him who can harm me,
And not in those of him who cannot call me to account.
Therefore yield me up thy pretty wings, O humming bird!

Sing for me in a prison, O lark!
Pay me thy rent, O widow! for it is mine:
Where there is reckoning there is no sin,
And where there is no reckoning sin is not.[35]

If, as I suggest, Butler was reflecting a common enough opinion in Victorian England, why were animal interests *in fact* not more respected? First, we might note that there was perhaps a greater sensibility to animals in late Victorian England, at least with regard to the issue of vivisection, than at any time since or before. But, more important, Butler's admonitions were not generally followed because of the human condition present in all peoples at all times. It is, as Butler makes so abundantly clear, a matter of power. He is reiterating the wise and well-known nineteenth-century admonition of Lord Acton that all power tends to corrupt and that absolute power corrupts absolutely. Many people do not do what they know justice requires. And this is the case because they have the power to avoid justice. Many more who believe that in principle the rights and wellbeing of animals matter either believe that other considerations are of even greater merit or are able so to rationalize their thoughts and their actions.

IV

My charges against the discipline for its too easy conclusions may seem unduly harsh, perhaps especially in my complaint that it is the most damning of plausible explanations that is frequently chosen and for no better reason than that it is the most damning. Certainly, the reader is entitled to demand that such charges be warranted by solid evidence. And I attempt to do so, where appropriate, throughout the book. But let me offer two representative examples at the outset to indicate what it is I am trying to overcome. We are often informed in the animal-rights literature that we distance ourselves from the animals we consume in order to avoid the guilt that we would feel if we acknowledged what it was we were consuming. Hence we disguise our predilection for eating cattle by calling their slain bodies "beef."[36] We delude ourselves further by calling pig "pork," calf "veal," deer "venison," and sheep "mutton." For example, James Serpell, who is certainly not one of the discipline's greater sinners but who has been followed by many who are, argues that: "Verbal concealment ... is commonplace. We talk about 'beef,' 'veal' and 'pork' rather than bull-meat, calf-meat, or pig-meat because the euphemisms, in every sense, are more palatable than the reality."[37] Of course, "distancing" is a common psychological device.

Of course, we are prone to euphemistic utterances. Yet "verbal concealment" and "distancing" are, quite simply, *not* the reason for the selection of the names for our meats. The authentic explanation is far less sinister – differently sinister, at any rate – and was provided almost two centuries ago by Sir Walter Scott in his novel *Ivanhoe,* set in the Middle Ages:

> "... how call you those grunting brutes running about on their four legs?" demanded Wamba.
>
>   "Swine, fool, swine," said the herd, "every fool knows that."
>
>   "And swine is good Saxon," said the Jester; "but how call you the sow when she is flayed, and drawn, and quartered, and hung up by the heels, like a traitor?"
>
>   "Pork," answered the swine-herd.
>
>   "I am very glad every fool knows that too," said Wamba, "and pork, I think, is good Norman-French; and so, when the brute lives, and is in the charge of a Saxon slave, she goes by her Saxon name; but becomes a Norman, and is called pork, when she is carried to the Castle-hall to feast among the nobles; what dost thou think of this, friend Gurth, ha?"
>
>   "It is but too true doctrine, friend Wamba, however it got into thy fool's pate."
>
>   "Nay, I can tell you more," said Wamba, in the same tone; "there is old Alderman Ox [who] continues to hold his Saxon epithet, while he is under the charge of serfs and bondsmen such as thou, but becomes Beef, a fiery French gallant, when he arrives before the worshipful jaws that are destined to consume him. Mynheer Calf, too, becomes Monsieur de Veau in like manner; he is Saxon when he requires tendance, and takes a Norman name when he becomes matter of enjoyment."[38]

Clearly, Scott was far more thorough and careful in his research than many modern scholars. Nor is Scott the sole historical source of this information. In his *Life of Chaucer* (1803), William Godwin praises the medieval poet for his revival of the English language, pointing out that the Anglo-Saxon words "cow," "sheep," and "pig" survived among peasants who tended the animals, while the animals as food became "veal," "mutton," and "pork," after the Norman lords who ate them.

What ought to surprise us, and indeed confound us, is why some modern scholars have not answered satisfactorily an obvious question that ought to have persuaded them to rethink the matter: If we change the names of foods to disguise what it is we are eating, why do we do it only for certain animals and not for chicken, goose, grouse, lobster, fish, quail, rabbit, hare, and the like? Why, in these instances, and not the others, were the food names the same as the names of the living beasts? The answer that only the most costly are changed will not do, as evidenced by grouse, plover, and partridge. Nor anyway would this affect the issue since we would still be eating the corpses of formerly sentient beings. Even if such animals are smaller and more biologically remote from us – the reason Jim Mason offers for the distinction[39] – the potential for guilt would remain the same. They are still sentient, self-directed beings and are commonly acknowledged as such. Moreover, how is it that the Old English "lamb," despite the animal's size, and similar biological propinquity to the human as cattle, escaped the fate of pig and calf? The answer appears to be that mutton – the French *mouton* – remained a generic term for the flesh of the sheep until the seventeenth century, when the significance of the animal's age to its tenderness crept into the language – long, of course, after the Norman power had waned. There was no thought to hide its nature from us. Indeed, the word "lamb" emphasizes youth and tenderness, which, if Mason were right, would have encouraged the feeling of guilt and would not have been introduced.

"Ham" – meat from the upper part of a pig's leg, salted or dried and smoked – prima facie would appear an exception to the rule. It is derived from the Old English *ham* or *hom*, denoting the back of the knee, which is in turn derived from a Germanic base, meaning "be crooked." However, while the anatomical usage is documented from the end of the first millennium, its alimentary usage appears to be a mid-seventeenth-century innovation, after Norman-French influences had waned. The reality remains that the Norman conquerors reserved certain foods for themselves and forbad them to the Saxons. Such comestibles bore Norman-French names. The remainder bore traditional Saxon names. "Pork," as Wamba said, is a traitor. And the treachery was then cast in perpetuity.

"Verbal concealment" played no role. Moreover, a little contemplation would have served to reinforce this conclusion. When, for example, we read in their creation myth of the evident pride the Cheyenne have in being meat eaters,[40] or when we discover in the memoirs of Samuel Rogers that the William, Dorothy, and Mary Wordsworth family were compelled "to deny themselves animal food several times a week" being "in such dire straitened circumstances"[41] – such examples could be repeated ad nauseam – we recognize that, generally speaking, animal food consumption was counted a mark of human achievement, of success, more to be boasted about than concealed, more to be commiserated than applauded when absent.

Jim Mason tells us that "Packers, processors and supermarket chains ...

reduce the carcasses to bloodless, shrink-wrapped packages that offer the consumer no clue as to their animal origins ... The various distance devices work in hand with misothery to keep animal exploitation from being emotionally and morally troublesome."[42] ("Misothery" is Mason's coinage for what he sees as a pervasive hatred of animals in the Western world.) To be sure, in modern grocery stores the meats are precut, prepackaged, and no longer resemble the animal corpses from which they are sliced. But this is not done, despite Mason's claim, via Serpell, to hide from us what they are. Mason is going to have a hard time explaining why a majority choose to order steaks in restaurants rare or medium rare (i.e., bloody) when they could so readily order them "bloodless." The choice would appear to accentuate rather than minimize the animal nature of the flesh that they are consuming. Certainly, where there are still butcher shops, the whole dead animal is hung for everyone to see – indeed, given prominence to entice the browser. It does not deter the customers. We still get to select living lobster and crab from the cramped tanks in which they are imprisoned. Their desirability, as measured by demand, has not waned one iota – even though modern sensibilities encourage some to insist that their still-living supper be chosen for them. Fresh fish for sale look scarcely any different from their still-swimming kin, and they are surrounded in ice to maintain the freshness and their similarity to the live being. They could be disguised, but it would deter rather than encourage purchase. Whole poultry is still recognizable as animal, even though it is dressed, for convenience not concealment, in a form that differs substantially from the living being. Not too long ago it was sold with the feathers, feet, and head still attached, and plucking the fowl was one of the most onerous of domestic tasks. The change has been for the convenience of the customer, and for the profit to the purveyor, not to hide the animal's nature.

To be sure, some have found it preferable that the nature of the animal food be disguised. And not merely in recent decades. Thus, in 1826, the literary critic and essayist William Hazlitt preferred not to have his sensibilities ruffled, declaring that "Animals that are made use of as food should either be so small as to be imperceptible, or else we should ... not leave the forms standing to reproach us with our gluttony and cruelty. I hate to see a rabbit trussed, or a hare brought to the table in the form which it occupied while living."[43] Clearly, Hazlitt was an exception, for the practices he abhorred continued unabated, without guilt of gluttony and cruelty. The concealment is thus less a physical concealment by those who purportedly wish to disguise from customers that they are eating flesh than it is an intellectual concealment by those authors who believe that people ought to be ashamed of the fact that they eat flesh.[44]

V

The second example is from Carol J. Adams's *The Sexual Politics of Meat: A Feminist-Vegetarian Critical Theory.* Despite a detailed and partly informative account of the vegetarianism of Joseph Ritson and its reception in the early nineteenth century and later, Adams's examination is ultimately more misleading than illuminating. Having rightly told us of the "ungenerous" description of Ritson's "incipient insanity" in the *Dictionary of National Biography,* she adds that the dictionary "exhibits a similar dismissive viewpoint in [its] biographies of other vegetarians."[45] Yet this is to misrepresent the dictionary and to misrepresent the prevailing view of vegetarianism. The representation is, again, designed to emphasize the tribulations of vegetarians rather than to seek a historically accurate account. In fact, the biographies of the vegetarians John Arbuthnot, William Buchan, George Cheyne, John Elliott, and George Nicholson are, in general, at least reasonably laudatory, in some instances emphatically so. Only in the instance of Dr. George Cheyne, who persuaded John Wesley to become a vegetarian, do we find a criticism of his vegetarianism, when we are told that he "carried his vegetarian views to great extreme, as when he maintains that God permitted the use of animal food to man only to shorten human life by permitting the multiplication of disease and sufferings, which should conduce to moral improvement." And here the criticism is of his religious views ostensibly derived from vegetarianism rather than of vegetarianism itself. Moreover, we are told that Cheyne's "literary and argumentative powers are generally admitted. All contemporary testimony gives a very favourable idea of his personal character. His reputation with the public was immense, and he was intimate with the most eminent physicians and persons of note in his time."[46] In addition, his biography is included in the 1838-40 publication of the *Medical Portrait Gallery of the Most Celebrated Physicians, Surgeons, etc., Who have Contributed to the Advancement of Medical Science,* going back as far as Aesculapius. The biography is neither "ungenerous" nor "dismissive." Further, it is worthy of note that the *Dictionary of National Biography*'s compilers were not always soft on those seen as the enemies of animal interests. Thus, for example, in the biography of the Scottish vivisector Dr. John Caverhill, we are told that "he conducted a large number of barbarous experiments on rabbits."[47]

Adams describes a caricature intended to ridicule Ritson, produced by James Sayer and published by Humphrey in 1803 immediately following the publication of Ritson's *An Essay on Abstinence from Animal Food as a Moral Duty:* "Large rats are eating carrots, a cow sticking her head in a window is munching lettuce from a large bowl. Parsnips, beets, and other root vegetables stand among the books on the bookshelves, juxtaposing this vegetarian food with written texts. Joseph Ritson's own *An Essay on Abstinence from Animal Food as a Moral Duty* lies opened to the title page ... In Ritson's left pocket sits *The Atheist Pocket Companion* ... A 'Bill of Fare' reads: 'Nettle Soup, Sour Crout, Horse Beans, Onions Leeks.'"[48] Adams goes on to mention, rightly, the depiction of Ritson in the cartoon as an excessive critic of his fellow antiquarians – he was well known for his writings on fairies, his well-founded argument that King Arthur was a historical figure, and his revival of the ancient legend of Robin Hood, much of which did not find immediate acceptance among his colleagues. She then sums up: "This is Ritson as his contemporaries saw him."[49] The clear implication is that his espousal of vegetarianism and atheism, along with his perceived undue sense of superiority, were what brought about his condemnation. But to what degree is this true of his espousal of vegetarianism?

It would certainly have added to the account and to the reader's ability to put the story in perspective if we had been told that James Sayer was a caricaturist in the pay of the Tory leader William Pitt the Younger, and that he had produced numerous other cartoons attacking prominent figures who did not possess a stellar conservative character, including one of the Whig leader Charles James Fox that accused him, both unjustly and unconvincingly, of the most heinous treachery. In fact, in this period, caricature was one of the most frequent forms of abuse, often, if not especially, from the right.[50] A renowned example by James Gillivray, entitled *The New Morality* and published by the Tory *Anti-Jacobin Magazine and Review* in 1798, lampooned Erasmus Darwin, Coleridge, Paine, Southey, Godwin, Priestley, the *Morning Chronicle* and *Morning Post* newspapers, and a few lesser individuals for their purported sympathy for the godless revolutionary spirit of France – not a whit less damning than the criticism of Ritson. Significantly, of those only Godwin was a vegetarian, and it was not for his vegetarianism that he was denounced. Indeed, his vegetarianism was almost certainly not known. Of course, it would be quite unwarranted to conclude that such was the light in which their contemporaries saw them. Some contemporaries, certainly. But decidedly not their contemporaries as a whole. To put the matter in context, it is worth mentioning that in opposition to Edmund Burke's *Reflections on the Revolution in France* – which had sold thirty thousand copies by 1797 and was a second Bible to the *Anti-Jacobin* – it is estimated that as many as fifty books saw the light of day, along with another 350 articles. It would appear that the godless revolutionary spirit of France, if that's what it really was – and in most instances it was not – lived in many!

Adams also mentions caustic critiques of Ritson's book in the *Edinburgh*

Review and the *British Critic*.[51] Yet it is difficult to evaluate such information unless we are also told – which we are not – that the *Edinburgh Review* had a reputation for vilifying those who did not share its Old Whig political opinions. ("Old Whig" was a term coined by Burke to identify his own views.) Its reviews of Wordsworth, Coleridge, and Byron – not a vegetarian among them – were every bit as vitriolic as its treatment of Ritson. The *British Critic* was founded to oppose the Dissenters and their opinions – we can only imagine what it thought of declared atheists! It was written and edited by conservative scholars and was instrumental in fomenting the anti-Jacobin reaction in the 1790s. One of its primary writers was Ritson's fellow antiquarian, and archenemy, the highly respected Thomas Percy. The magazine pilloried routinely anything of a controversial nature, including Mary Shelley's *Frankenstein,* which was said to possess "neither principle, object, nor moral." Its debunking of Ritson should neither surprise us, nor persuade us that this was a pervasive view of Ritson, nor convince us that the opposition to Ritson arose primarily because of his vegetarianism.

To provide a framework in which to view the reputation of Ritson and his animal cause, it is relevant to ask whether there were other views than those held by James Sayer and his ilk as well as those who funded them, whether opposition to Ritson was based primarily on his support for vegetarianism, and whether there was a more generous view of vegetarianism when not associated with Ritson. It would be churlish to ignore that Ritson's personality and his caustic and unrestrained writing encouraged detraction; he frequently termed his fellow antiquarians "fool" and "liar." He attacked Thomas Warton's scholarship in *Observations on Warton's History* (1782), a mordant critique of the Oxford professor of poetry's *History of English Poetry* (1774-81). He disputed the originality of Bishop Percy's *Reliques* and was less than impressed with Dr. Samuel Johnson's edition of Shakespeare. This was not a way to make influential friends. Yet we can find sympathy for him in the publisher Sir Richard Phillips, a man of sufficient stature and respectability – at least once he had recovered from a conviction for publishing Tom Paine! – that he was founder and editor of the *Leicester Herald* and sheriff of the City of London. The less respectable – he got Mary Wollstonecraft pregnant before they were wed – but decidedly accomplished, renowned, and revered William Godwin frequently dined with Ritson. He said that he had "acquired the friendship of many excellent persons." Ritson was one of the six he enumerated. Percy Bysshe Shelley and John Frank Newton respected him greatly. And in 1824 the admiring fellow antiquarian Joseph Haslewood wrote a tribute to him: *Some Account of the Life and Publications of the Late Joseph Ritson.* Haslewood aside, they were all vegetarians, and hence their sympathy was a *parti pris*. But as Keith Thomas observed, "From about 1790 there developed a highly articulate vegetarian movement"[52] – and he mentions a significant number of adherents. Ritson was not without allies, and his books sold well and were admired. To be sure, as Thomas also indicates, "Vegetarianism was a millenarian movement."[53] But not for this reason a small one. This was the age of perfectibility. It was an age in which many of the educated believed wholeheartedly that a change in political, social, and economic conditions would bring about a veritable utopia in which *all* ills would be cured. John Frank Newton even thought a change to a vegetarian diet would eventually bring about a customary lifespan of 150 years! The enemies of the radicals thought that instead of achieving the perfection they promised, they would rend asunder the very basis of traditional constitutional order on which the possibility for progress was based. Those who posed such a danger deserved the most complete disparagement. Compromise, conciliation, and half-heartedness were not the general nature of speculative thought in the revolutionary era.

In fact, while the Sayer caricature employs what might be considered the easy target of vegetarianism to ridicule Ritson – after all, as is now generally conceded, a number of his arguments and expectations were risible, including the fecund view that flesh eating rendered the consumer ipso facto fierce, and the book had just been published and offered the opportunity for reaction – the primary opposition to Ritson was to his Jacobinism. He liked to be called "Citizen Ritson." His home was adorned with pictures of Rousseau, Voltaire, and Paine. He was ferociously loyal to the principles of the French Revolution a decade after the Reign of Terror had shown its true nature and many of its most prominent early supporters had recanted. Even his account of Robin Hood was filled with revolutionary sentiment. None of this may have mattered too greatly but for the fact that there was a realistic fear in Britain that the French Revolution might be imported and for the reality that Britain and France were at war – a war in which Britain was not faring well. Thus for all intents and purposes, Ritson was viewed as a traitor to king, country, and fellow citizens. It is in this context that we may best understand the occasional vitriolic opposition to his espousal of vegetarianism.

It would be quite unwise to ignore Ritson's many detractors, and the venom heaped on his strident vegetarianism. But it would be equally unwise to imagine this to be predominantly a consequence of his attitude to animals or of his vegetarianism itself. If we turn to the experience of his contemporary, the printer and author George Nicholson, we hear a different story. In 1797 Nicholson published *On the Conduct of Man to Inferior*

Animals. This he expanded in 1801 into *On the Primeval Diet of Man; Arguments in Favour of Vegetable Foods; On Man's Conduct to Animals &c. &c.* The book was by and large a compendium of previous writings on vegetarianism and sensibility to animals from antiquity to the era of his contemporaries. Among the contemporaries, or near contemporaries, from whom he quoted passages in favour of vegetarianism, we find John Oswald, George Cheyne, John Arbuthnot, William Buchan, Sir John Elliott, James Graham, John Wesley, and John Stewart, not all of whom were radical innovators. Cheyne was a friend of Samuel Johnson, and his writings were roundly praised by that Tory dean of taste. Arbuthnot was physician to Queen Anne. Elliott was physician to the Prince of Wales, later George IV. And if Wesley roused the rabble it was in a different cause. Nicholson quotes numerous contemporary individuals who had written against cruelty to animals, especially against vivisection. But he also quotes from such magazines as *The Guardian, The Idler, Monthly Review, Monthly Magazine, Gentleman's Magazine,* and *Bell's Weekly Messenger,* journals that were the everyday bourgeois fare of those who could read and afford to buy them. (Printing taxes were extremely high and kept most publications out of the hands of even the relatively few members of the lower classes at this time who could read, although the age of the self-improving working man was dawning.) Moreover, on Nicholson's death, the *Gentleman's Magazine's* obituary was most respectful. The journal counted him "a man whose worth and talents entitle him to notice." Commenting that, in general, his writings have "already obtained the meed of praise from contemporary critics," the editors continued: "In a Treatise 'on the conduct of Man to inferior Animals' (which has gone through four editions) we have evidence of his humanity of disposition; and numerous Tracts calculated to improve the morals, and add to the comforts of the poorer classes, are proofs of the same desire of doing good. In short, he possessed, in an eminent degree, strength of intellect, with universal benevolence and undeviating uprightness of conduct."[54] We should certainly notice that the magazine was more impressed by Nicholson's humanitarianism toward animals and humans than by his arguments for vegetarianism per se, on which it was conspicuously silent. But we should also recognize that his radical espousal of vegetarianism did not prevent the praise from being offered. And we should not imagine complimentary obituaries a gratuitous feature of the *Gentleman's Magazine,* as the following caustic extract will indicate:

Died, April 4, [1789] at Tottenham, John Ardesoif, esq. a young man of large fortune, who in the splendour of his carriages and his horses was rivalled by few country gentlemen ... He was very fond of cock-fighting and had a cock upon which he had won many profitable matches; but he lost his last bet which so enraged him, that he had the bird tied to a spit and roasted before a large fire. The screams of the miserable animal were so affecting, that some gentlemen, who were present, attempted to interfere; which so enraged Mr Ardesoif, that he seized a poker, and with the most furious vehemence declared, he would kill the first man who interposed, but in the midst of his passionate asseverations, he fell dead upon the spot. Such, we are assured, were the circumstances which attended the death of this great pillar of humanity.[55]

Nor should we consider commendation from the *Gentleman's Magazine* for Nicholson and mocking denunciation of Ardesoif as "this great pillar of humanity" to be mere fellow-traveller backslapping among radical reformers. In *Vanity Fair,* perhaps the most popular Victorian novel about England in the Napoleonic era, William Makepeace Thackeray described the *Gentleman's Magazine* as one of the "standard works in stout gilt bindings"[56] that graced the bookshelves of a wealthy bourgeois home. Praise for Nicholson from the *Gentleman's Magazine* reflected not only the merit of its recipient but also the growing respectability of the causes he espoused.

In her 2003 Introduction to a new, and welcome, edition of Howard Williams's *The Ethics of Diet* of 1883, Adams repeats her claim about the *Dictionary of National Biography's* image of Ritson but now adds: "Regarding the vegetarian Sir Richard Phillips, *The Dictionary of National Biography* tells us that 'Tom Moore considered him a bore and laughed at his Pythagorean diet.'"[57] We are left with the impression that if you were a vegetarian you were in general a subject of ridicule. Adams did not bother to add the immediately following words from the *DNB:* "De Morgan credits his honesty, zeal, ability, and courage." Nor are we told, as is indicated in the same source, that he was "elected a sheriff of London in 1807" and that "he was knighted by the king" in 1808. Very strange treatment for someone who was apparently so risible! Further, she quotes the *DNB*'s description of William Lambe: "Lambe was accounted an eccentric by his contemporaries, mainly on the ground that he was a strict, though by no means fanatical, vegetarian." She does not tell us, as we can read in the same source, that he was elected to the College of Physicians, where "he held both the censorship and Croonian lectureship on several occasions between 1806 and 1828, and he was a Harleian orator in 1818." He may well have been regarded as eccentric for his diet. But, quite clearly, judging by the honours bestowed upon him, his eccentricity did not notably affect his reputation. Carol Adams has given us a significant glimpse into one part of the Georgian character. But it is only a glimpse and thus only a part – and ultimately a misleading one.

By 2000 Adams's misleading perspective had become orthodoxy. Without mentioning the source, since, presumably, it was now a commonly acknowledged fact, Steven G. Kellerman added to the authority of the myth, quoting the *Dictionary of National Biography*'s mention of Ritson's "incipient insanity" and implying that this was a common opinion of vegetarians. The subtitle of his article on the subject is "The Anti-Vegetarian Animus." He went further and referred to *The Cry of Nature* of John Oswald, telling us that "one cannot help assigning its kindly author to the category of persecuted animals."[58] On a broader and more careful inspection of the reputation of vegetarians of the period as a whole, we are driven to come to more balanced conclusions than those offered by Adams and Kellerman. Glorying in victimhood vicariously may have some psychological benefits. It does not have intellectual ones.

VI

To understand attitudes to animals and to vegetarianism in the period, we need to acquire a picture in the round, one that will allow us to recognize that at all times and places issues in the history of animal ethics are ever complex, with several sides to each story. A one-sided account will always deceive more than illuminate, leaving the reader with a faulty image. The history of ideas is the story of continuous battles fought among competing people with competing values, competing minds, and competing interests. It does not provide a unidimensional picture in any area of human contemplation, and this is certainly no less the case in our understanding of the human-animal relationship.

If the field of animal studies is to exist as a legitimate area of scholarship, it must first be judged by the standards of scholarship, not by its capacity to be useful for animal advocacy, even though most of us in the discipline are also animal advocates. Our capacity to persuade will depend ultimately on the rigour with which we have approached our discipline – on the degree to which our predilections have not interfered with our research or with our conclusions.

From the time of classical Greece to the eighteenth century, abating but not disappearing thereafter, the purported differences and purported similarities between human and animal souls constituted the primary basis on which human and animal rights were assessed. The status of the soul was the primary criterion for determining the right to ethical consideration. In the opening chapter we shall investigate the thorny questions of the nature of the soul with regard to its implications for the human-animal relationship.

I

In Quest of the Soul

I

I AM NOT ESPECIALLY CONCERNED with the question of whether nonhuman animals possess souls – or whether humans do for that matter. Of primary interest is the historical understanding of the soul and the significance to the status of animals of ascribing them souls or not. Donald Tannenbaum and David Schultz observed in *Inventors of Ideas* that the "philosophic idea of the psyche, or soul," arising in classical Greece, implied "that humans have a higher purpose in life than mere existence or the satisfaction of physical (bodily) appetites. The most basic questions of philosophy from then until now have been, what is this higher purpose, and how can it be achieved?"[1] And the answer has usually been determined to lie in those mental or moral attributes (customarily associated, at least until the last hundred years or so, with the concept of "soul") that distinguish the wiser from the less wise or the more saintly from the more secular humans, and a fortiori from nonhuman beings. If, as Plato argued in the *Republic* and elsewhere, the best should rule (these being the purveyors of the higher arts, who are conceived in terms of philosophical wisdom), then not only will some humans govern other humans, but humans as a whole will rule over animals as a whole – that is, over those who lack those attributes deemed necessary to the art of good governance, among which an appropriate "soul" was the primary requirement.

Religious usages of "soul" have tended to differ from the philosophical in that there are differing views of who constitute the best of people. Thus, for example, describing St. Simeon Stylites, the fourth-century ascetic monk to whose shrine near Antioch (the third city of the empire) so many medieval, and some later, pilgrims made the journey of homage, journalist and travel-writer H.V. Morton remarked that Simeon "believed that only by the complete humiliation of his body could his soul set itself free and fit itself to contemplate God." To this end he devised "all kinds of self-torture ... It is only with difficulty that we can even try to understand the mental attitude of the Fourth Century world. It was a world in revolt against materialism."[2] Animals were thus often relegated to the material world not because anyone thought that they were matter alone, but because they were incapable of those *highest* forms of spiritual activity that constituted the ultimate life of the soul. Since animals were neither speculative nor devout, they differed in some fundamental way from those who were capable of the most distinctively human kinds of activity. The very origins of philosophy and religion themselves, we may say, have argued against the inclusion of nonhuman animals as participants in the most significant of behaviours. The evaluative odds were stacked against them by a regard for the very concepts, questions, and assumptions of philosophy and religion as, generally speaking, exclusively human activities. If animals have souls, or psyches, these are not souls capable of the philosophical and spiritual heights achieved by those who are deemed the most worthy of beings – precisely because they engage in those special kinds of activity of which other species are incapable.

On the other hand, one could well argue that philosophy is also the source of conflict, dispute, and opposition. Philosophy makes enemies of one's fellow humans – an intellectual tribulation from which other species do not customarily suffer. Philosophy, some will say, leads us away from those things we know intuitively in our souls, while animals continue to know intuitively the truths of their souls – knowledge that humans have lost through their engrossment in the wonders of the intellect alone. The benefits of philosophy may not be as advantageous as those who philosophize usually imagine. Nor are the spiritual depths necessarily fulfilling. If the fullness of the soul requires self-flagellation in the manner of a Simeon Stylites, we may well wonder whether animals are superior beings in not so subjecting themselves. The failure of animals to engage in the highest of intellectual and spiritual activities may not be entirely to their detriment.

To the question of whether animals have souls at all, I can, initially, offer little more than the seemingly evasive response that much depends on what is meant by "soul." Suffice it to say for now, in the light of human-animal continuity, that if humans are deemed to be ensouléd, I can see no good reason for denying the souls of our fellow creatures. If, on the other hand, the human is deemed to be no more than what the early American bio-ethicist and Darwinian evolutionist J. Howard Moore called "a pain-shunning, pleasure-seeking, death-dreading organism," then prima facie so too should the nonhuman animal be no more than a pain-shunning, pleasure-seeking, death-dreading – or at least death-avoiding – organism.

Moore continued: "Man is neither a rock, a vegetable, nor a deity. He belongs to the same class of existences, and has been brought into existence, by the same evolutionary process, as the horse, the toad that hops in his garden, the firefly that lights its twilight torch, and the bivalve that reluctantly feeds him."[3] Nonetheless, even though he is writing of *The Universal Kinship* – "the kinship of all the inhabitants of the planet earth" – Moore does not disdain to tell us that "man is the most gifted and influential of animals"[4] even though he is closely related to all the other animals. But, as Plato had already pointed out in *Lysis* (221E-222D), to esteem another as one's kin (*oikeios*) is not the same as to esteem another as a being like oneself.[5] Kinship may be an important basis of ethical consideration. But it is not the only one. To proclaim another as kin is, at best, only an opening gambit in the question of who is entitled to ethical consideration. Even if all animals are related to us, they remain usually regarded as incapable of philosophical and religious endeavour, and are thus, in the view even of many who stress biological kinship, not possessed of the same kind of soul.

Writing also in the first decade of the twentieth century, Edward Clodd, an eminent English proponent of Darwinian theory, addressed the same issue of human-animal continuity:

> Is the mind different only in degree from that of the ape, the dog or the horse? Or does [man] alone possess an entity called the soul, which is *in* the body, but not *of* it? If this be so, it involves a series of breaks in Evolution and of intervals in man's development when, by supernatural intervention, he, the mortal, is endowed with an immortal spirit. For this intervention which must have taken place millions of years ago, when he passed from the protohuman to the human stage, is necessarily repeated in the case of every individual who has been, or who will be, whether he opens his eyes on this earth only to close them in unconsciousness that he has ever breathed, or lives a hundred years.
>
> Man's belief in his exceptional place in nature is a relic of the anthropocentric theory which assumed that the universe was made for him; the vast whole being thus subordinated to an infinitesimal part![6]

Yet, if humanity's arrogant anthropocentrism is no more than a relic, no longer justifiable as a consequence of the theory of evolution, if the interests of the rest of nature may not be "subordinated to an infinitesimal part," on what possible basis may humans deem it ethically appropriate to conduct invasive research on nonhuman animals for exclusively, or primarily, human ends, incarcerate them in what many deem prisons (euphemistically named "zoos") although they have committed no crime, or indeed do anything to them that would be deemed ethically inappropriate to do to humans? If there is nothing that sets humans apart, there can be no grounds for the subordination of "the vast whole ... to an infinitesimal part." Whatever Clodd may tell us, however, few evolutionists – and, for that matter, few others – have drawn the conclusions that Clodd's passage seems to imply: that there are no exclusively human attributes and that, accordingly, there are no natural grounds for treating humans *in principle* differently from other species. Even the animal liberationist Peter Singer asserts that "the right to life of a human being with mental capacities very different from those of the insect and the mouse" is greater than that of species with lower mental capacities.[7]

Evolution does not in fact appear to have provided a different answer to the age-old question of whether animals are for human use. Of course, most evolutionists would eschew teleological language – such as Aristotle's notion that man is intended by Nature to use animals for his ends – although Darwin himself in *The Descent of Man* noted how difficult it was to avoid such language and indicated that when he employed it, the words were to be read as metaphor. Yet his *practical* conclusion would be no different from the conclusion drawn by those who do use such language intentionally. Darwin would still use animals, and feel entitled to use animals, for human ends, as he did in killing pigeons, albeit reluctantly, for his scientific purposes.[8] Presumably, Darwin would not have felt entitled to kill humans for his research. And if he had believed that all species were intrinsically alike and thus of the same moral status, he would not have felt entitled to kill other animals either. And since he did kill them, and felt entitled to do so, it is difficult to escape the conclusion that, for Darwin, humans are somehow essentially different from, indeed superior to, all other species.

I am convinced that both human and nonhuman animals are more than their evolutionary biology. Nonetheless, as both Moore and Clodd imply, the working evolutionary assumption must be that humans and animals are alike in every respect unless and until relevant differences are convincingly demonstrated. And this, in turn, implies that humans and animals are entitled to similar consideration and treatment unless and until different entitlements and the appropriateness of different treatment are convincingly demonstrated – which is not to say that it is of necessity impossible to demonstrate such differences. Moreover, this would appear to be the *logical* implication of the Great Chain of Being as well as evolutionary theory – although, admittedly, *in fact* many (though far from all) who subscribed to the idea of the chain found a special place for the exclusively human soul, which must, as Clodd indicates, be at the very least put in question by the idea of human-animal continuity. The reality, however, is that if the great-chain advocates deemed humans entitled to special consideration as a consequence of something unique, or at least significantly superior, about the human constitution, so too do almost all the evolutionists. The difference is that the evolutionists do not usually tell us explicitly what it is. The question that the evolutionists must answer – almost all of whom support invasive experimentation on animals, and in the later nineteenth century in significantly greater degree than did their adversaries[9] – is that if animals are very much like us in all respects, if there is nothing that sets the human apart, then on what possible ground is it morally appropriate to conduct invasive experiments on animals that we would not permit on humans? If there is a special attribute that sets us apart, we are entitled to be told what it is. On the other hand, if we are told that other animals are quite different from us, on what basis can the invasive research be of any value? None of this is to suggest that there are of necessity no answers to such questions – only that nowhere are we told by such thinkers as Darwin, Moore, and Clodd what they might be.

The Great Chain of Being, or *scala naturae* (scale of nature), was conceived in classical Greece, notably in Plato's *Timaeus* and Aristotle's *De anima,* and systematized in the Hellenistic period by Plotinus, Porphyry, and Macrobius, the last named coming closest to giving the idea its later usage. The concept of the great chain came to the fore in the early Middle Ages, from which time it dominated Western thought until the later nineteenth century – and one still finds more recent profound instances, such as in Elijah Buckner's *The Immortality of Animals* (1903), where he calls it "the scale of existence," or "the scale of perfection," or even "that mighty chain of living beings,"[10] and in the mid-twentieth-century words of John Steinbeck: "One merges into another, groups melt into ecological groups until the time when what we know as life meets and enters what we think of as non-life."[11] According to the classical doctrine, everything in nature is arranged in a hierarchy from high to low, everything having its appropriate niche – from God to the various ranks of angels, to humans, and (via the higher mammals, birds, fish, crustaceans, and reptiles) to the lowest insects, to vegetation, and to rocks. In the words of Immanuel Kant: "Human nature occupies as it were the middle rung of the Scale of Being ... equally removed from the two extremes."[12] Those of a humanitarian disposition – such as Wilhelm Gottfried von Leibniz, John Ray, and Priscilla Wakefield – used the doctrine to stress the interdependence of nature, the worthiness of animals for their own sakes, and our own similarities and responsibilities to our fellow creatures. The nature poet John Clare, who

wrote on behalf of ill-treated badgers, donkeys, and food animals, emphasized the common origins of those who constituted the links of the chain:
"For dogs as men are equally / A link in nature's chain / Form'd by the hand
that formed me."[13] In 1838 William Drummond warned: "Neither let it be
imagined that any animal is so insignificant as not to form an important
link in the chain of being" even though, "as we descend in the scale of
being, the sense of pain diminishes till it altogether ceases"[14] – until, in the
words of the pre-Darwinian evolutionist Jean-Baptiste Lamarck, we reach
the "apathiques," those beings that are "deprived of all sensation, even that
of their very existence."[15] "Gradation," as the Cambridge mathematician
and astronomer Sir John Herchel observed in his *Discourse on Natural Philosophy* of 1830, "is the most prevalent principle in the great scheme of creation ... The human ... is very little above the highest intellectual quadruped.
We may thus trace intellect, however restricted in development, as it passes,
diminishing in degree through the whole of the encephalous animals."[16]
Moreover, humans are not always seen as being at the apex of the scale, for
as Joseph Hamilton wrote in *Animal Futurity* (1877): "the animals might
be greatly raised in the scale of being, as we know men will be."[17] One
could not be at the apex of the scale, as many modern commentators have
imagined the scale to be employed, and still be raised in that scale. Others, to be sure, used the concept to emphasize human superiority near the
apex of the scale, or at least above all the other animals, and claimed the
consequent entitlement to use animals as mere instruments of human purposes. The historian of ideas Arthur Lovejoy described the great chain as
"one of the half-dozen most potent and persistent presuppositions in Western thought."[18]

Now, of course, it is customary to insist that the theory of evolution
contains no such idea as a scale or a chain or a ladder; it is more a branch
system – in the words of Peter Bowler, "an ever-branching tree," for which
"a good classification system would be based on a correct identification of
the crucial points at which the branchings took place."[19] True enough. Yet
Clodd and Moore, like, as we shall see, Darwin in *The Descent of Man*,[20]
tell us that the difference between species is one of degree, whereby, logically, some must have more and some less of some relevant quality. Customarily, we read that it is not through Darwin but through the German
morphologist Ernst Haeckel's evolutionary studies that "Evolution became
a ladder rather than a tree, a linear sequence of stages through which life
had advanced towards the human form."[21] The ladder is a view emphatically
denied by the Darwinians. Yet, despite the denials, a scale, a hierarchy, a
chain, a ladder, is implied, at least in certain respects, in the work by Darwin and Darwinians such as W. Lauder Lindsay, Clodd, and Moore as much

as in the studies by Haeckel. Moreover, even when students of evolution *know* that evolution is not a ladder, they often continue to *think* as though it were. Thus, to take three instances at random, the geneticist Cathy Tsilfidis of the University of Ottawa Eye Institute, who studies regeneration in newts, tells us that the "higher you move in evolution, the less regenerative ability there is."[22] Hers is not an unusual example of statements made in unguarded moments. The thought is not responsive to the theory. Moreover, the thoughts of the founder of socio-biology, Edward O. Wilson, were not made in an unguarded moment. Throughout his work he writes of the "lower" and "higher" animals. "Optimize" and "maximize" are among his most frequently repeated terms. And he does not leave the impression that he is referring alone to greater complexity. Most strikingly in an 1876 address given at South Kensington Museum, T.H. Huxley, one of the greatest of Darwin's supporters, is reported to have proclaimed in a decidedly linear fashion that evolutionists "trace back the dog and the man, and find that at a certain stage, the two creatures are not distinguishable the one from the other."[23] Unfortunately, Hamilton's reporting of Huxley is not entirely convincing. He says also of Huxley that he had lately revived the hypothesis that "animals are mere automata."[24] Certainly, Huxley thought them less sentient than humans, but it is improbable that he thought animals without all feeling, and unfortunately, Hamilton provides us with no evidence for his bizarre assertion. In addition, in his 1879 *Mind in the Lower Animals,* the Darwinian evolutionist W. Lauder Lindsay has a chapter entitled "Evolution of Mind in the Ascending Zoological Scale."[25] Other chapter headings include "The Moral Sense ... in Animals," "Education of Animals," and "Animal Motives."[26] But he does not imagine that he is venturing into uncharted waters, for he says on the first page of his Introduction: "the subject of *Mind in the Lower Animals* is one that has from time immemorial been regarded, if not studied [scientifically], from the most different points of view."[27]

In *The Dreaded Comparison* Marjorie Spiegel remarks that "Darwin himself believed that evolutionary history is no basis for deciding 'who is better than whom.'"[28] In her frequently insightful *Kindred Brutes: Animals in Romantic-Period Writing,* Christine Kenyon-Jones comments on "Darwin's struggle to de-centre man from the universe" and enlists in her cause Sigmund Freud's view that "the researches of Charles Darwin and his collaborators and forerunners" were "one of the three severe 'blows' or 'wounds' that 'the universal narcissism of men, their self-love' had suffered from science." One immediately wonders whether Freud means, by reference to the forerunners, to extend the source of these "blows" back to Thales, Anaximander, and Anaximines among the evolutionary pre-Socratics or merely to Maupertuis, Bonnet, Goethe, Herder, Erasmus Darwin, Lamarck, Geoffroy Saint-Hilaire,[29] and the like among his more immediate successors, all of whom proclaimed some version of evolutionary theory that would have had a similarly detrimental effect on the concept of the exclusivity of humanity. Even if their science was flawed, Freud summarized the process as one in which, Kenyon-Jones writes, "the Darwinite researches had 'put an end to this presumption on the part of man.'"[30] But what, *in fact*, changed with Darwin? In what manner did this presumption decline, never mind come to "an end," even (if not especially) among the Darwinians?

Michael W. Fox observed that "As the concept of human superiority is, as Charles Darwin emphasized, logically and ethically untenable, then the only grounds for contending that it is humankind's right to exploit animals are based on custom and utility."[31] If so, then what other than evolutionary history accorded Darwin the right to determine who was higher? For he assuredly did determine some to be higher. Darwin, as well as Clodd, determined invasive experimentation on animals an appropriate human activity, for the moral acceptability of which some mode of human-animal qualitative distinction has to be assumed. It would appear that, for both evolutionists and those who subscribe to the great-chain theory, man is at or near the apex – perhaps beneath the angels, if their existence is contemplated, but well above the living, breathing beings we experience – and is entitled to special consideration, although other animals are also entitled to some, perhaps significant, but decidedly less, consideration. To be sure, Darwin and Clodd would be entitled to claim that there are grounds other than evolution on which human superiority may be predicated. And, if so, they must acknowledge a qualitative chain of being in the same manner as those who subscribe to the chain idea itself. Certainly, the science of natural selection is quite distinct from the theory of the great chain, but the moral implications derived from natural selection appear very similar to those of the chain.

Despite numerous other sound and significant historical insights in *The Dreaded Comparison*, Marjorie Spiegel eulogizes Darwin's understanding of the human-animal relationship – a view in which she follows most other writers on the history of animal ethics. She concludes that "from the misconstrued concept that humans are evolutionarily better than animals it easily followed (to those who were predisposed to that position) that whites could be evolutionarily superior to blacks."[32] She imagines this interpretation to be alien to Darwin himself. She is apparently unaware that racist utterances, seemingly to follow in Darwin's mind from his evolutionary interpretations, appear with chilling frequency in the *Descent*.[33] To take but a few of myriad examples, Darwin tells us of "savages," in contrast with Caucasians, that their "mental characteristics are ... very distinct; chiefly as it would appear in their emotional, but partly in their intellectual faculties";[34] "The strong tendency in our nearest allies, the monkeys, in microcephalous idiots, and in the barbarous races of mankind, to imitate whatever they hear, deserves our notice";[35] "these idiots [of arrested development] somewhat resemble the lower types of mankind";[36] by contrast with many whites, "Some savages take a horrid pleasure in cruelty to animals, and humanity is an unknown virtue";[37] "Judging from the hideous ornaments and the equally hideous music admired by most savages, it might be argued that their aesthetic faculty was not so highly developed as in certain animals, for instance, the birds";[38] "the fact of the races of man being infested by parasites which appear to be specifically distinct might fairly be urged as an argument that the races themselves might be classified as distinct species"[39] (although, in the final analysis, Darwin demurs on the grounds that "it is hardly possible to discover clear distinctive grounds between them");[40] "The Esquimaux, like other Arctic animals, extend round the whole polar regions."[41] Despite the taxonomical accuracy, the racism is clear. Darwin would not have made a similar comment about the English, like other European animals, inhabiting a temperate zone. Spiegel does refer, as though making a significant distinction between Darwin and some of his colleagues, to "the sentiments of the racist contingency of scientists such as A.R. Wallace who concluded that 'natural selection' could only have endowed the savages with a brain a little superior to that of an ape."[42] Not only did Charles Darwin and Alfred Russel Wallace share the same view of the "savage" brain, but whereas Darwin was emphatic in his support of almost unrestricted vivisection,[43] Wallace was a total abolitionist – although, unfortunately, more because of the practical destruction of the human character that he charged vivisection involved than out of concern for the animals: "I myself am thankful to be able to believe that even the highest animals below ourselves do not feel as acutely as we do; but that fact does not in any way remove my fundamental disgust at vivisection as being brutalising and immoral."[44] It is, for Wallace, the practitioner, and not the patient, who is brutalized.

Of course, Charles Darwin (as well as Alfred Russel Wallace) was doing no more than subscribing to the dominant (although far from universal) racist views of his time and culture. But to imagine that Darwinian evolution involves some idea of equality among humans or between humans and animals is a faulty judgment; "the mental faculties of man and the lower animals do not differ in kind, although immensely in degree," Darwin insists.[45] Indeed, Darwin acknowledges that nonracist views were becoming popular and he explicitly opposed them, remarking that "I have entered into the above details on the immorality of savages, because some authors have recently taken a high view of their moral nature, or have attributed most of their crimes to mistaken benevolence."[46] The animal-sympathetic Unitarian reverend William Drummond would have been one of those whom Darwin had in mind. In Drummond's view, from "the Africans, another race whom many take pride in degrading as an inferior species of mankind, Christians may learn the advantages of kind usage of cattle."[47] If the logic of evolution did not necessarily imply a pecking order, there was still a ladder on which Darwin thought he and his compatriots inhabited the highest rung.

A common difficulty with both evolutionary theory and the notion of a Great Chain of Being is that more often than not, they assume a human superiority over animals without attempting to demonstrate one, and, in the case of at least some evolutionists, they do so while apparently denying the very idea. Richard Milner pointed out that after Charles Darwin had drafted *The Expression of Emotions in Man and Animals* (1872) – the working title referred to the *Lower Animals* – he "resolved not to use the terms 'higher' and 'lower' in his description of animals" and thus "went through the manuscript striking out terms of rank."[48] James Rachels tells us that "such notions as 'higher' and 'lower' are very un-Darwinian ... There is no 'more evolved' or 'less evolved' in Darwinian theory."[49] Marjorie Spiegel refrains the theme, indicating that the "attempt to rank species reflects a chronic misinterpretation of Darwin's evolutionary theory."[50] In fact, a number of Darwinians from Herbert Spencer to Ernst Haeckel and J. Howard Moore have insisted that there are such ranks, and, as I have indicated in the case of Dr. Tsilfidis, modern scientists will often continue to *think* in such terms even though they *know* that evolutionary theory denies it. Moreover, the subtitle of *The Origin of Species by Means of Natural Selection* casts a ready doubt on the denial of the ladder: *or the Preservation of Favoured Races in the Struggle for Life*. More important, not only did Darwin use "higher" and "lower" with great frequency in *The Descent of Man* (1871), but he continued to do so in the second edition of the *Descent* (1874), which was revised and published two years after *The Expression of Emotions* had seen the light of day. His decision to strike out "terms of rank" appears to have been without more than momentary effect. Throughout the second edition of the *Descent*, we continue to read of "the lower animals," "the higher mammals," "animals very low in the scale," and "the organic chain." At one point, Darwin writes of "the ascending organic scale."[51] The idea of a *scala naturae,* in a manner very similar to that of those who subscribed to the great-chain doctrine, is central to

Darwin's mode of thought in the *Descent*. If he removed distinctions of
rank from *The Expression of Emotions,* he did so in the interest of a techni-
cal, scientific point – one that, if he had repeated it for the *Descent,* would
have rendered the whole book meaningless or at least incomprehensible.

The philosopher Mary Midgley, a pronounced advocate of the importance of evolution to the recognition of our ethical responsibilities to other
animals, writes of the "advanced species," of the "seemingly intelligent
actions of the higher animals,"[52] and observes: "As nervous systems grow
progressively more complex throughout the animal kingdom, it is perfectly
reasonable on disinterested grounds to suppose that the capacity both for
suffering and enjoyment also expands. There are many detailed cases in
which we can check this. For instance, we can see how social birds and
mammals are upset by solitude, or by the removal of their young, which
would have no effect at all on simpler creatures, and the power to remember and anticipate trouble, which is a specialty though not a monopoly of
humans, can increase its output."[53] Fair enough. Indeed, Midgley appears
to have captured the unexpressed underpinnings of Darwin's ethical conceptions with regard to animals. But the advocates of the role of evolution
in elevating the status of animals cannot emphasize the distinction between
the "advanced species" and others and then in the next breath refuse to
acknowledge the centrality of "higher" and "lower" to Darwinian thought.

As well as subscribing to a scale of increasing animal complexity, the
discourse of the *Descent* is self-consciously predicated on the idea of evolutionary moral *progress:* "the standard of morality has risen since an early
period in the history of man,"[54] we are told. Darwin adapts the ideas of
W.E.H. Lecky to that end. And, on the basis of evolutionary adaptation,
he believed that "man in the future will be a far more perfect creature than
he now is."[55] Moreover, the fact that Darwin believed it a matter of vital
importance to fight the 1875-76 attempt in the British Parliament to impose significant restrictions on vivisection – in the *Descent* he had written
in favour of vivisection on dogs when "fully justified by an increase of our
knowledge"[56] (note "knowledge," not medical advances) – indicates that
he thought very clearly in terms of human superiority, while many of his
anti-evolutionist adversaries were fighting the opposing cause on behalf of
the animals. This is not to argue that he was *necessarily* wrong to do so. It
is to indicate that, for Darwin, evolutionary theory did not in practice
eliminate distinctions of rank or qualitative assessments of relative animal
merit. Darwin, and indeed most of his followers, subscribed to the underlying moral principles of the Great Chain of Being at the same time
as they developed the theory of evolution by natural selection. If they did
not imagine that the possession of a distinctively human soul permitted

preferential treatment for humans, they must have believed that the differences of degree between humans and other animals were of sufficient magnitude that they amounted in effect to a qualitative difference, one that permitted the use of animals for human purposes. In other words, there is no distinction in principle between the Darwinian and traditional Christian categorization of relative animal value. For both Christians and Darwinians, humanity held dominion over other species. The relevant questions would always be to what degree this dominion required a concomitant responsibility and what dimensions this responsibility would assume.

Differences of degree imply a ladder. Darwin and his fellow evolutionists thought of the human-animal relationship very much in terms of a scale. And sometimes a scale with breaks large enough to constitute a qualitative difference. Thus T.H. Huxley, who some thought to possess a finer evolutionary mind than did Darwin, says:[57] "I have endeavoured to show that no absolute structural line of demarcation ... can be drawn between the animal world and ourselves; and I may add the expression of my belief that the attempt to draw a psychical distinction is equally futile, and that even the highest faculties of feeling and intellect begin to germinate in the lower animals. At the same time, no one is more certain than I of the vastness of the gulf between civilised man and the brutes; or is more certain that whether from them or not, he is assuredly not of them."[58] Certainly, Huxley, like Darwin in the *Descent,* found no problem in writing of "the lower animals." Indeed, Huxley continued to refer to them as "brutes"[59] – even though the term had gradually become unacceptable to some of those of animal sympathy. Nor did he find any difficulty in postulating civilized humans as distinctly superior to the brutes. Indeed, as David N. Livingstone has observed, "On the significance of human beings as a new evolutionary departure [James] McCosh sided with such individuals as [Alfred Russel] Wallace, [John Wesley] Powell, [Edward Drinker] Cope, and [Thomas Henry] Huxley in emphasizing that the advent of human intelligence inaugurated a new phase in evolutionary history. As [McCosh] put it, 'Man modifies Natural Selection, by bringing things together which are separated in physical geography ... The intellectual comes to rule the physical, and the moral claims to subordinate both.'"[60] A teleological evolutionist and professor of metaphysics, the spiritualist codiscoverer of evolution by natural selection, a neo-Lamarckian evangelical, a neo-Lamarckian palaeontologist, and the coiner of "agnosticism" and Darwin's greatest publicist concurred that evolution set humans apart! And thus the intellectual and the moral become the exclusive province of humankind, or almost exclusive; Darwin acknowledged on one occasion a moral sense in other species but counted it the factor in which humans and animals differed more completely than any other.[61] On another occasion he wrote that the "*Imagination* is one of the highest prerogatives of man,"[62] and on yet another that of "all the faculties of the human mind ... *Reason* stands at the summit."[63] Moreover, mankind "manifestly owes this immense superiority [over the other animals] to his intellectual faculties."[64] These faculties dominate the "lower" elements of the instinctive and the passionate. Such evolutionary philosophy returns us to the classical Greek conceptions of Plato in the *Gorgias* and the *Republic,* wherein the soul is divided into parts and the rational (which includes the moral) pursues justice by dominating the honour-loving spirited part, which in turn has precedence over the pleasure-loving passions, thereby creating harmony among them.

In a manner of speaking, for such Wallace-, Huxley-, and Darwin-compatible evolutionists, humanity, encouraged by the forces of evolution, has triumphed over the forces of evolution. It would appear that while not talking about the soul – indeed, sometimes eschewing such language as essentially meaningless – early evolutionary commentators thought of the human-animal relationship in terms similar to those of Plato. They saw humans possessing the most significant psychic parts (i.e., the rational and the moral) in far greater degree than other animals and hence being entitled to superior consideration. The major difference was that Plato thought these characteristics to be aspects of the soul. The Darwinians thought them aspects of the mind. And, as we shall see, sometimes these apparently competing concepts are synonymous.

Certainly, the idea of a significant human superiority over the animals has tended to prevail in Western thought. Thus, writing in 1918, just over half a century after the publication of *The Descent of Man,* and commenting on the role of the Sophists in Plato's *Republic,* Ernest Barker tells us that "Callicles may be regarded as using in advance the doctrine of 'tiger-rights' – to borrow Huxley's term – which many modern thinkers have also used, but which is fundamentally inapplicable to the world of human life."[65] Yet it is, we might suggest, more applicable to humans than Barker indicates and less applicable to other animals in general than Barker imagines. However, Huxley, Wallace, and for all intents and purposes, Darwin would side with Barker. Human mental evolution has allowed humans to escape the bonds of evolutionary law. In possessing faculties greatly beyond those of other animals, the human is of a different order, at least in that the moral laws applied to nonhumans differ greatly from those applied to humans.

In the 1890s James Mark Baldwin, an American psychologist, observed that the theory of natural selection could be developed to allow for animals' choice of a new lifestyle and that natural selection would then favour those who possessed the physical characteristics more suitable to the new habit. Both the anarchist Russian prince Peter Kropotkin in *Mutual Aid: A Factor of Evolution* (1902) and the South African general and statesman Jan Smuts in *Holism and Evolution* (1926) emphasized the role of cooperation in evolutionary development. And in the 1960s the Hungarian-born author Arthur Koestler was arguing that living beings can direct their own evolutionary development along purposeful lines.[66] Most people, evolutionists included, thought of humans as far more capable of choice, cooperation, and self-direction than other species, although other species possessed these abilities in lesser degree. Rousseau had long since argued that the distinguishing human essence was the capacity for self-development ("self-perfection" was his term). As James Wiser explained, for Rousseau: "unlike the animals, the humans were capable of responding to their own historical experience and changing themselves accordingly. In short, humankind is truly an historical species. The animals, according to Rousseau, are constant throughout time. As such, they do not evolve in light of their species experience. Humankind, on the other hand, does."[67] Thus what many see as the *logical* implications of evolutionary theory about human-animal similarities played little role in the ethical implications drawn from the theory by the evolutionists, which, in turn, followed some traditional lines of argument. Clodd's claim that "Man's belief in his exceptional nature is a relic of the anthropocentric theory"[68] may well be valid. But the reality did not change with the advent of evolutionary theory. The evolutionists themselves continued to think in the same terms as Rousseau. Animals mattered. Humans mattered a great deal more. And it was superior reason, a more refined moral sense, the capacity for self-improvement and self-direction that guaranteed humans the greater consideration even when the principles were not elucidated – and they never were elucidated by prominent evolutionists. They were the same characteristics that had always persuaded most of humankind from the pre-Socratics to the utilitarians to think of their own superior worth – in the words of John Stuart Mill, it is "better to be a human being dissatisfied than a pig satisfied." Even though the "immortal soul" was no longer mentioned as the exclusive attribute guaranteeing humans superior consideration, secularist evolutionists had managed to employ concepts that performed the same role.

At the end of the seventeenth century (a century and a half before "natural selection"), John Locke told us that: "There are some Brutes, that seem to have as much Knowledge and Reason as some that are called men; and the Animal and Vegetable Kingdoms are so nearly join'd, that if you will take the lowest of one, and the highest of the other, there will scarce be perceived any great difference between them; and so on until we come to the lowest and most inorganic parts of Matter, we shall find everywhere that the several *Species* are linked together, and differ but in almost insensible degrees."[69] The sense of "kinship" with the animals is no greater in Darwin than in Locke. Yet, if some brutes approximate men, they would appear to be few and far between. If there are "almost insensible degrees" between individual species, overall they constitute a gulf between the least and the most developed. In the words of wildlife storyteller Ernest Thompson Seton, "we and the animals are kin. Man has nothing that the animals have not at least a vestige of."[70] Yet it is a vestige, a part, of which, in the logic of Seton's statement, humans clearly have more. The Darwinian evolutionist John Howard Moore tells us explicitly of the human that in "important respects he is the most highly evolved of animals."[71] A hierarchy is indicated. Yet there is an inherent problem in the notion of a hierarchy, whether propounded by the advocate of the chain or by the evolutionist. It encourages us to imagine that other species become worthy, become valuable, become entitled to moral consideration, the more they possess the attributes of human beings, the more they approximate *our* abilities, *our* emotions, *our* rationality. Humans are the standard against which other animals are measured. Such comparisons encourage us to ignore that different species possess what John Rodman called "their own character and potentialities, their own forms of excellence, their own integrity, their own grandeur."[72] Would such categories as philosophical reason and reflective morality have ever been chosen as the relevant factors if it were not the primary possessors of such faculties who were making the selection?

It is important to contemplate the echolocation capacities of bats and dolphins and what they mean for their lives. The displays of the bower bird, the communicative versatility of the honeyguide bird, the cognitive map-charting of the honeybee, the engineering of the beaver, the loyalty of the wolf, and the camouflage of the assassin bug, not to mention the flight capacity of birds in general, should indicate to us that, to the extent that a hierarchy is conceptually appropriate, as it *might* be said to be in the above instances of the bower bird and the wolf, the human being stands far from the apex, and, in the instances of the bat, the dolphin, the honeybee, the honeyguide, and birds in general, the nonhuman animals engage in activities inconceivable for humans. In these latter instances, no hierarchy exists. The animals are, quite simply, in a league of their own. If in the evolutionary age we can continue to regard animals as appropriate instruments of human use – eating them, experimenting on them, using them to draw the plough are examples of human use – despite what some consider the logical but unpractised implications of evolutionary theory, it should scarcely surprise us that earlier ages sought a means whereby humans could conceive of themselves as entitled to greater consideration. The exclusive human *soul* may be thought to have offered itself as a convenient justification for such usage. The concept was indeed sometimes so used. But it is, however, a far from simple concept.

II

It is instructive to inquire about the historical usage of the term "soul." Chalmers' edition (1824) of the Reverend Henry John Todd's "corrected and enlarged quarto edition" (1818) of Dr. Samuel Johnson's famed *Dictionary of the English Language* (1756) provides us with significant insight into the variety of usages by a number of the more important writers in British literary history. Thus under "Soul" we read:

> (*ſaul*, Sax.; *saal*, Icel.; *seele*, Germ.) The immaterial and immortal spirit of man. [*Richard*] *Hooker*. Intellectual principle. [*John*] *Milton*. Vital principle. *Ibid*. Spirit; essence; quintessence; principal part. *Ibid*. Interiour power. *Shakespeare*. A familiar appellation expressing the qualities of mind. *Ibid*. Human being. [*Joseph*] *Addison*. Active power. [*John*] *Dryd*[*en*]. Spirit; fire; grandeur of mind. [*Edward*] *Young*. Intelligent being in general. *Milton*.[73]

It is the glory of the English language that its multifarious roots provide the potential for both subtlety and nuance often absent from other languages. Unfortunately, the variety of roots also encourages obfuscation, inspires us to imagine as meaningful distinctions that are merely verbal. Sometimes we find one concept expressed by different terms. Thus, in the above citations, Hooker, Milton, and Young move us toward the concept of "spirit." If we turn to "Spirit" in the same dictionary, we read:

> (*spiritus*. Lat[in]) Breath, wind. [*Francis*] *Bacon*. (*Esprit*. Fr[ench]) An immaterial substance; an intellectual being. [*John*] *Locke*. The soul of man. *Eccl*[*esiastes*] xii. An apparition. *St. Luke*, xiv. Temper; habitual disposition of mind. *Milton*. Ardour; courage; elevation; vehemence of mind. *Shakespeare*. Genius; vigour of mind. [*Edmund*] *Spenser*. Intellectual powers distinct from the body. [*Edmund Hyde, 1st Earl of*] *Clarendon*. Sentiment; perception. *Shakespeare*. Eagerness, desire. [*Robert*] *South* [ey]. Man of activity; man of life, fire and enterprise. *Shakespeare*. Persons distinguished by qualities of the mind. [*Gilbert*] *White*.[74]

St. Luke's apparition persuades us to take a look at "Ghost," which is described as the "soul of man. [*John*] *Pearson*. A spirit appearing after death.

Dryden," and "The third person of the adorable Trinity, called the Holy Ghost. *The Apostle's Creed.*"[75] In *The Immortality of Animals* (1903), Elijah Buckner tells us that the "words 'soul,' 'mind,' 'spirit,' 'ghost,' and 'eternal life,' are among those most frequently used ... to express the immaterial part of man ... but as they all convey the same idea it is of no consequence which is employed."[76] Yet it is, in fact, of considerable consequence, for the nuanced differences among the competing concepts have competing implications. Nor did the view of Buckner stand uncontested. Writing in *The Veterinarian* in 1840, W.C. Spooner had already indicated a decided opposition to the identification of the soul and the spirit.[77]

While immortality is only occasionally mentioned in these instances of the use of soul, spirit, and ghost, where it is mentioned the application is either explicitly or implicitly to humans alone. And, where it is not mentioned, often the immortality is implicit rather than an explicit characteristic of human destiny alone. While this is far from convincing evidence that immortality of the soul or spirit was thought to be an exclusively human prerogative, it is a prima facie indication that human immortality was at the very least a primary concern, certainly of greater concern than any potential immortality of nonhumans.

The multiplicity of uses – some of which are far from any concern with immortality – indicates the complexity of the concept. In the 1951 edition of the *Concise Oxford Dictionary,* the first definition of soul given refers to the "immaterial part of man." Immortality is not included, although the example offered of its usage refers to immortality. In the 2001 edition, the first definition is more explicit: "the spiritual or immaterial part of a human, regarded as immortal." The second of the 1951 definitions is more generic but equally common: "One's moral or emotional nature or sense of identity," followed by "emotional or intellectual energy or intensity." However important the idea of immortality might be to some uses of the concept, it is quite clear that the idea also has both meaning and vitality for those who would deny all possibility of immortality, whether as an attribute of the animal or the human soul.

In *Cassell's Latin Dictionary,* under "anima" one finds that "animal" is derived from – and that, in one of its senses, it means – *"the vital principle, the soul* (anima, *physical*; animus, *spiritual*)," which may appear rather confusing but which, the dictionary informs us, in Cicero's usage, can also mean *"the rational soul,"*[78] suggesting a spiritual rather than a physical entity. And this despite the fact, which the dictionary does not tell us, that the Epicureans tended to use "animus" exclusively for the rational soul. Moreover, the confusion is exacerbated if one turns to the entry under "soul," for there we are told that "anima" refers to "the living principle," whereas "animus" refers to "the emotional nature."[79] What should be clear is that the etymology of "soul" is far from clear with respect to the term's meaning.

The Roman satirist Juvenal applied the categories of *anima* and *animus* distinctively to human and animal by telling us that "Tantum animas, nobis animum quoque"[80] – that is, while the other species possess *anima,* humans have, beyond that, *animus.* At the turn of the nineteenth century, William Gifford translated this quotation and the previous three lines of Juvenal's *Satire XV* as:

> We, from high heaven, deduce that better part,
> That moral sense denied to creatures prone
> And downward bent, and found with man alone:
> For he who gave this vast machine to roll,
> Breathed life in them, in us a reasoning soul.[81]

To be sure, Gifford's translation takes us just a little further than Juvenal's original. But the animal-human distinction is clear enough: animals have a lower order of soul than do humans, the distinction being that their ordering principle is instinct, whereas ours is reason. Of course, one of the more important questions that arises from this distinction is: What is the consequential difference of status between humans and animals? What does it imply for immortality? Cicero stated unequivocally that whatever is sentient, has knowledge, and is discerning, is celestial and divine, and for that reason eternal.[82] The question yet remains: Which animals fit the category? For Cicero, the answer would appear to be humans alone.

Historically, the rational soul was often conceived of as the equivalent of the immortal soul. Even if, technically at least, "anima" refers to the seemingly oxymoronic physical soul – I follow Plotinus in believing that "if the Soul were a corporeal entity there could be no sense-perception, no mental act, no knowledge"[83] – in the medieval era "anima" was the customary term for the soul in general, too, as indicated by the contemporary entitling of Aristotle's originally unnamed work on the soul in its various dimensions as *de anima* and by the similar usage throughout, for example, Jean Bouridan's fourteenth-century Latin work, entitled (in French; it has no Latin title) *Le traité de l'âme* (Treatise on the soul).[84] Amid the complexities, and perhaps even confusions, one thing seems reasonably clear: the soul was thought to possess both immaterial and material dimensions, thus incorporating ideas of spirit, mind, psyche, heart, and brain – although often enough the brain was referred to as the material "seat of the soul." There is nothing in such usage that in principle denies either a mortal or an immortal soul to nonhuman animals. Nonetheless, one might be inclined to say that usage is so varied, so obfuscate, that any attempt to discern consistency and meaningfulness is doomed to failure. Since the concept has, however, played such a vital role in intellectual history, the task is necessary. Moreover, the editors of Funk and Wagnalls English dictionary did a commendable job in the mid-twentieth century of trying to provide some clarity. In addition to the customary definitions, the dictionary offers the following under "soul":

> *Synonyms:* mind, spirit. The *soul* includes the intellect, sensibilities, and will; beyond what is expressed by the word *mind,* the *soul* denotes especially the moral, the immortal nature; we say of a dead body, the *soul* (not the *mind*) has fled. *Spirit* is used especially in contradistinction from matter; it may in many cases be substituted for *soul,* but *soul* has commonly a fuller and more determinate meaning; we can conceive of *spirits* as having no moral nature; the fairies, elves, and brownies of mythology might be termed *spirits,* but not *souls.* In the figurative sense, *spirit* denotes animation, excitability, perhaps impatience; as, a lad of *spirit;* he sang with *spirit;* he replied with *spirit. Soul* denotes energy and depth of feeling, as when we speak of *soulful* eyes; or it may denote the very life of anything; as, the *soul* of harmony. Compare MIND.[85]

And when we make the suggested comparison, we read, again following the definitions:

> *Synonyms (noun):* brain, consciousness, disposition, instinct, intellect, intelligence, reason, sense, soul, spirit, thought, understanding. *Mind* includes all the powers of sentient being apart from the physical factors in bodily faculties and activities; in a limited sense, *mind* is nearly synonymous with *intellect,* but includes *disposition,* or the tendency towards action, as appears in the phrase "to have a *mind* to work." The *intellect* is that assemblage of faculties which is concerned with knowledge, as distinguished from emotion and volition. *Understanding* is distinguished by many philosophers from *reason* in that *reason* is the faculty of the high cognitions of a priori truth. *Thought,* the act, process, or power of thinking, is often used to denote the thinking faculty, and especially the *reason* ... As the seat of mental activity, *brain* is often used as a synonym for *mind, intellect, intelligence. Sense* is used as denoting clear mental action, good judgment, acumen; as, he is a man of *sense,* or he showed good *sense. Consciousness* includes all that a sentient being perceives, knows, thinks, or feels, from whatever source.[86]

Unfortunately, not all the historical figures who have written of animals and souls have possessed the rigour and clarity of the Funk and Wagnall editors. Nonetheless, since the editors are not advancing philosophical positions per se but are explaining the *usages* of language – which are ever shifting – we shall continue to bear their points in mind in attempting to discern the meanings inherent in historical statements. What should strike us immediately is the perceived relationship of soul to moral nature, mind, and reason. It should not come as a surprise, then, that the early writers who distinguished humans from other species wrote not just of the soul, or even of the immortal soul, but primarily of the rational soul, which implied for them the thinking soul with a conscious moral sense. Yet we can at the same time see the source of the confusions: the references to soul, employing ideas of sentience and will, where the human-animal distinctions are far less clear. Moreover, those who thought of an exclusively human rational soul did not deny animals their capacity for any reason but only for a particular speculative and philosophical kind of reason. And those who insisted that animals were rational wrote not of their capacity for philosophical speculation but of instrumental reason. Those who denied and those who affirmed animal reason would often talk past each other. What they were affirming and what they were denying were not common conceptions.

The abundance and diversity of ways in which concepts have been employed should alert us to a significant difficulty of analysis. When one person claims that animals have souls, she may mean something significantly different from what is meant by another person employing the same words. Indeed, it is quite conceivable that one person who claims that animals lack souls may be in substantial agreement with another who claims they are ensouled and vice versa. Clearly, caution is needed in any discussion of animal souls. We must constantly be asking what was meant and implied in particular cases. Was the person writing of a rational, a conscious, an eternal, a moral, or a sentient soul or writing of all, or some, or none of these? When the Swiss-American anti-Darwinian naturalist Louis Agassiz wanted to proclaim an immortal soul for animals in 1864, he claimed that "instinct is immortal," thus conceding human rational exclusivity, but indicating its irrelevance to the issue. By contrast, when Joseph Hamilton wanted to make his case for *Animal Futurity* in 1877, he suggested "that the various distinctions between reason and instinct that have been set up, do not serve to show that reason is peculiar to man, nor instinct peculiar to brutes. The two faculties may be as distinct as they have generally been described, but they are dove-tailed and blended together in the intelligence of both the human and the brute species."[87] Of course, the historical figures often pontificated on animal souls without making any of the distinctions that could help us to know with greater confidence what was meant. The problem is, however, somewhat reduced by the fact that what is often of the greatest interest is whether humans and nonhuman animals are deemed to share the same type or different types of soul, the same type or different types of spiritual attributes,[88] or whether one is ensouled and the other a purely material existence. And that is usually somewhat easier to discern.

I I I

In pre-Socratic Greek thought the idea that animals possessed souls was generally uncontroversial since the soul was often taken simply as that which animates, whether that be the breath (*pneuma*) or something else. As Buckner explains, "The word soul in its original signification stood for the principles which govern life both in man and the lower animals. It is true that the modes of explaining it were various. Sometimes it was regarded as the mere harmony of the bodily functions, and sometimes as a distinct entity of higher ethereal nature, but no essential distinction was made between the soul of man and the soul of the lower animals until a comparatively recent date."[89] However, the soul was also often regarded as the immortal aspect of an otherwise corporeal being. Already in the fifth century BC we find Alcmaeon, as later reported by Diogenes Laertius, telling us that "the soul is immortal and ... it moves continuously like the sun."[90] In general, although not always consistently, the soul was not seen to be entirely separate from material existence. The Greek word for the soul is *psyche,* which also means breath in some instances and mind in others; and the word is also used to denote the butterfly. Given the Homeric notion of the soul as a shadow of the body (*soma*), it is an application full of enchanting poetic imagery. In Pythagoras's conception, at least as described by later Pythagoreans – we will meet more of Pythagoras presently – the soul is related to the body; however, while the *psyche* has immortality (*athanathon*), the *soma* is mortal. The soul of humans is divided into three parts: intelligence (*nous*), passion (*thymos*), and reason (*phren*).[91] *Phren* is literally the midriff, but refers, almost indiscriminately, to the heart, will, or mind. It was seen as the seat of the highest mental faculties – which serves as a healthy reminder that just as Greek philosophy was in its infancy, so too was Greek knowledge of physiology quite negligible – although it is also worth noting that our continued use of "heart" is no more physiologically enlightened, even if the meaning remains relatively clear. Only *phren,* as the highest faculty of reason, was considered unique to human beings, *nous* and *thymos* being possessed by all animals. (It is a matter of some confusion that later Greeks did not use *nous* and *phren* in the same way.) Of course, this immediately raises the question of how transmigration of souls between humans and animals – of which more anon – was possible if there was something significantly different between human and animal souls. Paradoxical, too, is that, according to some Pythagoreans, vegetables were also ensouled – even though they possessed neither *phren,* nor *nous,* nor *thymos.* On the whole, however, the distinction between animal and plant held firm; Plato, for example, tells us that animals possessed a distinguishing self-consciousness (*Timaeus,* 77 BC), a view followed by Porphyry and also by some of the Stoics, including Hierocles.

It is important to remind ourselves that, as Martin West wrote: "Early Greek philosophy was not a single vessel which a succession of pilots briefly commanded and tried to steer towards an agreed destination ... it was more like a flotilla of small craft whose navigators did not all start from the same point or at the same time, nor all aim for the same goal; some went in groups, some were influenced by the movements of others, some travelled out of sight of each other."[92] Philosophy, West tells us, "meant originally something like 'devotion to uncommon knowledge' – but it did not acquire a specialized sense or wide currency until the time of Plato."[93] What we now call philosophy was an essentially unclear, unco-ordinated, and fumbling – if nonetheless often brilliant – myth-replacing adventure of the speculative minds of Ionia and Magna Graecia. We should not expect, nor do we get, cohesion and clarity or always even meaning-fulness. Religion, myth, and the beginnings of philosophy at first inter-mingled – West says Pythagoras "seems to have been part philosopher, part priest, part conjuror"[94] – and gave us concepts of spirit, soul, and mind reflecting a concern with immortality but also infused with the notion of the quintessence of the self (the more important of the individual attri-butes that constitute a personality), especially the higher mental disposi-tions of character, but always seen from a human standpoint. Animals came to be measured against those essentially anthropocentric conceptions, as indeed, they still are.

This anthropocentrism does not, however, arise primarily as a direct sequel to any conscious attempt to elevate the status of humans and den-igrate the status of animals. The fact that we tend to see things from the human perspective – just as the wolf sees things from the wolf perspective and the giraffe from the viewpoint of the giraffe – is the natural conse-quence of our being human animals, evolved through natural selection to adapt ourselves to our environment in the interest of our individual selves,

our immediate kin, and our species. What we see will inevitably, in the
first instance, reflect our interest as humans. This is far from the whole
story of what it is to be human. But it is an intrinsic part of the human
condition, the core of our animal condition, without which the central
elements of our behaviour are incomprehensible. Evolutionary forces did
not develop us to be rational, ethical beings engaged in the task of determining intellectually the appropriate relationship between our individual
selves and the remainder of our fellow creatures. We evolved to accomplish
certain tasks useful for our individual, kin, and species survival. Those, such
as the reductionist selfish gene theorists and the socio-biologists,[95] who
postulate this as the whole of the human character miss a great deal. But
those who ignore our animal nature and imagine we make decisions and
determinations in an intellectual and moral vacuum, from which our evolutionarily adaptive animal needs are excluded, miss a great deal more.
The human animal may well be more than its evolutionary development.
But it is most certainly also defined by this development. To try to explain
the human animal and its ethics entirely in philosophical terms is to make
the mistake of imagining humans as ethereal beings who have escaped all the
impulses, orientations, and commands of primordial individual, kin, and
species interests.

What is beyond our evolutionary development, although informed by
this development, is our desire to acquire knowledge for its own sake, not
just for its instrumentality, and our occasional willingness to sacrifice our
immediate basic needs to this goal. Moreover, and again informed but not
determined by our evolutionary development, is the idea of life as a self-
conscious, risk-taking adventure, as exemplified so well in the Homeric,
the Samurai, the Arthurian, and the Norse legends. Life is a journey of
self-discovery. Humanity's break with evolution as the sole driving element
of behaviour is immortalized in the famous inscription on the Temple of
Delphi: "Know thyself, and thou will know the Universe and the Gods."
Through self-discovery you will elevate yourself above your *mere* animal
self and come to understand your environment in a manner appropriate
to deities.

These intellectual adventures bring humanity to a self-conscious concern with being the best one can be, with being an accomplished person,
with acquiring what Machiavelli and Hobbes called "glory," and thus, in
fact, with becoming, by measuring oneself against one's fellows, competitive. We pursue the development of our own character, our own self-
identification, by distinguishing ourselves from others in terms of our
courage, our cunning, our reason, and our concern for the interests of our
familial kin, this being the basis of our self-conscious moral sense. As we

develop our social self-awareness, we come to see ourselves in relation to those to whom we are most closely bound by "blood" – for mutual kin protection is one of our species survival needs. Next we include in our conceptions of "like us" those who share the same language, and in our conceptions of "other" we include those who lack the capacity to communicate to us in our tongue. Thus the Greeks classified those who spoke in other tongues as "barbarians" – that is, those who spoke like sheep: "baa, baa." The Makritare of the rainforest of the Orinoco speak the So'to language. So'to means twenty, the number of digits possessed by humans. All non-So'to-speaking peoples, including their immediate neighbours the Kariña, are regarded as eighteen-clawed predatory cats and hence as beneath entitlement to full ethical consideration. Those who share one's religion are likewise deemed to belong, while those who do not are "infidels" (i.e., unbelievers) or pagans and heathens thought to be unenlightened, uncultured, unhallowed by the deity. Yet, while those who are not like us are "other," they are not completely beyond the pale, a phrase initially used pejoratively in Ireland in the eighteenth century to refer to those who were beyond the effective jurisdiction of English-controlled Dublin and hence outside the bounds of acceptable behaviour – as determined by English norms, of course. It is an instructive part of Judaic moral law that, wisely, it takes consideration for fellow Jews as more likely by nature than consideration for non-Jews. Thus in the Talmud we read that "it is forbidden according to the law of Torah to inflict pain upon any living creature. On the contrary, it is our duty to relieve the pain of any creature, even if it is ownerless or belongs to a non-Jew."[96] Likewise, "When horses, drawing a cart, come to a rough road or a steep hill, and it is hard for them to draw the cart without help, it is our duty to help them, even when they belong to a non-Jew, because of the precept not to be cruel to animals."[97] The authors of the law found it important to emphasize consideration for non-Jews, whose interests might otherwise be disregarded, as well as for animals. While our instinctive communitarian consciousness, our particularism, encourages us to give preference to those who are more closely related to us, our intellectually informed moral condition encourages us to believe that all those who share similar relevant properties, and to the extent that they do so, are owed a measure, albeit a lesser measure, of obligation, including both other humans and animals. Exemplifying the Victorian belief in the inevitability of progress, W.E.H. Lecky wrote, "At one time, the benevolent affections merely embrace the family, soon the circle expanding includes first a class, then a nation, then a coalition of nations, then all of humanity, and finally its influence is felt in the dealings of man with the animal world ... there is such a thing as a natural history of morals, a defined and regular order, in which our feelings are unfolded."[98] Even then, however, the natural inclination is to put the interests of one's own "blood," or genes, first. The selfish gene theorists have a point – even if it is one they grossly exaggerate.

In reality, most cultures have traditionally believed that cruelty to others is acceptable, provided, and to the extent that, it can be justified by some more immediate kin interest. The effect of self-conscious contemplation has been to diminish the extent to which our refusal to acknowledge the rights of others can be readily legitimized. Yet there remains the drive to acknowledge those who share our attributes as most deserving of consideration – hence the wars fought over tribe, religion, and nation. And as a species we identify ourselves against the "other" by what we see as the defining characteristics of humanity – intellectual reason, language, religion, a moral sense – those factors that, we believe, differentiate us from other species. Indeed, while the practice has been in decline since Karl Marx, it was once customary for philosophers to define the human essence in terms of whatever it was that distinguished us from other beings. In this manner, such diverse minds as Aristotle, Cicero, Augustine, Aquinas, Descartes, Hobbes, Hooker, Burke, Rousseau, Wollstonecraft, David Hume, Feuerbach, John Stuart Mill, and Karl Marx all sought the definingly differentiating human factor. For Cicero, Aquinas, Wollstonecraft,[99] and Mill[100] the essentially human characteristic was the possession of reason, for Hume only the degree of reason, for Descartes consciousness, which implied sentience, for Augustine and Hobbes individuality, for Hooker speech, for Burke religion, for Rousseau choice, for Feuerbach "the divine trinity" of "Reason, Will, and Affection," and for Marx creative labour. What such scholars shared in common was the belief that humans could be most fulfilled precisely through those characteristics that other species lacked, thus drawing the divide ever greater between our conception of what it is to be human and what it is to be animal. Whatever it was that made us unlike the other animals was the characteristic we should develop to make us more completely human and fulfilled.

Certainly, while most acknowledged that we owed some responsibility to other beings, animals received a relatively low level of consideration. As the Confucian philosopher Mencius (c. 372 BC to c. 289 BC) expressed it: "The superior man feels concern for creatures, but he is not benevolent to them. He is benevolent to the people but he does not love them. He loves his parents, is benevolent to the people, and feels concern for creatures."[101] The message seems to be, and it comes close to being the universal historical human message, that our primary duty is toward our immediate kin, secondarily to humanity in general, and finally toward the animal realm.

Moreover, the implication would appear to be that we should consider the interests of animals to the extent that they do not conflict with our responsibility toward humans in general and our kin in particular. What in most cultures and among most philosophers has been at issue is the relative weight that should be given to each of these. All sentient beings matter, but some matter more than others. The question remains one of degree – and, indeed, which animals are to be ranked above which other animals? Alternatively, we have to ask whether this question is itself biased. On what grounds can questions of degree be relevant if we cannot know the degrees of what substances, what categories, what concepts are most appropriately in question? It is easy to recognize that as humans we tend to choose those categories in which humans excel. It is easy to see how we can do this as particularist, communitarian animals. It is very difficult to see on what criteria we can do this if we are attempting to use some kind of universal standard in which humans and animals are treated prima facie as equally worthy of consideration. Of course, we do in fact use human standards almost all the time. In fact, we choose the human standards without any greater justification than that we are human and that human attributes are the most important and most naturally appealing *to us*. I shall return to this topic in the concluding chapter.

IV

Aristotle is an appropriate starting point for an understanding of the animal soul, for it is he who is customarily portrayed as the primary pernicious philosopher who determined the perceivedly low status of animals in the Western tradition. Thus, for example, Peter Singer noted that "Aristotle holds that animals exist to serve the purposes of human beings ... It was the views of Aristotle, rather than those of Pythagoras, that were to become part of the later Western tradition."[102] Jim Mason observed that Aristotle "was a strong advocate of dominionism. For him, nature was a hierarchy of beings, and man, having superior mental capacities reigned at the top."[103] And James Rachels queried, "In what way, exactly, do we resemble the Almighty? The favoured answer, throughout Western history, has been that man alone is rational. Aristotle, expressing the Greek view of the matter, had said that man is the rational animal, and differs in this respect from all other creatures."[104] Painted in *very* broad strokes, this overview is not far off the mark. But the strokes are so broad that all the nuances and all the nay-sayers, who are many and significant, are lost. There is far more than Aristotle to the Western tradition. Indeed, there is far more than one "Greek view of the matter." Moreover, Aristotle's so-called dominionism is far more subtle than one would expect on the basis of these assertions. And animals were not so "irrational" as this picture would suggest – even when they were referred to as "the irrational animals"![105]

Aristotle opened the *Metaphysics* with the assertion: "All men by nature have a desire to know. A sign of this is the joy we take in our senses, for quite apart from their usefulness we love them for their own sake."[106] He ascribes sensation to all animals and memory, intelligence, and the capacity for learning only to some. But it is philosophical reason, language, and the desire to know for its own sake that, for Aristotle, set humankind apart. Yet many traditions have questioned whether knowledge should be pursued for its own sake – on occasion, Buddhists, Confucians, Hindus, Jews, and Christians have all repudiated such a view. Their common fear has been that, in pursuing knowledge as an end in itself, humans would alienate themselves from their true natures and become more dissatisfied. Almost two thousand years after Aristotle, as other aspects of Aristotelianism were coming into disrepute, Thomas Hobbes, whose influence in the later seventeenth and early eighteenth centuries was paralleled only by that of Descartes, made a similar point: "Man is distinguished not only by his reason; but also by this singular passion from other animals ... which is a lust of the mind, that by a perseverance of delight in the continual and indefatigable generation of knowledge, exceeds the short vehemence of any carnal pleasure."[107] Neither Aristotle nor Hobbes denies animals the attributes of consequential reasoning, sentience, or memory, at least some animals. Aristotle and Hobbes are in accord that by nature humans are of a philosophical disposition, while other animals are not. And like Aristotle, Hobbes believes, despite some significant qualms about language leading to conflict, strife, and war, that the capacity for language has produced a superior life. Without it, "there had been amongst men, neither Commonwealth, nor Society, nor Contract, nor Peace, no more than amongst Lyons, Beares and Wolves."[108] Others, Rousseau, for example, are inclined to the view that humanity's speculative capacities are cultural acquisitions, with which we adorn ourselves as we divorce ourselves from our natural heritage.

It is, however, notable that even those who claim an inherently different speculative nature for humans and animals do not necessarily draw the same conclusions from it. Aristotle believes that our capacities for reason and speech make us – despite the acknowledgment of some admirable traits in other animals – an essentially superior species with the potential at least for essentially superior motives, such as the pursuit of justice and morality. Yet for Hobbes, human and animal impulses to action were essentially the same. Notoriously, Aristotle argued that "all animals must have been made by nature for the sake of man."[109] Far less well known – because such knowledge would interfere with the notion of a pernicious Western animal antipathy – is Hobbes's question: "I pray, when a lion eats a man and a man eats an ox, why is the ox made more for the man than the man for the lion?"[110] Humans and animals were sufficiently similar, despite the exclusive lust for knowledge among humans, that there was no logical justification for the view that animals were for human use. For Hobbes, the divisions between humans and animals are less marked than for Aristotle, for reason and language, while distinctly superior human attributes, are used to promote self-interested rather than noble ends. One of the purposes of language is to enable us to lie and calumniate.[111] Hobbes was a nominalist who insisted (not always consistently) that concepts like justice and rights have no real meanings but are merely tools developed to pursue the animal ends of those who develop them: "one man calleth *Wisdome,* what another calleth *Feare;* and one *Cruelty* what another *Justice;* one *Prodigality* what another *Magnanimity* ... such names can never be true ground for any ratiocination."[112] Nature makes us prey for some animals, predators to others. Moreover, "It is not prudence that distinguishes man from beast. There be beasts that at a year old observe more, and pursue that which is for their good more prudently, than a child can do at ten."[113] For Aristotle, there is something divine in our nature that distinguishes us from other creatures. According to Aristotle, the divine mind is characterized by the *reflexive* character of its thought, a capacity almost universally lacking in nonhuman animals. This self-reflection underlies self-consciousness; and for Aristotle, while animals are certainly conscious, they are not self-conscious, an opinion that many naturalists believe to have been largely born out by ethological study, the few exceptions to the lack of self-consciousness including chimpanzees and dolphins. For Hobbes, we are needs-satisfying, pleasure-pursuing animals like all the rest. Indeed, despite Hobbes's occasional adulation of human reason, more often than not, he sees humans not as the rational but as the rationalizing animal. Reason is merely an instrument for achieving the ends determined by the passions: "as oft as reason is against a man, so oft will a man be against reason."[114] Indeed, "life is but a motion of limbs ... I put for a general inclination of all mankind, a perpetual and restless desire of power after power, that ceaseth only in death."[115] This power is necessary to pursue successfully one's animal ends. There is, for Hobbes, nothing intrinsically special about humanity, other than a greater sense of individuality, at least not in terms of an altruistic morality, not even the possession of an individual immortal soul.[116]

In reality, the classical thinkers allotted humans a greater measure of divinity than they did other species, but they did not deny animals their divinity entirely – thus Aristotle's equivocation in *On the Parts of Animals*: "For of all living beings with which we are acquainted, man alone partakes of the divine, or at any rate partakes of it in fuller measure than the rest."[117] If it is then a rational soul that makes the human divine, what of the animal?

> If now this something that makes up the form of a living being be the soul, or part of the soul, or something that without the soul cannot exist (as would seem to be the case, seeing at any rate that when the soul departs, what is left is no longer a living animal, and that none of the parts remain what they were before, excepting in mere shape, like the animals that in the fable are turned into stone); if, I say, this be so, then it will come within the province of the natural philosopher to inform himself concerning the soul, and to treat of it, either in its entirety, or, at any rate, of that part of it which constitutes the essential character of an animal; and it will be his duty to say what this soul or this part of the soul is; and to discuss the attributes that attach to this essential character.[118]

For Aristotle, animals are more than mere material beings, for they possess a soul, their essence, without which the corporeal entity cannot exist, and, on the death of the animal, the soul departs the body. Indeed, Aristotle tells us that there is "something natural and something beautiful" in "every kind of animal" and that each animal "body must somehow or other be made for the soul."[119] Some animals are suited "for the development of courage and wisdom," and "some are of a more intelligent nature" and some less.[120] For Aristotle, clearly, animals possess souls, even if these souls differ, at least by degree, from human souls. The soul is still the animal's quintessence.

If humans "partake of the divine" in greater degree than other animals, although divinity is not entirely excluded from animals, what, for Aristotle, is it that sets humans apart and accords them this participation in divinity? According to Aristotle, the human soul performs three functions: the life of nutrition and growth, the life of sense perception, and the life of reason, each with its own standard of excellence. The first type of life is shared with both the animal and vegetative world, the second with the animal world, and the third is possessed exclusively by humans. Aristotle does not mean that animals lack all capacity to reason, but that they are incapable of speculative abstractions, of understanding "universals." In Aristotle's view, "the good of man is an activity ... in conformity with excellence or virtue, and if there are several virtues in conformity with the best and most complete."[121] These are the intellectual virtues. And nonhuman animals, for Aristotle, cannot achieve the intellectual or moral heights of which only humans are capable. Indeed, it would not be unreasonable to conclude that Aristotle has stated what has remained a primary, although far from sole, Western conception of the human-animal relationship. And, because of its godlike qualities, it is the rational soul, or the rational part of the soul, that is more customarily deemed capable of immortality. Moreover, it is in this that, for Aristotle and many who followed him, humanity achieves its ultimate distinctiveness. Yet in those societies where animals were deemed divine, their fate was certainly no better. They were killed and eaten precisely because they were divine. They were sacrificed and eaten precisely because to do so was to participate in the divine.

Aristotle elaborates his position in the first five chapters of the *Politics* in a manner, on the whole, less than sympathetic to animals (although he is almost equally unsympathetic to women and slaves):

The reason why man is a being meant for political association, in a higher degree than bees or other gregarious animals can ever associate, is evident. Nature, according to our theory, makes nothing in vain; and man alone of the animals is furnished with the faculty of language. The mere making of sounds serves to indicate pleasure and pain, and is thus a faculty that belongs to animals in general; their nature enables them to attain the point at which they have perceptions of pleasure and pain, and can signify those perceptions to one another. But language serves to declare what is advantageous and what is the reverse, and it therefore serves to declare what is just and what is unjust. It is the peculiarity of man, in comparison with the rest of the animal world, that he alone possesses a perception of good and evil, of the just and the unjust, and of other similar qualities ... he differs from animals, which do not apprehend reason, but simply obey their instincts.[122]

Later, he adds, "Animate beings other than men live mostly by natural impulse, though some are also guided to a slight extent by habit. Man lives by rational principle too [as well as by natural impulse and habit]; and he is unique in having this gift."[123] Animals possess a measure of intelligence and the capacity for calculative reason, but they lack the capacity for this philosophical principle, which is humanity's (or, at least, free, male Greeks')[124] most divine element. The "goods of the soul" are "fortitude, temperance, justice [and] wisdom"[125] – and animals may be said to possess fortitude in abundance, a smattering of wisdom, a touch, but no more, of temperance, and no sense of justice. In sum, philosophical rather than calculative reason and the possession of language rather than "the mere making of sounds" are necessary for the articulation of ideas, and the possession of a moral sense determined by deliberation rather than by instinct is what sets humans apart. Despite all this (and it is a great deal) the essential "animality" of humanity is acknowledged: if humans are divorced from law and justice, they become the lowest of the animals. And animals possess souls, of a kind, intelligence, of a kind, and sentience and courage in significant order. Moreover, they are self-directed, "for just as human creations are the products of art, so living objects are manifestly the products of an analogous cause or principle, not external but internal."[126] While the distinctions are not always clear, and nowhere nearly so emphatic as many commentators have imagined, in general it is safe to say that Aristotle distinguishes the rational souls of humans from the sentient souls of animals. Aristotle has set the stage for future discussions without giving the definitive answers that are sought. Aristotle's "soul" consists largely in the spiritual elements of the mind rather than in the capacity for eternal life. But the important elements are those in which humans alone excel. Yet there would be others – the theriophilists – who would respond, as we shall see later,[127] that the human acquisition of the rationality principle had certainly not made humankind happier and that animals are perfectly well equipped, better equipped perhaps than humans, to fulfill themselves as the beings that they are.

Sometimes the very use of the word "soul" misleads us when we are studying Greek thought. Indeed, I make it a practice, whenever I read a translated sentence in Greek philosophy that contains the word "soul," to repeat the sentence to myself, substituting the word "mind." Very often the implication is quite different, and the meaning in fact not only clearer, but probably closer to what was meant, although the rather nebulous conception of "life force" might sometimes be even closer. If we think of the term "psychology," we readily recognize it as being derived from the Greek word *psyche* and thus understand that psychology is the *logos* – in this instance "study" – of the *psyche*. Thus it is the study of the mind. If we were to translate the term as "the study of the soul," we would err significantly, as the soul is more closely the province of theology and something decidedly different. Bearing in mind the kinds of problems that these distinctions imply, we will probably come somewhat closer to understanding what the Greeks were grasping toward and what their later European followers intended. Indeed, in his study *The Immortality of Animals* (1903), Elijah Buckner is wise enough to remind us often of the relationship of the soul to mind. Frequently, when he refers to soul, he adds immediately "or mind."[128] And sometimes when he refers to mind, he adds "or soul."[129]

Further, he avers that "mind is synonymous with soul," that "the mind or soul never does and never can exist without thinking," and that "matter and soul [or mind] are the only constituents of the universe." Nonetheless, he observes aright that "there seems to be considerable confusion about the different terms used to express the immaterial parts."[130]

Following Aristotle, the Stoics, in general, although far from universally, took the view that animals were intended for human use. Animal exploitation was deemed acceptable because animals were "irrational creatures," a fact that was supposedly proven by the proposition that human rationality presupposes its antithesis: bestial irrationality.[131] Autobulus, in Plutarch's *De sollertia animalium,* responded squarely, arguing that there are many "irrational" objects in the world – that is, objects that do not possess "souls." "Everything that is soulless, since it has no reason or intelligence, is by definition in opposition to that which, together with a soul, possesses reason and understanding."[132] For Plutarch, there are no good grounds for drawing the Stoic conclusion that reason in one kind of being implies the absence of reason in all other kinds of being. Reason in one kind of being, even that which possesses it most fully, Autobulus argues, implies reason in all beings that possess the relevant capacity for reason, which is sentience. No creature is merely sentient since sentience implies a reasoning mind through which to experience sentience. Therefore, the Aristotelian distinction between the rational and sensitive soul is without value. To have sentience implies the capacity to reason.[133] Those who later concluded that animals possess immortal souls were greatly impressed by the arguments of Plutarch.

In the late-medieval era it was Aristotle who played a greater role than other previous philosophers in influencing the European mind, both in general and with regard to the human-animal relationship. This is reflected in the writings of Albertus Magnus, Averroës, Maimonides, and above all St. Thomas Aquinas, who was the great influence on later Christian thought and who always referred to Aristotle as "the Philosopher." Indeed, much of St. Thomas's writing is taken directly from Aristotle – today, perhaps unfairly given the temper of earlier times, we would excoriate the practice as blatant plagiarism!

Aristotle's followers, the Peripatetics, claimed the differentiating essence to lie in the fact that animals possessed sentient, or sensitive, souls, whereas humans possessed rational souls. In the article on "Rorarius" in his *Historical and Critical Dictionary* of 1697, Pierre Bayle accosts the Peripatetics, even though, like Bayle, they too are avowed opponents of the Cartesian view of animals as automata. He explains that sentience alone is insufficient to explain animal behaviour. Like Plutarch he understands that sentience implies a level of reason. He takes the case of a dog who, having been punished by his master, will not repeat the activity when his master stands before him with a raised stick in his hand:

> if this dog's action is accompanied by knowledge, then the dog must necessarily reason; he must compare the present with the past and draw a conclusion from this. He must remember both the blows he has received and why he received them. He must know that if he leaped to the dish of meat that strikes his senses, he would commit the same action for which he had been beaten; and he concludes that in order to avoid being beaten again, he ought to abstain from this meat. Now is that not definite reasoning? Can we explain this situation by simply supposing a soul that is capable of feeling, but not of reflecting on its actions, but not of recalling past events, but not of comparing two ideas, but not of drawing any conclusion?[134]

Yet are the Peripatetics, as here portrayed by Bayle, truly representative of Aristotle? Not only, for Aristotle, are some animals more intelligent than others, and some less, implying that intelligence is not a characteristic absent in animals, but, as he wrote in the *Historia Animalium* (588, A:8), "just as in man we find knowledge, wisdom and sagacity, so in certain animals there exists some other natural potentiality akin to these." Indeed, not just the Peripatetics, but, historically, the anglophone tradition in general followed this precept carefully, employing the word "sagacity" for animals where "reason" would be employed if the discourse were about humans, although with the intimation of "wise in a *practical* way" in order not to be required to bestow the favoured category of reason on the bestial kind. Thus, for example, Richard Eden in *Decades of the New World* (1555) says: "Are there many of such sagacitie and industrye in beastes of greater quantitie?" In 1646 Sir Thomas Browne, in his *Pseudodoxia Epidemica* (Vulgar prophecies), suggested that "Why they placed this invention on the Bever ... might be the sagacitie and wisdom of the animall." And in 1705, in *Voyage round the World*, Daniel Defoe wrote of "Black cattle ... by a natural sagacitie, apprehensive of being swept away into the flood."[135] Writing toward the close of the eighteenth century, the Quaker Priscilla Wakefield was clearly troubled by the seemingly artificial distinctions between reason, sagacity, and instinct:

> The distinction between reason and instinct are difficult to ascertain: to define their exact limits has exercised the ingenuity of the most profound philosophers, hitherto without success. Nor can the learned agree as to the nature of that wonderful quality, that guides every creature to take the best

means of procuring its own enjoyment, and of preserving its species by the most admirable care of its progeny. Some degrade the hidden impulse to a mere mechanical operation; whilst others exalt it to a level with reason, that proved prerogative of man. There are, indeed, innumerable gradations of intelligence, as of the other qualities with which the animal kingdom is endowed ... reason and instinct have obvious differences; yet the most intelligent animals in some of their actions, approach so near to reason, that it is really surprising how small the distinction is.[136]

Wakefield has no problem endowing animals with intelligence, nor does she fail to observe how close their actions come to reason, but she is unwilling to call it "reason." And on no better ground than that it is "the proved prerogative of man," although she does not tell us wherein the proof lies. While it is clear that she herself cannot distinguish between reason and the quality possessed by the animals, she prefers not to disturb the convention and refers to the latter as *"Extraordinary Sagacity."* Intriguingly, since it is difficult to determine the import, in 1839 William Karkeek, writing in *The Veterinarian,* observed that the "elephant exceeds [humankind] in strength, and, perhaps, in some cases, even in sagacity."[137]
Karkeek continued:

the ape can climb more easily, and the stag-hound run more swiftly. Among other animals, the eagle can float on the sunbeam, and sail through the blue depths with not a downy feather ruffled by the fierceness of the storm; the nautilus can spread on the wide ocean its sails, and plough securely the tempestuous wave; the fish dives to the unexplored halls of Neptune; the corallinae erect an island on the foaming billow; the bee constructs a mathematical figure, teaching man one of the finest lessons in architecture; the ant erects for itself majestic halls and palaces, displaying a knowledge, so to speak, of the intricacies of political economy, commerce, government and legislation; and all these, with man, exist as he exists, and vanish as he vanishes from the sphere of observation. *Why then should we, who are on a level with or beneath other animals in some respects, on the other hand be so unwilling to acknowledge the justice of their very humble comparative intellectual pretensions on the other?*

Whatever the views of the later Peripatetics, in the final analysis Aristotle himself, like Karkeek so many centuries later, *seems* to take the view that animal and human mentalities are different in degree but not in kind. Samuel Taylor Coleridge appears, also many centuries later, to be following the Aristotelian line when he tells us that animals possess understanding

while humans possess reason. This did not prove a hindrance to Coleridge's description of an ass as a "Poor little Foal of an oppressed race! ... I
hail thee *Brother*."[138] Nor did it prevent the conclusion to "The Rime of
the Ancient Mariner": "He prayeth well, who loveth well / Both man and
bird and beast. / He prayeth best who loveth best / All things both great
and small; / For the dear God who loveth us, / He made and loveth all."[139]
It is clearly possible to suggest that there are significant differences between
human and nonhuman animals without losing respect for them as the
beings that they are. Certainly, the question of difference of degree, or kind,
becomes the predominant, but not sole, consideration of future discussions of the human-animal relationship. Are we, or are we not, alike? Are
we, or are we not, kin? And, of course, any realistic answer is a complex
one. We are alike in many respects but differ in some. We are kin to all but
more closely related to some than to others and most closely related to our
conspecifics. If this is one path of Aristotelian influence, and one that
shows a significant measure of compassion for animals, the other path, via
the Peripatetics and Aquinas, is the one that we have heard far more of –
whereby the distinction is made clear that the human possesses a rational
soul, which, for this very reason, is capable of immortality, and that animals possess souls that feel but lack reason as a constitutive part of their
being. Hence they are entitled to little ethical consideration.

In each human there is a constant, albeit usually subconscious, tension
between the natural animal self and the cultural self. Indeed, there is a
struggle between our conception of ourselves as super-animals possessing
an intellect with a self-conscious moral sense and our rootedness in an animal nature responsible for our primordial needs, passions, lust. The very
possession of a self-conscious moral sense provides us with a sense of ourselves as the best of animals while we struggle with a fear that we are the
worst of all. Despite Aristotle's general elevation of the human over the
animal, he still concludes that:

> Man, when perfected, is the best of animals; but if he be isolated from law
> and justice he is the worst of all. Injustice is all the graver when it is armed
> injustice; and man is furnished from birth with arms [such as, for instance,
> language] which are intended to serve the purposes of moral prudence and
> virtue, but which may be used in preference for opposite ends. That is why,
> if he be without virtue, he is a most unholy and savage being and worse than
> all others in the indulgence of lust and gluttony. Justice [which is his salva-
> tion] belongs to the polis; for justice, which is the determination of what is
> just, is an ordering of the political association.

When Aristotle claims that man is a "zōōn politikon" – a political animal – he does not mean that humans are social animals (a common late-
twentieth-century mistranslation), as are bees and other gregarious animals,
but that humans are beings designed to live in a polis, which is, for all
intents and purposes, a moral communal order necessary for the fulfillment of higher human needs. When, however, humans do not use their
gifts toward the end of moral prudence as promoted by life in the polis,
they become not only less than completely human but lower than the other
species. The other species follow their natural promptings. If they fail to
follow the moral prudence for which their powers, such as reason and language, fit them, humans act in a manner destructive of humanity's very
nature.

Fyodor Dostoevsky preaches the same message in *The Brothers Karamazov* when he insists that: "people sometimes speak about the 'animal'
cruelty of man, but that is terribly unjust and offensive to animals, no animal could ever be as cruel as a man, so artfully, artistically cruel. A tiger
simply gnaws and tears, that is all he can do. It would never occur to him
to nail people by their ears overnight, even if he were able to do it."[140]
Dostoevsky concludes that there is "a beast hidden in every man, a beast
of rage, a beast of sensual inflammability at the cries of the tormented
victim, an unrestrained beast let off the chain."[141] Torn as we are between
culture and nature, if the bonds of culture are loosed, we are lower than
all the other brutes, which do no more than follow their natural instincts.

Mark Twain went into the issue in greater depth – and with greater
humour. In his short story "Man's Place in the Animal World" (1896),
Twain engages in what may be read, and usually is, as Swiftian satire. It
can also be read as stark realism. Twain does not merely take the common
classical-era theriophilic view that other animals live the superior life. He
suggests that they are the superior beings. The human capacities may be
greater, but they are used for inferior ends.

> I have been scientifically studying the traits and dispositions of the "lower
> animals" (so called,) and contrasting them with the traits and dispositions
> of man. I find the result profoundly humiliating to me. For it obliges me to
> renounce my allegiance to the Darwinian theory of the Ascent of Man[142]
> from the Lower Animals; since now it seems plain to me that this theory
> ought to be vacated in favor of a new and truer one, this new and truer one
> to be named the *De*scent of Man from the Higher Animals.
>
>   In proceeding toward this unpleasant conclusion I have not guessed or
> speculated, but have used what is commonly called the scientific method ...

Some of my experiments were quite curious. In the course of my reading I had come across a case where, many years ago, some hunters on our Great Plains organized a buffalo hunt for the entertainment of an English earl – that, and to provide some fresh meat for his larder. They had charming sport. They killed seventy-two of those great animals; and ate parts of one of them and left seventy-one to rot. In order to determine the difference between an anaconda and an earl – if any – I caused seven young calves to be turned into the anaconda's cage. The grateful reptile crushed one of them and swallowed it, then lay back satisfied. It showed no further interest in the calves, and no disposition to harm them. I tried this experiment with other anacondas; always with the same result. The fact stood proven that the difference between an earl and an anaconda is, that the earl is cruel and the anaconda isn't; and that the earl wantonly destroys what he has no use for, but the anaconda doesn't. This seemed to suggest that the anaconda was not descended from the earl. It also seemed to suggest that the earl was descended from the anaconda, and had lost a great deal in the transition.

A number of similar stories follow, with the inevitable interpretations.

These experiments convinced me that there is this difference between man and the higher animals; he is avaricious and miserly; they are not ... the cat is innocent, man is not ... Indecency, vulgarity, obscenity – these are strictly confined to man; he invented them. Among the higher animals there is no trace of them. They hide nothing; they are not ashamed. Man, with his soiled mind, covers himself ... man is the Animal that Blushes. He is the only one that does it – or has occasion to ... Man cannot claim to approach even the meanest of the Higher Animals ... I find the defect to be THE MORAL SENSE. He is the only animal that has it. It is the quality *which enables him to do wrong* ... It seems a tacit confession that heavens are provided for the Higher Animals alone. This is matter for thought, and for serious thought. And it is full of a grim suggestion: that we are not as important perhaps, as we had all along supposed we were.[143]

Twain's conclusion indicates that he intends the argument to be taken seriously – even though he is providing us with a hearty laugh at the same time. Deflating human hubris has always provided sensitive humans with a chuckle as well as an engagement with reality. Indeed, it is itself a mark of human hubris that some may prefer to read Twain's diatribe solely as satire. In fact, Twain is sufficiently serious to suggest that, on the evidence of human performance, it is the human alone who is denied entrance into heaven. All animals *except humans* possess immortal souls. And if one insists that Twain is merely jesting – and no doubt his tongue is partly in cheek – the same cannot be said for the physician Elijah Buckner, writing within a decade of Twain. He argues in *The Immortality of Animals* that on account of their lack of a moral responsibility for their behaviour, animals will certainly enjoy a heavenly destiny, whereas humans will have to earn it.[144] To be sure, some may suggest that Twain and Buckner are a little too kind to the nonhuman animals and point to the manner in which a cat toys with its mouse prey – unconvincingly explained away by Christopher Smart as playing "with it to give it a chance"[145] – or whales toy with their seal prey, or lions display animosity toward hyenas, or chimpanzees conduct themselves cruelly toward their immediate relatives – so convincingly demonstrated by Jane Goodall in *Through a Window: My Thirty Years with the Chimpanzees of Gombé*.[146] But this would show only that animals share some of the worst of human dispositions, and not to the degree displayed by humans. What should be noted, however, is that if we offer the answer that cats, whales, lions, and chimpanzees behave so because it is in their very nature, then we are suggesting an intrinsic difference between humans and nonhuman animals. *They* are incapable of using reason and moral judgment to inform their behaviour. *We* are superior in being so capable. On the other hand, this is precisely Twain's point. We are so capable and other species are not. But most of the time we fail to live up to our capacities and behave in a manner inferior to all the other animals. Aristotle, Dostoevsky, and Twain share a similar conception of human-animal differences. If there is a disagreement among them, it is that Aristotle believes humans reach their potentialities more often than Dostoevsky and Twain imagine.

Most, of course, have regarded human language and reason as that which allows us to pursue truth, understanding, and morality, which together are seen to elevate us above the other species. Yet Twain does not stand alone in questioning this common assumption. That animals do not have a complex language in the manner of humans may be said to provide them with advantages. Thus, given what Thomas Hobbes says about one of the primary purposes of language – the telling of lies, which leads to conflict and, in turn, to war[147] – the lack of a complex language may be seen to hinder the animals from obscuring the truth both from themselves and from their neighbours.[148] And, given James Joyce's view that language obscures the thoughts and dulls the senses,[149] animals may be seen to remain more in touch with their true selves than do humans. Animals understand themselves better than do humans. It would perhaps be unwise to rely too much on such speculations, but it would be more unwise entirely to ignore them.

What this should all suggest to us, against the too easy generalizations customarily to be found in the discipline, is that the status of animals, including that of their souls, is a very complex matter and always has been, that it is not consistent throughout the Western tradition but is ever changing, and that at any given time there have been significant disputes about this status among competing philosophies and factions. Moreover, there is a great discrepancy between human potential and human reality, which might be best expressed by a slight alteration to the famous opening of Charles Dickens's *A Tale of Two Cities,* reflecting on the human rather than the age: "It was the best of animals, it was the worst of animals, it was the animal of wisdom, it was the animal of foolishness, it was the animal of belief, it was the animal of incredulity, it was the animal of Light, it was the animal of Darkness, it was the animal of hope, it was the animal of despair, it had everything before it, it had nothing before it, they were all going direct to Heaven, they were all going direct the other way."[150] None of this should persuade us to forget that, on the whole, humans have judged their conspecifics far more leniently than the evidence warrants. But we have done so for no better reason than that we have been doing the judging – no doubt bonobos and porpoises would have reached different conclusions – and have been choosing categories of comparison that redound to humanity's favour. Historically, this was often accomplished by ascribing to humans a soul of a different order than that ascribed to animals. Slightly less often, however, the distinction was denied.

V

Whether animals possess souls would appear to be a question that one might address meaningfully only once one has determined what is meant by "soul." And we have no such clear determination. Moreover, confusion is increased if the writers do not specify whether they are writing of the soul per se or of the soul with a potential for immortal life. Unfortunately, there has been no consistency in the usage. For some, as will become apparent, no more than life itself is necessary for the existence of a soul, for others sentience, for others the possession of reason (and here there is no consistency between those who think instrumental reason adequate and those who require the capacity for philosophical reason). On occasion, a soulless being might be deemed one incapable of formulating concepts, or of self-reflection, or of the possession of an aesthetic sense. With such an array of potential meanings, or at least usages, before us, it might be thought judicious to ignore the question altogether. Yet our interest is less in the question of whether animals possess souls – immortal or otherwise – than in understanding how the question has been answered historically and the implications of those answers for understanding the human-animal relationship. In the final analysis, the question is important because it allows us insights into how our ancestors formulated the distinctions between human and nonhuman animals, with consequences both for their status and for their treatment.

Let us at the very outset of the quest again note, as we did in the Introduction, that the customary interpretations of the animal-rights literature require more careful evaluation. In *Reckoning with the Beast* (1980), James Turner observed how after 1800 most people came to recognize their kinship with animals but still fought to find a mark of distinction: "If nothing else, they had immortal souls."[151] However, this perennial claim that the Western tradition, especially in its Judeo-Christian dimensions, denies immortal souls to animals had not gone uncontested in the earlier animal-sympathetic literature. To be sure, we may encounter the view expressed by Joseph Hamilton in *Animal Futurity: A Plea for the Immortality of the Brutes* (1877) that a belief in animal immortality is "heterodox in religion," but when he follows that in the next paragraph with a partial list of the heterodox, the idea no longer strikes us as quite so heterodox: "Plato, Leibnitz, Charles Bonnet, Montaigne, Pascal, Gautier, Madame de Stael, Hallam, Bayle, Lewes, Leland, Theodore Parker, Bishop Butler, Samuel Rogers, Leigh Hunt, Southey, Foster, Abraham Tucker, John Wesley, Dr. Abercrombie, Henry Rogers, Henry More, Thomas Carlyle, Adam Clarke, Charles Kingsley."[152] He also quotes a Dr. Whitby to similar effect at a later point.[153] Perhaps the perennial "common man" might have doubts about animal immortality. But many a reputable author had opted for the spiritual view.

Writing in *The Veterinarian* in 1839/40, William Karkeek mentions a number of the above but adds the philosophers J. Prichard, Dugald Stewart, Ralph Cudworth, Thomas Brown, and David Hartley, the theologians Warburton and Barclay, the lawyer Matthew Hale, and a few less prominent figures, including Grew, Euler, Wardlaw, and Crousaz (apparently "a foreign writer of some eminence"),[154] although it is not always clear whether some of these writers are said to claim a future existence for the "brute creation" or only that animals have the kinds of minds necessary for eternal existence.

In *The Immortality of Animals* (1903), Elijah Buckner also lists a few of the more prominent of those who have claimed an afterlife for the animals, and here the futurity is explicit, only three of whom are also to be found on Hamilton's list. He mentions, among the ancients, the fourth-century

BC Greek comic playwright Menander;[155] Philo-Judaeus (or Philo of Alexandria), a pronounced influence on Clement and Origen and subsequent Christian theologians; St. Paul; and the Christian theologian Tertullian, who had a profound impact on ecclesiastical language. Among the moderns, he offers the seventeenth-century Dutch Rabbi Manasseh, a primary influence on Baruch Spinoza; John Wesley, the vegetarian founder of Methodism; Dr. Edward Bouverie Pusey, Hebrew scholar, theologian, and leader of the Oxford Movement after Newman's translation to Rome; Joseph Butler, Bishop of Durham; Canon Basil Wilberforce, Archdeacon of Westminster; the renowned Swiss-American naturalist and glaciologist Louis Agassiz, a Stoic opponent of Darwinism; the esteemed Methodist theologian Dr. Adam Clarke; the impressive bluestocking mathematician Mary Somerville; the Unitarian anti-vivisectionist Frances Power Cobbe; the president of the American Humane Education Society George Angell; Rev. Thomas T. Carter, Canon of Christ Church, Oxford; and the Reverends Robert Eyton, Joseph Cook, H. Kirby, and Dr. John Fulton.[156] In the mid-nineteenth century, the New York City University theologian Rev. Dr. George Bush, an authoritative Old Testament (or, more appropriately, Hebrew Bible) commentator, observed that the "phrase 'living soul' is repeatedly applied to the inferior order of animals," indicating that it would appear to apply equally to human and animal alike.[157] A formidable list to confound those of easy scholarship! And, as we shall see – this being a topic that I shall consider in some depth in the third chapter – it is far from a comprehensive list, even of the prominent. Indeed, could one find anywhere as extensive and half so impressive a list by any writer of any degree of repute of those who stated firmly that nonhuman animals do not enjoy an afterlife?

The Maryland anti-vivisectionist campaigner Henry O. Haughton introduced Buckner's book with the claim that "as all have the same origin and experiences in life, as all die the same death, all shall share the same destiny."[158] Haughton goes on to affirm, unfortunately without naming his source, and without providing any evidence, that "More than one hundred and seventy English authors, lay and clerical, uphold [the immortality of animal souls] and have written in its support."[159] Not to be outdone, Buckner claims: "one could continue to mention hundreds of noted divines, authors, and scientists, of modern times, as well as ancient writers, who have expressed themselves as believing in the immortality of animals."[160] In 1697, as I have noted, Pierre Bayle wrote the *Historical and Critical Dictionary*. It included, in the article on Gomez Pereira, an account of numerous contributors to the topic up to the seventeenth century, and in the article on Hieronymus Rorarius, he examined the views of several more recent commentators. Rorarius had written in 1544-47 a remarkable two-volume tome entitled *Quod animalia bruta saepe Ratione utantur melius Homine* (That brute animals are often rationally superior to humankind). In 1728 Georgius Heinricius Ribovius brought out a new edition,[161] to which he appended a lengthy annotated history of the opinions regarding the nature of beasts, indicating that, from the sixteenth to the early eighteenth century, philosophers and theologians were preoccupied with the theme and its relation to questions of grace, salvation, and immortality. Less than half the writings mentioned by Ribovius were sympathetic to the materialist view of animals. Distraught at the findings of Bayle and perceiving the same defects in the conclusion drawn by far too many others that there was little difference between the soul of man and the soul of animals, David-Renaud Boullier tried to stem the tide in 1728 with his *Essai philosophique sur l'Âme des Bêtes*.[162] Later in the century, a few more followed him. In 1749 Jean Antoine Guer published at Amsterdam a two-volume octavo compilation of an *Histoire critique de l'âme des bêtes, contenant les sentiments des philosophes anciens, & ceux des modernes sur cette matière*, in which he listed numerous authors on both sides of the question.[163] In 1838 the Reverend William Drummond wrote that, compared with the Greco-Romans and Natives, "Christians generally entertain a more exalted notion of the joys of the hereafter, and though one class would confine them to the contracted circle of their own denomination, another would extend them not only to all of their own species, but to the whole of animated being – a doctrine which is at least more accordant with the benevolent spirit of Christianity, than that which would sentence the majority of the human race to everlasting burnings."[164] Writing in the early 1860s, Ezra Abbot compiled a list of over two hundred contributions to the topic, mostly favourable, and referred the reader to further relevant bibliographies containing much that he had not cited.[165] As the biographer of the soul, William Rounseville Alger remarked in 1860, "the conflict is still thick and hot."[166] Even if these numbers were exaggerations, they would still be enough to demand that the subject be revisited and that miserly authority be brought to task.

2

Peripatetic Souls

I

TWO CONCEPTIONS OF THE NATURE of the soul that have played a part, if on the whole a peripheral part, in the history of Western thought but that serve as a significant introduction to the Western idea of "soul" are the transmigration of souls and the universal soul – although in reality they are merely different emphases of one creed. If peripheral, they are nonetheless important in order to show the relationship of Western thought to non-Western thought, to demonstrate further the complexity of the idea of the soul, to indicate a primary source of the more common Western conception of the soul, and, most significant, to portray the importance of the idea of the soul to our understanding of the relationship of human to animal as a means of association and dissociation, of common belonging, and of distancing. The idea of the transmigration of souls is one of the most important concepts through which the relative status of different species is expressed.

The doctrine of the transmigration of souls, or metempsychosis – Greek for "change of soul" – is associated primarily with the religions of India, especially Hinduism and, in a rather different form, Buddhism. It exists also, in less complex forms, in a number of Aboriginal societies, including those of Indonesia, Amazonia, and Australia. Among the Aztec warriors of Central America, the favoured rebirth was as a hummingbird, sipping nectar forever. A belief in reincarnation was also present among the Druids of Celtic France, although often here the new existence was in a mythological form. In Hinduism, the individual soul enters a new existence after bodily death, sometimes that of a nonhuman animal. *Karma*, or the force generated by an individual's prior moral conduct, determines the destiny of the soul, the more admirable behaviour ensuring a home among the birds or higher mammals, if a further life as a human is excluded, the more pernicious being reborn as a worm, a moth, or a louse. There is a decided hierarchy of animals as well as humans – birds, elephants, horses, lions, and tigers standing at the animal apex. According to *The Laws of Manu*, reincarnation as an animal always meant reincarnation in a lower form, lower than that of a human – that is, as a bird or an elephant if one was deserving – but "A priest-killer gets the womb of a dog, a pig, a donkey ... A priest who is a thief (is reborn) thousands of times in spiders, snakes, and lizards."[1] Those individuals who are especially devout may achieve release (*moksa*) from the otherwise eternal cycle of life and death. Writing in 1903, with, for the time, a typical sense of cultural and religious superiority tinged with paternalistic sympathy, Elijah Buckner opined that the doctrine "shows a noble and humane device of the ancients to deter man from indulging in sordid and mean passions. To teach that the souls of men after their separation from the body should pass into the form of such animals as they most resembled in their dispositions, then to endure the horrors and suffering which they were guilty of inflicting, was a wholesome doctrine in those dark days."[2]

In the formal Buddhist doctrine – which, we may say, is understood by as few Buddhists as there are Christians who understand the Trinity – the individual is an illusion, a grouping of elements that, on death, reverts to the original primal stream of being (*samsara*), when desire, the cause of the transmigratory cycle, ceases. So, too, are the soul and transmigration, as such, illusory. The most devout Buddhists, those who abandon all desire, are able to achieve a union with creation and thus to escape the cycle.

Despite these esoteric theological differences, in practice the doctrines of Hinduism and Buddhism are much closer to each other. Thus in the *Lotus Sutra* of Mahayana Buddhism, those who slander the Sutra are destined for a damnable life as a scabrous jackal, "a camel, donkey, pig or dog."[3] The more worthy enter the bodies of the perceivedly more admirable animals or remain as humans. In Chinese Buddhism a department of Hell is dedicated to assigning new animal bodies to the souls of the dead. The cultural archaeologist and anthropologist Nicholas Saunders has recounted that one "Buddhist tale relates how a priest was about to slaughter a goat for a ritual and saw that the creature was chuckling. 'Why are you chuckling? I am about to kill you!' the priest enquired. The goat replied that it was going to be reborn as a frog. 'And that makes you happy?' asked the priest. 'No. I am laughing because in my last life I was a priest.'"[4]

In both its Buddhist and Hindu dimensions, the doctrine is also seen to express kinship with all other animal forms. Thus in the *Lankavatara Sutra* of Mahayana Buddhism, we read that "In the long course of samsara, there is not one among living beings with form who has not been mother, father, brother, sister, son, or daughter or some other relative. Being connected with the process of taking birth, one is kin to all wild and domestic animals, birds, and beings born from the womb."[5] However, the Buddhist story of the priest, the goat, and the frog should remind us that the possession of a kin relationship did not hinder there being a decided hierarchy of animals. The human was well ahead of the goat, who was in turn ahead of the frog. Nor did it prevent the goat from being sacrificed entirely to satisfy a human purpose. Much is often claimed for the consequence of the idea of kinship that is quite unwarranted, in Aboriginal, Oriental, and Western writings alike. Certainly, few have considered St. Thomas Aquinas as sympathetic to animals. Yet in the *Summa contra Gentiles,* he wrote, "The Word also has a kind of essential kinship not only with the rational nature, but also universally with the whole of creation, since the Word contains the essence of all things created by God."[6] If we are suspicious that the idea of kinship, as employed by Aquinas, may have few positive implications for our consideration of the interests of other species, we should be equally cautious in interpreting the implications of the idea of kinship on other occasions and in other cultures.

In its Western form, the idea of metempsychosis, rather less of a rarity than many might imagine, bears some similarity to its Hindu counterpart, and it has been suggested by some – an idea derided by others – that both the Indian and the Greek forms of the doctrine are derived in common from Egyptian sources.[7] More customary now is to suggest that India was the immediate source of the Greek idea, and some, influenced by Ovid, have continued to maintain that Pythagoras, who first introduced the idea in the West, visited India and studied under the Brahmins.[8] The doctrine is not to be found in the earliest Hindu scriptures (the *Ṛg Veda*) but was formalized – no doubt from prior oral legends – in the *Upanishads* (c. 600 BC), at about the same time as the inauguration of Greek philosophy in the Ionian colonies of Asia Minor and several decades prior to the birth of Gautama Siddhartha (Buddha). From the *Upanishads,* we learn that departed souls who fail to pass the moon on their heavenly journey return to earth in the form of rain and enter into the bodies of whatever animals are deemed to correspond in character to the behaviour of the souls' possessors in their previous incarnations – again *karma*. Indeed, if Pythagoras himself, rather than just his followers, subscribed to the doctrine – and all the circumstantial evidence indicates that he did, even though we have no incontrovertible evidence of the thoughts of Pythagoras – he must have adopted the idea shortly after its scriptural development in India. Whatever its source, or if it is an indigenous product, it played a significant part in the thought of pre-Socratic Greece, both in the Orphic tradition – which derived from Thracia on the western shore of the Black Sea – and among

the Pythagoreans, many of whom were Greeks from central Italy via Asia
Minor. The Orphic tradition was once thought to derive directly from the
poetry of Orpheus. More common today is the belief that Orphic poetry
was composed in Pythagorean times (i.e., sixth century BC),[9] perhaps even
in part by Pythagoras himself, and that both are derived from older oral
traditions in which transmigration, a vegetarian diet, and ritual purification
to provide a smoother path for the eternal soul were preached. If there is
a consistent difference, it might best be conceived in the idea of the Orphic
cult as a predominantly mythological system, whereas Pythagoreanism was
striving to become more intellectual and philosophical. Certainly at some
point, probably around the fifth century BC, the two originally distinct
traditions fused, with Orpheus's lyre becoming the symbol of Pythagorean
music conceived in mathematical form.

In *Pythagoras and the Pythagoreans,* Charles H. Kahn has illuminated
the early role of the soul, which proved such a significant later part of the
two most important sources of traditional Western thought: Greek philosophy and Judeo-Christian religion.

> [The] network of connections between music, mathematics, and celestial
> phenomena ... constitutes one of the two fundamental principles of Pythag-
> orean thought. The other cluster of ideas is the conception of the soul as
> immortal and hence potentially divine, since in the Greek tradition death-
> lessness is the distinctive attribute of the Gods. In Pythagorean thought,
> immortality is conceived both in terms of the transmigration of souls (with
> the related notion of kinship between all living beings) and also in the pos-
> sibility of purification and escape from the cycle of rebirth, from the bondage
> of bodily form. (It is this conception of the afterlife that is common to the
> Orphic and Pythagorean traditions.)[10]

And, we might add, common to the Indian tradition as well. Here, with
Pythagoras, at the very outset of Western philosophy and a striving toward
a distinctive form of mental engagement, we encounter one of the dilem-
mas that is to confuse, perhaps confound, Western minds ever after in their
search for a compelling ethic. Are we to accept ourselves as the beings that
we are by instinctive nature, with the needs that we have by instinctive
nature – that is, as our *animal* selves? Or are we to glory in our acquired
characteristics, which orient us to the stature of gods and hence to immortality? In the fifth century BC, the atomist Democritus argued intellectually
what had once been a part of reigning mythology: that primitive humans
had been no more than animals, living in caves, and eating the fruits of
nature. They later invented language, domesticated (sometimes seen as a

euphemism for "captured and enslaved") other animals, and constructed cities, thus making human beings creatures of culture rather than nature. For most, of course, this was human progress. But for some it raised the question of whether there was a Golden Age of humanity before we became corrupted by the pursuit of reason, civilization, and "progress," whether there was an age in which we were more fulfilled by remaining as the animals we were in our human origins. Therewith lie the horns of a continuing human dilemma. Throughout the Middle Ages, the predominant notion was of a continuing fall from grace. From the Renaissance onward, the prevailing view was of the increasing capacity to solve human woes. But at all times there have been tugs in each direction.

A traditional Confucian proverb advises that it is unwise for persons to know more than is necessary for their daily living. As previously indicated, both Buddhism and Hinduism warn of the dangers of pursuing knowledge for its own sake. Of the pre-Socratics and classical thinkers, Hesiod, Empedocles, Virgil, Ovid, and Plutarch were among those who subscribed to the idea of the Golden Age – which, according to Tacitus, writing in the *Annals,* in the first century, was an era of "equality," a "time without a single vicious impulse, without shame or guilt ... [when] men desired nothing against morality."[11] Even Plato, in both the *Statesman* and the *Protagoras,* subscribed to the idea of a descent from the Golden Age of Cronus to a state of fallen humanity, paralleling the tradition of the Bible, which was then being composed in the lands of the Middle East. Prior to humans assuming power over the animals, Plato tells us, they conversed with each other and lived in harmony,[12] just as they do in Genesis and the Cheyenne creation myth.[13] The Chinese version of Kwang-tse, from around the fifth century BC, is most explicit with regard to the relationship between knowledge and virtue: "In the age of perfect virtue, men attached no value to wisdom ... They were upright and correct, without knowing they were upright and correct, without knowing that to be so was Righteousness: they loved one another, without knowing that to do so was Benevolence; they were honest and loyal-hearted, without knowing that it was Loyalty; they fulfilled their engagements, without knowing that to do so was Good Faith."[14] Moreover, it was an age when "men lived in common with birds and beasts."[15] Philosophy, it is being suggested, is unnecessary, and potentially dangerous, for our minds are inadequate to discern the ultimate truths. In the Golden Age, we understood what was good for us instinctively or intuitively in the same manner that other species know what is good for them instinctively or intuitively. When we employ our reason to discern our morality, we are led away from the eternal truths – from the intuited common sense – that lie within us.

In Genesis 3 we read of "The Fall":

Now the snake was the most subtle of all the wild animals that Yahweh God had made. It asked the woman, "Did God really say you were not to eat from any of the trees in the garden?" *2* The woman answered the snake, "We may eat the fruit of the trees in the garden. *3* But of the fruit of the tree in the middle of the garden God said, 'You must not eat it, nor touch it, under pain of death.'" *4* Then the snake said to the woman, "No! You will not die! *5* God knows in fact that the day you eat it your eyes will be opened and you will be like gods, knowing good from evil." *6* The woman saw that the tree was good to eat and pleasing to the eye, and that it was enticing for the wisdom that it could give. So she took some of the fruit and ate it. She also gave some to her husband who was with her and he ate it. *7* Then the eyes of both of them were opened and they realised that they were naked. So they sewed fig-leaves together to make themselves loin-cloths.[16]

Eve is not only the symbol of disobedience. She is also the symbol of philosophic wisdom. They go hand in hand. She is aspiring to the wisdom of the gods, as did Plato's philosophers. One may accept one's destiny from nature, and obey the commands of nature, or one may seek it in knowledge and speculation. The god Eve is disobeying is primordial Nature. As long as Nature is obeyed, humans and animals alike share the same soul. Once the fruit is eaten, an essential difference from our fellow beings is assured. Eve is removing us from the realm of the animals and making us "as gods, knowing good from evil," enticed by the allure of "wisdom." No longer will we obey our instinctive, our intuited, morality. Instead, we will attempt to discern it by reason. We will become as gods ourselves. It is no wonder that the Enlightenment tended to capitalize the word "Reason" – or, rather, that the cessation of the tradition of capitalization of nouns and some other significant words (normal until the seventeenth and, in some instances, the eighteenth centuries and continued today in the German language) did not apply to God, the Bible, the names of countries and cities, personal and familial names, titles, "State" – and "Reason."

It is notable that Eve can converse with the snake – at least before consuming the forbidden fruit, before becoming a being whose quintessence lay in Reason. In this, Genesis 3 resembles the version of the pre-rational age described by Kwang-tse, when "men lived in common with birds and beasts." We are still in the period that Chaucer later described as "those far off days ... [when] / All birds and animals could speak and sing,"[17] a condition that is repeated in most medieval myths, including those of the *Mabinogion* and the *Fairy Tales* collected by the brothers Grimm, many of which were still being recounted well into the nineteenth century. According to the creation myth of the Cheyenne:

> In the beginning the Great Medicine created the earth, and the waters upon the earth, and the sun, moon, and stars ...
> In this beautiful country the Great Medicine put animals, birds, insects, and fish of all kinds. Then he created human beings to live with the other creatures. Every animal, big and small, every bird, big and small, every fish, and every insect could talk to the people and understand them. The people ... went naked, and fed on honey and wild fruits; they were never hungry. They wandered everywhere among the wild animals, and when night came and they were weary, they lay down on the cold grass and slept. During the days they talked with the other animals, for they were all friends.

Later, "the Great Medicine blessed them and gave them some medicine spirit to awaken their dormant minds. From that time on they seemed to possess intelligence ... they clothe[d] their naked bodies with skins of panther and bear and deer ... They were no longer able to talk to the animals, but this time they controlled all other creatures, and they taught the panther, the bear, and similar beasts to catch game for them ... He ... gave them corn to plant and buffalo for meat ... No matter how fierce or wild the beast, it became so tame that people could go up to it and handle it."[18] In like manner, the Pima of southern Arizona considered reason the distinguishing human characteristic, as did the K'iche Maya of Guatemala in describing humans in the *Popol Vuh* as "the creature with reason endowed." For each, reason distinguished us from the other animals.

The similarity of the Cheyenne story to Genesis is remarkable. Both are a part of a universal history of humankind's consciousness rather than distinctive legends of the Cheyenne and the Jews. There is but one human animal. The order of creation is substantially the same (Genesis 1:passim). Humankind is originally vegetarian (Genesis 1:29 and 2:17), animals are provided for humans as companions and helpers (2:18-19), animals and humans converse (3:1-5), humans become rational beings, recognize their nakedness, and clothe themselves (3:5-13). They come to control and domesticate other animals (1:26), and God permits them to become carnivorous (9:3-4). For both Genesis and the Cheyenne, the acquisition of reason produces a very different kind of human being, and not necessarily one for the better. No longer are we at one with the animals. The coyote has betrayed us – for the Kalahari Bushmen it is the hare, and among the Bassari and in both the Mesopotamian *Epic of Gilgamesh* and Genesis the betrayer is the snake, as seen in the following passage from Genesis 3:

14 Then Yahweh God said to the snake,
  "Because you have done this,
  Accursed be you
  of all animals wild and tame!
  On your belly you will go
  and on dust you will feed
  as long as you live.
15 I shall put enmity
  between you and the woman,
  and between your offspring and hers;
  it will bruise your head
  and you will strike its heel."

Both Adam and Eve are expelled from the garden of Eden (3:23) – from the primitive perfection of the Golden Age – the man being required for the first time to toil arduously for the fruits of the earth (3:17-19). Previously, in the Golden Age, by implication in Genesis and explicitly in Hesiod's Greek version, "all good things were theirs, and the grain-giving soil bore its fruits of its own accord in unrestricted plenty, while they at their leisure harvested their fields in contentment and abundance."[19] The creation story of the Cheyenne and the Jews is, to all intents and purposes, the identical story.

The Scythians, the Hyperboreans, and occasionally the Ethiopians were the "noble savages" of the ancients – chosen as such no doubt because they were sufficiently unknown that they could be readily mythologized. From the sixteenth century onward, these peoples were replaced by the newly "discovered" inhabitants of the New World. Yet the Amerindians were not so naive about their own reality. For the Cheyenne, for example, in telling their legends, the Golden Age is long past. No longer are they at one with the animals. According to the myth of "How the Buffalo Hunt Began,"[20] the buffalo and the coyote were the eternal enemies, the former because they hunted and ate humans until the humans defeated them in a race with the help of the magpie and the hawk. Thereafter, it was the buffalo who became prey. The coyote sided with the buffalo in the race and hence was thereafter hunted for his fur alone. The "Creation Myth" had already pilloried wolves as, along with the white people, "the trickiest and most cunning creatures." In the Western tradition, too, the wolf remains a perennial enemy, even though Kant confirmed the principle while denying its universality: "The more we come in contact with animals, and observe their behaviour, the more we love them, for we see how great is their care for their young. It is then difficult for us to be cruel in thought even to a wolf."[21] It is true, according to Italian myth, that wolves raised Romulus and Remus and, according to the German *Heldenbuch* (first edited in the fifteenth century), that Wolfdietrich was raised similarly by wolves. But the general reputation was not thereby salvaged. The serpent of Genesis is also derided unjustly. And to such a degree that the novelist Joseph Conrad decided to take revenge on the snake's behalf. Whereas in the Garden of Eden, a corrupting serpent intrudes into Paradise, in *Nostromo* it is corrupting humans who invade what Don Pepe calls "the very paradise of the snakes" to devastating effect.[22] Later in the novel, Charles Gould remarks that "It is no longer a Paradise of snakes. We have brought mankind into it, and we cannot turn our backs upon them to go and build elsewhere."[23] Humanity must deal with the problems it has created. As previously noted, for the African Bushmen it is the hare who plays a role similar to that of the serpent in Genesis. The hare is the beguiler who deprives humans of eternal life,[24] just as the serpent does in Genesis. The Mahayana Buddhist slander of the snake shows considerable similarity to that of Genesis 3:14-15:

> [Being reborn in] ... the body of a serpent
>> long and huge in size,
>> measuring five hundred yojanas,
>> deaf, witless, without feet,
>> slithering along on his belly,
>> with little creatures
>> biting and feeding on him,
>> day and night undergoing hardship,
>> never knowing rest.
> Because [the maligner of the holy text] slandered this sutra,
> this is the punishment he will incur.[25]

Humans and animals were no longer at one, either in the Bible or in Aboriginal and Oriental thought. The commonality of human experience in different cultures that are centuries and even continents apart is both startling and pleasing for those who recognize an essential human equality and acknowledge the intrinsic similarity and equality of Western, Eastern, and Aboriginal culture. Of course, this should not give the impression that the snake was broadly disrespected. The Hindu god Vishnu rode the cobra of wisdom. And both the Bible and deists like Voltaire chose the serpent as the symbol of wisdom, although it was often not an unadulterated wisdom, as when, in Samuel Richardson's *Pamela* (1741), the eponymous heroine is described as being "as innocent as a dove, yet ... as cunning as a serpent."[26] Still, in the mid-sixteenth century, the French army surgeon Ambroise Paré said, in apparent repetition of contemporary lore derived from the Hebrew Bible, that the serpent was eminent for wisdom,[27] and he offered none of Richardson's limitations. The Egyptian pharaohs wore the cobra in their crowns as a symbol of royalty and divinity. For many pre-Columbian civilizations, the serpent symbolized transformation and cosmic rebirth because of the shedding of its skin. And in the European *Fairy Tales* collected by the brothers Grimm, there are three "Tales of Snakes" in which, in turn, a child and snake enjoy a friendly relationship; the essential animal equality of human and snake is stressed; the snake has gifts to impart to us if we treat it, and its gifts, with respect.

A decided hierarchy of animals was being established. But it was no simple matter. Some animals were generally higher than others, some higher in some respects and lower in others. Hierarchy might be influenced by fear of the animal, respect for the animal's accomplishments, or disdain for the animal's weakness, including its perceived ugliness – as Shakespeare so misrepresents the toad as "ugly." It was above all, however, through the acquisition of reason, in both imitation and defiance of the gods, that we came to dissociate ourselves from animal life and to count ourselves among the godlike creatures of reason, among whom immortality was the defining attribute.

II

All this leads us to Plato, who, instead of fearing the acquisition of reason, of whose perils we are warned in some of the older traditions, grasps it as the acquisition of human excellence. In part, Pythagoras is his inspiration. He subscribed to Pythagoras's doctrine of the transmigration of the soul between different species and also believed that there is something divine in the human make-up. Indeed, some (e.g., Moderatus) have seen the whole Platonic school as a mere plagiarism of Pythagoras. They "have taken for themselves the first fruit of Pythagorean thought."[28] In contrast with the doctrine of Genesis, which warns us of the dangers of becoming "as gods," of changing from instinctive to rational creatures, Plato's view, following the injunctions of Pythagoras, is that "Knowledge is unsurpassed by any law or regulation; reason, if it is genuine and really enjoys its natural freedom, should have universal powers."[29] Elsewhere he insists that "no city nor individual can be happy except by living in company with wisdom under the guidance of justice."[30] As we have seen, in Plato's view the human soul consists of three parts: a rational part, which pursues truth; a spirited part, which pursues honour; and the passions, which seek pleasure. Each part possesses its own virtue. Wisdom corresponds to reason, courage to honour, and moderation to the pleasure-seeking passions. Justice is achieved when there is a "harmony" among the three elements, which acknowledges the rule of philosophical reason, the subordination of both honour and the passions to reason, and the subordination of the passions also to honour. Plato would appear to consider nonhuman animals primarily creatures of the appetites, incapable of honour and philosophical reason, which are the exclusively human capacities. Moreover, "When there is implanted in the soul of men, a right opinion concerning what is honorable, just and good, and what is the opposite of these – an opinion based on absolute truth and settled as an unshakable conviction – I declare that such a conviction is a manifestation of the divine in a race which is of supernatural lineage."[31] Unlike other animals, humans possess "a manifestation of the divine" in their natures since they pursue truth through reason. They are "of supernatural lineage," a view still being repeated in the nineteenth century by Spooner and Manthorp, for example, in the pages of *The Veterinarian*.[32] Plato's view was paraphrased almost two millennia later in the famous expression of Cervantes, "Where truth is, God is, truth being an aspect of divinity." Among animals, only humans pursue "the truth." And in the early nineteenth century, Beethoven made music the elevating principle: "I know well that God is nearer to me in my art than to others, I consort with him without fear." His friend Bettina von Arnim said "he treated God as an equal." He possessed a heavenly art to which no other species – and for that matter no conspecifics – could aspire. Yet, despite his music, if Beethoven was at all godlike, it was as an Old Testament autocrat, although his idiosyncrasies have also been termed a "divine madness." His behaviour toward family, servants, and associates alike made him what Aristotle would have called "a most unholy and savage being" and in that respect lower than the other animals. Thomas Carlyle, who was a revered Romantic historian, an opponent of the French Revolution with a disposition derived from a medieval sense of honour and order, and a valiant opponent of the unconscionably modern animal experimentation, noted that "divineness ... does not come from Judea, from Olympus, Asgard, Mount Meru, but is in man himself; in the heart of everyone born of man – a grand revolution, indeed, which is altering our ideas of heaven and earth to an amazing extent in every particular whatsoever."[33] The godhead lies within each human soul. But Carlyle was himself no benevolent god. Tennyson said of Thomas and Jane Carlyle's quarrelsome marriage that it was ideal. Any other solution would have made four people unhappy.[34] Carlyle's reputation was that of a tyrant.

As Nietzsche observed, the gods lived no more. We had instead become our own gods – a point reiterated in H.G. Wells's 1922 novel, *Men as Gods*. But human aspirations to divinity are not always endearing. In losing our animality, we lose something valuable, and not all of what we gain is worth possessing, to which the lives of Beethoven and Thomas Carlyle bear witness, even though a certain kind of divinity lived within them. But it was the cantankerous divinities of mythology not the purveyors of peace and harmony that they emulated. Nevertheless, in elevating themselves above the animals and becoming as gods, many such minor divinities came to believe that they had acquired the same moral obligation to the lower animals as God had for humans. If we were no longer animals in the same manner that others were animals, this did not give us the right to use them at will. Instead, it imposed on us responsibilities toward those over whom we had "dominion." In Carlyle's case, the animals were treated as beings on whom the cruelties of vivisection should not be permitted. In Beethoven's case, divinity permitted him to treat all those beneath him with contempt. Becoming godlike allows, in principle at any rate, the equal likelihood of treating the lower animals with respect or disdain. A belief in the inferiority of the animal soul has not of necessity denied it ethical consideration.

Despite his elevation of the human to a quasi-godlike status, Plato has humans share their souls with other animals. Indeed, on death, most take on the souls of other creatures – which is difficult to comprehend if only the human shares in the divine, if only the human is of a supernatural lineage. In the *Phaedrus* Plato describes the cavalcade of gods and purified souls who have escaped the transmigratory cycle of life and death, which is achieved, he remarks in the *Theaetetus,* by imitating the gods as far as possible, a view that in turn is examined more closely in the *Phaedo*. And, of course, only humans can imitate the gods because only they possess philosophic reason. Indeed, this apparent contradiction persuaded Iamblichus, writing in the fourth century against the transmigration conclusions of Porphyry and Origen, who were following Pythagoras and Plato, to conclude that only human souls were of an exclusively rational nature and that since a rational soul cannot become a nonrational soul, transmigration between humans and nonhuman animals was not possible.[35]

In fact, as Richard Sorabji has demonstrated in *Animal Minds and Human Morals,* there were myriad discussions in Greek philosophy about the relationship of the soul to reason (*logos*), to rational intuition (*nous*), to nature (*physis*), and the like,[36] and there were almost as many divisions as there were philosophers. There was little consistency and not a great deal of clarity. It would appear, however, that Plato conceives of the lower

humans as capable of sharing souls with nonhuman animals, but to the
extent that the accomplished humans rise to the fulfillment of the human
capacity for philosophy and wisdom, they are able to escape the confines
of entrapment in the bodily cycle and reach the realm of the gods. Likewise, for both Hindus and Buddhists, the more one becomes godlike the
more one is able to escape the limitations of the eternal cycle of bodily life
and death – that is, the more one is able to escape the confines of human
animality. As we have noted, those Hindus who are especially devout
achieve *moksa* – release from the cycle. Eschewing all bodily desire, the
Buddhist rejoins *samsara* – the original primal stream of being. Platonic,
Hindu, and Buddhist conceptions of the transmigration of the soul, and
the capacity for escape from the cycle, share a great deal in common, most
of which involves shedding one's animal nature, divorcing oneself from
one's kinship with the animals – although the West stresses reason, albeit
allied with virtue, whereas the East emphasizes devoutness. Indeed, the
Hindu scholar Basant K. Lal argues that: "The Hindu recommendation
to cultivate a particular kind of attitude toward animals is not based on
consideration for the animals as such but on consideration about how the
development of this attitude is a part of the purificatory steps that bring
men to moksa ... Hinduism in all its forms teaches that we have no duties
to animals and thus implicitly denies that they have rights."[37] Of course,
this is not an uncontested view – indeed, I regard it as a significant exaggeration, as it ignores the authentic respect sometimes accorded to animals
– but it reflects the fact not only that there are varieties of interpretation
in all cultures, but also that the kinship implied by transmigration does
not necessarily result in a benevolent relationship with other species or in
their being granted a higher status.

   In Book 10 of the *Republic,* Plato engages in a scathing critique of Homer
and the traditional poets, determines that they promote illusion rather than
reality, and concludes: "Did he and they pass on to posterity a kind of
Homeric way of life after the manner of Pythagoras? Such a legacy is
counted among Pythagoras's greatest achievements. Even now his successors use the term *Pythagorean* to denote a certain way of life that many of
our contemporaries look upon with respect."[38] When, in the *Philebus* (16c),
Plato writes of "the ancients, our superiors, who dwelt nearer to the gods,"
it is assumed that he is referring to Pythagoras as well as to the inhabitants
of the Golden Age. The mathematical curriculum of the *Republic* is derived
directly from the Pythagorean quadrivium (arithmetic, geometry, astronomy, and music), whose purpose it is to orient the mind away from the
sensory pursuits in favour of the study of abstract forms. Thus Plato's character of Socrates tells us: "the task is useful in the quest for the beautiful

and the good."[39] In following Pythagoras and damning Homer, Plato places himself on the side of abstract reason and scientific knowledge against myth, poetry, and tradition as legitimate forms of knowledge. Indeed, he would have the Homeric verses proscribed. The essence of humanity is our philosophical spirit, which most clearly distinguishes us from the animals. Nonetheless, Plato has no difficulty in accepting the flight of the human soul to the animal and vice versa. If, as Kahn says, "in the Greek tradition deathlessness is the distinctive attribute of the gods,"[40] through the transmigration of souls, both humans and animals would appear to be deathless, and hence both partake in a measure of divinity. Yet this is not the message one receives in general in Plato's dialogues. Thus in the *Meno* Plato uses the Pythagorean doctrine of transmigration to pursue "an epistemology of innate ideas and a priori knowledge," which he elaborates in the *Phaedo,* insisting that the basis of "recollection is a prenatal acquaintance with eternal Forms"[41] – well beyond the capacities of nonhuman animals to have acquired in their transmigrations. Plato tells us further in the *Phaedo* that "thought is best when the mind is gathered into herself and none of these things trouble her – neither sounds nor sights nor pain nor any pleasure – when she has as little as possible to do with the body, and has no bodily sense or feeling, but is aspiring after being." The philosopher is to pursue "absolute justice ... absolute beauty and absolute good," and "he attains to the knowledge of them in their highest purity who goes to each of them with the mind alone, not allowing when in the act of thought the intrusion or introduction of sight or any other sense in the company of reason, but when the very light of the mind in her clearness penetrates into the very light of truth in each."[42] The body is animal, and the mind is that in which humans can lose their animal nature. Clearly, Plato sees a vast difference between the human capable of philosophy and the nonhuman animal incapable of such intellectual elevation. It would be strange if there were such an obvious paradox in the writings of one of the West's greatest philosophers – the possibility of transmigration of souls between humans and animals yet their possession of quite different types of souls. In fact, the paradox is greatly diminished if we recognize that Plato's distinction is not just between humans and animals, but between philosophical humans, ordinary humans, and animals, the latter two seeming at least to share more in common in relevant respects than the first two. The philosophical human, the spiritual human, is able to escape the cycle of life and death, to become in essence a god. Other humans continue to share their souls with animals, for they remain essentially animal and fail to achieve the godliness of the philosopher kings, whose minds embody justice.

In later centuries of European history, the emphasis on the benefits of education may well be understood as the emphasis on losing our animal qualities and making us more fit to commune with the gods. With the democratization of education, the plebeians came to be seen as essentially human in the same way – almost anyway – as were the upper classes. Until then the societal elites regarded lower-order humans as essentially of a different species, just as had Plato regarded those who did not belong to the guardian class as essentially inferior in nature.

Metempsychosis is so central a part of Plato's thinking that he discusses it in the *Phaedo, Republic, Phaedrus,* and *Meno.* It is even implied in the judgment myth toward the close of the *Gorgias.*[43] The lengthiest treatments – in the *Phaedo* and the *Republic* – provide us with confusingly different accounts. In the *Phaedo* Plato tells us: "if we must have acquired this knowledge [of the ideal equal] before we were born, and were born having it, then we also knew before we were born and at the instant of birth not only the equal or the greater or the less, but all other ideas; for we are not speaking only of equality absolute, but of beauty, good, justice, holiness, and all which we stamp with the name of essence in the dialectical process, when we ask and answer questions. Of all this we may certainly affirm that we acquired the knowledge before birth." All higher truths are innate and live as Forms within the soul. Thus "our souls must have existed before they were in the form of man – without bodies, and must have had intelligence ... our souls must have had a prior existence."[44] Yet, surely, such ideas are not present in the minds of nonhuman animals or, for Plato, in those of the lower human classes either, as he explains in the myth of the metals in the *Republic.* All share the soil in common, but some are of gold, some of silver, and some of bronze. Their natures are essentially different, the lower classes being incapable of understanding the Forms and hence needing to have their societal roles explained to them not by higher forms of reason but by myth: the "noble lie." Plato tells us that along with the enlightened souls, there are also the corrupt, who come from the lower classes, for in Plato's view there is a unity of virtue and reason:

> These must be the souls, not of the good, but of evil, who are compelled to wander about such places [as tombs and sepulchres] in payment of the penalty of their former evil way of life; and they continue to wander until the desire which haunts them is satisfied and they are imprisoned in another body. And they may be supposed to be fixed in the same natures which they had in their former life.
>
> What natures do you mean, Socrates?
>
> I mean to say that men who have followed after gluttony, and wantonness,

and drunkenness, and have had no thought of avoiding them, would pass
into asses and animals of that sort. What do you think?

I think that exceedingly probable.

And those who have chosen the portion of injustice, and tyranny, and
violence, will pass into wolves, or hawks and kites: – whither else can we
suppose them to go?

Yes, said Cebes; that is doubtless the place of natures such as theirs ...
Even among them some are happier than others; and the happiest both in
themselves and their place of abode are those who have practised the civil
and social virtues which are called temperance and justice, and are acquired
by habit and attention without philosophy and mind.

Why are they the happiest?

Because they may be expected to pass into some gentle and social nature
which is like their own, such as that of bees and ants, or even back again
into the form of man, and just and moderate men spring from them ... But
he who is a philosopher or lover of learning, and is entirely pure at departing,
is alone permitted to reach the gods.[45]

There is much here that corresponds to the doctrines of Hinduism (just
as Plato's class system of philosopher kings, warriors, and artisans in the
Republic bears considerable similarity to the Hindu caste system).[46] The
ultimate purpose of the soul is to escape from the circle of life and death
(to achieve *moksa* for Hindus, reach nirvana for Buddhists). Prior to this,
in successive lives the soul resides in the bodies of animals according to
desert in the former life. Thus, for Plato, there is a hierarchy of animals in
which the predators and the supposedly slothful are placed toward the
bottom and the gentle and social animals toward the top. (It should be
noted that whereas humans are distinguished by their individual philosophical achievements or lack of them – giving rise, for Plato, to moral
achievements or lack of them since, as noted, he believes in a unity of the
true and the just – animals are distinguished by *species* characteristics.)
But only the very best of humans – the philosophers, the Brahmins, the
Bhikku – can hope to reach the gods. The animal possesses an immortal
soul – or, rather, the immortal soul possesses the animal, as it does the
lower-class human, but only in the sense of enduring pain forever. Heaven,
the Buddhist nirvana, is beyond the animal's grasp as well as beyond that
of the lower human orders. There would appear to be two kinds of immortality: (1) an eternal continuity in the bodies of different beings of
different status; (2) a permanent existence as pure spirit. Thus there would
appear to be the material, the material combined with the spiritual, and the
purely spiritual. In most versions, the animals are predominantly material

with an element of the spiritual, but only the best of humans can aspire to the purely spiritual. In the Platonic, Hindu, and Buddhist traditions, animals are not divorced from humans as such but from the superior class of humans – those who have approached the gods, those who have the capacity for immortality in its purely spiritual sense.

The version of metempsychosis in the *Republic* is far more respectful of animals than that of the *Phaedo,* even with a hint of theriophily. Having praised Pythagoras and philosophy over Homer and poetry, Plato tells us:

> The spectacle, [the unnamed prophet] said, was worthy to behold, in what manner the several souls made choice of their life; for it was both pitiful and ridiculous and wonderful to behold, as each for the most part chose according to the habit of their former life. For he told that he saw the soul which was formerly the soul of Orpheus making choice of the life of a swan, through hatred of womankind, being unwilling to be born of woman on account of the death he suffered from them. He saw likewise the soul of Thamyris, making choice of the life of a nightingale, and he saw likewise a swan turning to the choice of a human life, and other musical animals, in like manner, as is likely. And that he saw one soul, in making its choice, chuse the life of a lyon, and that it was the soul of Telamonian Ajax, shunning to become a man, remembering the judgment given with reference to the armour. Then next he saw the soul of Agamemnon, and that this one, in hatred also of the human kind, on account of his misfortunes, exchanged it for the life of an eagle ... And that in like manner the souls of wild beasts went into men, and men again into beasts.[47]

In this version, both humans and animal get to choose the body in which their soul is next to be housed. And it is less past merit that occasions the choice than past experience. Swans and songbirds choose human form. Some humans, on account of their unfortunate experiences as humans, choose a life as an animal in whose form they will not experience what harmed them as humans. This version of metempsychosis would appear to have no Buddhist or Hindu counterpart, although it has some similarity to earlier Egyptian conceptions.

Here there is no explicit hierarchy of animals or of humans over animals. Nonetheless, it would be inconsistent with Plato's views in general to imagine that animals may now aspire to the divine. Yet the obvious question remains: When the soul present in a human enters into the animal, does it lose its capacity for philosophy and an intellectual appreciation of wisdom, beauty, goodness, and justice? Even though these may be confounding questions, it is certainly a matter worthy of our contemplation that the genius acknowledged by consensus as the greatest of the classical minds – even Aristotle acknowledged Plato's intellectual primacy, although he deemed his brilliance the very source of his errors in the practical sciences – believed in a theory that, in general if not entirely, the majority in the later Western tradition, both Christian and secular, has rejected out of hand. Moreover, the very positing of the transmigration of souls when the soul possessed by the human contains characteristics unavailable to animals may cause us to wonder whether transmigration was more a literary and theological device than a serious theory that related to the nature of animals. Certainly, this would appear probable of the version reported in the *Republic,* if less likely of that in the *Phaedo.*

As we have noted, in the fourth century, Iamblichus pointed out against Porphyry and Origen, who accepted the transmigration of souls, that if animal souls were not fully rational in the way that human souls were, then no transmigration could take place. Indeed, the suggestion has often been made that neither Pythagoras nor Plato believed in transmigration as fact. It was a way of telling a story in which humans were understood to be *like* kites or bees without ever *becoming* kites or bees, just as the myth of the metals is to be read not as though some people are *actually* made from different metals, but as though this is a useful allegorical way of understanding matters that may be beyond the grasp of the less philosophical of humans. It is *as though* people were made from different metals. This may well be so. But the story is far more illuminating if we recognize that Plato has a conception of animals and humans in which certain types of humans may be said to differ as much from some of their conspecifics as from non-human animals. However, whatever Pythagoras and Plato may have meant, the later history of metempsychosis in the Western tradition is very much a sideline, even if an occasionally very enlightening one.

A particularly enlightening version of metempsychosis is offered to us by Menander, a poet of the "New Comedy" and a near contemporary of Plato. It is one in which the animal is treated as superior to the human, but perhaps only to demean the characters of Menander's associates:

If one of the gods should say to me, "Crato, when you die, you shall immediately be reborn, and you shall be whatever you wish, a dog, a sheep, a goat, a man, a horse; for you must live twice. This is your destiny. Choose whatever you wish." "Anything," I think I should immediately say, "make me anything rather than a man." This animal alone wins success unjustly and does ill. The best horse has better care than the others. If you are a good dog, you are much more highly honored than a bad dog. A well-born cock lives on a particular kind of food, the low-born fears his better. A man, even if he is good, well-born, well-bred, has no advantage among the present generation. The flatterer does best of all, the sycophant comes next, and the malignant man is third. To be born an ass would be better than to see one's inferiors living more splendidly than oneself.[48]

Enlightening as it is, Menander appears to offer this version of metempsychosis as a literary device through which to make a telling point. It does not suggest a serious subscription to the notion of the transmigration of souls, although it reflects the fact that such notions were common and readily believed. Whether any of Plato's successive variations on the idea of transmigration are to be taken seriously is difficult to discern. Nevertheless, given the frequency with which they arise in Platonic writings, it is clear that metempsychosis was at the very least an important metaphor for Plato's thought and a central way of ascribing relative status to different species.

III

How closely does either of Plato's versions of metempsychosis correspond to that of Pythagoras? In reality we have very little idea. Nothing written has come down to us from Pythagoras – despite the popularity among theosophists, mystics, and others of the neo-Pythagorean (and essentially pseudo-Pythagorean) *The Golden Verses of Pythagoras*.[49] As Daniel Dombrowski pointed out: "Despite the fact that there has been as much preserved about Pythagoras as about any figure in antiquity, we still have great difficulty knowing who he was ... The reports about him from late antiquity ... are in conflict, and later Pythagoreans, who existed centuries after the master, are understandably open to the charge of distortion and hyperbole."[50]

The most complete version of "Pythagoras" is the one provided to us by Ovid in *Metamorphoses,* which was composed in Rome around the time of the birth of Christ, over a half-millennium after Pythagoras's death. It is the source of the most commonly understood, if not necessarily most accurate, version of Pythagoras's espousal of the doctrine of the transmigration of souls and of his consequent eschewing of the consumption of flesh – with the result that until the term "vegetarianism" was coined in 1842, followed by the founding of the Vegetarian Society in 1847, those who elected not to consume flesh were customarily known as "Pythagoreans."[51] Biographies of Pythagoras were written by Iamblichus, Porphyry, and Diogenes Laertius, but, again, it is nigh impossible to know how they correspond with the life of the historical figure.[52]

"O my fellow-men," Pythagoras is reported by Ovid to have taught, "do not defile your bodies with sinful foods ... Alas what wickedness to swallow flesh into our own bodies, to fatten our greedy bodies by cramming in other bodies, to have one living creature fed by the death of another! In the midst of such wealth as earth, the best of mothers, provides, nothing forsooth satisfies you, but to behave like the Cyclopes, inflicting sorry wounds with cruel teeth! You cannot appease the hungry cravings of your wicked gluttonous stomachs, except by destroying some other life!"[53] Of course, the Orphics and the advocates of the Golden Age were vegetarian before Pythagoras, but it was he who, in Ovid's version of events, provided the doctrine with its ethical foundation:

> All things change, but nothing dies; the spirit wanders hither and hither, taking possession of what limbs it pleases, passing from beasts into human bodies, or again our human spirits pass into beasts, but never at any time does it perish. Like pliant wax which, stamped with new designs, does not remain as it was, or keep the same shape, but yet is still itself, so I tell you that the soul is always the same, but incorporates itself in different forms. Therefore, in case family feelings prove less strong than greedy appetite, I warn you, do not drive souls that are akin to yours out of their homes by impious killings, do not nourish blood with blood.[54]

For the Ovidian Pythagoras, whoever first ate flesh put humankind on the early steps to a life of crime. Who with any humanity could ignore the cries of kids and calves and slit their throats? Who could kill and eat the ox who had worked one's own land? Ovid's depiction of the animal-embracing Pythagoras became the image accepted by posterity.

Yet, if this is an accurate depiction, in light of Plato's rejection of Homer in favour of the wisdom of Pythagoras, would one not expect Plato also to have espoused vegetarianism – at least to the extent that Pythagoras was a complete vegetarian, for Porphyry says of him that "Fish he ate rarely"?[55] But then, apparently lacking blood, fish were not seen so readily to belong to the category of animals, as did mammals and birds, a distinction that continued among many later "vegetarian" groups, including such Christian sects as the medieval Cathars. However, both Porphyry and the anonymous biographer whose study was preserved by the Byzantine patriarch Photius also claimed that Pythagoras and his followers ate sacrificial flesh on occasion.[56] And in his learned account of *The Presocratic Philosophers*, Jonathan Barnes casts doubt on the idea that Pythagoras was vegetarian at all.

Certainly, there is no evidence that Plato was himself a vegetarian, and if he had been so, it is likely that he would have announced it as a part of his ethical system. Plato accepts that in the Golden Age – from which his own age appears to be acknowledged as a regression – humans were vegetarian and lived a fulfilled life. Yet in his own age he condones hunting and butchering as well as the raising of livestock for consumption, and the "culinary art" is said to consist in the seasoning of meats, which are deemed wholesome and recommended to athletes.[57] Nonetheless, Daniel Dombrowski has made a persuasive case that Plato's *Republic* is conceived as a return to the principles of the Golden Age in which vegetarianism will be the accepted practice.[58] Likewise, Howard Williams has numbered Plato among the antiflesh dietary reformers.[59] And certainly Socrates's statement in the dialogue about the division of labour leads to this conclusion: "To feed them they will make meal from barley and flour and wheat; some they will cook, some they will knead into fine flat-cakes and loaves." And when Glaucon questions whether they should not also have relishes at their feasts, Socrates replies: "I forgot that; they will have something more, salt, of course, and olives and cheese, onions and greens to boil, such as they have in the country. And I suppose we shall give them dessert, figs and chickpeas and beans, and they will toast myrtle berries and acorns before the fire, with a drop to drink, not too much."[60] All wholesome nonflesh foods, of course, even if dairy products are included. And when Glaucon asks how Socrates would feed the citizens if he were founding a city of pigs rather than the republic, he replies with disdain that they could continue to consume the luxurious foods they now consume. Yet it seems clear that Plato, through the character of Socrates, is concerned with the moral and physical health of the citizenry, not with the protection of animals, especially as the term the "city of pigs" is used with such disdain for pigs. His opposition is to "fine food ... and ointments and incense and pretty girls and cakes, all sorts of each." His intent is for the city not to exceed "the bare necessities," that which epitomized the Golden Age. One can be confident of this from what Socrates has said when discussing the appropriate size of the city: "Yet it would still not be so very large, even if we were to add oxherds and shepherds and the other herdsmen, that the farmers might have oxen for the plow, and the builders draught-animals to use along with the farmers for carriage, and that the weavers might have fleeces and skins."[61] There is no objection to the use of animals for human ends and no objection to the use of the skins of slaughtered, or at least dead, animals. Moreover, what could be the purpose of keeping sheep if not for their flesh and wool? It would be strange if goat and pig skin were used for raiment and protection from the elements and the flesh of the domesticated animal were not consumed. If Plato does intend a vegetarian city, it is, unlike the image we have of Pythagoras, not out of respect for animals but to maintain an ascetic citizenry. Nonetheless, if Plato is truly a consistent follower of Pythagoras, Howard Williams's interpretation is certainly worthy of consideration: "The obligation to abstain from the flesh of animals was founded by Pythagoras on mental and spiritual rather than on humanitarian grounds. Yet that the latter were not ignored by the prophet of *akreophagy* is evident equally by his prohibition of the infliction of pain, no less than of death, upon the lower animals, and by his injunction to abstain from the bloody sacrifices of the altar."[62] Perhaps the same was true of Plato, although the evidence does not encourage such a belief. Still, perhaps the inconsistency of domestication of sheep and a vegetarian diet was merely an oversight of detail – an offence of which Plato was frequently guilty.[63]

We cannot know whether Ovid's view of Pythagoras is historically accurate. But we have indications in the writings of some of his followers and those who cited them. Thus in his third-century *Life of Pythagoras*, Porphyry confirmed Ovid's account – although his primary source might well have been Ovid himself! – stating that it "became very well known that [Pythagoras] said, first, that the soul is immortal, then that it changes into other kinds of animals; and further that at certain periods whatever has happened happens again, there being nothing absolutely new, and that all living beings belong to the same kind. Pythagoras seems to have been the first to import these ideas into Greece."[64] Likewise, in the first century, Diodorus had affirmed in *Universal History* that "Pythagoras believed in metempsychosis and thought that eating meat was an abominable thing, saying that the souls of animals entered different animals after death."[65] Around the same time, the Roman Stoic philosopher Seneca reported that Pythagoras "held that all beings were interrelated, and there was a system of exchange between souls which transmigrated from one bodily shape to another. If one may believe him, no soul perishes or ceases from its functioning at all, except for a tiny interval – when it is being poured from one body into another."[66] In the third century, Diogenes Laertius reported in *Lives of the Philosophers* the statement of Xenophanes, a contemporary of Pythagoras, that "once when [Pythagoras] passed a puppy that was being whipped they say he took pity on it and made this remark: 'Stop, do not beat it; for it is the soul of a dear friend. I recognized it when I heard the voice.'"[67] Xenophanes intended this anecdote as an affront, but the belief in the principle of metempsychosis was affirmed. Sextus Empiricus, of third century or so vintage, reported in *Against the Mathematicians* on what was known as the Italian school – that is, those Greeks who, like Pythagoras, had migrated to Greek colonies in Italy (Magna Graecia): "Pythagoras and Empedocles and the rest of the Italians say that we have a fellowship not only with one another and the gods but also with the irrational animals. For there is a single spirit which pervades the whole world as a sort of soul and which unites us with them. That is why, if we kill them and eat their flesh, we commit injustice and impiety, inasmuch as we are killing our kin. Hence these philosophers urged us to abstain from meat."[68] While this may sound closer to the idea of the universal soul, of which more later, than the transmigration of souls, it is clear that the poet Empedocles not only believed in metempsychosis, but claimed, according to Hippolytus in *Refutation of All Heresies,* that in earlier lives:

> already have I once been a boy and a girl
> and a bush and a bird and a silent fish in the sea
> He said that all souls change into every sort of animal.[69]

Moreover, according to Aristotle in the *Rhetoric,* Empedocles believed that "there is by nature a common justice and injustice" between humans and animals:

> This is what Empedocles says about not killing animate creatures; it is not
> the case that this is just for some and not just for others.
> But a law for all, through the broad
> air it endlessly extends and through the boundless light.[70]

Unlike Plato, most Pythagoreans appear to have derived their transmigration ideas from their respect for animals and their refusal to eat them out of a sympathy for them. For Apollonius of Tyana (fl. AD 100), on the other hand, the avoidance of flesh derived from his asceticism (which included abstinence from sex and wine), which was a part, he believed, of the appropriate path to accessing the spiritual world, a view reminiscent of both Buddhism and Hinduism. Yet metempsychosis and respect for animals seems to have played little role in his vegetarianism.[71] Apollonius, we might suggest, was closer to Plato than to Pythagoras. Xenocrates, Plato's disciple, and later head of the Academy in the fourth century BC, was concerned that the consumption of animal food would contaminate humans and assimilate the eater to the souls of irrational beasts – although he did also believe that animals deserved protection.[72] This might be taken as further evidence, if not confirmation, of Plato's attitudes. It was the contemporaneous head of the Aristotelian Lyceum, Theophrastus, who was, apparently, the first to emphasize the anatomical and psychological features shared by humans and nonhuman animals, sense perception and feeling especially – a view not incompatible with that of Aristotle himself –

and to suggest, going much further than Aristotle, that respect for animal life is based upon the concept of community (*oikeiôsis*) as the principle that associates all animals, thus beginning a longstanding argument, principally between the Stoics and their adversaries. At least some of the Stoics denied that humans belonged to the same community as other animals, arguing that humans therefore could not owe the animals just treatment. In Theophrastus's view, however, we have a right to kill dangerous animals for self-protection, but only in the same manner that we have a right to protect ourselves against criminals.[73] Otherwise, the animals have a right to their lives. Moreover, there is no suggestion that Theophrastus subscribed to the idea of the transmigration of souls. Respect for animals seems to have derived just as readily from the notion of anatomical similarity as from the idea of metempsychotic kinship.

Whereas, in general, metempsychosis was taken less than seriously by many of the pre-Socratics – as we have noted, Xenophanes poked fun at it – Plato's myths gave the doctrine respectability. Certainly, we find transmigration espoused by such neo-Platonists as Porphyry (apparently) and Plotinus (more obviously), who found the doctrine helped to solve the problem of the apparent evil of animals eating each other.[74] The idea was used likewise in both the Islamic and Jewish traditions for similar ends.[75] If the souls of animals continued after bodily death, and were later re-housed, then the quintessence, the soul, of the animal had not been killed.

IV

If such ideas played less of a role after the advent, and increasing dominance, of Christianity, they did not disappear entirely. In the fourth century, Arnobius wrote of "the hidden mysteries" whereby "the souls of wicked men on leaving their bodies, pass into cattle and other creatures"; thus, if this is true, he says, "it is even more clearly shown that we are allied to them, and not separated by any great evil, since it is on the same ground that both we and they are said to be living creatures, and act as such."[76] Origen accepted the possibility of transmigration into plants as well as animals.[77] Nonetheless, generally speaking, Christianity espoused a much more personal conception of survival and salvation. Still, transmigration did not disappear. During the Middle Ages, both the vegetarian Albigensians (Cathars) of the Languedoc in France and the Bogomils in the Balkans embraced the doctrine – often human to human transmigration, occasionally human to animal. The historian Keith Thomas sees in the medieval era and the Renaissance "hints of popular belief in the transmigration of souls" – and, given the difficulty of uncovering popular, rather than official, formal religious or intellectual beliefs in the distant past, "hints" may be evidence of something more widespread. Indeed, one of the primary justifications for speculation in historiography is the importance, and equal difficulty of discovering, the ideas of "the people." As Karl Popper wrote, "what really happens within the realm of human lives is hardly ever touched upon ... The life of the forgotten, of the unknown individual man; his sorrows and his joys, his suffering and his death, this is the real content of human experience down the ages ... all the history which exists, our history of the Great and the Powerful, is at best a shallow comedy."[78]

Thomas also acknowledges neo-Platonist scholars as postulating "the movement of the universal soul into every kind of animal creation," thus suggesting that "even the beasts had the divine spark within them."[79] Nor did the conception of the divine spark in animals diminish. The Anglican priest Thomas Young assured his readers in 1798 that other species possessed the divine "spark of life," and he seems to assume that this would find a ready acceptance among his readers.[80] And if this was the view of the scholars, how much more would it be the view of those who lived cheek by jowl with the animals?

In the Renaissance, the poet and divine John Donne (1572-1631) wrote *Metempsycosis*, a poem reflecting the fact that, although the transmigration of souls was not widely held, it was not entirely ignored. In the poem, Donne bemoaned the ills done to fish, while they in turn hurt neither humans nor beasts. In a letter of 5 February 1649 to the Cambridge Platonist Henry More, René Descartes mentioned disparagingly those who are given "to the superstitions of Pythagoras"[81] – an unlikely comment if there were no Pythagoreans around. The Water Poet John Taylor wrote "The Olde, Olde, Very Old Man" in 1635 to commemorate the life of Walter Parr, who died in that year, reputedly at the age of 152, declaring that "He was of old Pythagoras' opinion." Perhaps this referred to his vegetarianism alone, although this was certainly not the case when, ironically in light of their doctrines, J. Barnard announced in *The True Life of Heylyn* [Healing] of 1683 that "the Soul of St. Augustine (say the Schools) was Pythagorically transfused into the corps of Aquin."[82] The vegetarian Anabaptist Thomas Tryon wrote *Pythagoras: His Mystick Philosophy Reviv'd* around 1684, but the philosophy revived referred first and foremost to the eschewing of flesh foods rather than to the transmigration of souls. However, the pantheistic antinomian Ranters, who throve around the time of the Commonwealth, allegedly held that "when we die we shall be swallowed up into the infinite spirit, as a drop into the ocean, and so be as we were; and if ever we be raised again, we shall rise a horse, a cow, a root, a flower and such like."[83] In the late seventeenth century, John Dryden's translation of Ovid's *Metamorphoses* was widely read and maintained public interest in Pythagorean doctrine. Indeed, we find it expressed in John Wilmot's "Satyr" of 1675, where the author gets to choose "What Case of Flesh, and Blood, I pleas'd to weare."[84]

A few years later John Gay, in *Trivia* (1716), wished the doctrine to be true:

> If, as the *Samian* taught, the Soul revives,
> And, shifting Seats, in other Bodies lives,
> Severe shall be the brutal Coachman's Change,
> Doom'd in a *Hackney* Horse the Town to range.[85]

("Samian" refers to Pythagoras since he was born at Samos in Asia Minor; "Hackney" refers to a kind of coach, and a "Hackney Horse" to the horse that pulls such a coach.)

In similar vein, the eighteenth-century novelist Henry Fielding, when faced with an instance of cruelty to a horse by a coachman, described how: "My son Tom told me, as we pursued our walk that he had a facetious [agreeable] acquaintance ... who professed the Pythagorean principles, and affirms that he believes [in] the transmigration of souls. This gentleman, as Tom informed me, comforts himself on all such occasions with a persuasion that the beasts he sees thus abused have formerly been themselves ... coachmen; and that the soul of the then driver will in his turn pass into the horse and suffer the same punishment which he so barbarously inflicts on others."[86] Writing a decade after Fielding, the British parliamentarian and commissioner of the Board of Trade, Soame Jenyns, took the doctrine quite seriously in *Free Inquiry into the Nature and Origin of Evil,* bemoaning the failure of his contemporaries to do so:

> But the pride of man will not suffer us to treat this subject [of the transmigration of souls] with the seriousness it deserves; but rejects as both impious and ridiculous every supposition of inferior creatures ever arriving at its own imaginary dignity, allowing at the same time the probability of human nature being exalted to the angelick, a much wider and more extraordinary transition, but yet such a one as may probably be the natural consequence, as well as the reward of a virtuous life; nor is it less likely that our vice may debase us in the servile condition of inferior animals, in whose forms we may be punished for the injuries we have done to mankind when amongst them, and be obliged in some measure to repair them, by performing the drudgeries tyrannically imposed upon us for their service.[87]

The anonymous author of *Adventurer* 5 (1752) described a dream of transmigration in which a flea tells how it suffered in various incarnations as a
puppy, a bullfinch, a cockchafer, a lobster, an earthworm, and a pig. No
doubt the purpose of the exercise was less to convince the readers of the
reality of metempsychosis than to instill in them the fact of the unbear-
able and unjustifiable cruelties meted out to each of these animals in turn.[88]
Nonetheless, metempsychosis was treated as a readily recognized creed, as
indeed was Pythagoreanism in general. In the Foreword to *The Pythagorean
Sourcebook and Library,* Joscelyn Godwin mentions the indebtedness of
the astronomers Copernicus, Galileo, and Kepler to the Pythagoreans for
their heliocentric astronomy and observes that "Largely as a consequence
of the work of ... two great 'romantic' Pythagoreans, Thomas Taylor (1758-
1835) and Fabre d'Olivet (1767-1825) ... the nineteenth century swarms with
semi-Pythagoreans, great and small, but all of rather limited effect on the
world in general."[89]

There is more than a mere hint of transmigration in William Wordsworth's great ode "Intimations of Immortality from Recollections of Early
Childhood":

> Our birth is but a sleep and a forgetting:
> The Soul that rises with us, our life's Star,
>   Hath had elsewhere its setting,
>   And cometh from afar:
>   Not in entire forgetfulness,
>   And not in utter nakedness.

Nor is it without relevance to our respect for our fellow animals:

> Ye blessèd Creatures, I have heard the call
>   Ye to each other make; I see
> The heavens laugh with you in your jubilee;
>   My heart is at your festival,
>   My head hath its coronal,
> The fulness of your bliss, I feel – I feel it all.[90]

William Blake mentioned his previous existences on several occasions,
commenting, for example, on his once having been Socrates and on another
occasion mentioning "the books and pictures of old, which I wrote and
painted in ages of Eternity before my mortal life"[91] – although he never
mentioned having previously been an animal. Charles Baudelaire hints at
reincarnation in "Former Life" of *Les Fleurs du Mal.*[92] In *Joseph Andrews,*

Henry Fielding writes of "The Lawyer, [that is, the author, who] is not only alive, but hath been so these 4000 years [that is, since creation]."[93] Oscar Wilde's iconoclastic literary mother, Lady Jane, "claimed her aquiline look came from having been an eagle in a previous existence."[94] The theosophists numbered Robert Browning among the reincarnationists. And certainly the theosophists themselves, led by Helena Petrovna Blavatsky and Annie Besant, subscribed to an idiosyncratic version of the doctrine. W.B. Yeats was a member of the Theosophist Society for a time (1887-90) and maintained a belief in metempsychosis even afterward, to which *A Vision* (1926) is witness. In *Finnegans Wake* James Joyce embraced a cyclical view of history, and in *Ulysses* he made frequent allusion to the doctrine of metempsychosis, Bloom telling his wife, in the Calypso episode: "They call it reincarnation. That we all lived before on the earth thousands of years ago ... They say we have forgotten it. Some say they remember their past lives." Whether Joyce believed in the doctrine seems unlikely, but there is no doubt he took the matter very seriously.

Raised in an Orthodox Jewish family in Poland before the Second World War, Isaac Bashevis Singer reports the then continued existence of the doctrine. He heard and read that the souls of the dead were reincarnated in cattle and fowl and that killing them with a kosher knife and saying the blessing with fervour served to purify these souls.[95] Fellow Jew Saul Bellow is emphatic, if a trifle opaque, in telling us in "Cousins," via Ijah Brodsky, speaking for the author: "We enter the world without prior notice, we are manifested before we can be aware of manifestation. An original self exists or, if you prefer, an original soul." In *The Bellarosa Connection* Bellow refers to this innateness and tells us that it is "a tricky word 'innate,' referring to the hidden sources of everything that really matters." If, on the whole, in the Western world the doctrine was taken with a pinch of salt by the exalted, it probably lived a healthy life among the peasants. And even the exalted sometimes recognized the value of the retribution inherent in the creed.

V

The world-soul, anima mundi, or universal soul can be found in different forms in different societies. In the West it was first given explicit literary form in Plato's *Timaeus* (29/30), where we are told that "this world is indeed a living being endowed with a soul and intelligence ... a single visible living entity containing all other living entities, which by their nature are all related." The central idea is that each individual soul is a part of a greater collective soul, which is itself an organism and which, like each individual soul, has breath and motive and is animated by a deeply rooted vital force that is the basis of justice and reason. If one wishes to seek a precursor, one can perhaps be found in Pythagoras's contemporary Anaximines, who regarded cosmic air as a live substance, analogous with the human soul, with the implication that the soul is not entirely separated from the material world but is a natural part of it. As Martin West indicates, it "is tempting to see here an affinity with the Upanishadic doctrine of a universal wind or breath with which both the unchanging life-soul of the world and the individual self are identical, by which living things and worlds are held together, and which the whole universe obeys."[96] Whether the analogy holds for Anaximines, Plato's doctrine, as in so much else, certainly bears considerable similarity with the Upanishadic world of Hinduism. The Hindu idea of the world-soul is expressed in the *Chandogya Upanishad* (3.14):

> All this universe is in truth Brahman. He is the beginning and end and life of all ...
> There is a Spirit that is mind and life, light and truth and vast spaces. He contains all works and desires and all perfumes and all tastes. He enfolds the whole universe, and in silence is loving to all ... This is the Spirit that is in my heart, greater than the earth, greater than the sky, greater than heaven itself, greater than all these worlds.
> This is the Spirit that is in my heart, this is Brahman.
> To him I shall come when I go beyond this life.[97]

Many later Western philosophers took up the idea. The Stoics, who flourished in Rome from around a hundred years before Christ to a couple of centuries after and who had a considerable influence on Christian ethics, described a "World-Soul," a kind of immanent God, that made all humans brothers and worked in conjunction with reason and providence. However, although there were exceptions among the Stoics, the majority excluded animals from this community since they did not, it was said, possess the reason that was the foundation of the sense of community pertaining to all human beings. The Dominican heretic Giordano Bruno (1548-1600) was not so exclusive. His pantheistic conception of the world-soul conceived of all creation as one life, animated by God. In *Cause, Principle, and Unity* he described "the universal intellect" as "the innermost, most real, and essential faculty and the most efficacious part of the world soul. It is the one and the same thing, which fills the whole, illumines the universe, and directs nature in producing her species in the right way."[98] If, for Bruno, nonhuman animals do not participate in the rational life, they are nonetheless themselves the product of universal reason and thus are kin with all species. The great German polymath Johann Wolfgang Goethe (1749-1832) conceived of the earth as an ensouled organism. The world-soul permeated and gave purpose to this organism within which all individual life interacted. He wrote of animals as "brothers" and as "ends in themselves," which were an integral part of the world-soul. Both Henry More (1614-87) and Johann Gottfried Herder (1744-1803) saw nature itself as a kind of world-soul and animals as being ensouled. Indeed, More affirmed that they who "think that [animals] are not all made for themselves [are] ... ignorant of Man and the Knowledge of Things ... [the animals] enjoy their own beings."[99] Herder stressed the cultural and organic superiority of humans over other species, but he nevertheless wrote, against what was a common belief, that humans possessed "in the most perfected form all the powers and capabilities of other species."[100] This is patently untrue, he adds. All species are perfected in their own way, and if humans are superior in some ways and superior overall, animals are superior in other respects consistent with their own natures. Friedrich von Schelling (1775-1854) used the term "world-soul" as a unifying principle of the organic and inorganic. In *Ideen zu einer Philosophie der Natur* (1797),[101] he held that nature could not be subordinated to reason and that the principle of mind is already inherent in nature. In accord with the prevailing conceptions of the Great Chain of Being, he postulated different degrees, or levels, within this organic unity, which was a history of progress toward harmony from a previous fall. Animals as a part of nature participated in the world-soul.

While Ralph Waldo Emerson had great respect for the animal realm, animal participation is conspicuously absent from his conception of the "Over-Soul." Thus he refers to "that Unity, that Over-Soul within which every man's particular being is contained and made one with all other; that common heart of which all sincere conversation is the worship, to which all right action is submission; that overpowering reality which confutes our tricks and talents, and constrains everyone to pass for what he is, and to speak from his character and not from his tongue, and which evermore tends to pass into our thought and hand and become wisdom and virtue and power and beauty."[102] Nonetheless, even if Emerson does not entirely contradict this in *Nature,* he offers there an image in which the human-animal relationship is far closer:

Neither does the wisest man extort [Nature's] secret, and lose his curiosity by finding out all her perfection. Nature never became a toy to a wise spirit. The flowers, the animals, the mountains, reflected the wisdom of his best

hour, as much as they had delighted the simplicity of his childhood ...
Nature stretches out her arms to embrace man, only let his thoughts be of
equal greatness ... every animal function from the sponge up to Hercules,
shall hint or thunder to man the laws of right and wrong, and echo the Ten
Commandments ... How much industry and providence and affection we
have caught from the pantomime of brutes ... Each creature is only a modification of the other; the likeness in them is more than the difference and
their radical law is one and the same.[103]

The third sentence from the end indicates all of animate life as a part of
the same moral system, the penultimate sentence indicates what we have
to learn from other species, and the final sentence appears to inform us
that all species have a place in the "Over-Soul." The message appears to be
one already expressed in the Book of Ecclesiastes (3:19-21) of the Hebrew
Bible: "For the fate of human and the fate of animal is the same: as the
one dies, so the other dies; both have the selfsame breath. Human is in no
way better off than the animal – since all is futile. Everything goes to the
same place, everything comes from the dust, everything returns to the dust.
Who knows if the human spirit mounts upward or if the animal spirit goes
down to the earth?" A few words from d'Alambert in the *Encyclopédie* might
be added: "Everything in nature is connected. Nature isn't at the surface,
it's in depth."[104] Nor should we imagine that the concept of the universal
soul or universal spirit is restricted to those who have been mentioned. It
is also to be found in the works of Jakob Boehme, Emanuel Swedenborg,
Henry David Thoreau, John Ruskin, John Muir, Joseph Conrad, Marcel
Proust, and Albert Camus, among others.[105] The nineteenth-century French
poet Stéphane Mallarmé said that he could be understood only as "an
aptitude of the universal spirit." In the final analysis, the idea of the universal soul is about the relationship of the divergent elements of nature to
each other. It can be used, like kinship in transmigration, like the various
ranks within the Great Chain of Being, or like the human-animal similarities and differences in the theory of evolution, to elevate the animals or
to denigrate them. It is not the system itself, or the concepts themselves,
that provide answers. Most depends on the disposition of the person who
uses the system. But most of those who have used the concept of the universal soul have done so to draw us closer to animated nature.

In at least one respect the idea of the world-soul, or universal soul, can
be used to the great detriment of animal interests, as we have already seen
with metempsychosis. Thus, if the soul, or spirit, of the animal is regarded
on death as returning to the universal soul before being born again, the
killing of an animal for food (or, indeed, for any other purpose) can be

justified in the belief that one is not "really" killing the animal. Thus, for example, when the Netsilik Inuit catch a seal, they sprinkle water in its mouth to satisfy the thirst of its spirit, which is a part of the universal seal spirit. The Inuit believe we are all a part of a universal soul with corresponding obligations and rights to each other. To kill a seal does not harm the seal, provided that one reveres and appeases the seal spirit as a part of the universal spirit. The Inuit place mittens on the carcass of the seal in the hope and expectation that the seal's spirit will report its favourable treatment, will be content to "offer" itself again as food on a future occasion, and will encourage other seals to make themselves available as prey.

It is understood by the Inuit that there is something *in principle* unacceptable about the killing of other beings. Yet at the same time, for the Inuit, meat eating is a necessity in the inhospitable clime they inhabit, where anything more than the most meagre horticulture is impossible. The Inuit must thus *atone* for their behaviour and become *at one* with the spirit of the animal they have killed. The logic of the situation, in relation to their beliefs, requires that they engage in exculpatory behaviour (*at-one-ment*). In order to justify their behaviour to themselves, they must employ terms that excuse the killing by treating it as a nonkilling: the spirit of the animal returns to the universal spirit (soul), whence to be born again. The essence of the reverential behaviour is an apology for behaviour that can be excused only by necessity and that requires an elaborate myth in order to render the objectively harmful acts acceptable to those who commit them. After all, whatever the myth may suggest, the animal in fact dies. And if life matters, what is done to the animal is a transgression. Moreover, if the idea is in principle satisfactory, then one must inquire why it does not apply equally to human lives, which are also a part of the universal spirit.

Whatever conclusions one may wish to draw from such a tangled web, it should at least be clear that they cannot be simple ones. In the hands of the Pythagoreans – and of the Brahmins, although not of Hindus in general – the kinship implied by the transmigration of souls led to vegetarianism – in some instances out of respect for the animals and in others for health reasons or in order not to defile one's own body with the corpses of other creatures. In Plato's case, there would appear to have been a willingness to continue flesh consumption in the existing society, and if vegetarianism were to become the norm of the new republic, it would have little to do with the interests of animals but be practised out of concern for health, simplicity, and asceticism. On the other hand, it is intimated in the later references to transmigration in the Western tradition that some have been attracted to the doctrine not because it is convincing metaphysically, but because it is a convenient way to find a retribution for the evils perpetrated on animals in this life. And the view that scholars like Pythagoras and Plato used the doctrine more as a convenient metaphor than as a metaphysical description of reality has a ring of truth, even if we lack sufficient evidence to be fully convinced. Howard Williams was one of those who was convinced that such was the case. In *The Ethics of Diet* we are told that "we may presume that by [the doctrine of metempsychosis] Pythagoras intended merely to convey to the 'uninstructed,' by parable, the sublime idea that the soul is gradually purified by a severe course of discipline until finally it becomes fitted for a fleshless life of immortality."[106]

Transmigration in all its forms may suggest a human-animal kinship, but it has nonetheless always indicated a decided hierarchy among animals, with humans at the apex. All animals may be related, but this does not suggest that they share anything like an equal worthiness or that they are necessarily the objects of great respect. In the Hindu and Buddhist traditions, domestication is seen to denigrate animals (with the exception of cattle), and freedom is seen to elevate them, as well as the birds, who are the freest of all and, as they most obviously possess a faculty beyond human accomplishment, the highest of all. Majesty, too, as with elephants, lions, and tigers, is seen as a mark of favour, whereas the simple animals – the louse, the moth, and the worm – constitute the nadir. Likewise, other things being equal, those animals seen as beautiful are favoured over those judged to be ugly. And, again, other things being equal (and often they are not), the friendly are favoured over the harmful. This often leads to the contradiction of certain species being treated as both awesome and frightening (in India both tigers and cobras fall into this category). But then, as Plato has Socrates proclaim in the *Euthyphro,* "where reverence is, there is fear,"[107] a maxim adopted by Hobbes in his insistence that all reverence, all worship, is based in fear.

The Hindu and Buddhist conceptions do not differ significantly in general from the customary Western conception of the animal hierarchy, although they do differ from that of Plato, which may be taken as something of an exception, but only something, for the republics of the insects and the consequent organizational capacities of their members were frequently praised. For Plato, in the *Republic,* the gentle and social animals, such as the ant and the bee, constitute the animal heights, whereas the aggressive animals, such as the wolf, the hawk, and the kite, are (along with the ass, perhaps for its stubbornness) the most despised. There are elements of this in Western thought generally. The bee has been the object of admiration from the poetry of Virgil through the prose of Montaigne to the verse of Ralph Waldo Emerson; and the wolf was constantly feared until the twentieth century (even where it was extirpated!), but the elements of insignificance often accorded the insects and those of awe generally accorded predators are entirely ignored in Plato's scheme in the *Phaedo*. Elsewhere in Plato it is the songbirds that are elevated – a far more customary practice. So much so that when, in the early nineteenth century, we find the poet John Clare extolling the dowdy sparrow rather than the radiant swallow, the wren rather than the customarily eulogized nightingale, as well as the badger rather than the bear, and the rights of the fly over the purity of one's ale, the reader is a little taken aback, although it is not an unpleasant surprise.[108] The message is extended by John Steinbeck in *Cannery Row*. Those whose souls fly to the gods can look after themselves; all others share a common need and love: "Our Father who art in nature, who has given the gift of survival to the coyote, the common brown rat, the English sparrow, the house fly and the moth, must have a great and overwhelming love for no-goods and blots-on-the-town and bums, and Mack and the boys. Virtues and graces and laziness and zest. Our Father who art in nature."[109] The philosophers and the priests approach the gods. The meek and the powerless, both human and animal, inherit the earth and become the salt of the earth. Later in *Cannery Row,* Doc tells us that "All of our so-called successful men are sick men, with bad stomachs, and bad souls, but Mack and the boys are healthy and curiously clean. They can do what they want. They can satisfy their appetites without calling them something else."[110] Mack and the boys are animals – but are better for it than those who aspire to be gods. Their souls are animal souls. Plato and Steinbeck share the same conception of a human hierarchy – but their heroes and demons are reversed.

In modern animal-welfare and animal-rights discussions, the talk is customarily of the wellbeing and rights of animals in general versus those of humans. Perhaps greater understanding might be achieved if animals were far less homogenized and if one were to talk more often of the wellbeing and the rights of humans, of zebras, of crocodiles, of pythons, of sharks, of dolphins, of whales, of hyenas, of deer, of lice. After all, if the human is to be treated as a distinct species, with species-specific needs, aren't other species worthy of similar treatment? When George Bernard Shaw soliloquized on the rights of "a poor innocent microbe" in *Too True to Be Good,* he wrote in jest. But he was also asking us to distinguish between the microbe and the dog. We might be surprised when we encounter the view of Cornelius Lapide, who claimed in the sixteenth century, continuing a long tradition of thought that had begun with Empedocles in the fifth century BC, that "lice, flies, maggots and the like were not created directly by God but by spontaneous generation, as lice from sweat"[111] – a view repeated by the reputed biologist John Needham in the eighteenth century

when his experiments seemed to confirm it. The belief that such animalcules were so despicable that God would not have deigned to create them
did not finally disappear until the results of the research of Louis Pasteur
became known. But at least we are not as astonished at such views as we
would be if Lapide had accorded the lowliest status to polar and panda
bears, to chimpanzees and mustangs. And we are pleasantly surprised when
we encounter William Cullen Bryant's paean "To a Mosquito" (1826),
being far less surprised at his sympathy for the hunted bird in "To a Water-
fowl" (1815).

In all traditions, the human is given special treatment, and the human
capacity for reason is always an important part of the human-animal distinction. In the West, however, even greater emphasis has been laid on reason, albeit of a special type – that is, philosophical reason, intellectual
speculation, and scientific hypothesis. The relationship of this type of reason to the possession of soul, or at least a special kind of soul, has played
a significant role in Western thought. Indeed, in the seventeenth and eighteenth centuries, there was what Ben Lazare Mijuskovic has called a "unity
and identity of thought and soul."[112] If animals seemed to possess the relevant characteristics of thought and consciousness, they were deemed ensouled. If they lacked these attributes, they were deemed soulless, or at least
bereft of the relevant kind of soul – an immortal soul. But if there was a
"unity and identity of thought and soul" in the Renaissance and the Enlightenment – most of the time anyway, although perhaps less frequently
in France, as we shall see in Chapter 5 – there was also a significant element of this conception in previous centuries, too. Nor did it die out
afterward. From October 1839 to August 1840, *The Veterinarian* published
an amiable dispute between two veterinary surgeons, William Karkeek and
W.C. Spooner, on whether the possession of reason was sufficient for
immortality, each agreeing that animals were rational but disagreeing
about the implications of rationality for everlasting life.[113] Nonetheless, as
we shall see in greater detail in the next chapter, the customary claim that
the Western tradition has denied animals the possession of souls, or
rational souls, or immortal souls, and that as a consequence animals have
been ill-considered and ill-treated in the West, is misleading.

In perhaps the most famous book on the immortality of the soul, one
that went through ten editions between 1860 and 1878 – and perhaps second only in readership to Plato's work on the topic – William Rounseville
Alger advised his readers that:

Physical death is experienced by man in common with the brute. Upon
grounds of physiology there is no greater evidence for man's spiritual survival

through that overshadowed crisis than there is for the brute's. And on grounds of sentiment man ought not to shrink from sharing his open future with these mute comrades. Des Cartes and Malebranche taught that animals are mere machines, without souls, worked by God's arbitrary power. Swedenborg held that "the souls of brutes are extinguished with their bodies." [*Outlines of the Infinite*, ch. 2, sect. 4, 13.] Leibnitz by his doctrine of eternal monads, sustains the immortality of all creatures. Coleridge defended the same idea. Agassiz, with much power and beauty, advocates the thought that animals as well as men have a future life. [*Contributions to the Natural History of the United States*, vol. 1, pp. 64-66.] The old traditions affirm that at least four beasts have been translated to heaven; namely, the ass that spoke to Balaam, the white foal that Christ rode into Jerusalem, the steed Borak that bore Mohammed on his famous night-journey, and the dog that wakened the Seven Sleepers.[114] To recognize, as Goethe did, brothers in the green-wood and in the teeming air, – to sympathize with all lower forms of life, and hope for them an open range of limitless possibilities in the hospitable home of God, – is surely more becoming to a philosopher, a poet, or a Christian, than that careless scorn which commonly excludes them from regard and contemptuously leaves them to annihilation.[115]

To be sure, Alger continued his narrative by observing a "vast" distinction between "the dying man and the dying brute," and we are left with a clear impression one can be more confident of an immortal soul, an eternal life, for humans than for animals. Nonetheless, the point remains, a belief in animal futurity is "more becoming to a philosopher, a poet or a Christian" than a belief that animal life is extinguished at death. The reality is that most in the Western tradition have ascribed some kind of soul to animals, many have ascribed rational and immortal souls to animals, and some of those who have denied animals the possession of immortal souls have still thought it important that animals be treated well.

VI

In general, although without any degree of precision or consistency, we may say that soul, or spirit, was regarded as indivisible, immaterial, rational, self-directed, and immortal. Matter was, by contrast, divisible, incapable of self-movement, lacking the capacity for thought, and perishable. Hence the debate about animal souls tended to revolve around questions of whether they were capable of self-control or were instinctual beings, whether they could think and reason in the same manner that humans could think and reason, and whether they were spiritual or solely material entities.

In the minds of many, answers to the question were of more than philosophical moment. If animals were not moral agents – that is, were not morally culpable for their behaviour – yet were deemed to possess souls and consequent immortality, then the fear was that moral pandemonium would ensue. People could be dissuaded from inflicting harm on others by the promise of eternal damnation and the consequent promise of a life hereafter should they obey the injunctions of their superiors, both ecclesiastical and lay. They could be persuaded to accept the injustices perpetrated on them in this life if they could be persuaded of Matthew 19:24: "it is easier for a camel to pass through the eye of a needle than for someone rich to enter the Kingdom of Heaven." Thereby should wrongs be righted, iniquities avenged. They could be persuaded that heaven would be their just reward for willing submission in the here and now. Yet if animals, who possessed neither a moral sense nor the capacity for choice, held the same destiny before them, then heaven could be achieved without acting morally and submissively. If heaven was not a special reward for the deserving, then there was little point in belonging to the deserving. In such circumstances, the heavenly odds were stacked against the beasts. In the opinion of many, to have ensouled the animals would have been to trivialize the plight of humans.

3

A Natural History of Animal Souls

I

THERE IS A COMMON ASSUMPTION in the literature on animal ethics and in books on world religions that Aboriginal and Oriental belief systems ascribe souls to animals, whereas the Western tradition does not, and that accordingly the Western attitude to animals, and the status assigned to them, is distinctly less considerate than that of others. In the West alone, animals are seen to be for human use, and in the West alone humans are seen to believe that they are entitled to exploit animals. It has thus become customary, where traditions are compared, to find the Aboriginal and Oriental experience lauded and the Western tradition castigated, especially, but not solely, in its Judaic and Christian dimensions.[1] This orientation is exemplified by Christopher Key Chapple, who writes: "noninjury to animals holds a prominent place in both [Jaina and Buddhist] traditions. Animals are regarded to be none other than our very selves ... By contrast, the status of animals in most cultures, especially those arising from Europe and America, regards animals to have no such kinship relation. The book of Genesis and the writings of Aristotle, Aquinas, Descartes and others have justified the position that animals exist solely for human exploitation."[2]

In like vein, we are told by Peter Knudtson and David Suzuki, in their survey of the wisdom of a substantial variety of indigenous cultures from around the world, that unlike Westerners, Native peoples recognize "the Kinship of All Life." In contrast to Western oppressive attitudes, "the Native Mind is imbued with a deep sense of reverence for nature. It does not operate from an impulse to exercise dominion over it ... The vocabulary of Native knowledge is inherently gentle and accommodating towards nature rather than aggressive and manipulative."[3] Even the best of commentators have bemoaned that until very recent times, there has been nothing in the Western tradition to match the classical humanitarianism of Plutarch or, especially, Porphyry without their seeming to notice that there is an even longer hiatus from the Hindu *Upanishads* and the Buddhist *Dhammapada* to any modern humanitarian expressions, perhaps Gandhi excepted.

One is struck by the tendentiousness and the one-sidedness of the examples chosen to represent the Western tradition. Why merely the "dominion" (*rādâ*)[4] of Genesis 1 and no mention of the equal creation from the topsoil of Genesis 2? Why the assumption that *rādâ* implies that "animals exist solely for human exploitation"? What is the evidence? Both modern Biblical studies and numerous historical interpretations have insisted that the implication of *rādâ* is that although humans are given authority over the animals, they also owe them a significant responsibility. It would be difficult to recognize this from a reading of the relevant recent, and some not so recent, animal-rights or ecology literature. Thus Edward Payson Evans, Lynn White Jr., William Leiss, John Passmore, Peter Singer, Keith Thomas, Roderick Frazier Nash, Jim Mason, and very many others have imagined that God's granting of dominion constituted a licence for humans to use animals as they wished without any consideration for their welfare.[5] In reality, the Genesis story is a complex one – indeed, it has more than one author, with rather different tales to tell. In the more recent P narrative (Genesis 1), humans are told to exercise dominion – supreme authority – over other species. The older J narrative (Genesis 2) presents a quite distinct conception of the human-animal relationship. Here, as Theodore Hiebert has indicated, "J conceives of this relationship in terms that are more communal. As humans and animals are made alike from the topsoil, they possess no distinct ontological status, both being referred to simply as living beings."[6] Now certainly, it is the P narrative that is more commonly referred to, although Buckner in *The Immortality of Animals* (1903), for one, emphasizes the human-animal equality of Genesis 2.[7] Still, far more people are, and historically have been, aware of the grant of Genesis 1:26, which reads, in its best-known King James version, as the awarding to humankind of "dominion over the fish of the sea and over the fowl of the air, and over the cattle, and over all the earth, and over every creeping thing that creepeth upon the earth."[8] And certainly the awarding of *rādâ* is the granting of a significant power that may be used despotically. Nonetheless, an implication of the term is that humans were intended to share not only some of God's prerogative, but also, as Andrew Linzey has explained, "his moral nature," acting toward animated nature as God did toward humans, bringing order to chaos, and bringing blessings and goodness rather than tyrannical mastery to the world.[9] Certainly, in Genesis 1 animals are subordinated to humankind, and humans are entitled to their use. But, as the detailed research of Elijah Judah Schochet has demonstrated, domesticated animals were to be regarded as "the delicate tool," as instruments for human use, to be sure, but to be used with feelings of compassion for, and kinship with, animals.[10]

Moreover, since the grant of dominion occurs before the right to eat flesh is conferred, this dominion does not even include the killing of animals for food. And Adam's naming of the animals (Genesis 2:20) – naming was an act denoting special significance – indicated the close relationship between human- and animalkind. Indeed, in the pastoral era, husbandman and animal lived their lives in a close proximity we find difficult to imagine in the modern era. Further, for his protection of the animals, Noah was granted the title of *zaddik* – a man of charity.

Perhaps more importantly, what we might call limited dominion has been the customary historical interpretation of the grant. Thus in *The Primitive Organization of Mankind* (1677), Lord Chief Justice Sir Matthew Hale wrote that the "end of man's creation was that he should be the viceroy of the great God of heaven and earth in this inferior world, his steward, villicus, bailiff or farmer of this godly farm of the lower world." It was thus that humankind was "invested with power, authority, right, dominion, trust and care, to correct and abridge the excesses and cruelties of the fiercer animals, to give protection and defence to the mansuette and useful ... to preserve the face of the earth in beauty, usefulness and fruitfulness."[11] And he adds: "I ever thought that there is a certain degree of justice due to the creatures as well as from man to man."[12] In *Seasons* (1728), poet and former divinity student James Thomson explained the human role as "The Lord and not the Tyrant of the world."[13] In his *Self-Interpreting Bible* (1776), the staid Biblical traditionalist John Brown declared it an "honourable dominion over the creatures."[14] In a chapter on Biblical lessons in *An Essay on Humanity to Animals* (1798), Thomas Young tells us that the interpretation to be drawn from "passages of scripture" is that "it is the will of God that we should abstain from cruelty, and cultivate humanity, towards the brute creation."[15] "Gentle dominion" were the words of the radical publisher George Nicholson in *On the Primeval Diet of Man* (1801).[16] In *An Essay on Abstinence from Animal Food as a Moral Duty* (1802), the even more radical (and atheist) antiquarian Joseph Ritson declared that the "dominion" of Genesis was instituted "for the sake of authority, protection, and the glorious offices of benevolence and humanity."[17] The British parliamentary leader in the animal cause, Lord Erskine, remarked in a debate in 1809 on the second reading of an unsuccessful animal-welfare bill that "the dominion granted to us over the world is not conceded to us absolutely. It is a dominion in trust; and we should never forget that the animal over which we exercise our power has all the organs which render it susceptible of pleasure and of pain."[18] In *The Rights of Animals* (1838), the Unitarian-Christian Irish preacher William Drummond argued that "man's dominion over [the animals] is a delegated trust, which he is required to use with discretion and lenity." Moreover, "though [God] gives man the privilege to use [animals] as his wants and necessities require, he gives no authority to abuse the privilege, and convert liberty into lawless licentiousness ... the charter given to man invests him with the privilege to *reign,* not with authority to *tyrannise* ... Man's powers have their limits, and animals have their rights ... They were formed for their own enjoyment of life ... it is a dominion of justice and mercy." Nonetheless, he concedes that the contrary "obnoxious" belief, born of "arrogance and self-conceit," is sometimes "maintained by philosophers, and sanctioned by divines."[19] In the pages of *The Veterinarian* for 1839, William Karkeek described the doctrine on human usage of animals at will as "preposterous and absurd" and quoted the poet Robert Southey in his cause from *The Doctor,* where Southey deems it "monstrous" and "tyrannical."[20] He concludes that "it really does appear almost unnecessary to refute such an assumption."[21] Nonetheless, he quotes further from Southey: "Made for thy use, indeed, when so may seem to have been made for thy punishment and humiliation!"[22] In her popular novel *Agnes Grey* (1847), the devout Anne Brontë insisted that the Genesis passage, together with other Biblical texts, implied a significant responsibility to what she called our fellow "sentient creatures," although her commentary also showed how the issue was a cause of constant dispute between the compassionate and the selfish.[23] The veterinarian and former Nonconformist minister William Youatt treated the matter at some length in *The Obligation and Extent of Humanity to Brutes* (1839). Having claimed the "dominion" of Genesis to be a "delegated authority," he asks:

> How far does this delegated authority extend? It reaches to the accomplishment of every benevolent purpose: and by no casuistry can it be understood as sanctioning cruelty. Has he who may be accused of habitually or too frequently inflicting unnecessary pain on those who are indebted for life to the same parent with himself, never thought of the interest which that all-wise Being must feel in the welfare of his offspring; and the indignation with which he must behold every instance of wanton cruelty inflicted on his children, and they are all his children, throughout the whole circle of animated existence?[24]

In November 1847 the London *Times* reported the words of a Clerkenwell magistrate on sentencing a youth to a fine of twenty shillings or ten days' imprisonment for cruelty to a horse: "I will teach you that poor animals were sent here for man's use and not his abuse."[25] In light of such evidence, if the claim is made that the Biblical and Christian view is that "animals exist solely for human exploitation," we need to be offered a great deal more than unsubstantiated assertion. We would have to be shown how the passages just quoted were quite unrepresentative of the tradition. Nowhere have I ever encountered even a half-hearted effort at such a demonstration.

Again, if this is not mere self-indulgent rationalization, why think only of the Aristotle who argued that "animals must have been made by nature for the sake of men" and ignore the Aristotle who wrote that "in all of [the animals] there is something natural and beautiful" and that "just as in man we find knowledge, wisdom and sagacity, so in certain animals there exists some other natural propensity akin to these"? Why ignore the Aristotle who wrote of animals as possessing souls, intelligence, nobility, wisdom, beauty, and practical reason?[26] Moreover, why choose Aristotle as the sole representative of the classical Western tradition? One could have selected Theophrastus, Plutarch, Porphyry, or Sextus Empiricus, who, in outright contradiction of Chapple's claim, emphasize our kinship with, and similarities to, other species. Why are we not reminded that Plutarch claimed that there were only three general precepts in earliest Athenian law: "Honour your parents; worship the Gods; hurt not animals"?[27] Why are we not told of Oppian's view in *Halieutica* that "The hunting of dolphins is immoral, and that man can no more draw nigh to the gods as a welcome sacrificer nor touch their altars with clean hands, but pollutes those who share the same roof with him, who willingly devises destruction for the dolphins." In quoting this passage from the fifth-century-BC poet, Antony Alpers observes: "Looking over all the known facts, I rather think that Oppian's view of cruelty to dolphins, even though he is alone among Greek writers in expressing it so fervently, may have been fairly widely held."[28] Claudius Aelianus's famous tale of the dolphin and the boy of Leonidas would lend credence to this view.[29]

Why treat Augustine and Aquinas as the sole representatives of earlier Christianity? Important as Augustine and Aquinas are – and not a whit as one-sided as the manner in which they are customarily presented – a good case could be made that the animal-respecting Arnobius, Athanasius, and Lactantius, or Leonardo da Vinci, St. Francis, and St. John of the Cross, among many other medieval and Renaissance figures, were entitled to a passing mention if a not too inaccurate picture were to be painted. Why

Descartes – when, among anglophones, no more than a handful of commentators appear to have embraced his view of animals as automata and, among francophones, rather more found his automatism a readier ground for ribald derision than for applause? Even the Catholic Church placed his works on the Index. The Christians Jean de la Fontaine, Henry More, Pierre Bayle, Pierre Gassendi, the Duchess of Newcastle, Samuel Johnson, Joseph Addison, Alexander Pope, John Locke, John Ray, Étienne de Condillac, Charles Bonnet, and Jonathan Swift, among many others, and the deists Voltaire and Viscount Bolingbroke, among many others, castigated the Cartesian understanding of animals. Indeed, Bolingbroke thought that only rarified philosophers could make such a naive error. Descartes had indicated that animals were like complex watches made by God but, like watches, that they lacked consciousness. Bolingbroke was confident that, despite Descartes, the common man would continue to be able to tell the difference between the town bull and the parish clock. Almost as damning was the short question, in a list of many objections to Cartesianism by the Reverend William Drummond: "Is it grateful for being wound up, or grieved if suffered to run down?"[30] On reading Mary Midgley's brilliant decimation in *Beast and Man* of the Cartesian view of animals,[31] I wonder who there could possibly be who needs to be convinced.

Why claim that the Western tradition denies our kinship with animals when St. Basil of Caesarea, Johann Wolfgang Goethe, Samuel Taylor Coleridge, Walter Scott, and the Brontë sisters, among many others, expressed this kinship so vociferously and clearly? If the Eastern tradition alone regards animals "as none other than ourselves" – and there are grounds for disputing the claim as a generality about these traditions – why ignore the words of William Wordsworth in "The Ruined Cottage" (1795): "All things shall live in us and we shall live / In all things that surround us"? What of Jesse Chambers' observation of D.H. Lawrence that "With wild things ... he was in primal sympathy – a living vibration passed between him and them"?[32] Why not refer to the egalitarianism of Percy Bysshe Shelley in *Queen Mab* (1812): "All things are void of terror: man has lost / His terrible prerogative, and stands / An equal amidst equals"?[33] The reality is that the Western tradition in general, and the Christian tradition in particular, like other traditions, but perhaps even more decisively, is a complex concurrence of competing ideas. It is an ongoing unresolved debate. It cannot be captured in unidimensional terms.

There are inherent dangers in treating cultures or religions in terms of their proclaimed ideals and assuming these ideals to be exemplified in practice, as do so many, including Chapple as well as Knudtson and Suzuki. What if one were to claim that the Judeo-Christian tradition exemplified the desire to create a harmonious community of all species? Isaiah 11:6-9 could be offered as evidence, where "the wolf will live with the lamb, the panther lie down with the kid ... The cow and the bear will graze, their young will lie down together. The lion will eat hay like the ox ... No hurt, no harm will be done on all my holy mountain." In similar vein, in Hosea 2:18, God affirms that: "I shall make a treaty for [the people] with wild animals, and the birds of heaven and creeping things of the earth; I shall break the bow and the sword and warfare, and banish them from the country, and I will let them sleep secure." One may wish to add that the principle is secured in Romans 8:21, where "the whole creation itself might be freed from its slavery to corruption and brought into the same glorious freedom as the children of God." What if one were to announce that Westerners are altruistic and peace-loving? Matthew 7:12 might be offered as evidence, where we are instructed: "So always treat others as you would like them to treat you." The claim could be bolstered by citing Matthew 5:39-44, where we read as the statement of Jesus Christ: "I say this to you: offer no resistance to the wicked. On the contrary, if anyone hits you on the right cheek, offer him the other as well ... if anyone requires you to go one mile, go two miles with him. Give to anyone who asks you ... love your enemies and pray for those who persecute you." Of course, no one would accept, or would be expected to accept, these exhortations as accurate descriptions of the reality of Western or Christian culture. Despite Biblical admonitions, few would imagine mercy, fidelity, and charity to be universal, or even common, Christian practices. Indeed, if they were representative of Western cultural history, there would be no need to promote them so vociferously. Many would suggest that the very need for the injunctions arises precisely from the failure to achieve the peaceable kingdom. Why, then, do we glibly assume that the statements of, say, the *Dhammapada,* the *Acaranga Sutra,* or the *Popol Vuh* are representative of the reality of the Oriental and Native worlds? Moreover, why are the noble ideals of these traditions customarily reported to us and the statements and practices disparaging of animals ignored?

Now it would be churlish to deny that there is much beauty, wisdom, truth, and justice in Oriental and Aboriginal pronouncements on appropriate attitudes to animals. To take but a few examples, in the *Bhagavatam* we read that "The perfect devotee of the Lord is one who sees Atman [the principle of life] in all creatures as an expression of the Supreme Being and all beings as dwelling in the Supreme Spirit."[34] According to the *Jaina Sutra,* "with the three means of punishment, words, thoughts, and deeds, you shall not injure living things."[35] The *Acaranga Sutra* instructs us that "All beings are fond of life; they like pleasure and hate pain, shun destruction and like to live. To all life is dear."[36] The *Dhammapada* announces that "A man is not great because he is a warrior and kills other men; but because he hurts not any living being he in truth is called a great man."[37] Further examples could be offered ad libitum.

While references to such proclamations in the Eastern religions occur frequently in animal-rights literature, one rarely encounters any acknowledgment that pronouncements of a not too different nature are to be found in abundance in the Judaic tradition. As we noted earlier, one of the principles of the Torah – *Bal Taschit* (do not destroy) – plays a similar role to that of the originally Jaina concept of *ahimsa* (the principle of noninjury to living beings) in Jaina, Buddhist, and Hindu thought. Moreover, "it is forbidden according to the law of Torah to inflict pain on any living creature." Further, Jewish texts tell us that "Jews must avoid plucking feathers from live geese, because it is cruel to do so";[38] "Rejoicing cannot occur at an animal's expense";[39] "Animals are not penned up on Shabbat";[40] "One who prevents an animal from eating when at work is punishable by flagellation";[41] "Thou thinkest that flies, fleas, mosquitoes are superfluous, but they have their purpose in creation as a means of a final outcome ... As the Holy One, blessed be He, created in His world, He did not create a single thing without purpose."[42] A number of benevolent principles are to be found in the Hebrew Bible, including: "The upright has compassion on his animals, but the heart of the wicked is ruthless" (Proverbs 12:10); "Yahweh is generous to all, his tenderness embraces all his creatures" (Psalm 145:9); and the powerful statement of the traditional King James version: "He that killeth an ox is as if he slew a man" (Isaiah 66:3). Notoriously, there are no such statements in the New Testament. The best that may be said is that the New Testament is silent on the matter because the principles had already been laid out in the Hebrew Bible and nothing needed to be added. Despite the New Testament silence, however, there are a considerable number of compassionate and respectful statements to be found among early Western thinkers, both explicitly in traditional myth and explicitly in Christian pronouncements. Overall, it would be fair to say that the traditional religions of India have expressed a rather more consistent and profound concern for animals than did the early Judeo-Christian tradition, although with nowhere near the degree of difference expressed in so much of the literature concerned more with ideology than with truth, more with cultural self-flagellation than with objectivity. If we are shocked by D.H. Lawrence's repulsion at what he called the "barbaric substratum of Buddhism,"[43] we should at least recognize that one of our most sensitive literary figures found a dimension in Buddhism that we no longer care to recall.

Since Aboriginal values are usually expressed in the form of traditional tales, explicit ethical pronouncements are rare (and where one finds them, one is entitled to question whether they have been inserted by those with an axe to grind, for this is not the traditional customary Aboriginal form of expression).⁴⁴ Nonetheless, they are often implicit. Says Bill Neidjie, elder of the Bunitj clan of the Australian Kakadu people: "People look for food, animal look for food. Lizard look, bird look. We all same ... Eagle our brother, like dingo our brother."⁴⁵ In the Cheyenne myth of "How The Buffalo Hunt Began," cooperation with the friendly animals is stressed.⁴⁶ In the African Bushman legends, the similarities between human and animal are emphasized.⁴⁷ Despite occasional aberrations – as in the Cheyenne despisal of the wolf and the coyote and in the Bushman antipathy to the hare – the notion of a human-animal similarity is common, at least in respect of the idea that all animals face the same trials and tribulations, especially with regard to self-preservation. Their view corresponds to that of Ecclesiastes 3:19, where "the fate of human and the fate of animal is the same: as the one dies the other dies; both have the self-same breath. Human is no way better off than animal," although the message occurs with rather greater consistency in Aboriginal legends.

We have already encountered a number of statements in the Eastern scriptures where the disparagement of animals is disturbing. The disturbance is not decreased when we read in the Buddhist *Lotus Sutra* that:

> he will enter the Avichiheu [hell]
> [who slanders this Sutra] ...
> Though he may emerge from hell,
> he will fall into the realm of beasts,
> becoming a dog or a jackal,
> his form lean and scruffy,
> dark discolored, with scabs and sores,
> something for men to make sport of ...
> Camel, donkey, pig, dog –
> these will be the forms he will take on.
> Because he slandered this Sutra
> this is the punishment he will incur.⁴⁸

In the *Tibetan Book of the Dead* we are warned about the dangers of rebirth as "the lowest of the low, like beasts."⁴⁹ In the ten levels of existence of Buddhist cosmology, "beasts or beings with animal nature" belong to the third level from the bottom, beneath the level of the ogres and the demons, and well beneath the realm of human beings – who are not seen to be "beings with animal nature"! According to Burton Watson, scholar of Buddhism, the lowest four levels, in which the animals are included, represent the "evil paths: the lowest, most painful and undesirable levels of existence."[50] In the Buddhist view, a Bodhisattva, one who has escaped the eternal cycle, "has by then gained five advantages, he is no more reborn in the state of woe, but always among gods and man; he is never again born in poor or low-class families; he is always male, and never a woman; he is always well-built and free from physical defects; he can remember his past lives, and no more forgets them again."[51] In this statement of the Buddhist view, humans themselves are decidedly hierarchical – the wealthy, the high-born, males, and the well-formed are in and of themselves superior beings – and the animals are an altogether inferior existence, as their categorization below the level of the ogres and the demons so graphically demonstrates. Yet it is quite rare to encounter any acknowledgment in Western animal-oriented literature of the classism, sexism, and speciesism of the Buddhist tradition.

Certainly, such miserable animal existences as those described in the *Lotus Sutra* must have souls in order to participate in transmigration. But, equally clearly, the possession of souls, immortal souls at that, has done little to elevate their status. Indeed, a convincing case can be made that many of those in the Western tradition who denied animals the possession of immortal souls had a higher sense of the animals' worth than those who expressed these particular Buddhist conceptions. In reality, there are, of course, other Buddhist conceptions – expressed most admirably in the *Dhammapada* – but just as we must recognize a complexity in the Buddhist tradition, so too, and a fortiori, there is a complexity in the Christian tradition that goes much further than the customary comments about "dominion" and the lack of animal souls.

If we inquire about the extent to which animal treatment conforms to the more favourable norms of religion, philosophy, and myth, we will be compelled to note that animal sacrifice has been practised in Hindu temples for millennia and that it is still not completely eradicated – human sacrifice was finally prohibited during the Raj – and this by contrast with the divine injunction, some two and three-quarters millennia old, from Isaiah 66:3-4: "Some slaughter a bull, some kill a human being, some sacrifice a lamb, some strangle a dog, some present an offering of pig's blood ... all these people have chosen their own ways and take delight in these disgusting practices ... They have done what I regard as evil, have chosen what displeases me." There are, to be sure, other Hebrew Bible passages that tell a different story, and in practice animal sacrifice did not disappear for centuries. Indeed, the dicta of the Old Testament are confusing. Leviticus

17, in effect, requires sacrifice, restricts the slaughter of animals to sacrifice, and deems it a heinous offence to kill animals otherwise. On the other hand, Deuteronomy 12:15 allows the slaughter of animals for food without the rites of sacrifice. Again, Exodus 12 commands sacrifices, while Amos 5 pours scorn on the sacrificial festivals: "I take no pleasure in your solemn assemblies. When you bring me burnt offerings ... your oblations, I do not accept them."

Certainly, Jesus found it necessary to repeat the condemnation of sacrifice found in Isaiah.[52] In Matthew 9:13 and 12:7, he repeats the admonition of Hosea 6:6: "Mercy is what pleases me, not sacrifice."[53] According to John 2:13-16, he drove the doves, oxen, and sheep out of the temple partly because he abhorred the use of the temple as a market place and partly because of the misuse of the sacrificial animals by the moneychangers.[54] In the second century, the theologian Clement of Alexandria identified what sacrifice had become in Old Testament days – not a purification and an offering but merely an elaborate justification for meat eating.[55] In reality, the practice ceased among the Jews only when the Romans destroyed the temple in Jerusalem, the site of the sacrifices, in the year 70 – it had replaced the one built by Solomon in the tenth century BC.[56] The original Jerusalem temple had become the sole site of sacrifice in 622 BC, on the orders of King Josiah, so as to encourage the solidarity of a pilgrimage to Jerusalem, of which animal sacrifice to God was an integral part.[57] But if flesh eating was restricted to the consumption of sacrificed animals and if animals were sacrificed only in the Jerusalem temple, those Jews who lived a distance away are likely to have eaten meat only rarely. In the Roman Empire generally, sacrifice was finally proscribed by the emperor Theodosius toward the end of the fourth century, over sixteen hundred years before its gradual decline in the Hindu world. And gradual is the decline. In the last few days of December 2003 and the first few days of January 2004 – a week in all – tens of thousands of goats, sheep, and buffalo were ritually sacrificed in eastern India by devout Hindus at religious festivals.

Cockfighting continues in both Hindu and Muslim India. It was outlawed in Britain in 1835, with most other Western jurisdictions following in time, although three US states continue to permit the barbarous practice. And a World Cup of Cockfighting was held in the Dominican Republic in 2002. It has also taken place in the Philippines. Some two hundred sloth bears are poached from their dens each year in India to endure a deplorable life as dancers. They are trained painfully, suffering inter alia two nose piercings with a carpet needle without anaesthetic. Neither state nor *sadhus* concern themselves overly to inhibit the practice. Such practices have long been outlawed in most Western jurisdictions.

We are often told of the fine sensibilities to animals of the Jaina. The anthropologist Hope MacLean tells us that they "refrain from all killing, even of a mosquito or an ant that crawls in their path."[58] What she does not tell us is that the manner in which they accomplish this apparently benevolent act results in the death of twice as many animalcules as would be killed if the Jaina did not practise their restraint. A person is hired to walk ahead of the Jain priest, brushing the path so that there will be no insects on which to tread. Of course, this results in at least as many insect deaths as, probably more than, would have occurred if the priest had walked boldly forth. Moreover, if, as is customary, the sweeper walks to the side as he brushes, his steps will kill as many insects as would the Jain if he had received no assistance in his purity – and this in addition to the killings via the brush. Certainly, such behaviour must encourage us to give due consideration to Basant K. Lal's claim that religious attitudes of this kind in India are "not based on consideration for the animals ... [but are] a part of the purificatory steps that bring men to moksa."[59] Lal is referring explicitly to Hindu practices, but the principle applies more broadly. Clearly, it is the Jain's purity that is at stake, not the interests of the insects.

The rightly revered Gandhi complained bitterly of the treatment of the supposedly sacred cow: "How we bleed her to take the last drop of milk from her. How we starve her to emaciation, how we ill-treat the calves, how we deprive them of their portion of milk, how we castrate them, how we beat them, how we overload them."[60] On another occasion he remarked: "The ideal of humanity in the West is perhaps lower, but the practice of it very much more thorough than ours. We rest content with a lofty ideal and are slow or lazy in its practice. We are wrapped in deep darkness, as is evident from our paupers, cattle and other animals."[61] If the cow is sacred in India, sanctity does not appear a quality worthy of possession. Certainly, Gandhi makes it abundantly plain that, whatever the underlying principles of attitudes to animals on the Indian subcontinent, it most assuredly cannot stand as a model to be emulated.

The anthropologist Göran Burenhult tells us of the Toraja of the highlands of the Sulawesi in Indonesia that: "To the Toraja, the water buffalo is the most sacred of all animals, and it is also the principal symbol of wealth. Buffalo are bred for one single purpose: to be sacrificed during grand funeral ceremonies."[62] The *sacred* animal is bred for one sole purpose – its own death! The buffalo's carotid artery is cut with a long knife and the *sacred* animal bleeds to death. As many as 250 buffalo may be killed for a single funerary ceremony for an important person. Whatever *sacred* may mean for the Toraja, it must be the very antithesis of what it meant for the devout Albert Schweitzer, for whom "life itself is sacred." According to Schweitzer, "a thinking man feels compelled to approach all life with the same rever-ence he has for his own." For William Blake, "all life is holy." For John Ruskin in *Unto This Last,* which Gandhi said transformed his thinking, "there is no Wealth but Life." The anti-vivisectionist (and poet) Robert Browning proclaimed "the wild joys of living." D.H. Lawrence announced that "for man, as for flower and beast and bird, the supreme triumph is to be most vividly, most perfectly alive."[63] He would have felt constrained to ask of the Toraja breeder, "How could he betray the great privilege of life?" For the Toraja – and countless other societies – what is *sacred* is put to death, neither to furnish necessary food, nor raiment, nor shelter but to appease the spirits – and not the spirits of the buffalo! Neither the possession of an immortal soul nor sanctification is of any benefit to the animal. Indeed, both are decidedly detrimental. It is thus difficult to give credence to the frequent assertion – advanced, for example, by Peter Singer – that the fail-ure of Christianity to allow animals an immortal soul has had a "very neg-ative influence on the way in which we think about animals" and has resulted in their abominable treatment in the West, which he contrasts with other "religions, especially in the East, which have taught that all life is sacred."[64]

To be sure, in the Western tradition, examples of the view that animals are expendable are rife. Thus, for example, in John Steinbeck's *The Grapes of Wrath,* Jim Casy quotes from Blake and, clearly speaking for the author, proclaims that "all that lives is holy." Yet when we discover that Steinbeck lived at times on the earnings of his good friend Ed Ricketts, who collected, with Steinbeck's assistance, marine specimens for vivisection in schools and cats for vivisection elsewhere, that he described a frog hunt in *Cannery Row* without any apparent concern for the lives of the frogs, and that in *The Log from the Sea of Cortez,* he exulted in nature – quite incorrectly of course, relatively few species being carnivorous – as that in which "Every-thing ate everything else with a furious exuberance,"[65] we are entitled to interpret his proclamation on the holiness of life as less than convincing. He describes the wonders of the Sally Lightfoot crabs with awe: "little crabs, with brilliant cloisonné carapaces ... They have remarkable eyes ... They are beautiful."[66] Yet they are there to be plucked. For Steinbeck, nature is a Darwinian struggle for survival, in which the fittest not only succeed, but deserve to do so. But if we interpret Steinbeck's claim of the holiness of life as inauthentic, we must apply the same criterion to the claims of the Toraja regarding the sacredness of the lives of their buffalo and to the claims of the Hindus regarding their holy cows.

In earlier times, North American Indians would drive whole herds of buffalo over the cliffs to their destruction – up to nine hundred at a time according to archeological evidence. In the excavation of kill sites, it has been found that, as Simon Davis reports, "Many of the limb bones were whole, some were semi-articulated, and spinal columns often intact. This suggests mass killings and substantial waste."[67] If Mark Twain could make ribald and telling remarks about a fictional English earl and his entourage killing seventy-two buffalo and eating only one, it would appear he could have found historical, rather than imaginary, instances somewhat closer to home, although even then it would have been becoming politically incorrect to do so.[68] Davis reports also that in "South America nearly 80 percent of its large mammal fauna became extinct" in earlier centuries and that in his opinion, Native "people played the dominant role in bringing about the extinctions" through overhunting.[69] In Madagascar fourteen species of lemurs have been hunted to extinction by the Native population. Once, the largest lemur species was the size of a gorilla. Now the largest species is the size of a human infant. When Madagascar was first populated around the fifth century, the Native immigrants rapidly brought about the extinction of six species of giant elephant birds.[70] One could continue at length in such a vein. On the other hand, one could report many instances of the most profound animal consideration. And one could do the same for the West and Christianity, as well, of course, as reporting countless instances of consummate cruelty. The reality is that all cultures are complex and contain both horrendous animal disregard and commendable sympathy. At best, the respectful edicts of the philosophies are reflections of the ideal. They bear little comparison with how animals are treated in practice. Like the Oriental ascription of souls to animals, the Aboriginal tradition appears to have been no more successful in warranting respect and considerate treatment for them.

II

One of the most commonly offered explanations for the purported elevation of the status of animals in non-Western cultures and the degradation of the animal in Christianity is that, whereas other cultures ascribe souls to animals, Christianity does not. For example, Gary Kowalski tells us that "The word 'animal' comes from a Latin root that means 'soul.' To ancient thinkers, soul was the mysterious force that gave life and breath to the myriad of the earth's creatures ... Later, theologians restricted the possession of a soul to human beings."[71] One might question the accuracy of Kowalski's implication about the significance of the root of "animal." While the root is indeed "anima," in the usage of the Epicureans and many who followed them, "anima" referred to the sensitive soul alone, while "animus" referred to the rational and immaterial soul. Moreover, ascribing life and breath to animals was not in and of itself to accord them the status Kowalski imagines. Thus, as we noted earlier, Juvenal in *Satire XV* distinguished humans from animals in that God "Breathed life in them, in us a reasoning soul."[72] For Juvenal, and a number of the ancients, the possession of life and breath was not equivalent to the possession of a rational or immaterial soul.

One might question, too, Kowalski's account of the role of theologians in general, given the research of Ezra Abbot to which we have already alluded.[73] Indeed, only materialists, who would deny humans and animals alike the possession of immaterial souls, and Cartesians, who would ascribe souls exclusively to humans, would deny the animal any kind of soul – although certainly many, including theologians, have suggested a grand distinction between the human and the animal soul. But, initially, one might question the assumption behind Kowalski's statement: that the ascription of souls to animals ensures them greater respect and more favourable treatment. To the contrary, for example, the Italian Renaissance churchman and saint, Cardinal Robert Bellarmine (1542-1621) avowed that the very lack of an immortal soul made kindness toward animals obligatory because they could not be recompensed in heaven for their sufferings.[74] In *De la Sagesse* (On wisdom) of 1601, Pierre Charon deemed animals inferior to humans in some respects, yet almost equal to them in reason, and superior in others, for he regarded them as possessing a considerable degree of virtue, generally greater than that of humans. Nonetheless, they did not possess immortal souls. But this was seen as no detriment either to their worthiness or to the respect in which they must be held. As we shall see, this was in essence the view held by the court physician Marin Cureau de la Chambre, the Catholic priest Pierre Gassendi, and the fabulist Jean de la Fontaine. Indeed, Gassendi held the view in the mid-seventeenth century that animals were rational, that it was unnatural to eat them, and that, even then, they lacked immaterial, heaven-destined souls.

In the second half of the eighteenth century, we encounter a fervent French moral outrage against the perpetrators of animal cruelty in the writings of Anne-François-Joachim de Fréville, Friedrich Melchior von Grimm, Joseph Lavallée, Louis-Sébastien Mercier, and Pierre Louis Moreau de Maupertuis, without any one of them concluding that animals possessed immortal souls. The French outrage at the mistreatment of animals was only a whit less, if not quite as broad, as that experienced in Britain in the same period. And their first attempt at animal-welfare legislation (in 1804, inspired by Lavallée) came only four years after the first bill was introduced into the British Parliament, although success had to await the *Loi Gram-mont* of 1850 – twenty-eight years behind the British. If the French, with some notable exceptions, were loath to grant animals immortal souls, they were still concerned to hinder cruelty toward them.

The Catholic poet Alexander Pope, whose verse and prose showed a consistent appreciation for the suffering of animals, expressed the view on at least one occasion that animals would perish, but he believed that, as a consequence, they were entitled to more considerate treatment in the here and now. In 1713 he wrote: "The more entirely the inferior creation is submitted to our power, the more answerable we should seem for our mismanagement of it; and the rather, as the very condition of their nature renders these creatures incapable of receiving any compensation for their ill-treatment in this."[75] If there were no heaven for animals, the injustices could not be remedied. Hence there was an even greater obligation to treat animals well now. As late as 1839, we can find the Cornish veterinarian William Karkeek repeating the message: "if, indeed, the beasts do perish, that very circumstance ought to produce greater kindness toward them."[76] But he was not so sure that they perished, even though it should be noted that W.C. Spooner, responding to Karkeek in 1840, thought Karkeek to be in a decided minority.[77]

In fact, Pope, too, was not always so convinced that animals lacked eternal souls. The anecdotist Joseph Spence reported a conversation with Pope some thirty years later, around 1743 or 1744, in which, initially, Spence had speculated that dogs possessed the faculty of reason:

> [POPE:] So they have to be sure. All our disputes about that, are only disputes about words. Man has reason enough only to know what is necessary for him to know; and dogs have just that too.
> [SPENCE:] But then they must have souls too; as unperishable in their nature as ours?
> [POPE:] And what harm would that be to us?[78]

As Pope rightly recognizes, in the customary discussion of the question of animal reason, the disputants frequently talked past each other – their disagreements were "only disputes about words." But what should also be clear from the conversation is that the answer to the question of whether animals possessed immortal souls was no one-sided matter in the mid-eighteenth century or, indeed, later. The matter was, in fact, hotly debated. Equally significant is the indication that the answer to the question revolved around the issue of whether animals were perceived to be rational. The implication of the conversation is that the possession of the faculty of reason is both a necessary and a sufficient condition for the possession of an immortal soul. Thus everything depended on what was meant by reason.

By consensus, the most significant explicitly Christian general study of animal ethics in the eighteenth century was Humphry Primatt's *The Duty of Mercy and the Sin of Cruelty to Brute Animals* (1776). Primatt was the Anglican vicar of Swardeston in Norfolk and a Cambridge graduate – it is remarkable how many of the advocates of animal interests from the seventeenth to the nineteenth centuries were Cambridge graduates. He argued his case for ethical treatment of animals largely based upon Biblical pronouncements and what he regarded as Christian principles, centring his hypotheses on the issue of pain and suffering: "Superiority of rank or station exempts no creature from the sensibility of pain, nor does inferiority render the feelings thereof the more exquisite. Pain is pain, whether it be inflicted on man or beast, and the creature that suffers it, whether man or beast, being sensible of the misery of it while it lasts, suffers *evil;* and the sufferance of evil, unmeritedly, unprovokedly, where no offence has been given, but merely to exhibit power and malice, is cruelty and injustice in him that occasions it."[79]

Primatt was following in the footsteps of the Reverend Richard Dean, author of *An Essay on the Future Life of Brutes* (1767): "Brutes have sensibility; they are capable of pain; feel every bang, and·cut or stab, as much as man himself, some of them perhaps more ... For a man therefore to torture a brute shows a meanness of spirit. If he does it out of wantonness, he is a fool and a coward; if for pleasure he is a monster."[80] And Dean was a successor to Christopher Smart, who, in mid-century, had begged "God" to be "merciful to all dumb creatures in respect of pain."[81] John Hawkesworth, writing an adventure story in the *Adventurer* (no. 17, 1752), also allowed for the centrality of pleasure and pain to animal ethics. Having kicked his dog, a potentate is visited by the spirit who resides in his magic ring: "'Amurath,' said he, 'thou hast offended against thy brother of the dust; a being who like thee, has received from the ALMIGHTY a capacity of pleasure and pain: pleasure which caprice is not allowed to suspend, and pain which justice only has a right to inflict.'" In number 32 of the *Adventurer,* published in the same year, we again find commentary on the significance of animals' pain: "To take pleasure in that by which pain is inflicted, if it is not vicious is dangerous; and every practice which, if not criminal in itself, yet wears out the sympathising sensibilities of a tender mind, must render human nature proportionately less fit for society."[82] These passages were written some thirty-seven years before the far more famous, but certainly no more significant nor profound, statement by the secular utilitarian Jeremy Bentham: "The question is not, can they *reason*? Nor, can they *talk*? but, can they *suffer*?"[83] When Hilda Kean writes of the concern with pain in animal-welfare issues as "the Benthamite nostrum,"[84] she ignores many more worthy forbears. So too does Richard Sorabji. In commenting on pain in *Animal Minds and Human Morals,* he tells us that "Bentham shifted the ethical question."[85] The question was in fact shifted long before Bentham.

Humphry Primatt does not, however, reach the same optimistic conclusion with regard to animal immortal souls as does Richard Dean (and, as far as I can ascertain, Hawkesworth did not address the issue directly). To be sure, Primatt acknowledges that animals have souls, referring to a body "being animated by the soul of a brute or the soul of a man,"[86] but he does not award this soul a heavenly destiny. Indeed, he offers something similar to Pope's animadversions regarding recompense for injury.

> [Eternal life] is the hope and confidence of a *man*. And a most comfortable reflection it is indeed to the virtuous and innocent sufferer who knows that he has an almighty Patron and Avenger, who will finally cause that the malice and wickedness of his enemies here shall at length promote the degree of his glory hereafter. But what hope, what glimpse of a recompense hereafter, awaiteth the afflicted *brute*? An hereafter for a brute, a recompense for a beast, has a strange sound in the ears of a man. We cannot bear the thought of it. Injustice itself is a virtue in the judgment of partiality; and in the pride of our heart we rather say: let man be happy, though all creation groan. Yet it is a truth that ought not to be concealed, that God is a righteous judge; and it is a presumption in Man to determine the limits of the divine goodness. However, as we have no authority to declare, and no testimony from heaven to assure us, that there is a state of recompense for suffering brutality, we will suppose there is none; and from this supposition we rationally infer that cruelty to a brute is an injury irreparable.[87]

For Primatt, as for Pope before him, the animal's lack of an immortal soul, if that be the reality, requires our greater consideration for it: "His present life (for any thing we know) is the whole of his existence; and if he is unhappy here, his lot is truly pitiable; and the more pitiable his lot, the more base, barbarous, and unjust in man, must be every instance of cruelty towards him."[88] For Primatt, even if animals lack immortal souls, our responsibility to apply the golden rule still holds: "do you that are a man so treat your horse, as you would be willing to be treated by your master, in case that you *were* a horse ... let this be your invariable rule, everywhere, and at all times, *do unto others as, in their condition, you would be done unto.*"[89] The fact that animals have sentience is quite sufficient to ensure that they are entitled to ethical consideration.

Even though, like Pierre Gassendi before him, Emanuel Swedenborg denied immortal souls to animals, he nonetheless advocated vegetarianism and influenced the Reverend William Cowherd and his Bible Christian Church (founded 1809) to abstain from flesh. Whether they were ensouled or not, we were not entitled to use animals as food. In *The Rights of Animals* (1838), William Drummond is adamant that *if* "this life is their all," their condition, *therefore,* "should be rendered as comfortable as possible, and no injury offered them by which it can be curtailed."[90] The veterinarian and former Nonconformist minister William Youatt, writing in 1839, took up the same theme:

> The Gospel which promises us another life, says not one word with regard to [animals] ...
>
> I admire the intelligence and the moral qualities of the inferior animals – I find each perfect in the situation in which he is placed – but I see plainly the limit which he has never yet passed. All has reference to the present state of being, and to his compartment of it. He lives his day, and he passes of the stage ...
>
> ... [in heaven] *Man will there live again:* but what other forms of existence will surround him has not yet been revealed, and it becomes us not to be wise above that which is written.[91]

Whereas Pope and Primatt clearly hope for animals in heaven, and Primatt is unwilling to accept more than is promised in the scriptures, Youatt accepts the lack of an immortal soul for animals more or less with equanimity. Nonetheless, in Youatt's view, and this is the crux of the matter, whether possessed of immortal souls, animals are entitled to be treated humanely not for the sake of the humans who practise benevolence but for the sake of the animals themselves, who are entitled to just treatment precisely because it is just: "The claims of the lower animals to humane treatment, or at least, to exemption from abuse, are as good as any that man can urge upon man. Although less intelligent, and not immortal, they are susceptible of pain ... they have as much right to protection from ill usage as the best of their masters have."[92]

While, for many, the ascription of rationality to animals, or a moral sense, which was deemed an integral part of rationality, was sufficient for the ascription of an immortal soul, this was not so for Youatt. Despite "the intelligence and moral qualities of the inferior animals," the lack of a Biblical authorization was sufficient to deny the immortality of the animal soul. Nonetheless, animals were entitled to earnest moral consideration. The purported lack of an immortal soul does not seem to have been of necessity inimical to the earthly interests of animals.

III

Despite the evidence to the contrary, as I have indicated, the purported failure of the Christian tradition to ascribe immortal souls, or even souls of some other kind, to animals has been widely regarded as a symptom of Christianity's supposed failure to recognize the worthiness of other species. Thus, for example, Peter Singer opens his discussion of the Biblical and Christian tradition in his extremely influential *Animal Liberation* with the caustic claim: "To end tyranny we must first understand it." What follows is a decided unwillingness to attempt to understand it. He determines instead to vilify it. He goes on to assert that: "Christianity brought into the Roman world the idea of the uniqueness of the human species, which it inherited from the Jewish tradition but insisted upon with still greater emphasis because of the importance it placed on the human being's immortal soul. Human beings, alone of all beings living on earth, were destined for life after bodily death. With this came the distinctively Christian idea of the sanctity of all human life."[93] For Singer, the Christian sanctification of human life meant the denigration of other life forms. Even the erudite Mary Midgley is convinced that "the church often and explicitly insisted that all plants and animals must be viewed merely as objects given to man as his instruments, that to have any sort of regard for them in themselves was sinful and superstitious folly."[94] Yet Singer's assertion flies in the face of quite contrary claims avowed by some Christians. And if there is any validity to Midgley's claim, there must have been a lot of wicked and credulous inanity among the faithful. Thus, for example, in *The Rights of Animals* (1838), William Drummond proclaims as, to him, an obvious fact that "no one, though but superficially acquainted with that divine religion, can for a moment question that humanity to animals is a duty accordant with its whole spirit; and that every act of cruelty, though to a reptile or a fly, is in direct hostility to its nature. Christianity is throughout a religion of mercy – of mercy not limited to any tribe or nation, nor to the sphere of rationality itself, but extending to the extreme limit of life and sense."[95]

Moreover, "it is virtual atheism to affirm that [Christ] cares not how wantonly and cruelly they are destroyed."[96] Prima facie, most of us are likely to consider Drummond's claim an exaggeration, certainly of practice if less so of principle. But it surely indicates to us that Singer's reproval should not go unanswered. One wonders whether Singer gave any consideration to the possibility that the sanctification of one life form might encourage the sanctification of others. One wonders whether Singer gave any consideration to the human exclusivity of other societies when, for example, in the traditional story of "The Coming of Gluscabi" of the Abnaki, only the human forms himself out of the dust left over from the deity's creation of all other life forms; or when, in the Cheyenne creation myth, only humans are granted intelligence by the Great Medicine. With regard to the Christian view, initially one might note that it was the preexisting Roman Stoic tradition that, along with Judaism and certain aspects of the Greek tradition, inspired Christianity to acknowledge the sanctity of the individual and to recognize that all humans possess a certain inherent equal dignity, rather than Christianity alone being the source of the conception.[97] Following this implication of an anti-animal stance at the very origin of the Christian tradition, Singer refers to "the Christian doctrine that animals do not have immortal souls"[98] as though it were an orthodoxy of the religion as a whole rather than an opinion shared by some influential Christians and rejected by some others. Unfortunately, Singer does not stand alone. If the mortality of the animal soul is not a part of Christian orthodoxy, it appears to be a part of the orthodoxy of those recent animal-rights commentators who have mentioned the topic. We can find such pronouncements on the animal soul in the writings of many, including many of the more admirable scholars: Angus Taylor, Barbara Noske, and Randy Malamud, along with William Leiss and Jim Mason, and many others.[99] It is certainly important to acknowledge that numerous Christians, including some highly influential Christians, among them Augustine, Aquinas, Descartes, Malebranche, and Bossuet, denied immortal souls to animals, the last three denying them souls of any kind. And even in the late twentieth century, the Dominican Ambrose Agius, in his otherwise admirable little book *God's Animals*, maintained the view that humans and animals had a separate destiny.[100] Like most aspects of the Western tradition, it is a complex story of argument and counterargument rather than the achievement of an invariable doctrine, despite numerous attempts in that direction, especially in the Roman Catholic world.

It is customary to assert that the greatest influences on Christian thought with regard to the status of animals were Aristotle and those Stoics and Epicureans who denied we had any *communitas* with animals and hence owed them few obligations – and certainly not "justice." And a modestly good case can be made for it. Unfortunately, however, more often than not, it is imagined that the contrary emphasis in classical thought played little or no role. In fact, to downplay the competing ideas in classical thought is to neglect a strain that played an important – if, for some considerable time, a rather less significant – role. We may find this strain exemplified when Simplicius tells us that Diogenes of Apollonia (fl. c. 430 BC) believed that "Humans and other animals, inasmuch as they breathe, live by the air. And this is for them both soul and intelligence ... The souls of all animals are indeed the same."[101] Plutarch (c. AD 46-120) emphasized the similarities in reason, sentience, and virtue between humans and other species[102] – and Plutarch was one of the most widely read of scholars from the sixteenth to the very late nineteenth centuries, and not without numerous enthusiasts until well into the twentieth. He had a profound effect on English literature, not least in that his works supplied the historical material, via Sir Thomas North's translation of his *Parallel Lives,* for Shakespeare's *Coriolanus, Julius Caesar, Antony and Cleopatra,* and *Timon of Athens.* While Plutarch does not discuss immortal souls as such, the implication is clear: if humans possess them, there can be no good ground for denying them to other species. Seneca (c. 3 BC-AD 65) – also widely read until the late nineteenth century, if not to the same degree as Plutarch, although no one had a greater influence on Renaissance tragedy – was, at least for a time,[103] imbued with Pythagorean philosophy, believing in the transmigration of souls and drawing vegetarian implications therefrom,[104] none of which can have escaped the notice of later centuries.

In his systematization of the idea of the Great Chain of Being, the vegetarian Plotinus (c. AD 205-70) argued that "the animals and plants share in reason and soul and life" but to varying degrees.[105] Soon plants were dropped from the human, animal, plant trilogy. Porphyry (c. AD 233-304), a student of Plotinus far more persuasive than the master, especially against the more pernicious of the Stoic doctrines, made an appealing case for human-animal continuity. However, we should not fail to notice that his book *On Abstinence from Animal Food* was not translated into English until 1793 – a French translation had appeared in 1747 – and thus did not have a great influence on pre-utilitarian thought, although equally we should not disregard that most of the educated were well acquainted with the classical Greek language and frequently cited the original works in their writings. Porphyry observed that:

> we do not consider kinship limited to humans and gods but maintain that there is also a communal relationship with the animals ... we do not wish to treat them as beings of another kind, excluded from our community ... we declare that all people as well as all animals are of the same kind, because the base of their bodies are by nature the same. In so speaking I do not refer to the primary elements, since they also pertain to plants. I refer instead to

the skin, the flesh, and to the dispositions present in all animals, and to much else besides, since the souls they possess are no different from ours by nature; I refer further to the correspondence of the appetites, to animal passions, to their reasoning as well, and, above all, to their feelings. And, as with the body, so too certain animals have a complete soul, while others have it in lesser degree for all animals, however, the natural principles are the same ... red blood courses in the veins of all animals, and they all demonstrate in common the sky as father and the earth as mother ...

Since animals are allied to us, it should appear, according to Pythagoras, that they are allotted the same soul that we are: he may justly be considered impious who does not abstain from acting unjustly toward his kindred ...

He ... who admits that he is allied to all animals, will not injure any animal ... since justice consists in not injuring anything, it must be extended as to every animated nature.[106]

With such a context of ideas – and only a few of the more significant and better-known examples have been given here – the case for immortal animal souls and for a recognition of our obligations to animals cannot have been without some persuasive merit to early Christians.

Not surprisingly, then, the writings of a number of early Christian scholars reflect profound indebtedness to the more sympathetic classical thought with regard to the status of animals. The Jewish thinker Philo-Judaeus (Philo of Alexandria) is said to have influenced the early church profoundly, not least in his assertion that as a future heavenly reward, "There is no doubt but that [in the] hereafter dumb animals will be divested of their ferocity, and become tame and gentle after the manner of other creatures whose dispositions are subdued to harmony and love."[107] Thus, for example, the Christian theologian Tertullian observed at the turn of the third century that: "There shall be an end of death, when the devil its chief master, shall go away into the fire which God has prepared for him and his angels; when the manifestations of the sons of God shall release the world from evil, at present universally subject to it; when the innocence and purity of nature being restored, animals shall live in harmony with each other, and infants shall play without harm with animals once ferocious; when the father shall have subdued His enemies to His Son and put all things in subjection under his feet."[108] So also asks Arnobius (fl. fourth century), who taught rhetoric at Sicca Venria in North Africa during the reign of the emperor Diocletian: "Will you, laying aside [all] partiality, consider in the silence of your thoughts that we are creatures quite like the others, or separated by no great difference? For what is there to show that we do not resemble them? Or what excellence is in us, such that we scorn to be ranked as creatures?" Having shown the bodily similarities between humans and animals, he continues:

> But we have reason, [one will say], and excel the whole race of dumb animals[109] in understanding. I might believe this was quite true, if all men lived rationally and wisely, never swerved aside from their duty, abstained from what is forbidden, and withheld themselves from baseness, and [if] no one thought folly and the blindness of ignorance demanded what is injurious and dangerous to himself. I should wish, however, to know what this reason is, through which we are more excellent than all the tribes of animals.
>
> [Is it] because we have made for ourselves houses, by which we can avoid the cold of winter and heat of summer? What! Do not the other animals show forethought in this respect? Do we not see some build nests as dwellings for themselves in the most convenient situations; others shelter and secure [themselves] in rocks and lofty crags; others burrow in the ground and prepare for themselves strongholds and lairs in the pits which they have dug out? But if nature, which gave them life, had chosen to give them also hands to help them, they too would, without doubt, raise lofty buildings and strike out new works of art.
>
> Yet even in those things they make with beaks and claws, we see that there are many appearances of reason and wisdom which we men are unable to copy, however much we ponder them, although we have hands to serve us dextrously in every kind of work.[110]

Clearly, for Arnobius, animals and humans share their attributes in common, including reason. And if there are any respects in which humans might be thought to excel, it is not difficult, Arnobius surmises, to imagine those animal capacities with which humans cannot compete. In some respects nonhuman animals are superior. Arnobius raises the Pythagorean doctrine of metempsychosis, if without full conviction, and concludes, with greater conviction that "it is more clearly shown that we are allied to them, and not separated by any great interval, since it is on the same ground that we and they are said to be living creatures and act as such." But what does Arnobius mean by "living creatures," and how does it relate to the concept of "soul"? The theologian Dr. George Bush (1796-1859), of the City University of New York, whom we met briefly in the first chapter, made the following point in his commentary on Genesis: "The phrase 'living soul' is repeatedly applied to the inferior order of animals. It would seem to mean the same when spoken of man that it does when spoken of beasts, viz., an animated being, a creature possessed of life and sensation, and capable of performing all the physical functions by which life is distinguished, and we find no *terms* in the Bible to distinguish the intellectual faculties of man from the brute creation."[111] Elijah Buckner made this point in 1903 – a point that has become far more common in recent years but that, in fact, had already been made explicit, if equivocally, in 1839 by William Karkeek in the pages of *The Veterinarian:*[112]

> It is acknowledged by all the best Greek and Hebrew scholars today that, in every passage of Scripture where the Hebrew word *nephesh* or the Greek word *psyche* is used, it should be translated soul, and when *nephesh chayah* is used it should be translated living soul. This is admitted by the marginal reading found in many old English Bibles. In Genesis 2:7, when the divine writer speaks of Adam, the translation is correct, as it reads in the Hebrew, *nephesh chayah*, which translated into English means living soul, but there are nine more passages in Genesis where the same Hebrew words are used, but as they refer to lower animals the true meaning has been perverted by the English translation.[113]

Moreover, the commonality of human and animal is stressed in the use in Genesis on four occasions of the term "all flesh" (*kol basar*) to indicate a similar treatment for humans and animals alike.[114] If we turn to the Book of Wisdom, the evidence is perhaps even stronger. There we read: "how could any thing have endured, if it had not been thy will? or been preserved, if not called by thee? But thou sparest all; for they are thine, O Lord, thou lover of souls" (Wisdom 11:26). Clearly, souls of a similar nature, as well as flesh of a similar nature, are allotted to all creatures in common.

Despite the acknowledgment of "all the best Greek and Hebrew scholars" of a century ago, no great advance appears to have been made. Lewis G. Regenstein observed in 1991 how "the Hebrew word for 'soul' (*nephesh*) and 'living soul' (*nephesh chayah*)" are used interchangeably of humans and animals in Genesis. "But," when referring to animals, "most modern Bibles substitute the word 'life' for 'living soul,' or use 'living creatures.'"[115] Arnobius, writing in Latin, uses the same term to apply to humans and animals alike. As living creatures in common, with a shared reason, both humans and animals must, for Arnobius, be understood as *nephesh chayah* – living souls. They could be distinguished from plants in that vegetation possessed neither blood nor breath.

Many who followed Arnobius struck a similar chord. Lactantius (c. AD 240-320), a student of Arnobius, who wrote the first systematic account in Latin of the Christian attitude to life, allowed that human exclusivity lay alone in having a religious nature, a view followed in essence by Edmund Burke as late as the eighteenth century. Pliny had said that elephants were religious, too (a story still being repeated as credible by Pierre Gilles as late as the sixteenth century), but then he also described three-acre whales, merciful lions, and an octopus that climbed trees! In fact, the notion of animals worshipping God was very common in medieval thought (e.g., in the writings of Saint Kieran and Saint Francis of Assisi) and continued later – for example, in the words of William Karkeek writing in *The Veterinarian* and citing John Wesley in his cause.[116] However, for Lactantius, apart from the natural proclivity for worship among humans, "the other things, even those which are thought proper in man, are found in other animals also," including speech, laughter, love, foresight, and understanding. Can anyone doubt, he asks, that "there is reason in them when often they delude man himself?"[117]

Athanasius (c. AD 293-373), the chief defender of Christian orthodoxy against Arianism – the denial of the divinity of Christ – also ascribed reason to animals and insisted that "the omnipresent and perfectly holy word of the Father is present in all things and extends His power everywhere illuminating all things visible and invisible. Containing and enclosing them in Himself, He leaves nothing deprived of His power, but gives life and protection to everything everywhere, to each individually and all together."[118] For Athanasius, all creatures, both singly and in common, are under God's protection. The similarities rather than the differences between humans and animals are stressed.

Saint Basil, the fourth-century bishop of Caesarea, is even more emphatic in telling us that God has "promised to save both man and beast" and that animals "live, not for us alone, but for themselves, and for [God]."[119] They are, moreover, "our brothers ... to whom Thou gavest the earth as their home in common with us."[120] That is, animals are, in part, ends in themselves who are akin to us and share the same kinds of souls that we possess. By degree of contrast, the third-century Christian scholar Origen denied rationality to animals. Nonetheless, as Richard Sorabji informs us, he "sometimes accepts the transmigration of souls into animals, and the eventual salvation of all souls, evidently even those that were in animals."[121] Although rationality and immortality are often related, the instance of Origen once again indicates that it is not a universal rule. Often at issue was what was meant by rationality, whether a calculative ability would suffice, whether the capacity for philosophical speculation, or understanding "universals," was a prerequisite. And the answers were not always easy to discern. Indeed, the requisite information for a confident judgment is often absent from the evidence and the arguments. And this is because the thinkers of the time dwelt on the questions that mattered to them in their historical and cultural context and were often not concerned to raise the issues that we, many centuries and mind patterns away, often presume, or at least wish, they must have had in mind. Moreover, just as in the Oriental and Aboriginal traditions, when we read that the animals are our brothers, we can be reasonably confident that it does not mean that we are to treat them as we would our siblings. It does mean, however, as it does in the Oriental and Aboriginal traditions, that we recognize a close relationship between humans and other species.

IV

If rationality and immortality were once fairly commonplace characteristics attributed alike to human and animal, this began to change in the latter part of the fourth century with the great influence of Saint Augustine. In *On the Catholic and Manichean Ways of Life,* Augustine tells us in Stoic vein that "we see and appreciate from their cries that animals die with pain. But man disregards this in a beast, with which, as having no rational soul, he is linked by no community of law."[122] Rationality and immortality are treated effectively as coterminous. And it is only toward like rational souls that we have reciprocal moral obligations. Only those with rational souls constitute the moral community. To provide a context in which Augustine's arguments may be better understood, it is important to recognize that here Augustine is arguing against the Manichaean way of life – Augustine had formerly belonged to the sect for nine years, and thus there was an element of self-justification in his remarks as well as an attempt to convince his fellow Catholics that he had distanced himself from his disreputable past. Manichaeism included a vegetarian diet, at least for the elect, or *Primates Manichaeorum,* although not for the "Hearers," the lower orders to whom Augustine himself belonged. At this point in the argument, he is concerned primarily with justifying the legitimacy of the consumption of animal food – although, in general, Augustine is also concerned to warn against the Manichees' excessive asceticism. Indeed, vegetarianism was not the rarity we might imagine it to have been even among Christian sects. In *Of the Morals of the Catholic Church,* as I noted earlier, Augustine indicates that, among Christians, vegetarians were "without number."[123] Augustine was in a battle for flesh-eating orthodoxy against such early Christian thinkers as Clement of Alexandria, who called Christian communal meals *agapes* – literally love feasts – for being "in essence heavenly food," claiming that such meals should not include the smell of roasting meat.[124] Hence in *The City of God,* Augustine remarks: "We do not apply 'Thou shalt not kill' to plants, because they have no sensation; or to irrational

animals that fly, swim, walk, or creep, because they are linked to us by
no association or common bond. By the Creator's wise ordinance they are
meant for our use, dead or alive. It only remains for us to apply the commandment 'thou shalt not kill' to man alone, oneself and to others."[125]

There would appear to be an element of rationalization in Augustine's
argument. If the ground on which we are entitled to kill plants is that
"they have no sensation," one is entitled to wonder why the sensation criterion does not apply to animals as well as plants. Moreover, since animals
"die in pain," as Augustine acknowledges, even if animals "are meant for
our use, dead or alive," one has to wonder why "man disregards this in a
beast." Surely, since, in the discussion of plants, "sensation" has been acknowledged as a criterion of ethical consideration, we ought to have a moral
obligation to diminish this suffering in the death of animals, even if the
animals "are linked to us by no association or common bond." The seemingly obvious ethical consideration with regard to pain and suffering appears
to escape Augustine. Having "no rational soul," animals appear to possess
neither immortality nor the status of beings for whom ethical consideration is appropriate.

One should note, however, that this dismal lack of consideration for
animals is remedied to a degree elsewhere in Augustine's writings. Thus,
for example, in *Confessions* he condemns animal sacrifice, treating it not
merely as something anti-scriptural, but as a pronounced "evil,"[126] and insisting that, if all beings are not equal, nonetheless "each is good in itself,
and all very good together, because our 'God made all things very good.'"[127]
When we consider the continuation of animal sacrifice both in India and
among many Aboriginal groups to the present time, Augustine's condemnation might be considered a matter of comparative significance. Moreover, in *The City of God,* he wrote that "living creatures show their love of
bodily peace by their avoidance of pain, and by the pursuit of pleasure to
satisfy the demands of their appetites they demonstrate their love of peace
and soul. In just the same way, by shunning death they indicate clearly
how great is their love of peace in which soul and body are harmoniously
united."[128] Animals are deemed to possess seemingly oxymoronic corporeal souls, their sentience is acknowledged, they are seen as seekers after
peace and harmony, but lacking "rational souls" – immaterial souls – they
lack the capacity for immortality.

Even if Augustine shows a partial sympathy for animals, there is little
doubt that it is, in large part, because of Augustine that the early Christian theological emphasis on human-animal similarities was replaced by a
harsher doctrine. With the exceptions of such sects as the Bogomils in Bos-
nia and Bulgaria and the Cathars (or Albigensians) in Languedoc in the

early years of the second millennium, who included both vegetarianism and, apparently, immortality of animal souls in their doctrines, the *theology* of Christianity turned away from a recognition of animal worthiness, at least as a generality, although there remained numerous anomalies within this theology. However, even though the theology tended to turn away from a concern with animal wellbeing, there is an abundance of evidence that the culture was kinder to animals than was the dogma. Certainly, medieval myth – Carolingian, Franco-Arthurian, and *Mabinogion*, among others – allowed sometimes for a common human-animal language, sympathy, and soul. Medieval myth may be exemplified by Piers the Ploughman, who averred that "the thing that moved me most, and changed my way of thinking, was that Reason ruled and cared for all the beasts, except only for man and his mate; for many a time they wandered ungoverned by Reason."[129] And Andrew Linzey has estimated that more than two-thirds of the medieval saints demonstrated "a practical concern for, and befriending of, animals."[130] Another theologian, Stephen H. Webb, has confirmed that "one of the criteria for sainthood seems to be the compassionate treatment of animals."[131] In medieval religious myth, Saint Jerome is portrayed as an ethical vegetarian, Saint Columba as ordering his monks to care for an exhausted crane, and Saint Neot, Saint Hubert, and Saint Aventine as saving hares and stags from hunters. Saint Goderic of Finchdale is reported to have rescued birds from snares, Saint Carileff to have protected a bull from the royal hunt, and Saint Monacella, Saint Anselm, and Saint Isidore each to have saved a hare from the hounds. Saint Roch has his dog and Saint Anthony his pig. Saint Kevin, Saint Guthlac, and of course Saint Francis of Assisi are reported to have had a mysterious rapport with animated nature. Saint Thecla, a first-century follower of Saint Paul, was purportedly befriended by lions who were sent by her enemies to devour her and who later defended her against a bear; Saint Sergey was protected from his enemies by a bear; Saint Blaise, patron saint of cattle, gave sanctuary to wild beasts; Saint Gall, patron saint of birds, shared his dwelling with a bear; Saint Harvey charmed animals with his music and is invoked to protect flocks and herds. In *Beasts and Saints,* Helen Waddell has translated medieval accounts of an old monk's generosity to his ox and to a lion visitor, the compassion of a holy man for a penitent wolf, Saint Macarius and a grateful hyena, the kindness of a wild ass to the abbot Helenus, Saint Malo's compassion for a swineherd and his sow, Saint Cuthbert's companionship with the sea otters, Saint Benno encouraging the frogs to pray croakingly to God, Saint Ciarin (Kieran) ordaining a boar as his first monk, and many others of like vein. Moreover, Kieran preached vegetarianism to the boar and to other carnivorous animals that joined his flock. It matters little whether the stories reported of these saints are true. We can be quite confident that some, perhaps many, of them are not. More likely, they are exaggerations. Of significance is that the prevalence of these accounts reflects that a mark of great Christian worthiness was the caring for our fellow creatures, especially when they were oppressed. As Nietzsche explained concerning the mark of the saint: "In him the ego is melted away, and the suffering of his life is, practically, no longer felt as individual, but as the spring of the deepest sympathy and intimacy with all living creatures."[132] If saints are rare, and if their exploits are sometimes in part a product of the imagination, they epitomize for the people the life of honour and justice, which few lesser mortals are ever able to achieve.

Not all the stories were myths. Some were well-documented statements of the elevated. Thus asks Saint Isaac the Syrian (d. c. 700), successively abbot of Spoleto and bishop of Nineveh:

> What is a charitable heart? It is a heart which is burning with love for the whole creation, for men, for the birds, for the beasts ... for all creatures. He who has such a heart cannot see or call to mind a creature without his eyes being filled with tears by reason of the immense compassion which seizes his heart; a heart which is softened and can no longer bear to see or learn from others of any suffering, even the smallest pain being inflicted upon a creature. That is why such a man never ceases to pray for the animals ... [He is] moved by the infinite pity which reigns in the hearts of those who are becoming united with God.[133]

And again, if the eloquence of a Saint Isaac is something of a rarity, he is nonetheless but one in the panoply of saints who displayed their profound sense of belonging with their fellow creatures.

The renowned Saint Francis (1182-1226) admonished his flock, informing them that "all the creatures under heaven, each according to his nature, serve, know, and obey their Creator better than you."[134] His first biographer, his contemporary, Thomas of Celano, described him as a man of "very great fervor and great tenderness toward lower and irrational creatures ... he called all creatures *brother,* and in a most extraordinary manner, a manner never experienced by others, he discerned the hidden things of nature with his sensitive heart."[135] His disciple Saint Bonaventure tells us that when "he considered the primordial source of all things, he was filled with even more piety, calling creatures, no matter how small, by the name of brother or sister, because he knew that they had the same source as himself."[136] Only philosophical theologians expressed themselves on esoteric matters such as the status of animal souls and how many angels could dance on the head of a pin – and Saint Francis was no philosopher. But there can be little doubt that even if he acknowledged animals as less rational than humans, he saw them as his close relatives to whom he owed fraternal obligations. Given the multitude of stories told of him and his affection for animals, it is difficult to imagine that he thought of them as having a separate destiny, especially since he saw them as worthier worshippers of God than humans. Nonetheless, it is sometimes suggested that if Saint Francis continued to eat animals, as appears likely, at least on special occasions, his use of "brother" and "sister" was mere subterfuge. Perhaps. But if this is the case, so is every such use by members of Aboriginal societies – which are universally omnivorous. Again, the Western and non-Western experiences share more in common than is usually conceded.

Convinced a priori that the Oriental tradition must be more compassionate than Christianity, the Australian philosopher John Passmore commented on a fourteenth-century Japanese writer who opposed the caging of wild birds that such a sensibility "could certainly not be matched by European writers of the same period."[137] In fact, in the fifteenth century, as we shall see, Leonardo made a practice of buying caged birds in the Florentine market place and releasing them. W.E.H. Lecky remarked in 1869 that the thirteenth-century Saint James of Venice "was accustomed to buy and release birds with which Italians used to play by attaching them to strings, saying that he 'pitied the little birds of the Lord,' and his tender charity recoiled from all cruelty, even of the most diminutive animals."[138] And in the very same century as the remarks of the Japanese essayist, Chaucer, in the "Manciple's Tale" of *The Canterbury Tales,* observed that the fine treatment offered by Phoebus to his captive bird is no adequate replacement for liberty: "But God knows, none can compass in such cases / The power to restrain a thing which nature / Has naturally implanted in a creature / ... Caged in a cage of gold however gay, / That bird would rather twenty thousand fold / Be in a forest which is rough and cold, / Feeding on worms and other wretched trash."[139] In fact, the idea that the bird would prefer freedom to imprisonment in a cage is a common theme of medieval literature, including the *Roman de la Rose.* Passmore's remarks indicate that there is a lengthy tradition to be overcome of ignoring the realities of Western history. Of course, this did not mean that the nefarious practices stopped. In the early nineteenth century, Lord Erskine purchased seven robins from a boy who had just caught them, released them in the garden of his London home, and commemorated the occasion by writing a poem titled "The Liberated Robins": "Now harmless songsters, ye are free / Yet stay awhile and sing to me."[140] But neither did the practices stop in the Orient.

Lecky continued his comments on Saint James of Venice by observing that medieval monastic farming "represents one of the most striking efforts made in Christendom to inculcate a feeling of kindness and pity towards the brute creation."[141] Indeed, Ruth Burtt has testified that: "The records of the monasteries often show a marked consideration for their working beasts on the part of monastic landlords and this is notably so in the large estates owned by the great abbeys ... Thus 13th century rules laid down by the Abbey of St. Peter, Gloucester, expressly state the servants of their manors are to take care of the oxen, making sure that the manger in each stall contains a night's food for each team, and they are to plough without injuring or distressing the beasts."[142] And Walter of Henley's thirteenth-century husbandry manual recommended that working animals should not be "overloaded or overworked, or overridden or hurt."[143] It would appear that those who concentrated on good works rather than elevated thoughts had more interest in treating animals well than in determining esoteric points of theological abstraction. Even when animal rationality was denied, we sometimes encounter a heartfelt concern. Thus Richard de Wyche, who became bishop of Chichester in 1244, bemoaned the lives of food animals, declaring: "Poor innocent creatures: if you were reasoning beings and could speak you would curse us. For we are the cause of your death, and what have you done to deserve it?"[144]

None of this care and concern for animal wellbeing persuaded the greatest medieval theological doctor, Saint Thomas Aquinas, to reverse Saint Augustine's pronouncements on the nature of the animal soul, much to the lasting detriment of Roman Catholicism and to the despair of many Roman Catholics – although recent years have witnessed a measure of rethinking. Pope Pius IX (1846-78) had prevented the opening of a proposed animal-welfare shelter in Rome on the grounds that it would divert attention from human-animal welfare concerns. And the *Catholic Dictionary* of 1897 stated unequivocally, and downright immorally, that "The brutes are made for man who has the same rights over them which he has over plants and stones." This contradicted the earlier constant recognition of a limited "dominion." In the same era, the anti-vivisectionist, vegetarian, theosophist physician Anna Kingsford had declared, on witnessing a barrage of animal cruelty while in Rome: "Argue with these ruffians, or with their priests, and they will tell you, 'Christians have no duties to beasts that perish.' Their Pope has told them so."[145] As late as 1984, Pope John Paul II avowed "it is certain animals are intended for human use."[146] Thus it came as a pleasant surprise when in 1987 the papal encyclical *On Social Concerns* advocated "respect for the beings which constitute the natural world." Even more pleasing was Pope John Paul II's declaration at a public audience in

1990 that "also animals possess a soul" – and he certainly spoke in far friendlier and more compassionate terms than had been customary about the human-animal relationship. Still, it cannot be ignored that the Catholic Church has always allowed animals their souls. It is quite inappropriate to claim, as does the modern theologian Stephen H. Webb, that in "the Middle Ages, theologians taught that animals have no souls."[147] Prior to the Cartesians in the Renaissance, no theologians taught that animals lack souls. It was always the particular nature of the soul that was at issue – a soul as such was acknowledged by all. It has always been "rational souls," "immortal souls," not souls as such, that have been at issue, and Pope John Paul II did nothing to clarify the matter.

In line with Plato and Aristotle, who wrote of humanity's godlike character through reason, Aquinas remarked in the *Summa Theologica* that mankind's capacity "to reason makes him like the angels" – and here "mankind's," not "humankind's," is strictly meant, for in Aquinas's view, following Aristotle, "woman is a misbegotten" man, and "nothing misbegotten or defective should have existed in the original creation."[148] And unlike the many commentators we have already met,[149] he reads the dominion over animals of Genesis to mean "making use of them without hindrance."[150] To be sure, vegetation, animals, and humans all possess souls but, respectively, of a vegetative, sensitive, and rational nature. In the Thomist view, "it is clear that the sensitive soul does not possess any process of its own exercised by itself alone; but all activities are of a composite nature. *From which it follows that since the souls of brute beasts do not function by themselves alone, they do not subsist.*"[151] Animals, in other words, are not self-directed beings. And the question of what is said to "subsist" is determined by reference to Aristotle's notion of *sophia,* philosophical reason, which, not being possessed by animals, denies them the capacity to approach the gods. Hence they cannot be immortal. Now Aquinas, like Aristotle, does not deny animals reason per se: "All animals by their natural instinct have a participation of prudence and reason."[152] It is sufficient for Aquinas's argument that rationality is not the active principle of animals – although where he derives the idea that it is *in fact* rather than *in potential* the active principle in humankind is difficult to discern. Again, Aquinas insists on kindness to animals, which leaves one in some doubt about the interpretation of dominion over animals "as making use of them without hindrance." Perhaps it is only the fact and not the degree of being unhindered to which Aquinas refers: "God loves all things that exist ... to love something is nothing else than to will its good"; "It is God's custom to care for all His creatures, both the greatest and the least. We should likewise care for creatures, whatsoever they are in the sense that we use them in conformity with the divine purpose, in order that they might not bear witness against us in the day of judgment." Sometimes, when Aquinas writes of kindness to animals, the appropriate motive for this kindness appears to be the improvement of the character of the one who is being kind. On other occasions, it would appear that one should be kind to animals for the sake of other humans. In this instance, the motive is to avoid being judged harshly in one's quest for eternity. Two serious problems with Aquinas's analysis immediately come to mind: If it is not wrong to be cruel to animals in and of itself, and for the sake of the animals themselves, why should their testimony matter or be offered? And, if the animals are going to be there at judgment day to offer a possible complaint, they must already in some manner have escaped a full and final death at their earthly demise. Aquinas's analysis leaves far more perplexing questions than it offers answers.

Aquinas's conclusions should scarcely surprise us, for those of his mentor Albert von Böllstadt (Albertus Magnus), who was one of the foremost students of animal life of the Middle Ages, show an even greater lack of sensitivity. As much influenced by Aristotle's ethology and metaphysics as was Thomas, Albert tells us after quoting Aristotle that pygmies stand one ell tall, give birth at three, and die at seven. Albert asks *"utrum pygmaei sint homines?"* (whether pygmies are human) and delivers an emphatic "no" in response. They possess language, thought, and the ability to make things but have no moral community and lack philosophy, a moral code, and art – indeed, they are afraid of pictures. Ultimately, however, the reason they are not human is that they lack immortal souls. How do we know that they lack immortal souls? In tautological manner, the answer appears to be because they are not human. How do we know? The implied answer is that they lack the capacity for philosophy and morality. The reality is that Albertus Magnus had never seen, addressed, or studied a pygmy. His answers were predicated neither on evidence nor on sound argument but on his prior assumptions that only "normal" humans possessed immortal souls. Perhaps more important, we can see already foreshadowed in Albert's description of the pygmy a mode of thought that produced the view preponderated among the Schoolmen that humans possess an immortal soul, as evidenced by the possession in humans of a kind of reason superior to that of the animals, one affording an understanding of universal ideas; that they have the capacity for abstract language, which is not found in the communicating sounds of the animals; and that they possess both an aesthetic and a moral sense, which are beyond the abilities of other species. Since animals act primarily, if not solely, from instinct, they have none of the aptitudes for which reward or punishment would be appropriately assessed. While they have feelings, and a small degree of intelligence, they lack the higher gifts of humankind. Hence they possess only a corporeal, sensitive, and mortal soul.

However much Albert and Thomas may have influenced the more or less customary view of the church, we should not imagine that they necessarily represent medieval Christianity in general, sometimes not even the formal view of the church itself, nor later Catholicism with regard to sensibilities to animals. Thus, in the century following the Thomist writings, we find the visionary and advocate of church reform Saint Catherine of Siena (1347-80) claiming that "the reasons why God's servants love His creatures so deeply is that they realise how deeply Christ loves them. And it is the very character of love to love what is loved by those we love."[153] The founder of the Order of the Most Holy Saviour, Saint Bridget of Sweden (c. 1300-73), wrote *Revelations,* which was formally approved by the Council of Basle in 1431. There she announced: "Let a man fear, above all, me, his God, and so much gentler will he become towards my creatures and animals, in whom, on account of me, their Creator, he ought to have compassion for to that end was rest ordained on the Sabbath."[154] The import of even these compassionate comments pales against that of the British ecclesiastical moral treatise *Dives et Pauper* – in essence, "The Wealthy and the Ordinary" – since, as Henry Chadwick has pointed out, "*pauper* means a person of modest means rather than someone without food, roof, or clothing; Ovid defined him as 'a man who knows how many sheep he owns.'"[155] The treatise is probably of late-fourteenth-century vintage and certainly composed no later than 1410. It discussed inter alia certain features of the mass, the ten commandments, and church customs, concluding:

> When God forbade man to eat flesh, he forbade him to slay the beasts in any cruel way, or out of any liking for shrewness. Therefore, He said, "Eat ye no flesh with blood (Gen. IX), that is to say, with cruelty for I shall seek the blood of your souls, at the hands of all beasts." "That is to say: I shall take vengeance for all the beasts that are slain out of cruelty of soul and a liking for shrewness." For God that make all hath care of all, and He will take vengeance upon all that misuse His creatures. Therefore, Solomon saith, "that He will arm creatures in vengeance on their enemies." (Sap. V); and so men should have thought for birds and beasts and not harm them without cause, in taking regard they are God's creatures. Therefore, they that out of cruelty and vanity behead beasts, and torment beasts or fowl, more than is proper for men's living, they sin in case full grievously.[156]

There can be no doubt that in each of these instances, it is the interests of the animals and the prevention of cruelty to them that is at issue – a clear indication of a Christian recognition that animals mattered in and for themselves, although there is no pronouncement on the status of their souls.

In fact, in the medieval era there was an increasing interest in animals for their own sakes, rather than merely as instruments of ethical instruction. It has been customary to regard the bestiaries as a mere means of socialization, in which the animals are no more than characters on the stage playing their moralizing roles for the edification of the audience. But, as Ron Baxter has shown in *Bestiaries and Their Users in the Middle Ages,* if this is how bestiaries began, editions produced in the later part of the medieval era showed a decided interest in animals for their own sake: "they made it more of a reference book and less of a lecture script."[157] People in general were beginning to think about animals rather than merely through them. And while the religious hermit Richard Rolle of Hampole's fourteenth-century "The Nature of the Bee"[158] is still dependent on ethological inaccuracies derived from Aristotle and is still largely concerned with presenting a moral about how we should follow the animals in our behaviour, it is difficult to read the piece without recognizing an authentic respect for the birds and bees of the story.

If Aquinas believed that Genesis grants humankind the authority "of using [animals] without hindrance," it is very far from clear that he spoke with general Christian authority or even that he consistently espoused this view in his own writings. And certainly there is no evidence for Peter Singer's broadly shared view that the claimed Christian denial that animals possess immortal souls had a "very negative influence on the way in which we think about animals." On the basis of the available evidence, it would appear that early Christianity viewed animals with no less sympathy than adherents of other religions, other cultures, other climes.

V

In fact, many Christians not only rejected the Thomist view of the human-animal relationship, but also disagreed with him fundamentally on the issue of the immortality of animal souls itself. Around the time of the Reformation, the animal-soul story became far more complex and interesting. But not from the Protestant side alone. Thus immediately prior to Luther's posting of the ninety-five theses on the door of the castle church at Wittenberg, Leonardo da Vinci was writing in his *Notebooks* of "the spirit of the sentient animals"[159] – and although his words are consistent with the notion that this sentience denotes only a sensitive soul, the implications take us very much further. Certainly, since he is writing of the capacity for self-direction driven by the spirit, his conception of the animal soul is very different from that of Aquinas. His *Notebooks* are replete with statements against cruelty to animals, and hints at his own apparent vegetarian persuasion are to be encountered there, too, as is his disgust at human hubris. To take a characteristic example of his humanitarianism alone, he complains, writing about bees: "And many will be robbed of their stores and their food, and will be cruelly submerged and drowned by folks devoid of reason. O justice of God! Why dost thou not awake to behold thy creatures thus abused?"[160] Clearly, if rationality is the criterion of immortality, it is the human variety of which Leonardo despairs.

He has much to say about lack of consideration for animals in farming. Writing of "sheep, cows, goats, and the like," he bemoans the fact that "from countless numbers will be taken away their little children and the throats of these shall be cut, and they shall be quartered most barbarously."[161] In discussing "asses that are beaten," he exclaims to Nature: "I see thy children given up to slavery to others, without any sort of advantage, and instead of remuneration for the good they do, they are paid with the severest suffering, and spend their whole life in benefiting their oppressor."[162] If Leonardo does not tell us explicitly that animals have a heavenly destiny, he makes it abundantly clear that many humans do not deserve one. What, however, appears to be of the greatest significance with regard to Leonardo's respect for animals is that his first biographer, Giorgio Vasari, writing in 1550 of the lives of several Renaissance artists, not only comments on this respect very favourably, but clearly expects his readers to rejoice in this respect and consideration along with the writer. Leonardo, he remarks,

> was so pleasing in conversation that he attracted to himself the hearts of men. And although he possessed, one might say, nothing and worked little, he always kept servants and horses, in which latter he took much delight, and particularly in all the other animals, which he managed with the greatest love and patience; and this he showed when often passing by the places where birds were sold, for, taking them with his own hand out of their cages, and having paid them what was asked, he let them fly away into the air, restoring them to their lost liberty. For which reason nature was so pleased to favour him, that, wherever he turned his thought, brain, and mind, he displayed such divine power in his works, that, in giving [the animals] their perfection, no one was ever his peer in readiness, vivacity, excellence, beauty, and grace.[163]

Nor was Vasari alone in commending Leonardo's attitude to animals with admiration and awe. In 1515 the traveller Andrea Corsali wrote to Giuliano di Medici – a patron of Leonardo, the Florentine ruler to whom Machiavelli first dedicated *The Prince,* and the brother of Pope Leo X – telling him that "certain infidels called Guzzarati [Hindus of northwest India] do not feed upon anything that contains blood, nor do they permit among them any injury to be done to any living thing, like our Leonardo da Vinci."[164] Leonardo's vegetarianism is further exemplified when he remarks on the disgusting way in which people make their "gullet" into "a sepulchre for all animals" and adds "now does not nature produce enough simple (vegetarian) food for thee to satisfy thyself? And if thou art not content with such, canst thou not by mixture of them make infinite compounds, as Platina wrote, and other authors on feeding?"[165] Moreover, he castigated the owners of oxen who first use them and then eat them: "The masters of estates will eat their own labourers."[166] Christian opinion (as expressed by the church and heavily influenced by Saint Thomas) and Christian opinion (as expressed by the admiration for the compassion of a Leonardo or the animal-protecting saints) were as alike as chalk and cheese.

To put Leonardo's attitudes in context, two further matters need to be reported. First, his vegetarianism was not at all uncommon for the time and later, although usually more out of necessity than choice. Writing almost three centuries later, Marianne Stark observed in *Letters from Italy, between 1792 and 1798* that: "The most remarkable quality in the Florentine Peasants is their industry, for, during the hottest weather, they toil all day without sleep, and seldom retire early to rest: yet, notwithstanding this fatigue, they live almost entirely on bread, fruits, pulse, and the common wine of the country."[167] Even in the mid-nineteenth century, we find Alphonse de Lamartine commenting on "the robust and healthy pastoral peoples, and even our industrious rural workers, who work more, live simpler and longer lives, and who eat meat perhaps ten times in their lives."[168] Around 1872 the British government conducted a study on the diet of agricultural labourers in Europe and determined that they ate very little flesh. Only in Pomerania and England were flesh foods commonly a part of the regimen. In Belgium and Scotland they consumed a little bacon but otherwise no meat. In Prussia they ate flesh only on feast days. In the Netherlands they fed on a little fish but otherwise no meat. In Bavaria, Italy, Sweden, and Russia, no meat was consumed, and this was similarly the case in Ireland, although they deemed it important to mention that there a "little whisky is taken."[169] In the sixteenth century, peasant life was no less sparse. Perhaps Leonardo's apparent vegetarianism reflected his respect for the simplicity of the peasant life almost as much as a concern for animals. Certainly, we know that meat was on his grocery list, and if this was to feed his employees and guests, at least we can be sure that he did not attempt to impose his diet on others.

Curiously, the paragraph of praise for da Vinci's consideration for animals in Vasari's *The Great Masters: Giotto, Botticelli, Leonardo, Raphael, Michelangelo, Titian* is followed by a passage in complete contradiction of it: "Leonardo carried to a room of his own into which no one entered save himself alone, lizards great and small, crickets, serpents, butterflies, grasshoppers, bats, and other strange kinds of suchlike animals, out of the number of which, variously put together, he formed a great ugly creature, most horrible and terrifying ... and so long did he labour over making it, that the stench of the dead animals in that room was past bearing, but Leonardo did not notice it. So great was the love he bore towards art."[170] If the story is accurate, and if the animals were not already dead when he acquired them, we must assume that Leonardo had a greater love for art than for animals, just as Darwin had a greater love for physiology than for animals; or that he considered animals such as mammals and birds to be of a far superior order to reptiles and insects and therefore deserving of ethical consideration, while the meaner beings were not (and bats have always had a bad press); or that he developed his deep and abiding respect for all animals after the event reported by Vasari. Perhaps it is merely one more indication of the contradictory nature of the human being.

VI

With the advent of Protestantism, there was an increasing belief that doctrinal matters were not solely a matter of traditional church orthodoxy, best left to the acknowledged theological Fathers of the Church. To be sure, in the early Protestant decades, there was no intention of instituting rival denominations with competing creeds. The earnest desire was to reform the church united. Nonetheless, there was an increasing sense that doctrine should be subject to more than pontifical or council declaration. Certainly, Martin Luther's view of the animal soul was something other than what one might have expected from Rome and closer to the more humanitarian views that seemed to prosper more readily away from the pressures of the traditional church hierarchy.

On being asked by a young girl called Thekla whether her recently deceased companion dog would now be destined for heaven – "Will Nix rise again at the last day?" – Luther offered a very different answer than would have come from an Augustine or an Aquinas. He told his entourage, "as he smiled at the grieving little girl," while they "trembled at such blasphemy, spoken from the innocent lips of a child":

We know less of what that other world is like than this little girl knows of the empires or powers of this world. But of this we are sure, the world to come will be no empty lifeless waste ... God will make new heavens and a new earth. All poisonous and malicious and hurtful creatures will be banished there, all that our sin has ruined. All creatures will not only be harmless, but lovely and joyful, so that we might play with them. The suckling child shall play on the hole of the asp, and the weaned child shall put his hand on the cockatrice's den. Why, then, will there not be little dogs in the new earth, whose skin might be as fair as gold, and their hair as bright as precious stones?[171]

Clearly, Luther's answer refers to the "new heavens and a new earth" of Isaiah 65:17, and in particular to verse 25 of the same chapter, where the "wolf and the young lamb will feed together; *the lion will eat hay like the ox,* and dust be the serpent's food" (italics in the original); also, one assumes, Luther is referring to Isaiah 11:8, where, in the King James version, the serpent and mythical cockatrice will lose their sting, or as *The New Jerusalem Bible* has it, "the infant will play over the den of the adder; the baby will put his hand into the viper's lair." There are poisonous and harmful animals – once a great deal more harmful to humans than they are now. It was human sin in the Garden of Eden that had ruined the originally benevolent nature of all species, and, on the day of judgment, God will render all unto their original Edenic nature, Luther argues. If Luther leaves the exact form unclear – as, of course, does Isaiah 65 – there can be no doubt that he accepted not only that animals may have an afterlife, but also that humans will occupy the same place in some kind of mutually beneficial association.

Nor was this the only occasion in which Luther mentioned animals in the afterlife: "You must not think that heaven and earth will be made of nothing but air and sand, but there will be whatever belongs to it – sheep, oxen, beasts, fish, without which the earth and sky or air cannot be."[172] If so significant a Christian as the founder of Protestantism could be so explicit about animals in the afterlife, how, one wonders, have so many managed to reach the conclusion that Christianity denies animals immortal souls? Is it poor scholarship, or a wilful denial of the reality of the Christian religion? Is it ignorance or prejudice?

We should not imagine the newfound confidence regarding animal immortality to be solely a Protestant matter. For instance, the Spanish mystic and poet Saint John of the Cross (1542-91) had no doubts about immortal animal souls. In "Spiritual Canticle" (1584), he wrote:

According to St. Paul, the son of God is the highest of His glory and the image of His substance. It must be known, then, that God looked at all things in the image of His Son alone, which was to give them their natural being, to communicate to them many natural gifts and graces and to make them finished and perfect, even as he says in Genesis in these words: God saw all the things that He had made and they were very good. To behold them and find them very good was to make them very good in the Word, His Son. And not only did He communicate to them their being and their natural graces when He beheld them, as we have said, but also in the image of His Son alone He left them clothed with beauty, communicating to them supernatural being ... in uniting Himself with man He united Himself with the nature of them all. Wherefore said the same Son of God ... I, if I be uplifted from earth, will draw all things unto me.[173]

If, for San Juan de la Cruz, there are spiritual distinctions between humans and other species, they can be but minor. The animal, too, is to be uplifted from earth and drawn unto the godhead.

By the turn of the seventeenth century, similar, if decidedly less mystical, sensibilities were increasingly common. In 1580 Michel Eyquem, seigneur de Montaigne, applauded the luminaries of the classical period who proclaimed that animals and humans shared reason, emotions, affection, and choice, avowing in concurrence with Lucretius in *De rerum natura,* "We are neither above nor below the rest; all that is under heaven ... shares the same law and the same fortune. / ... 'All things are bound by their own chains of fate.'"[174] Having suggested a common destiny, he appears to attribute the same kind of soul to the animals as he does to humans when he avers that we see "in our cruder works" – cruder than that of the animals, that is, in that we lack their abilities – "the faculties that we use; and that our soul applies itself with all its power; why do we not think the same of them?"[175] Humans, it would appear, had no fundamentally distinctive characteristics of soul, certainly no obviously superior ones.

The quasi-pantheist, yet presbyterian, John Milton (1608-74) held to a not dissimilar view, believing that God created all beings not out of nothing, but out of himself. In the *Treatise of Christian Doctrine,* he observed that, as a consequence: "There seems therefore no reason why the soul of man should be made an exception to the general law of creation ... For ... God breathed the breath of life into other living beings and blended it ... intimately with matter ... every living thing receives animation from one and the same source of life and breath ... Nor has the word *spirit* any other meaning in the sacred writings, but that breath of life which we inspire, or the vital, or sensitive, or rational faculty, or some action or affection belonging to these faculties."[176] At the very least, Milton is insisting that both humans and animals are *nephesh chayah* – living souls – even though his concept of "soul" may not be the traditionally spiritual one. Indeed, like Voltaire, and in line with much deist and Christian thought, Milton imagined that the Jews had no concept of an immaterial soul, and he saw his conceptions as consistent with Jewish thought. As Denis Saurat explains, in Milton's "cosmology, there is no place for death. Immortality is a direct consequence of the way in which the world is built, since 'no created thing can be finally annihilated.' Matter is divine and indestructible, and man has no soul that can be separated from his body. Therefore, every being is naturally and normally immortal."[177] The ontological status of animal and human were the same, as indeed Genesis 2 had proclaimed. Thus in *Paradise Lost* (1677, bk. 7, 387-88), we read: "And God said; let the waters generate / Reptile with spawn abundant, living soul." Even reptiles shared *nephesh chayah* with humans. Moreover, in the same epic poem (bk. 8, 261-66 and 369-75), Milton first has Adam rejoice in his animal companions, and is then critical, via the "Universal Lord," of human self-absorption and deems other species both worthy of fellowship and possessed of sufficient reason for the task:

> About me round I saw
> Hill, dale and shady woods, and sunny plains,
> And liquid lapse of murmuring streams; by these,
> Creatures that lived and moved, and walked or flew;
> Birds on the branches warbling; all things smiled
> With fragrance, and with joy my heart o'erflowed ...

Yet Adam, on reflection, wants more appropriate company. Raphael responds:

> What call'st thou solitude? Is not the Earth
> With various living creatures, and the air
> Replenished, and all these at thy command
> To come and play before thee? Knowst thou not
> Their language and their ways? They also know,
> And reason not contemptibly; with these
> Find pastime, and bear rule; thy realm is large.

All are rational; all have souls. But, to be sure, not all souls are equal: there is a scale of being from angel to human to animal. The angel says (bk. 5, 486-90):

                     whence the soul
Reason receives; and reason is her being,
Discursive or intuitive; discourse
Is oftest yours, the latter most is ours;
Differing but in degree, of kind the same.

And the animal soul, too, while of the same kind, differs also, and more, in degree.

Like Milton, a believer in the universal soul, the pantheist physician and kabbalist Robert Fludd (1574-1637), one of the most influential thinkers of the early seventeenth century – although now largely forgotten – claimed humans and animals to possess similar souls. Both Milton and Fludd follow the teachings of the *Zohar,* a thirteenth-century kabbalistic Spanish compendium of the nonorthodox Jewish traditions, in the belief that "all souls form one Unity with the essential soul,"[178] although these souls would appear to be material entities. Or, at least, it would appear that there are no essential distinctions between the body and the soul.[179] If the teachings were not always clear, of one thing we can be certain: for Fludd, whatever the soul might be, it is shared alike by human and animal, and on the final day, all will be resurrected. This seemingly confusing idea of the possibility of there being a purely material world in which humans and animals would, nonetheless, at least ultimately, survive death, in that all would be resurrected simultaneously at the judgment day, rings strange to modern ears.

Yet particularly in the sixteenth and seventeenth centuries, there were many factions in contention on the question of both the nature and fact of immortality. Even in Rome, at the Fifth Lateran Council in 1513, Pope Leo X was compelled to cut the Gordian knot of philosophy and theology with canonical authority and declare the soul immaterial and immortal, yet liable in the afterlife to both physical pleasure and pain. The complexities that serve to confuse were exacerbated, if from a different perspective, in the writings of the abbé de Villars in *Le Comte de Gabelis* (1670): "I shall not conceal from you the fact that the opinions of your modern philosophers concerning the nature of animal souls do not please me at all; I pass between them, and distance myself equally from those who take the purely mechanistic view and those who suppose them to be material souls, yet filled with knowledge and emotions. I believe that their souls are spiritual, that they possess reason as we do, even though they differ in their experience and their duties."[180] Despite this, the abbé does not accord the animals an immortal soul! The sixteenth and seventeenth centuries were witness to a variety of interpretations, some of which are difficult to categorize in any manner meaningful to later generations.

In fact, what is commonly missed in modern discussions of the histor-ical ascription of souls to animals, or the lack of it, is the context of the status of the soul itself. In the Renaissance centuries, there were many fol-lowers of the Spanish-Arabian medieval scholar Averroës, known as the "mortalists" – numbering among them Faustus Socinus, George Wither, Thomas Hobbes, and Richard Overton, along with Milton and Fludd, who insisted that the individual soul was naturally mortal and as corrupt-ible as (or one with) the body. In the words of the Cromwellian intriguer and political leveller Richard Overton: "the whole man ... is a compound merely mortal, contrary to the common distinction of soul and body ... the present going of the soul into heaven or hell is a mere fiction ... at the resurrection is the beginning of our immortality, and the actual condem-nation, and salvation, and not before."[181] They denied personal immortal-ity upon death, tried to accommodate Christianity to the new humanism, but still regarded themselves as devout and faithful Christians – indeed, in some instances, as the only genuine devout and faithful Christians. For most of the mortalists, resurrection was collective following the last day – hence, although mortal, the individual soul would inhabit the hereafter, and for a number of them this included animals as well as humans.

It should, then, not surprise us in the midst of such complexities that, following a similar but largely ignored exposition by a Spanish physician, Gomez Pereira, in 1554, a Descartes should arrive on the scene to declare, or be understood to declare, yet another refinement: that while humans were spiritual and ensouled, animals were material and not only devoid of an immortal soul, but also devoid of a sensitive soul and hence of sen-tience, or at least consciousness, altogether. For Descartes, the soul is an entity entirely separate from the body, whose essential attribute is thought. It can thus exist independently of the body because thought is a substance whose nature can be grasped without reference to material existences. (And since animals cannot think, they cannot have souls!)

There is in fact considerable dispute as to the niceties of the Cartesian position partly because Descartes was not always clear, partly because he appeared to maintain one position on one occasion and another on another, and partly because his position appears to have softened with age. Despite the recent contentious debate on the matter,[182] the most significant inter-pretation is perhaps that offered by Hester Hastings, writing in 1936:

> When Descartes said that the brute soul was seated in the blood of the heart, he made it a material soul. Since matter cannot initiate action or motion, another cause for the life and behaviour of animals had to be found. Des-cartes explained animal action as purely mechanistic, as involuntary action,

following upon external stimulus. He did not say, in the *Discours de la Méthode,* that beasts do not feel, but those who refer to his system later, whether partisans of it or opponents, constantly claim that in order to avoid the theological dangers of giving beasts a soul, Descartes made them *unfeeling automata,* possessing eyes in order not to see, ears in order not to hear, and so on. Although Malebranche was really the main preacher of the insensibility of beasts, Descartes was considered the originator of this pure mechanism. It seems evident that animals were looked upon as acting involuntarily, or mechanically, for one reason only: in order that their actions should not be explained as due to a spiritual soul. Brute souls and brute automata are inextricably linked.[183]

It is perhaps one of the less frequently recognized aspects of the study of animals as automata and the analysis of immortal animal souls that very many of those, at least in France, who rejected the concept of mechanistic animals did not simultaneously draw the conclusion that they possessed immortal souls. Thinking, feeling, judging beings they might be; immortal did not seem, for many, a necessary concomitant.

In France, Descartes was followed by Pierre Chanet, M. Des Fournelles (nom de plume of Géraud de Corderoy), Antoine Le Grand, and above all, Nicolas Malebranche, who proved very much more of a mechanist than the master. His animal automata

> eat without pleasure, they cry without pain, they grow without knowing it; they desire nothing, they fear nothing, they know nothing; and if they behave in a seemingly intelligent manner, it is because God, having made them thus to preserve them, has so formed their bodies that they avoid mechanically and fearlessly everything capable of destroying them. Otherwise, it should be said that there is more intelligence in a mere egg than in the wisest man, because it is certain that there are more different parts and that it produces more regulated parts than we are capable of understanding.[184]

Notoriously, Malebranche kicked his dog in order to be able to declare that the resultant cries were no more than mechanical responses. No pain was felt. Only a spring was sprung.

While Cartesian rationalist philosophy was in general greatly admired, it is remarkable how many expressed their conviction of the validity of its arguments *except* with regard to what they saw as the preposterous notion of animals as *bêtes machines.* For example, Marin Cureau de la Chambre, physician in ordinary to the king – and one who knew both human and animal anatomies from experience – argued in adamant opposition, in *Traité de la Connoissance des Animaux* (1648), that animals could reason and were ingenious. He even questioned the right of human dominion. But he did not offer the animals a heavenly home.

The Catholic priest Pierre Gassendi (1592-1655) was appalled at what he saw as Descartes's blindness where animals were concerned. In fact, long before Descartes published his *Meditations,* which denied animals the capacity for reason definitively, Gassendi had announced his position in the Preface to *Exercises in the Form of Paradoxes in Refutation of the Aristoteleans* (1624): "I restore reason to the animals; I find no distinction between the under-standing and the imagination." He went on to argue that it is simply a prejudice to deny animals the faculty of reason, that animals reason in the same manner as humans, and that "all knowledge" – whether human or animal – "is in the senses or derived from them" (2.6.2).[185] Thus when the *Meditations* appeared, Gassendi was prepared to do battle.

In the *Second Meditation,* Descartes observed a grand distinction between the human and the animal in the faculty of judgment. Only humans possessed the capacity to draw rational but unverified inferences from their experience. Thus, if looking down on a street from a window we saw only moving hats and clothes, we would correctly judge that they were worn by persons. A dog would be quite incapable of performing such a feat. Gassendi wrote privately to Descartes, expressing what he called his "Doubts." Descartes responded publicly, dismissing Gassendi's reservations. In return, Gassendi went into print with what he now termed his "Rebuttals":

> You deny that any dog has a mind and leave him only with an imagination, [but the dog] also perceives that a man, or his master, is hidden under the hat and clothes, and even under a variety of different forms ... Is it not true that if you think the existence of a mentality is evidenced by your realization that there is a man underneath when you see nothing but his hat and clothes, and if likewise a dog realizes that there is a man underneath when he sees nothing but his hat and clothes, is it not true, I say, that you should also think that the existence of a mentality like yours is evidenced by the dog?[186]

Yet, what might appear a surprise in light of the common historical identity of reason and immortality, Gassendi concludes that the animal soul is corporeal and hence mortal. Despite this, he considered humans and animals sufficiently alike that we have no entitlement to a carnivorous diet: "Can a use so noxious be called *natural*? Faculty is given by nature, but it is our own fault that we make such a perverse use of it." Moreover, "there is no pretence for saying that any right has been granted to us by [moral] law to kill any of those animals which are not destructive or pernicious to the human race."[187] Our only right to harm animals is in self-defence.

Again, we are presented with a paradox that the denial of an immortal soul is no hindrance to the recognition of the greatest ethical obligation on our part toward the animals. However, as for so many eighteenth- and nineteenth-century commentators who made persuasive cases for vegetarianism – Pope, Mandeville, Hartley, Goldsmith, Rousseau, Voltaire, Lamartine, Wagner, and Tennyson, for example – there is no evidence that Gassendi practised what he preached. Still, for a while, vegetarianism was at least a semi-respectable cause, both among those who accorded animals immortal souls and those who denied them.

If we find Gassendi's conclusions regarding the immateriality of the animal soul rather surprising in light of his views of animal rationality and our ethical obligations, the Cambridge Platonist Henry More (1614-87), a Fellow of Christ's College, who was equally distraught at Descartes's analysis, meets our expectations more readily. Writing to Descartes in 1648 to praise his work in general but to abominate his view of animals as automata, he complained bitterly of "the internecine and cutthroat idea you advance in [*The Discourse on*] *Method* which snatches life and sensibility away from the animals":

> But I beg you, most penetrating man, since it is necessary by this argument of yours, either to deprive animals of their sense, or to give them their immortality, why should you rather set up inanimate machinery than bodies motivated by immortal souls, even though that may have been the least consonant with natural phenomena so far discovered? In this, indeed, the most ancients judged and approved; take Pythagoras, Plato, and others. Certainly, the persistent idea is presented in all the works of Plato, and has given courage to all Platonists. Nevertheless, such a remarkable genius has been reduced to these straits, that, if one does not concede immortality to the souls of brutes, then all animals are of necessity inanimate machines.[188]

For More, despite the "remarkable genius" of Descartes, both the evidence of science and the philosophy of the historical masters conferred immortal souls to animals. Nor was More alone in his time. Keith Thomas has demonstrated that, at least in England, "a substantial proportion of commentators took the view that animals, like the rest of nature, would be restored to the perfection they had enjoyed before the Fall."[189] The contemporary philosophical originator of this view may have been John Smith (1618-52), along with Ralph Cudworth the earliest of the Cambridge Platonists, whose principle concerning the immortality of the soul was: *"That no substantial and invisible thing ever perisheth.*"[190] If animals possessed thought, they must possess immortality. And, for the Cambridge Platonists, it would appear, they possessed thought. Moreover, the complexity of the thought system was not relevant. Commenting on the possibility of animals occupying other terrestrial bodies after physical death, Cudworth stated: "This seemeth to be no more than what is found by daily experience in the course of nature, when the silkworm and other worms, dying, are transformed into butterflies. For there is little reason to doubt that the same soul which before actuated the body of the silkworm doth afterward actuate that of the butterfly; upon which account it is that this hath been made by Christian theologers an emblem of the resurrection."[191]

The clergyman John Norris of Pemberton (1657-1711) provides an illuminating account of Cartesian persuasiveness in the intellectual culture of the seventeenth century, along with a recognition that, however convincing the Cartesian arguments might have been to the contemporary mind, there is something in human experience that makes the whole scheme dubious and dangerous. Some of Descartes's followers appear to have thought of him as Aristotle thought of Plato – i.e., as the greatest of geniuses – but it is precisely the greatest geniuses who commit the greatest blunders. As Norris writes in 1701:

> To conclude now with a Word or two concerning the *Treatment* of Beasts. Tho' it is my Opinion, or if you will, my Fancy, that Reason does most favour that side which denies all Thought and Perception to brutes, and resolves those Movements of theirs which seem to carry an appearance of it (because *like* those which we exert by Thought) into mechanical Principles, yet after all, lest in Resolution of so abstruse a Question our Reason should happen to deceive us, as 'tis easy to err in the Dark, I am so far from incouraging any practices of Cruelty, upon the Bodies of, these Creatures, which the Lord of the Creation has (as to the moderate and necessary use of them) subjected our Power, that on the contrary I would have them used and treated with as much tenderness and pitiful regard, as if they had all that Sense and Perception, which is commonly (tho' I think without sufficient Reason) attributed to them. Which equitable Measure, they that think they really have that Perception, ought in pursuance of their own Principle, so much the more Conscientiously to observe.[192]

Norris's response is significant for several reasons. First, it makes clear what today we customarily doubt: that, despite well over a half-century of Cartesianism, "Sense and Perception ... is commonly ... attributed to" animals. That is, the Cartesians had, on the whole, failed to convince. Second, it indicates that the primary problem with the idea of animals as automata was not merely a metaphysical one but was readily understood to lie in the fact that it provided an apparently ethical justification for the invasive animal experimentation that was now a pastime of a whole host of amateur drawing-room scientists. Third, it demonstrates that intellectual speculations countering the ethical intuitions of our souls (or minds) need to be thought through again. And Norris is trying to think them through again, for, if they appear to convince his reason, they fail utterly to persuade his conscience. Finally, Norris's response affirms that even those who might be persuaded of the general validity of Cartesianism and who might have accepted the proposition that animals were for human use – "moderate and necessary use," Norris says – should deem it appropriate that animals be treated with tenderness and compassion.

The influence of Descartes was indeed most varied. And it was most voluminous. Apart from Pierre Bayle, however, to whom we shall return,[193] in the eighteenth century the only prominent francophone contributor to the debate who came out with a definitive and unequivocal stance on behalf of the immortal animal soul was the Swiss naturalist Charles Bonnet. He argued that:

> To imagine the animal soul to be mortal merely because the animal is not man is equivalent to imagining the human soul to be mortal merely because man is not an angel.
>
> The soul of the animal and the soul of mankind are equally indestructible by secondary causes. It requires an equally distinctive act of the divinity to annihilate the soul of a worm as to annihilate the soul of a philosopher. Indeed, what proofs may be offered for the annihilation of the souls of beasts? We are told that they are not *moral beings*. But is it not the case that moral beings are those capable of happiness? ... If we put the soul of a brute into the brain of a man, I do not know whether it will rise to the level of understanding universals.[194] I have nothing to say about that. It may well be that there are relative differences among souls in the same manner as we observe among bodies. But notice too the physical diversity which separates human souls.
>
> Why do you restrict the extent of DIVINE GOODNESS? It wishes to bring about the greatest possible amount of happiness. It should be acknowledged that, by degrees, it raises the level of the soul from that of the oyster to the sphere of that of the monkey, and from the soul of the monkey to that of the sphere of mankind.[195]

This singularity is not, however, witness to a general Cartesian victory. A dozen or so of the French publications of the time side with Descartes;

about as many ascribe a spiritual principle to the beasts without necessarily denying or affirming animal immortality; and there is an almost equal number in which animal rationality is stressed, while at the same time, the superior rationality of humans is also emphasized.[196] By 1730 we find the abbé Desfontaines writing that the Cartesian automatists are perceived as a rump of "tenuous men, who pay little attention to the voice of nature."[197] Morfauce de Beaumont in *Apologie des bestes* (1732) dubbed Cartesianism "a vain invention of the schools," and in *Traité des animaux* (1755) Condillac described it as "unacceptable to reasonable people."[198] In *Lettres philosophiques, sérieuses, critiques* (1733), Saunier de Beaumont described the Cartesians as *romanciers*. Nonetheless, he acknowledged humans as distinctly superior to animals because, as a consequence of our upright stance, our eyes are turned toward the heavens, our hands are designed for dexterous productivity, and only we can sit down to practise the arts – a view very similar to that of Xenophon two thousand years earlier. Despite these attributes, animals possess reason, as exemplified by the dreaming of dogs and the capacities of certain animals to learn tricks. He also repeats a number of the classical examples of animal sagacity, from the skills of the swallows and spiders to the faithfulness of the turtledove and the chastity of the crow – "the crow is so chaste that it has been said she would spend up to five hundred years without a mate, if her own were killed."[199] To be sure, we soon get naturalists like Johann Friedrich Blumenbach trying to correct the excesses of such tales,[200] but the greater concern with ethological accuracy did little to inhibit the increasing respect for animals; indeed, it encouraged a greater respect for animals in their own authentic natures.

Among those works that support a human-animal similarity while failing to acknowledge an immortal animal soul (and often enough a human one, too) are the writings of the deists and those even less drawn to traditional spiritual concepts. They are epitomized by the famous quip of Helvétius: "give a horse a hand and he'll be as clever as a man." But they did not take the issue of the immaterial soul very seriously. Thus, for example, we encounter Voltaire writing of the operation of the soul, both animal and human, as analogous with the operation of bellows.[201] In essence, like the French naturalist René Antoine Ferchault de Réaumur, the inventor of the temperature scale and author of a definitive study of insects, Voltaire considered disputation about the nature and existence of souls, both human and animal, to be futile and unprofitable. Julien Offray De La Mettrie has the most to say. As a mechanist, his view of the animal differs little from that of Descartes. But, for La Mettrie, humans are machines, too! *L'Homme-machine* (1745) – the title is a parody of Descartes's *bête machine* – is his best-known work. Here La Mettrie allows only for physical differences:

Between man and beast, the actions of whom derive from mechanical causes ... The excellence of reason does not depend on a big word devoid of meaning (immateriality), but on the force, extent, and perspicuity of reason itself ... The human body is a machine which winds its own springs. It is a living image of perpetual movement. Nourishment keeps up the movements which fever excites. Without food, the soul pines away, goes mad, and dies exhausted ... the diverse states of the soul are always correlative with those of the body ... How can human nature be known, if we may not derive any light from an exact comparison of the structure of man and of animals?

In general, the form and the structure of the brains of quadrupeds are almost the same as those of the brain of man; the same shape, the same arrangement everywhere, with this essential difference, that of all the animals man is the one whose brain is largest, and in proportion to its mass, more convoluted than the brain of any other animal.[202]

In *The Natural History of the Soul* he adds: "Matter, far from being indifferent, as it is supposed to be, to movement and to rest, ought to be regarded as an active, as well as a passive substance."[203] Both man and beast are matter, yet have reason, and are self-directed. Moreover, as La Mettrie makes clear in the *Traité de l'âme,* "the cries of animals are stronger than any sophisms of that philosopher [i.e., Descartes]. They have feelings and ideas, are organized in the same way as man ... The essence of the soul of man and animal is and always will be as unknown as the essence of matter and body."[204] Whether such statements are strictly compatible with a mechanistic philosophy may be doubted. What cannot be doubted is that animal and human souls are, for La Mettrie, in all essential respects the same.

That such arguments were causing a degree of consternation is evident. The cardinal Melchior de Polignac (1661-1741) had produced a Latin treatise against the rationalist teachings of Lucretius, which in 1747 was translated into French as *L'Anti-Lucrèce,* temporarily reviving the mechanistic view of nonhuman animals exclusively, which had been effectively in abeyance throughout the century. It was an attack on Bayle, Boyer d'Argens, and Guillaume-Hyacinthe Bougeant, who had been too readily granting the animals a soul, and on La Mettrie, who had maintained an essential animal-human equality in denying a meaningful soul to both. Most French writers of the period, in fact, hedged their bets, including both the comte de Buffon and Pierre de Maupertuis, the latter of whom was anyway claimed by the Prussians and accused elsewhere of adherence to the principles of Leibniz, which we shall shortly visit.

Plain and simple logic could not allow these writers to deny animals a soul, on the whole not even a moderately rational one, but ecclesiastical authority had not yet dimmed sufficiently, perhaps not yet at all, to suffer them to risk a damning public and private censure. It is perhaps no accident that Bayle was a Protestant Frenchman, Bonnet a Protestant Swiss from Geneva, just as Rousseau's animal-friendly conceptions were born in his hometown of Geneva, and Leibniz a Saxon Protestant, albeit one who tried in vain to reunite, first, Catholicism and Protestantism and, subsequently, the two main branches of Protestantism. Moreover, the seventeenth-century Dutch rabbi Manasseh, an influential teacher of the philosopher Baruch Spinoza, observed that "dumb animals will have a much happier state than they ever enjoyed here when they with man shall rise again."[205] The prospect for Jewish and Protestant animals seemed rather brighter than for Catholic animals.

Still, there was no strict dichotomy between the liberal Protestant and authoritarian Catholic view. In *L'Existence de Dieu démontrée par les merveilles de la nature,* the Dutch physician, mathematician, mayor, and Protestant theologian Bernard Nieuwentyt demonstrated an obstinate refusal to draw conclusions that his analysis would have compelled others to accept. He demonstrates reason among animals and raises the issue of the status of their souls:

> We shall not dwell on the matter; very strong arguments can be offered for and against ... their organs are the same; they take care in their work; they defend themselves against their enemies; they set traps with dexterity; they skilfully avoid eventualities that threaten them; they foresee the ambushes set by their foes; they convene together; they subject themselves to a consistent standard of law; they choose leaders; they employ a division of labour; they work for the common good; they love each other and are faithful in their pairings; they show sadness on departure from each other; all this is to be found among the bees, swans, cranes, beavers, pigeons, and turtles – which seem to indicate that a soul directs their behaviour.

Nevertheless, Nieuwentyt adds, "But we do not know how to relieve ourselves of the difficulties which are faced by this view" – the question of moral responsibility and just deserts. So we "shall not dwell on the matter."[206]

There can be no doubt that immortal animal souls fared far better in liberal and formally Protestant England than in conservative and formally Catholic France. In 1677 the diarist John Evelyn mentioned a sermon he had heard delivered in London by a Dr. Stradling. The sermon, delivered on Romans 8:22, was intent on "showing how even the creatures should enjoy a manumission and as much felicity as their nature is capable of, when at the last day they shall groan for their servitude to sinful men."[207] The animals, too, will be freed from slavery and find a place in heaven. The founder of the Society of Friends, the apparently (but perhaps not in reality) vegetarian George Fox, writing in about 1673, also found the basis of his apparent belief in animal immortality in Romans 8: "liberty is a natural right, and every natural creature would have his natural right, its liberty; and Christ gives liberty, and breaks the bonds asunder; and where the spirit of the Lord rules, there is liberty; but where it is quenched, there is the bondage, and not liberty, that bondage that causeth the whole creation to groan, which the creature waited to be delivered from into the glorious liberty of the sons of God, by Christ."[208] This would appear to be the first application of the doctrine of natural rights to animals, not long after the idea had been applied to humans in the mid-seventeenth century, and it is used to argue the case for their immortal souls.[209]

For some of the Protestants – for example, Elnathan Parr in *A Plaine Exposition upon the ... Epistle of Saint Paul to the Romanes* (1620) – the Pauline doctrine was to be interpreted as indicating that, as a sheer question of numbers, each species would be represented in heaven but not each individual animal. By contrast, the Protestant reformers John Bradford and William Bowling of Kent believed that each individual animal would be saved, the latter writing in 1646 that "Christ shed his blood for kine and horses ... as well as men."[210] To be sure, many continued to doubt these theological interpretations, but, certainly, around the time of the Commonwealth, they were in considerable vogue. In line with Bradford and Bowling, the revolutionary leveller Richard Overton, whom we have already met, declared in *Man's Mortalitie* (1643) that "all other creatures, as well as man, shall be raised from death at the resurrection." Moreover, he quotes with approval the sixteenth-century French army surgeon Ambroise Paré's view that "if we will diligently search into [animal] nature, we shall observe the impression of many virtues," together with "all [man's qualities] good and bad, and every faculty he has," in some of which "for his five senses [man] is by them excelled."[211] Clearly, humans were to be levelled to animalkind. To the extent that Overton is representative of the levellers, their extreme egalitarian ideas were not restricted to class alone. Those who concur with the famous quip of the Whig historian Thomas Babington Macaulay that the seventeenth-century "Puritan hated bear-baiting, not because it gave pain to the bear but because it gave pleasure to the spectators"[212] have, quite simply, failed to capture the spirit of the Commonwealth.

VII

Perhaps the most important work on animal souls came not from Britain but from a Frenchman, resident in Rotterdam, writing at the turn of the eighteenth century. It was composed by a skeptic – although a confessing one – of Huguenot background and faith. The author was Pierre Bayle, and the book was *Historical and Critical Dictionary,* published in 1697. The most significant article, from my perspective, was a lengthy one on Rorarius, Hieronymus – a sixteenth-century nuncio of Pope Clement VII at the court of the King of Hungary – who wrote two volumes, published in 1544 and 1547, in which he undertook to show "that brutes frequently make a better use of their reason than men."[213] In commenting on Rorarius, Bayle produced probably the most complete, erudite, and convincing discussion of the question of animal souls published to that date – and arguably the most profound even if we include the subsequent years. It is certainly beyond the scope of this book to reproduce all the arguments that he undertook to demonstrate the errors of the Aristotelians (whom he calls Peripatetics, as Aristotelians were known from Theophrastus on), of the Thomists (whom he calls Schoolmen), and of the Cartesians. The very premise of Bayle's article is of paramount interest as a reflection of the prevailing, if far from universal, intellectual context. He does not bother to discuss the question of immortality itself. It is quite sufficient in Bayle's view to demonstrate that animal and human reasoning are similar in kind because he takes for granted that the rational soul is the immortal soul. For Bayle, if the animal's mind can be shown to be rational, it follows ipso facto that its soul is immortal. A few snippets will suffice to show the timbre of Bayle's arguments.

> It is evident to anyone who knows how to judge things that every substance that has any sensation knows what it senses, and it would not be more absurd to maintain that the soul of man actually knows an object without knowing that it knows it than it is absurd to say that the soul of a dog sees a bird without seeing that it sees it. This shows that all the acts of the sensitive faculties are by their nature and essence reflexive on themselves ... the memory of beasts is an act that makes them remember the past and makes them aware that they are remembering. How then can anyone dare say that they do not have the power to reflect on their own thoughts or to draw inferences ... It is admitted [by those other than the Cartesians who deny immortal souls] that [the dog] senses bodies, that it discerns them, that it desires some of them and abhors others. This is enough. It is therefore a substance that thinks and thus is capable of thought in general. It can then receive all sorts of thoughts. It can then reason ...

[T]he Schoolmen are not able to prove that the soul of man and the soul of beasts are of a different nature. Let them say, and let them repeat thousands of times, "The Soul of man reasons and knows universals and virtue; and that of animals knows nothing about all this." We shall answer them: "These differences are only accidental and are no sign that there is a difference in the species of the subjects. Aristotle and Cicero at the age of one did not have more sublime thoughts than those of a dog; and even if they had remained in infancy for thirty or forty years, the only thought in their souls would have been sensations and childish passions for playing and eating. It is then by accident that they have surpassed the beasts; it is because their organs, on which their thoughts depend, acquired such and such modifications, which do not appear to the organ of beasts. The soul of a dog in the organs of Aristotle and Cicero would have lacked nothing for acquiring all the knowledge of these two great men."

Here is a very fine line of reasoning: such a soul does not reason and does not know universals; therefore it is very different from the soul of a great philosopher. For if this line of reasoning were just, then it would be necessary to say that the soul of small children is not of the same species as that of mature men.[214]

Bayle's views resemble fairly closely those of the German philosopher Wilhelm Gottfried Leibniz, whom Bayle admires and whose arguments he discusses, and critiques, at length – some half of the article is devoted to Leibniz's and Bayle's responses to each other. In 1695 Leibniz had published in the *Journal des Savants* an essay entitled "Système nouveau de la nature et de la communication des substances, aussi bien que de l'union qu'il y a entre l'âme et le corps" (New system of nature and the communication of substances, as well as of the union of soul and body). As the title indicates, Leibniz's article was concerned primarily with humans, but it was also of considerable relevance to the thorny issue of animal souls. Bayle provided a critique in the first edition of the *Dictionnaire,* to which Leibniz responded sympathetically in print in the *Histoire des Ouvrages des Savants* for July 1698, and, in kind, Bayle replied further in the second edition. This, in turn, provoked further argumentation and clarification from Leibniz, who was quite delighted to have a sympathetic critic who could help him elucidate his position.

Much of the form and language of Leibniz's argument appears arcane today, and we shall not discuss the argument's intricacies. After all, I am less concerned with its philosophical aspects and logical success than with understanding the role, and the degree of acceptance, of the belief in animals' immortal souls and its influence on the status of animals. In summary, then, Leibniz's argument is that "matter alone cannot constitute a true unity, and ... thus every animal is united to a form that is a simple, indivisible and truly unique being ... this form never leaves its subject; from which it follows that, properly speaking, there is neither death nor creation in nature."[215] Leibniz's hypothesis, Bayle tells us, "lead[s] us to believe": "(1) that God, at the beginning of the world, created the forms of all bodies and, hence, all the souls of beasts, (2) that these souls have existed ever since that time, inseparably united to the first organized bodies in which God placed them. This spares us having recourse to the theory of metempsychosis, which otherwise we would have had to turn to."[216] Leibniz believes, Bayle tells us, that a dog's soul, for example, "acts independently of its body; 'that it has everything from within itself by a perfect *spontaneity* with regard to itself ... That its internal perceptions arise from its own original constitution, that is to say, its representative nature (one capable of expressing the beings that are outside itself with respect to its organs), a condition which was given to it from the time of its creation and which constitutes its individual character'... all souls are simple and indivisible."[217] After further reflection, Leibniz concluded in the *Monadology* of 1714: "it may be said that not only the soul (mirror of an indestructible universe) is indestructible, but also the animal itself, though its mechanism may often perish in part and take off or put on an organic slough."

Generally speaking, at least in Britain, the philosophers left the arguments about the nature of the soul to the clergy and theologians. To be sure, the physician, mathematician, and philosopher David Hartley made a stab at it, full of sympathy for the animals but equally full of caution. In *Observations of Man* (1749), he remarked: "And if there be any Glimmering of the Hope of an Hereafter for them, if they should prove to be our Brethren and our Sisters in this higher Sense, in Immortality as well as Mortality, in the permanent Principle of our Minds, as well as the frail Dust of our Bodies, if they should be partakers of the same Redemption as well as our Fall, and be Members of the same mystical Body, this would have a particular Tendency to increase our Tenderness for them."[218] If he was attracted to the proposition, he was insufficiently attracted to exceed the methodological limits of his anatomically based philosophy. Nonetheless, he added that "the future existence of brutes cannot be disproved by any arguments, as far as yet appears."

Among the clergy and theologians, the story, at least in Britain, was rather different. As Keith Thomas has observed, despite there being some who were angered at the use of the Bible to prove animal immortality: "In the later seventeenth century many otherwise orthodox clergy regarded the issue of animal immortality as entirely open. Samuel Clarke told an acquaintance that he thought it possible that the souls of brutes would eventually be resurrected and lodged in Mars, Saturn or some other planet, while the physician Dr. Charles Leigh thought there was a 'spiritual immaterial being' in all living creatures. Ralph Josselin dreamed in 1655 that Christ was born in a stable because he was 'the redeemer of man *and beast* out of their bondage by the Fall.'"[219]

If many late-seventeenth-century clergy saw the matter of animal immortality as "open," by the eighteenth some were becoming much more confident. The most prominent Anglican theologian of the first half of the eighteenth century, Joseph Butler (1692-1752), successively bishop of Bristol, dean of Saint Paul's, and bishop of Durham, wrote his famous book *The Analogy of Religion, Natural and Revealed, to the Constitution and Course of Nature* (1736) to combat the increasing influence of deism in England by demonstrating the reasonableness of Christianity. Perhaps surprisingly, much of the opening part of the book dealt with the question of animal souls – of "the natural immortality of brutes," as Butler termed it. "[T]here can no probability be collected from the reason of the thing, that death will be their destruction ... there is nothing more certain than that *the reason of the thing* shows no connection between death and the destruction of living agents. Nor can we find any one thing throughout the whole *analogy of nature,* to afford us even the slightest presumption, that animals ever lose their living powers; much less, if it were possible, that they lose them by death."[220] It is certainly worthy of reflection that no other theological work had any greater influence for the next half-century and that Butler's espousal of the immortality of animal souls had no negative influence on his ecclesiastical career. He continued to be elevated throughout the remaining sixteen years of his life. Indeed, his bishopric appointments came after the *Analogy* was published, and the bishopric of Durham was one of the wealthiest and most powerful of the Anglican Church.

Two Anglican clergymen followed Butler in writing books to argue not only that animals possessed eternal souls, but also that we are obligated to treat them with the greatest consideration for their potential suffering. Thus in 1742 Reverend John Hildrop, Oxford D.D. and Anglican rector, wrote *Free Thoughts upon the Brute Creation,* in which he argued, predominantly on Biblical premises, that animals would have immortal life. "If death came into the world only after Adam's sin, then before it came, animals must have been immortal. If they be intended for God's glory (and so is every created thing intended) then He will not slay them. When man assumes that God will destroy, they conceive a wasteful deity; hence an imperfect one. This God of ours made all things to be happy." It is not animals but humankind who have demonstrated their unworthiness: "What is the rage of tigers, the fierceness of lions, the cruelty of wolves and bears, the treachery of cats and monkeys, when compared with the cruelty, the treachery, the barbarity of mankind?"[221] Reason and revelation, Hildrop remarks, "declare it to be a breach of natural justice, and indication of a cruel and unnatural temper, to abuse or oppress [animals], to increase the miseries, and aggravate the suffering of those innocent unhappy creatures, and to add, by our barbarity, to the weight of that bondage to which they are made subject by our disobedience, to put them to unnecessary labours, to punish them with immoderate severities, or withhold from them those necessary refreshments which their state and condition requires."[222]

In 1767 Richard Dean, a schoolteacher and Anglican clergyman in Middleton, just northwest of Manchester, wrote an *Essay on the Future Life of Brutes, Introduced with Observations upon Evil, Its Nature and Origins*, which he prefaced with the observation that the idea of animal immortality, which he set out to prove, was "not quite so novel as some folks perhaps have been inclined to imagine."[223] Yet, of course, if it was far more common than recognized, this meant, nonetheless, that the belief that animals lacked immortal souls was still somewhat in the ascendant, although rejected by an increasingly significant number of important and influential voices as well as by some of the less celebrated. One of the less celebrated to be unpersuaded by Dean was a certain James Rothwell, perhaps a neighbouring Anglican minister in Lancashire, who wrote a little book in response, *Letter to the Rev. Mr. Dean of Middleton*, in which he chastised Dean for allowing animals an afterlife without their having to follow the moral law, while humans were always put to the moral test to earn their way into heaven. Moreover, Dean had provided only three sources in support of his thesis – Saint Paul, Origen, and Tertullian – the first of whom he had misinterpreted and the other two of whom were heretics, one for proposing transmigration and the other for leaving the fold. Moreover, if animals were to live for eternity, where would there be an adequate supply of water in heaven for the fish, and if all are to be eternally happy, how is a louse supposed to amuse itself? Nonetheless, and this was now the eternal norm, even if they did not have an eternal life, they must always be treated with kindness and generosity.[224]

Yet, if Rothwell was unconvinced, perhaps it was in part because Dean's approach was a little novel. Whereas previous arguers for the animal immortal soul had sometimes tended to vilify humans to make their point, as Hildrop had done by and large, Dean was decidedly more concerned to elevate the status of animals, a view that was increasingly to become more common:

> If it be allowed that brute animals are more than mere machines, have an intelligent principle residing within them, which is the spring of their several actions and operations, men ought to use such methods in the management of them, as are suitable to a nature that may be taught, instructed and improved to his advantage, and have not recourse to force, compulsion and violence. Brutes have sensibility, they are capable of pain, feel every pang and cut or stab, as much as man himself, some of them perhaps more, and therefore they should not be treated as stocks or stones.[225]

Animals not only are creatures with "an intelligent principle," but are capable of improvement, a characteristic long proclaimed an exclusive attribute of humankind. Not only do they possess sentience in no less a degree than humankind, and perhaps more, but they are creatures of "sensibility," too – which implies that, if Dean is being careful with his words and means more than "sentience," he was according a lot more to animals than was normally claimed for them.

From the middle of the eighteenth century in Britain, arguments for the immortality of animal souls became common. In *Light of Nature Pursued* (1754), Abraham Tucker, writing under the pseudonym of Edward Search, queried: "Why should we think it an impeachment of His equity, if He assigns them wages for all they undergo in this important service ... of giving their labours, their sufferings, and their mortal lives? Or an impeachment of His power and His wisdom, if such wages accrue to them by certain stated laws of universal nature running through both worlds [i.e., that of the human and that of the animal]?"[226] The wages, of course, are the potential for eternal life.

The Scottish physician George Cheyne (1671-1743), friend of Alexander Pope and recipient of Samuel Johnson's admiration, who wrote widely on medical matters, preaching the merits of temperance and vegetarianism, declared in 1740 that "it seems utterly insensible that any creature ... should come into this state of being and suffering for no other purpose than we see him attain here ... There must be some infinitely beautiful, wise and good scene remaining for all sentient and intelligent beings, the discovery of which will ravish us and astonish us one day."[227] And Keith Thomas observed that in "the 1770s the Calvinist divine Augustus Toplady declared that beasts had souls in the true sense and he had never heard an argument against the immortality of animals which could not be equally urged against the immortality of man."[228] The poet Anna Seward, renowned as "the Swan of Lichfield" and whose father was canon residentiary of Lichfield cathedral, asked in her late-eighteenth-century verse: *"Has God deprived animals of immortal souls?"*[229] Her answer was a resounding "no." The

British parliamentarian Soame Jenyns, who, as we have seen,[230] was seriously tempted by the doctrine of the transmigration of souls, proclaimed the "certainty of a future state" for animals and asked: "Is not the justice of God much concerned to preserve the happiness of the meanest insect which he has called into being, as of the greatest Man that ever lives? are not all creatures we see made subservient to each others uses? and what is there in Man, that he alone should be exempted from the common fate of all created beings?"[231] He was adding a certitude to the musings of the anonymous author of *Adventurer* 37 (1753), who had declared that when he considered "the inequality with which happiness appears to be distributed among the brute creation, as different animals are in a different degree exposed to the capricious cruelty of mankind, in the fervour of my imagination I began to think it possible that they might participate in a future retribution; especially as mere matter and motion approach no nearer to sensibility than to thought, and he who will not venture to deny that animals have sensibility, should not hastily pronounce that they have only a material existence."[232]

In *Elements of the Philosophy of the Human Mind* (1828), the philosopher Dugald Stewart reported of Dr. Thomas Brown, professor of moral philosophy at the University of Edinburgh from 1810 to 1820, that he "believed that many of the lower animals have the sense of right and wrong; and that the metaphysical argument which proves the immortality of man, extends with equal force to the other orders of earthly creatures."[233] If legitimacy, and even orthodoxy, were once on the side of the nay-sayers, the tide would appear to have turned. More and more, those of sensibility could find no sufficient reason to exclude the animals from the afterlife.

Despite the increasing acknowledgment of immortal souls, a decided novelty in the argument for the negation deserves mention. It was offered from France by the chevalier de Suze in *Suite des erreurs et de la vérité* (1784). He allows an animal sensual superiority. Although animal pleasures are no more than sensual, they are, nonetheless, "more real and unvarnished, since they are neither corrupted, nor diminished, nor altered by reflection upon them; and whenever they are not enjoyed, it seems that animals have less need of them than does mankind, because they do not think about them."[234] Animals have the luxury of avoiding the futility of scholarship! Nor do they fear death, nor are they anxious about immortality, nor do they concern themselves with the success of their kin and posterity. If animals thereby enjoy so much happier a life, humans are entitled to be compensated for their comparative misery by immortality. Immortality is not a reward for a more moral life, but a compensation for a less pleasant one! Theriophily, indeed!

Perhaps the most significant contribution of the later eighteenth century, primarily because of the source and its rapidly growing influence, came from the vegetarian John Wesley, for a while a growing power within the Anglican Church as the founder of the evangelical movement of Methodism, which separated formally from the parent body in 1795, four years after the founder's death. In a sermon on Romans 8:29-30, "The General Deliverance" (1788), Wesley declared that "nothing can be more express":

> Away with vulgar prejudices, and let the plain word of God take place. They "shall be delivered from the bondage of corruption, into glorious liberty," – even a measure, according as they are capable – of "the liberty of the children of God" ...
>
> [T]he whole brute creation will, then, undoubtedly, be restored not only to the vigour, strength and swiftness which they had at their creation, but to a far higher degree of each than they ever enjoyed. They will be restored, not only to that measure of understanding which they had in paradise, but to a degree of it much higher than that, as the understanding of an elephant is beyond that of a worm. And whatever affections they had in the garden of God, will be restored with vast increase; being exalted and refined in a manner which we ourselves are not now able to comprehend. The liberty they then had will be completely restored, and they will be free in all their motions. They will then be delivered from all irregular appetites, from all unruly passions, from every disruption that is evil in itself, or has any tendency to evil. Thus in that day, all the vanity to which they are now helplessly subject will be abolished; they will suffer no more, either from within or without, the days of their groaning are ended.[235]

Wesley's vegetarianism, which he did very little to publicize, never became a significant aspect of Methodism. Indeed, perhaps more surprisingly, his proclamation of the immortality of animal souls did not become a generally acknowledged part of Methodist orthodoxy, even though Wesley's sermons are considered the doctrinal foundation of the creed. Nonetheless, Adam Clarke, premier Methodist theologian, publicist, friend of John Wesley, and Biblical commentator, added his voice to the cause:

> It does not appear that the animal creation are capable of a choice; and it is evident that they are not placed in their present misery, through either their choice or their sin; and if no purpose of God can be ultimately frustrated these creatures must be restored to that state of happiness for which they have been made, and of which they have been deprived through the transgression of man. Had not sin entered into the world, they would have had

much greater enjoyments without pain, excessive labour, and toil, and without death, and all those sufferings which have arisen from the fall. It is therefore obvious that the gracious purpose of God has not been fulfilled in them; and that, as they have not lost their happiness through their own fault, both the beneficence and justice of God are bound to make them a reparation. Hence it is reasonable to conclude that, as from the present constitution of things they cannot have the happiness designed for them in this state, they must have it in another.[236]

In light of such theological interpretations it should not surprise us that the Methodists were in the vanguard of the early-nineteenth-century movement to enshrine animal-welfare legislation in the constitution and culture of the country, but the Anglicans followed the Methodists closely, both in their animal advocacy and in the recognition of their souls. Thus Rev. Edward Bouverie Pusey, theologian, professor of Hebrew at Oxford University, and leader of the Oxford Movement after John Henry Newman's conversion to Catholicism, announced unequivocally: "All nature, having suffered together, shall be restored together. As to us, death is to be the gate of immortality and glory, so in some way to them creation includes all created beings, and all creation must include our nature too, in that one common groan and pang."[237] Rev. Thomas Thelluson Carter, the noted canon of Christ Church, Oxford, declared: "We may, moreover, connect with the resurrection of our Lord, the hope for restoration of the entire creation; for the whole world looks forward to that future state. As the whole world of creation around us suffers from the effect of the fall, so in some way they will know a resurrection and be transformed into a pure, more blessed, more beautiful, state. The lowest creatures are not to be destroyed, but, after their manner, according to their kind, will be restored, giving praise and glory to 'Him who created them.'"[238] Finally, to mention further only one of the more eminent of the Anglican divines, the anti-vivisectionist canon Basil Wilberforce, Archdeacon of Westminster, remarked that "these beautiful and useful forms of life, which are sometimes so cruelly tortured, are bound to pass over into another sphere, and that in the great eternal world men and animals should sink or swim together."[239]

The Lake Romantics, too, Anglicans all, were interested in the question of the animal soul. To take but two of numerous examples, the devout Tory Robert Southey wrote in his youthful poem "On the Death of a Favourite Old Spaniel" (1797): "But fare thee well! mine is no narrow creed, / And he who gave thee being did not frame / The mystery of life to be the sport / Of merciless man! there is another world / For all that live and move – a better one!"[240] The no less devout, but perhaps more conservative than Romantic Tory, Samuel Taylor Coleridge was somewhat less certain than Southey and, against the changing tenor of the times, more denigrating of humankind than elevating of animals. Discussing the Socinian creed, he commented:

> If it should be asked, why this resurrection, or re-creation is confined to the human animal, the answer must be – that more than this has not been revealed. But some have added, and in my opinion much to their credit, that they hope, it may be the case with the Brutes likewise, as they see no sufficient reason to the contrary. And truly upon *their* scheme, I agree with them ... Men are *on the whole* distinguished from other Beasts incomparably to their *disadvantage* by Lying, Treachery, Ingratitude, Massacre, Thirst of Blood, and by Sensualities which both in sort and in degree it would be libelling their Brother-beasts to call *bestial*, than to their advantage by a greater extent of Intellect. And what indeed, abstracted from the Free-will, could this intellect be but a more shewy instinct? of more various application indeed, but far less secure, useful, or adapted to its purpose, than the instincts of Birds, insects, and the like.[241]

If he is only half-convinced himself, although clearly wishing to believe, Coleridge acknowledges the complete confidence of a significant group of Christians less conformist than himself. Rather than raising animals to a common kind of reason with humankind, Coleridge reduces humans to the possession of a common kind of instinct with the animals. Moreover, human character is of a decidedly lower order than that of the animals. Coleridge's argument represents an exemplary case of theriophily, a topic to which we shall return in the next two chapters.

VIII

The second half of the nineteenth century was, on the whole, rather more concerned with providing practical remedies for animal abuse than with discussing the heavenly status of the beasts. But the acknowledgment of animals as fully ensouled became increasingly commonplace, at least in those instances where it was raised – even though in an increasingly secular world, it was less commonly at the forefront of people's minds. Still, the instances that arose were often highly instructive. The greatest French historian of the late-Romantic school, Jules Michelet, wrote in *La Bible de l'humanité* (1865): "Man does not deserve his salvation but by the salvation of all. The animal has its rights before God."[242] In *L'Oiseau* (1856) he observed: "Preconceived ideas apart, you cannot offend nature by restoring a soul to the beast. How much greater the Creator's worth if he created persons, souls, and will than if he constructed machines."[243] By "persons," Michelet makes it clear that he means animals, rather than only humans, who, he believes, have no right to reserve the application of the concept to themselves. Of course, the ascription of "personhood" to animals was unusual. By contrast, the removal of "personhood" from humans and their relegation to "animalhood" was rather less so. Thus, half a century later to be sure, Lenin issued a decree in January 1918 calling on the agencies of the newfound state – never destined to wither away – to "purge the Russian land of all kinds of harmful insects," which agencies immediately proceeded to do, with more power than authority, more vitriol than goodwill, more hatred than justice – indeed, with as much disdain as that with which the Malebranchian vivisectors treated animals. For Lenin, humans who lacked his vision, or came from the wrong segment of society, lacked "personhood." They were "insects." Years later, the novelist Alexander Solzhenitsyn listed a few of the groups whose "personhood" had been removed to be replaced by the category of "insect," under which title they would be condemned to death: "'former *zemstvo* members, people in the Cooper movements, homeowners, high school teachers, parish councils and choirs, priests, monks and nuns, Tolstoyan pacifists, officials of trade unions' soon all to be classified as 'former people.'"[244] Whereas Michelet and some of his fellow Romantics wanted to raise the status of animals to that of persons, Lenin wanted to reduce to the status of insects those humans who did not share the Bolshevik worldview. History is not uniformly progressive. Of course, for Lenin, neither insects nor humans possessed eternal souls nor souls of any kind worthy of consideration. For Michelet, animal and human souls were in all relevant respects identical.

In George Eliot's masterpiece, *Middlemarch* (1871-72), Dorothea Brooks criticizes the excessive interbreeding and pampering of Maltese pups: "I believe that all the petting that is given to them does not make them happy. They are too helpless: their lives are too frail. A weasel or a mouse that gets its own living is more interesting. I like to think that the animals about us have souls something like our own, and either carry on their own little affairs or can be companions to us ... These animals [i.e., the Maltese lap dogs] are parasitic!"[245] Eliot appears to believe that animals who live a natural, self-directed life are ensouled. Of those who are mere appendages, mere parasites, she is not so convinced. For Eliot, living a natural life, which involves procuring one's own livelihood, or performing a valued service, appears to be a prerequisite of the possession of a soul. Insofar as these requisites are fulfilled, the animal soul appears to correspond to the human soul. For Eliot, it would appear that it is valued labour that gives us a valuable life, and it is a valuable life that imparts the soul. It is unlikely that Eliot was specifically criticizing the view of Philo of Alexandria, but she certainly stands in diametric opposition. The first-century Jewish philosopher wished for a time when wild animals would be domesticated. He wanted them "gentle in emulation of the docility and affection for the[ir] master, like Maltese lapdogs who fawn with their tails, which they cheerfully wag."[246] Philo, Auguste Comte, and the London Zoological Society were in accord.[247] For Eliot, animals should be more self- than human-directed. It was this that entitled them to a soul. It was Arcadia, not Eden, for which she pined.

The profound mystic Edward Carpenter, an English cult figure in the late nineteenth century whose books went into a multitude of printings, wrote in *Towards Democracy* (1883): "Behold the animals. There is not one but the human soul lurks within it, fulfilling its destiny as surely as within you."[248] And the practical parliamentarian, the seventh earl of Shaftesbury, by general repute the greatest British social reformer of the century, told the House of Lords on the occasion of the 1876 debate over controlling vivisection that not only did animals have immortal souls, but they were more deserving of them than some humans. He added that "these ill-used and tortured animals are as much His creatures as we are; and to say the truth, I had, in some instances rather been the animal tortured than the man who tortured it. I should believe myself to have higher hopes, and a happier future."[249] For Lord Shaftesbury, the heavenly prospects for vivisected animals were decidedly rosier than for the vivisectors (or "vivi-dissectors," as William Drummond called them, striving for an elusive accuracy).[250]

Rev. John George Wood, author of a number of works of natural history, including the very popular *Illustrated Natural History,* wrote *Man and Beast, Here and Hereafter* in 1874. In the Appendix to *Animals' Rights Considered in Relation to Social Progress* (1892), the renowned animal advocate Henry S. Salt described Wood's book as "a plea for animal immortality, by a well-known naturalist. His plan is three-fold. First, to show that the Bible does not deny a future life to animals. Secondly, to prove by anecdotes, 'that the lower animals share with man the attributes of Reason, Language, Memory, a sense of moral responsibility, Unselfishness, and Love, all of which belong to the spirit and not to the body.' Thirdly, to conclude that, as man expects to retain these qualities after death, the presumption is in favour of the animals also retaining them."[251] In the book, Wood added a touch of theriophily: "I feel sure that animals will have the opportunity of developing their latent faculties in the next world, though their free scope has been denied to them in the short time of their existence in the present world. They surpass many human beings in love, selflessness, generosity, conscience, and self-sacrifice. I claim for them a higher status in creation than is generally attributed to them, and claim they have a future life in which they will be compensated for the suffering which so many of them have to undergo in this world."[252] It is notable that the attributes Wood ascribes to animals are very similar to those offered by several of the classical writers, by the theriophilists in general, and by Ambroise Paré in the sixteenth century, which were repeated by Richard Overton in the seventeenth century, discussed at length by George Nicholson and William Youatt in the earlier nineteenth century, and followed by Charles Darwin in the 1870s.

Frances Power Cobbe, Unitarian preacher, journalist, author of a book on Kantian philosophy, and a cofounder of the Victoria Street Society, later the National Anti-Vivisection Society, added a distinctive voice:

> I will venture to say plainly that, so far as it appears to me, there is no possible solution of this heart-wearing question save the bold assumption that the existence of animals does not end at death. It is absolutely necessary to postulate a future life for the tortured dog or cat or horse or monkey, if we would escape the unbearable conclusion that a sentient creature, unoffending, nay, incapable of giving offence, has been given by the Creator an existence which, on the whole, has been a curse. That conclusion would be blasphemy. Rejecting it with all the energy of our souls, we find ourselves logically driven to assume the future life of lower animals.[253]

One may feel that, in her insistence, she is grasping at straws. A burning desire for justice, not a theological interpretation or even a conviction that it will be so, appears to lead to her conclusion. But perhaps justice is a superior criterion.

The Scottish novelist, poet, and Christian fantasy writer George Macdonald (1824-1905) was described by W.H. Auden as "one of the most remarkable writers of the nineteenth century." C.S. Lewis remarked: "I have never concealed the fact that I regard him as my master. I fancy I have never written a book in which I did not quote from him." He was ordained a Congregationalist minister and became pastor of Arundel in 1850, a disastrous experience that resulted in his resigning his living after three years. He joined the Anglican Church as a lay member in 1860. He continued, however, to preach independently on occasion and to publish his Biblical commentaries. In 1892 he wrote *The Hope of the Universe*, which contained *Hope of the Gospel*, from which comes this extract:

For the earnest exploitation of the creature waiteth for the manifestation of the sons of God – Romans viii, 19.

Let us try, through these words, to get the idea in Saint Paul's mind for which they stand, and have so long stood. It can be no worthless idea they represent – no mere platitude, which a man, failing to understand it at once, may without loss leave behind him. The words mean something which Paul believes vitally associated with the life and death of his master ...

First then, what does Paul, the slave of Christ, intend by "the creature" or "the creation"? If he means the visible world, he did not surely, and without saying so, mean to exclude the noblest part of it – the sentient! If he did, it is doubly strange that he should immediately attribute not merely sense, but conscious sense, to that part, the insentient, namely, which remained. If you say he does so but by a figure of speech, I answer that a figure of speech that meant less than it said – and how much less would not this? – would be one altogether unworthy of the Lord's messenger.

First, I repeat, to exclude the sentient from the term common to both in the word creation or creature – and then to attribute the capabilities of the sentient to the insentient, as a mere figure to express the hopes of men with regard to the perfecting of the sentient for the comfort of men, were a violence as unfit in rhetoric as in its own nature. Take another part of the same utterance: "For we know that the whole creation groaneth and travailleth in pain together until now:" is it not manifest that to interpret such words as referring to the mere imperfections of the insensate material world, would be to make of the phrase a worthless hyperbole? I am inclined to believe that the apostle regarded the whole visible creation as, in far differing degrees of consciousness, a live outcome from the heart of the living one, who is all and in all: such view, at the same time, I do not care to insist upon; I only care to argue that the word creature or creation must include everything in creation that has sentient life.

Macdonald continues the argument in this vein for a further twelve pages before concluding, with implicit reference to Psalm 36: "To those whose hearts are sore for that creation, I say, The Lord is mindful of his own, and will save both man and beast."[254]

There is a continuous tradition from the Renaissance on, somewhat stronger in Britain than elsewhere, of a subscription to the immortal animal soul, maintained both by scholars of distinction and by those merely speaking from the heart. As the nineteenth century succeeded in softening manners in France, there, too, an occasional acknowledgment of the eternal animal soul could be heard. In addition, new clarion calls were now being heard loud and clear in the American voice. American clergy were impressed with the stance taken by the revered naturalist Louis Agassiz. Thus, for example, Rev. Joseph Cook, an influential Boston cleric who gave a series of popular lectures at Tremont Temple, which were published as *Monday Lectures on Scepticism, Biology, Transcendentalism* (1878), queried:

> Do not facts require us to hold that the immortal part in animals having higher than automatic endowments is external to the nervous mechanism in them as well as in man? What are we to say if we find that straightforwardness may lead us to the conclusions that Agassiz was not unjustifiable when he affirmed in the name of science, that instinct may be immortal, and when he expressed in his own name, the ardent hope that it might be so. Shall we, too, not hope that this highest conception of paradise may be the true one? Would it not be a diminution of supreme bliss not to have union with God through these, the most majestic of His works below ourselves?[255]

The apparently obscure reference to instinct possibly being immortal arises from the fact that, in many instances, as we have seen, the possession of reason was seen as the necessary, and sometimes the sufficient, condition of the potential for immortality. And even though animals may reason to a degree, it was not their ordering principle, as it is with humans. Thus animals were denied the possibility of an afterlife. Agassiz had argued to the contrary that even if animals were predominantly instinctive beings, they should not thereby be precluded from immortality. As a prelude to his discussion of instinct in *Contributions to the Natural History of the United States* (1857), Louis Agassiz had written:

> When animals fight with another, when they associate for a common purpose, when they warn one another in danger, when they come to the rescue of one another, when they display pain or joy, they manifest impulses of the same kind as are considered among the moral attributes of man. The range of their passions is even as extensive as that of the human mind, and I am at a loss to perceive a difference of kind between them, however much they may differ in degree and in the manner in which they are expressed. The gradations of the moral faculties among the higher animals and man are, moreover, so imperceptible, that to deny to the first a certain sense of responsibility and consciousness, would certainly be an exaggeration of the difference between animals and man. There exists, besides, as much individuality, within their respective capabilities, among animals as among men, as every sportsman, or every keeper of menageries, or every farmer and shepherd can testify who has a large experience with wild, or tamed, or domesticated animals.

This argues strongly in favor of the existence in every animal of an immaterial principle similar to that which, by its excellence and superior endowments, places man so much above the animals. Yet the principle exists unquestionably, and whether it be called soul, reason, or instinct, it presents in the whole range of organized beings a series of phenomena closely linked together; and upon it are based not only the higher manifestations of the mind, but the very permanence of the specific differences which character-ize every organism. Most of the arguments of philosophy in favor of the immortality of man apply equally to the permanency of this principle in other living beings. May I not add, that a future life, in which man should be deprived of that great source of enjoyment and intellectual and moral improvement which result from the contemplation of the harmonies of an organic world, would involve a lamentable loss, and may we not look to a spiritual concert of the combined worlds and all their inhabitants in pres-ence of their Creator as the highest concept of paradise?[256]

With these words, written by an adamant anti-evolutionist two years before Darwin's *Origin of Species* and fourteen years before Darwin's *The Descent of Man,* Agassiz accorded animals the very same attributes that Darwin accords them and that are regarded in so much of the animal-rights liter-ature as representing, when advanced by Darwin, a revolutionary change in understanding and, specifically, a break with the past via evolution. Nei-ther of these hypotheses can be maintained with any degree of rigour, as will be demonstrated in Chapter 8.

The New York Episcopalian minister renowned for his books on the Holy Land and on marriage law, Dr. John Fulton, remarked:

This is a redeemed world, with not one suffering creature that has been left out of Christ's all-embracing redemption. And more, I dare believe that Saint Paul was right when he looked upon this redeemed world, and seeing how its lower orders are groaning and suffering together with us, even until now, he was inspired to prophesy of a better time, when their redemption and ours shall be perfected together, and the glory that is to be revealed shall sur-pass all the present suffering. My own belief, which I do not pretend to under-stand, but in which I can nevertheless believe, the salvation of Christ is broad enough to include and does include the dumb creation.[257]

The American best known to animal advocates who took up the cause of the animals' everlasting soul was George Thorndike Angell, a founder of the Massachusetts Society for the Prevention of Cruelty to Animals (1867), the Bands of Mercy (1882), and the American Humane Education Society (1889), of which he remained a most influential president for two decades.

He asked plaintively: "Can there be any doubt that the Almighty, who has given them one life, has power, if He wishes, to give them another? In God's great universe, comprising as it does perhaps millions of worlds larger than our own, is there not room enough for all?"[258] Nor did only the pacific think so. According to Lytton Strachey, in his *Eminent Victorians,* General Gordon of Khartoum was adamant in his belief in the immortality of animal souls. And, of course, we have noted earlier the compilation of works on the topic by Ezra Abbot; the supportive proclamation by the dean of the study of the soul William Rounseville Alger;[259] and the words of Elijah Buckner, who devoted a whole book to the proof of animal immortality – a view to which Buckner adds: "The notion that animals were created only for the use of man is a weak and unwarrantable conceit. They were created long before man, and would have been better off, if mankind had never been made."[260]

IX

Émile Zola, who was decidedly anti-Catholic and far more socialist than Christian, represents an animistic viewpoint increasingly common in Western thought in the nineteenth century and later – a viewpoint subscribed to by both the secular and the devout. Responding to an unfriendly critic, Zola wrote: "You call *Germinal* 'a pessimistic epic of human animality.' So be it, I shall accept your definition, provided you let me enlarge upon this word 'animality.' You situate man in the brain. I situate him in all his organs. You isolate him from nature, I have him occupy the earth from which he comes and to which he returns. You enclose the soul in a human being, and I feel it to be there and everywhere else, in animals, in plants, in pebbles."[261] Thomas Hardy said he looked "upon all things in animate nature as pensive mutes."[262] E.M. Forster thought "as though inanimate nature had purposes and volitions of its own."[263] Edgar Allan Poe was fascinated by *androides,* inanimate entities, apparently imbued with spirits. Victor Hugo insisted that stones have souls, even though it was the "worst evil" that inhabited them. By contrast, William Wordsworth's *Excursion* expresses an animism that views the whole world as alive with intimations of the divine. His *Prelude* is a hymn to the universal spirit in which humans and animals are united: "I saw one life and felt that it was joy." "Indeed," wrote Coleridge in 1802, "we are all, *one Life.*"[264] If there are periods in Western thought in which great distinctions are made between humans and the rest of nature, there are also periods in which some of the great literary figures see all of nature as ensouled and as of one kin.

The prevalent thesis that Christianity and Western civilization have invariably denied animals immortal souls cannot be maintained. Especially, but not solely, in the English-speaking world, many Christians of different epochs, different denominations, and different dispositions have cheerfully proclaimed the ensouled beast – although many, too, have followed the authority of Augustine and Aquinas in allowing only a limited soul. And a few – rather more than a few in France – have even imitated Descartes in denying its existence other than as coterminous with the body. Nor has the affirmation of the immortal soul been restricted to a radical fringe. Rather, it has been espoused by some of the most prominent and influential national, cultural, and church leaders as well as by some eminent theologians and a few significant philosophers. Moreover, affirmation of animal souls in Asia and among Aboriginal peoples has not always served animal interests well. Nor has the denial of animal souls in the Christian West necessarily hindered ethical consideration. The issue of the relationship between the proclamation of the ensouled animal, the affirmation of distinctive cultural and religious attitudes, and the treatment of animals is a great deal more complex than is customarily allowed.

4

Return to Nature

THE GOLDEN AGE AND THE HAPPY BEAST

I

THE CONCEPTS OF "RETURN TO NATURE," "the Golden Age," and "the happy beast" share in common the underlying idea that there is something better about life in its original simplicity and that progress and civilization have served not to better the human condition but to destroy its very essence. In fine, our fundamental nature is seen to lie in our origins rather than in our development. An associated belief is that a thing, idea, or being can be understood fully only if it is understood through its origins. These are ideas that have played an important role in the evolution of Western thought with respect to the directions in which society has been moving and the relative status ascribed to different kinds of beings.

"A mouse is miracle enough to stagger sextillions of infidels," wrote Walt Whitman.[1] If mice were miracles in their natural selves, why should humans have thought it appropriate to deviate from the miracle of their own natural selves? Émile Zola observed that "Man partakes of the brute and the angel and it is precisely this mixture that constitutes what, by common accord, we call the human element, it is from the eternal struggle between body and soul that morality is born."[2] If humankind had not tried to reach the angels, it would have known its morality intuitively and would have avoided the painful struggle between body and soul, which is the lot of the human alone, for the body and soul of other animals are in accord. We have acquired reason, philosophy, education, and a self-conscious morality, which has led us away from our instinctive and surer knowledge. Samuel Butler's Erewhonians were wise beyond human history, for they "had no schools or systems of philosophy, but by a kind of dog knowledge did that which was right in their own eyes and in those of their neighbours."[3] It was when they began to abandon this way of life that the problems of civilization began, although it is notable that they remained "well armed with bows and arrows and pikes"[4] and that they continued to eat "supper of milk and goat's flesh"[5] or "fried flesh of something between mutton and venison."[6] To be sure, they had tried vegetarianism, but finding the arguments against harming vegetables just as persuasive as those against harming animals, they had abandoned a fleshless diet. Others thought of original human wisdom as that in which, as Ovid wrote, "content with foods produced without constraint, they gathered the fruit of the arbute tree and mountain berries and cornel berries and blackberries clinging to the prickly bramble thickets, and acorns which had fallen from the broad tree of Jupiter."[7]

This was the kind of thinking, always paradoxical, as we shall see, that persuaded many profound minds, from antiquity to modernity, to desire a return to the primitive satisfaction of our natural condition – even if there was little accord regarding the particulars of this primitive way of life. Nowhere is this clearer than in Hobbes's postulate of the state of nature as a "war ... of every man, against every man"[8] contrasted with Locke's far more peaceful idea that "men living together according to Reason, without a common Superior on Earth, with Authority to judge between them, is *properly the State of Nature*."[9] Neither Hobbes nor Locke was advocating a return to this original state, although Locke thought it had much to commend it if one could avoid breaches of the natural law, while Hobbes deemed it the very worst condition that humankind could endure. Locke thought of humans as, by nature, at least in part altruistic. Hobbes thought of humanity as, by irremediable nature, eternally self-interested. A civil society could serve to reduce the harm of human self-interestedness by the use of reason. Nothing would, however, serve to alter human motives. Enlightened self-interest was the best we could hope for. Naturally, it was only those who saw some idyllic way of life in our natural state who wished to return to it. But even those who found the natural state admirable disagreed about the conditions of this natural state.

The importance of the idea of the Golden Age in Western thought lies primarily in suggesting that it is when we lose as little as possible of our primordial natures that we are most fulfilled.[10] Its generality may be judged from Hélène Guerber's classical account of the legend of the Holy Grail, the sacred vessel variously described as the chalice of the Eucharist or as the dish of the Pascal Lamb but having pre-Christian origins suggestive of some primordial state of perfection:

> Of all the romances of chivalry the most mystical and spiritual is undoubtedly the legend of the Holy Grail. Rooted in the mythology of all primitive

races is the belief in a land of peace and happiness, a sort of earthly paradise, once possessed by man, but now lost, and only to be attained again by
the virtuous. The legend of the Holy Grail, which some authorities declare
was first known in Europe by the Moors and Christianised by the Spaniards
was soon introduced into France, where Robert de Borron and Chrestien de
Troyes wrote lengthy poems about it. Other writers took up the same theme,
among them Walter Map, Archdeacon of Oxford, who connected it with
the Arthurian legends. It soon became known in Germany, where in the hands
of Gottfried von Strassburg, and especially of Wolfram von Eschenbach, it
assumed its most perfect and popular form.[11]

This conception of a return to nature utopia is rooted, as Guerber indicated, "in the mythology of all primitive races" – or, at least, almost all.
Thus, for example, the Babylonian creation myth *Enuma Elish* differs in
that chaos and violence reign there rather than peace. Nonetheless, in *The
Myth of the Eternal Return,* the profound anthropologist of religion Mircea
Eliade has argued that, predominantly, primitive religions are inherently
concerned with returning to the past in order to overcome the inevitable
decay involved in the march of history, which removes us from the perfection of the creation of the gods.[12] Mythic beginnings provide security.
The reality of historical experience threatens with disharmony, uncertainty,
and despair.

    The idea of perfection in our origins, the state of our blessed immor-
tality, did not disappear with the diminution of the influence of myth after
the Renaissance. Associated with this idea is the notion that it is in the
acquisition of culture that we destroyed our original nature, which in turn
corrupted our virtue. Thus Immanuel Kant writes: "We live in an age of
discipline, culture and civilization, but it is still far from being a moral age.
Under the present conditions people can say that the happiness of the State
grows alongside the misery of the people. And there remains the question
of whether we might not be happier living in a primitive condition where
we would have none of our present culture. For how does one make people happy without making them moral and wise?"[13] On witnessing the rise
to power of the Nazis in the 1930s, Sigmund Freud felt compelled to argue
not only that civilization and culture were detrimental to human society,
but also that they threatened human existence itself, and while one might
wish to respond that Nazism is the negation, not the fulfillment, of civilization – indeed, the Nazis explicitly repudiated progress and "civilization"
much of the time – one must at the same time acknowledge, however
reluctantly, the accomplishments of Leni Riefenstahl, Martin Heidegger,
and even Albert Speer. Correspondingly, it is sometimes argued, since other

species have not acquired culture and have thereby remained true to their original natures, that there is some respect in which they are more fulfilled than are we. They remained pure, while humans became corrupted. Hence the idea of "the happy beast."

At the outset, however, it is important to emphasize that there is no agreement among the return-to-nature thinkers regarding the precise content of the Golden Age to which we should aspire to return. Indeed, there are two quite antithetical forms, which we might call the Edenic and the Arcadian. The "ideal type"[14] of Eden is rural, simple, peaceful, altruistic, symbiotic, innocent, loving (*agape*), cooperative, compassionate, meek, tender, egalitarian, and vegetarian – in short, the world of the angelic and the saintly. This is the Eden of Genesis before humans and animals became carnivorous and before the fruit of the tree of knowledge was eaten. The "ideal type" of Arcadia is rural, simple, industrious, adventurous, loving (*eros*), loyal, courageous, strong, honourable, respectful, hierarchical, hunting, and carnivorous – in short, the world of Pan, King Arthur, and "the noble savage."[15] We might say that the totem of the first is the unicorn,[16] and of the second the lion. And they are both in conflict with the cultured soul that delights in progress, books, learning, the arts, and the finesse of civilization as well (of course) with science, technology, and concomitant luxury.

To be sure, no single postulation of the Golden Age fits either of these "ideal types" completely; on the sixth day even the animals are made subservient to humankind. Nonetheless, as a prima facie indication of the former conception, we might offer the earliest written statement of the Golden Age theme: that of Hesiod in his eighth-century-BC *Works and Days,* where we are told (in a fusion of two earlier oral myths) of the Five Ages: the age of gold, the age of silver, the age of bronze, the age of heroes, and the age of iron, the last of which is our present degraded age of "toil and misery," of "constant distress." By contrast: "First of all the deathless gods having homes on Olympus made a golden race of mortal men. These lived in the time of Cronus when he was king in heaven. Like gods they lived with hearts free from sorrow and remote from toil and grief; nor was miserable age their lot, but always unwearied in feet and hands they made merry in feasting, beyond the reach of all evils. And when they died, it was as though they were given over to sleep."[17] It is notable that whereas, some four centuries later, Plato conceived of being godlike as employing the greatest strivings of the intellect, for Hesiod being godlike implied essentially a life of ease in which "All good things were theirs, and the grain-giving soil bore its fruits of its own accord in unstinted plenty, while they at their leisure harvested their fields in contentment and abundance." As M.L. West has written, "Every reference to a 'golden age' in Western literature and speech derives directly or indirectly from Hesiod."[18] In elaborating the idea, the Pythagorean poet Empedocles tells us that in that age "the altar did not reek of the unmixed blood of bulls, but this was the greatest abomination among men, to snatch out the life and eat the goodly limbs." Moreover, "All were gentle and obedient to men, both animals and birds, and they glowed with kindly affection towards one another."[19]

Such early Christian groups as the Montanists and Marcionites, later deemed heretical, feasted on milk and honey in their Eucharist in order to re-create the Edenic vision of Exodus 3:8, in which God came "to bring them ... to a country rich and broad, to a country flowing with milk and honey." This was a return to their vegetarian, although not vegan, past of a life without harm and without conflict. Milk was symbolically opposed to blood, and honey was seen as a foretaste of heaven. They represented a vision of the peaceable kingdom.[20]

A more recent conception of Eden is offered by the vegetarian Leo Tolstoy in a letter to his wife, Sonya, written on a visit to Ivan Turgenev's home, Spasskoye, in which he exults in the splendour of the forest that beckoned him to eternal rest: "Cool grass underfoot, stars in the sky, the perfume of flowering laburnum and limp birch leaves, trills of the nightingales, buzzing of beetles, the cry of the cuckoo, the solitude, the easy movement of the horse beneath me and a sensation of physical and moral health. As always, I thought of death. It seemed clear to me that everything would be just as good – although in another way – on the other side, and I understood why the Jews imagine paradise in the form of a garden."[21] Future heaven would resemble the Edenic past. And should one have mentioned to Tolstoy that usage of animals for human ends, the riding of horses, is scarcely an Edenic ideal, he would have replied that his relationship to his horse was symbiotic – a relationship of mutual benefit.

More recent still is the statement of Elijah Buckner in *The Immortality of Animals* (1903), where, in the original state,

the earth, teeming with every variety of useful productions, was the great storehouse of the Almighty, from which all living things were commanded to help themselves. They were all vegetarians, for they were commanded by God to live on nothing else. There was no necessity to destroy one life to support another ... In this primeval innocence, there was surpassing beauty in every animate and inanimate object, and every living thing in the heavens above and all that moved in the waters or upon the earth below, were at peace.

After the path of history has run its course: "In the final state of the world the ferocious and the carnivorous animals will change their destructive appetites and passions. They will eat vegetable food, and become gentle, and exhibit kindly dispositions, and all will be restored to primeval peace and happiness."[22] It is certainly a utopian picture, one that Buckner bases on Biblical authority while citing the interpretations of several prominent theologians in his cause.[23] Buckner's view reflects fairly closely that of John Wesley, as expressed in his sermon on Romans 8.[24] Wesley writes of both humans and animals living in a state of happiness in Eden. Both were immortal. Both were vegetarian. The only difference in principle between them was that humans possessed a religious sense. Both suffered through the fall from grace. But at the day of judgment, humans and animals will be restored not only to their original happiness, but to a happiness even superior to that of their original Edenic condition.

Sometimes there are even conflicts within the expression of the Edenic idea itself, as, for example, in the dispute between Linton and Catherine in Emily Brontë's *Wuthering Heights* (1847):

"One time we were near quarrelling. He said the pleasantest manner of spending a hot July day was lying from morning till evening on a bank of heath in the middle of the moors, with bees humming dreamily about among the bloom, and the larks singing up overhead, and the blue sky and bright sun shining steadily and cloudlessly. That was his most perfect idea of heaven's happiness; mine was rocking in the rustling green tree, with a west wind blowing, and bright white clouds flitting rapidly above; and not only larks, but throstles, and blackbirds, and linnets, and cuckoos pouring out music on every side, and the moors seen at a distance, broken into cool dusky dells; but close by great swells of long grass undulating in waves to the breeze; and woods and sounding water, and the whole world awake and wild with joy. He wanted all to lie in an ecstasy of peace; I wanted all to sparkle, and dance in a glorious jubilee.

"I said his heaven would be only half alive, and he said mine would be drunk; I said I should fall asleep in his, and he said he could not breathe in mine."[25]

In part, the disagreement is aesthetic, but perhaps also a touch of Arcadia, entirely absent from Linton's thoughts, is creeping into Catherine's Eden.

Likewise ambivalent is the image offered by Sir Ralph Brown in George Sand's *Indiana* (1831), where we read that "it is in the heart of beautiful unspoilt nature that we discover an awareness of His power, pure of all human profanation. So let us go back to the desert in order to be able to pray."[26] The rejection of civilization in favour of human origins is clear. But "beautiful unspoilt nature" becomes confused with the desert in which one's prayers alone may be purer. Perhaps the idea is that "unspoilt nature" – identified in the novel with Ascension and Madagascar – is too fulfilling to encourage one to pray. One prays only when one has unrequited needs or out of habit.

Yet this ambivalence is not some reflection of modernity's inability to recognize its own nature. It is already present in the Mesopotamian *Epic of Gilgamesh,* variously dated at from 3000 to 1700 BC. Here Enkidu is a primitive human who lives in accord with the animals, sharing in common with them a vegetarian diet. The temple girl, Shamat, and the symbol of civilization, Gilgamesh, escort Enkidu on an Arcadian adventure to prove him capable of human valour and human lust. When he returns to the animals temporarily, they no longer acknowledge him, and he no longer possesses their speed and strength. There is a mixture of Arcadia and Eden in both the human and the animal life.

A modern example might suit to exemplify the more fully Arcadian attitude. Jeffrey Myers examines the influence of the renowned English poster artist Mark Greiffenhagen's popular Victorian painting *An Idyll* (1891) on D.H. Lawrence:

> In *An Idyll,* which plays an important symbolic role in [Lawrence's] novel [*The White Peacock* (1911)], a swarthy Pan figure, of great vigor and vitality, bare-chested and clothed in animal skins, seems rooted in a meadow, where grazing sheep and olive trees are lit up by a setting sun. He is lifting a pale, Pre-Raphaelite young woman off her feet, which are covered with bright poppies and daisies. He presses her half-naked bosom against his own body, entwines his fingers in the thick auburn hair cascading down her blue garment to her buttocks, and kisses her cheek as she turns away and swoons limply in his muscular arms, fearful of her feelings.[27]

The passion of sex must always change an Eden into an Arcadia. However, Arcadia at its starkest is to be found in Juvenal's *Satire VI* (i), "The Golden Age," as rendered by John Dryden in the seventeenth century:

> In Saturn's reign, at Nature's early birth,
> There was that thing called chastity on earth;
> When in a narrow cave, their common shade,
> The sheep, the shepherds, and their gods were laid:
> When reeds and leaves, and hides of beasts were spread
> By mountain huswives for their homely bed,

And mossy pillows raised, for their rude husband's head.
Unlike the niceness of our modern dames
(Affected nymphs with new affected names),
The Cynthias and Lesbias of our years,
Who for a sparrow's death dissolve in tears,
Those first unpolished matrons, big and bold,
Gave suck to infants of gigantic mould;
Rough as their savage lords who ranged the wood,
And fat with acorns belched their windy food.[28]

While the primitive still thrives on acorns, and "the bounteous year / Her common fruits in open plains exposed, / Ere thieves were feared, or gardens were enclosed," nonetheless sheep are kept for their skins and for food. Most striking, however, is that Juvenal reproaches the citizens of the civilized world for their excessive sentimentality, "who for a sparrow's death dissolve in tears." It would appear that the rugged laws of nature are respected in this Arcadia more than the individual lives of animals – as, indeed, the rugged laws of nature must always be at the expense of the peaceful and the serene. The conflict comes alive again in the modern contrast between animal-rights theory, in which each individual animal is an end in him- or herself, and holistic ecology, which espouses ideas of environmental sustainability that, in principle, emphasize the health of the whole as the end in itself – although in inconsistent reality we usually find humans, in comparison with the rest of nature, treated as ends in themselves.

The Bible scholar and Genesis translator Stephen Mitchell took this Arcadian orientation to its extreme in remarking that: "I wouldn't want my paradise to have the lion and the lamb lying down together. I want my lion to be a lion, not a vegetarian – to love the taste of blood, to be able to use its teeth and claws. When we talk about paradise in the way prophets do – swords into plowshares, lions chummy with lambs – we're talking about a fantasy of safety and tameness; a zoo, not the factual world that God created."[29] In reality, according to Genesis 1:29, not only were humans originally created as vegetarians, but so too were all other animals: "to all the wild animals, all the birds of heaven and all the living creatures that creep along the ground, I give all the foliage of the plants as their food." In the Judeo-Christian tradition, whether one is an Edenist or an Arcadian is influenced predominantly by whether one's archetype occurs before or after the Flood, when flesh eating, once forbidden, is now permitted. For Mitchell, it is clearly postdiluvian.

While Eden is an object of beauty, serenity, and reverence, Arcadia relates more to the awesome, the sublime, and the majestic. Thus, when George

Eliot thinks through the animals in *The Mill on the Floss* (1860), she tells us: "If those robber-barons [of the medieval Rhine castles] were somewhat grim and drunken ogres, they had a certain grandeur of the wild beast in them; they were forest boars with tusks, tearing and rending, forever in collision with beauty, virtue, and the gentle uses of life; they made a fine contrast in the picture with the wandering minstrel, the soft-lipped princess, the pious recluse, and the timid Israelite ... these Rhine castles thrill me with a sense of poetry."[30] The "wild beast" evokes a primal respect bordering on fear but is viewed in awe. The tiger and the lamb are different animals with contrasting natures. The tiger and the lion represent Arcadia, the unicorn and the lamb represent Eden.

Sometimes the Arcadian concept has managed to confuse as much as to enlighten. Thus, under the caption of "Rural Harmony," Peter Bowler refers to Gilbert White's "Arcadian image of country people in harmony with their environment" in White's groundbreaking *The Natural History of Selborne* (1789) – the first detailed study of the fauna and flora of a parish (that of Selborne in Hampshire, of which White was the curate) – which "established his reputation as founder of our modern respect for undisturbed nature."[31] Yet White was not averse to planting four lime trees between his home and the butcher's yard opposite "'to hide the sight of blood and filth.' His action symbolized a growing effort not to abolish slaughter-houses, but to hide them from the public gaze."[32] And one is uncomfortable reading his description of a tortoise as "the most abject reptile and torpid of beings."[33] Moreover, while one might expect an Arcadian to consider subsistence hunting a necessary and admirable activity, sport hunting ought to be conceived differently. Yet not only was White himself an avid "sportsman," but he claimed that it was "impossible ... to extinguish the spirit of sporting, which seems to be inherent in human nature."[34] On the other hand, White shows not only great curiosity toward all kinds of animals, but also, despite the occasional aberration, a lack of repugnance for the usually despised animals. He is even "much taken with the [tortoise's] sagacity"[35] and describes "the wonderful spirit of sociality in the brute creation"[36] with considerable sympathy and respect. What this should suggest to us, once again, is that paradox is the nature of humankind. And the contrasting images of Eden and Arcadia are reflective of the centrality of this paradox to the human condition.

II

The attitudes of Empedocles looking back to Peace and of Lawrence looking back to Pan are antithetical, the reason being that we have competing

notions of what constitutes the essential human paradigm. There is a notion
of a gentle primitivism contradicted by a notion of a robust primitivism,
the feminine yin contradicted by the masculine yang. Are we essentially
rabbit or wolf, prey or predator, Cyparissus[37] or Pan? In reality, we are each
of the antonyms in frequent tension.

An example of the tensions existing side by side and unrequited is to be
found in Virgil's *Eclogues* (42-29 BC)[38] – for instance, in the idea of Eden
in *Eclogue V:* "The wolf intends no ambush to the flock, the nets / No
trickery to deer: Daphnis the good loves peace."[39] Similarly, in *Eclogue VI:*
"Then truly you could see Fauns and wild animals / Playing in rhythm."[40]
By contrast, in *Eclogue II* we find the Arcadian view:

> Pallas can keep her cities,
> But let the woods beyond all else please you and me.
> Grim lions pursue the wolf, wolves in their turn the goat,
> Mischievous goats pursue the flowering lucerne,
> And Corydon you, Alexis – each at pleasure's pull.[41]

And the peroration of *Eclogue X* tells us:

> The choice is made – to suffer in the woods among
> The wild beasts' dens, and carve my love into the bark
> Of tender trees: as they grow, so my love will grow.
> But meanwhile with the Nymphs I'll range on Maenala
> Or hunt the savage boar.[42]

Arcadia triumphs!

Nowhere are the tensions clearer than in the self-consciously contradictory and paradoxical – and *hence* illuminating – writings of Henry David
Thoreau. Nathaniel Hawthorne described Thoreau as "a young man with
much of wild original nature still remaining in him ... and nature, in return
for his love, seems to adopt him as her special child, and shows him secrets
which few others are allowed to witness."[43] But what "nature" is it that lies
within him, and what does "nature" disclose? Moreover, does he embrace
culture at the same time as embracing paradoxical concepts of original
nature? Can he espouse progress and the return-to-nature theme at once?
He certainly appears to when, in one extended breath, he embraces alike
the delights of the woods and the passing steam train: "the tantivy of wild
pigeons ... a fishhawk ... a mink ... reed-birds ... the rattle of rail road cars
... the whistle of the locomotive."[44]

In *Walden,* Thoreau tells us that "no nation that lived simply in all respects, that is, no nation of philosophers, would commit so great a blunder as to use the labor of animals." Yet it is not for the sake of the animals that he arrives at this position but out of fear that one "should become a horse-man or herds-man merely[45] ... husbandry is degraded with us, and the farmer leads the meanest of lives. He knows Nature but as a robber."[46] Oddly, then, this nation of simplicity is not only a "nation of philosophers," but is predicated, in common with Jean-Jacques Rousseau, Karl Marx,[47] and Herbert Marcuse,[48] on a concern with eliminating human one-dimensionality. Despite his love of the awe-inspiring steam locomotive and the world of functional specialization it signifies, he espouses a Rousseauian primitivism when he avers: "Simplicity, simplicity, simplicity! I say, let your affairs be as two or three, and not a hundred or a thousand, and keep your accounts on your thumb nail."[49] On the one hand, he comments that "whether we should live like baboons or like men is a little uncertain,"[50] and on the other hand, he observes: "What recommends commerce to me is its enterprise and bravery ... Commerce is unexpectedly confident and serene, alert, adventurous, and unwearied."[51] On the one hand, he is a lover of books and learning, of art and culture, and on the other, he tells us of the much admired Alek Therien that "his thinking was so primitive and immersed in his animal life, that, though more promising than a merely learned man's, it rarely ripened to anything which can be reported."[52]

There can be no doubt that Thoreau is well aware of the ambivalence – indeed, contradictions – in his thought and character, and, rightly, he knows it a part of the human condition. As an Arcadian, he writes:

> As I came home through the woods with my string of fish, trailing my pole, it being now quite dark, I caught a glimpse of a woodchuck stealing across my path, and I felt a strange chill of savage delight, and was strongly tempted to seize and devour him raw; not that I was hungry then, except for that wilderness he represented. Once or twice, however, while I lived at the pond, I found myself ranging the woods, like a half-starved hound, with a strange abandonment, seeking some kind of venison which I might devour and no morsel could have been too savage for me. The wildest scenes had become unaccountably familiar.

Yet Thoreau contrasts this immediately with an insight into a competing character from within: "I found in myself, and still find, an instinct toward a higher, or, as it is named, spiritual life, as do most men, and another towards a primitive rank and savage one. And I reverence them both. I love the wild not less than the good. The wilderness and adventure that are in fishing still recommend it to me. I like some days to take rank

hold on life and spend my day more as the animals do."[53] At issue is whether
to live as a proper animal and thus in line with nature – that is, to be "the
happy beast" – or to live as a super-animal and thus in line with culture.
Clearly, the animals that he has in mind are the predators, the Arcadian
cheetah rather than the Edenic gaur. Thoreau continues at some length
speaking in different voices and different characters, all a part of the contradictory whole that constitutes both the Thoreauian and the human
condition. Finally, Eden emerges:

> It may be vain to ask why the imagination will not be reconciled to flesh
> and fat. I am satisfied that it is not. Is it not a reproach that man is a car-
> nivorous animal? True, he can and does live, in great measure, by preying
> on other animals; but this is a miserable way, – as anyone who will go to
> snaring rabbits, or slaughtering lambs, may learn, – and he will be regarded
> as a benefactor of his race who shall confine himself to a more innocent and
> wholesome diet. Whatever my own practice may be, I have no doubt it is a
> part of the destiny of the human race, in its gradual improvement, to leave
> off eating animals.[54]

Thoreau never loses sight of the tensions in the human condition. There
is, for Thoreau, something admirable in the human as savage, as peace-
maker, and as thinker, but they constitute centrifugal forces within the
human breast, tearing the individual apart. "We are conscious of an animal
in us, which awakens in proportion as our highest nature slumbers. It is
reptile and sensual, and perhaps cannot be wholly expelled; like the worms
which, even in life and health, occupy our bodies. Possibly we may with-
draw from it, but never change its nature. I fear that it may enjoy a certain health of its own; that we may be well, yet not pure."[55]
     We may be well in our animal nature but not in our intellectual nature
at the same time. Already, on visiting the Jardin des Plantes in Paris[56]
in 1833 – over a quarter-century before Darwin published *The Origin of
Species* – Ralph Waldo Emerson had recognized that animals live within
us as a part of our evolutionary heritage: "I feel the centipede in me – cay-
man, carp, eagle and fox. I am moved by strange sympathies."[57] What
Thoreau is adding – which is perhaps implied but not explicit in Emerson
– is that it is not merely animals' biology that constitutes the development
of our being, not merely that we have sympathies with the animals as a
consequence of our kinship, but that their psychology lies within our nature,
too. So, too, writes Victor Hugo in *Les Misérables* (1862), reminiscent of
Plato in the *Republic*:[58] "It is our belief that if the soul were visible to the
eye every member of the human species would be seen to correspond to

some species of the animal world and a truth scarcely perceived by thinkers would be readily confirmed, namely, that from the oyster to the eagle, from the swine to the tiger, all animals are to be found in man and each of them exists in some man, sometimes several at a time."[59] However much we strive to become creatures of culture, intellect, self-conscious morality, and civilization, we can never escape the animality that must of necessity be an integral part of our souls, not merely to compete with our intellect, our morality, and our civilization, but to constitute an irremediable part of them. Testosterone and estrogen infuse the totality of our lives. The human condition cannot but be a paradox. Indeed, Cain, the first murderer, was also the builder of the first city.[60] Remus was victorious in his choice for the site of Rome. Romulus slew him. Civilization is founded on conflict and violence. The view of Thoreau's fellow Transcendentalist Margaret Fuller – whose early death by drowning denied the world of thought a great prospect – is far more appealing. She refers to "that spontaneous love for every living thing, for man and beast and tree, which restores the golden age."[61] Unfortunately, Thoreau's vision is far more convincing – as, indeed, is that of Aldous Huxley in *Island* (1962). Writing of the South Sea islands, he proclaims them "an Eden innocent unfortunately not only of Calvinism and capitalism and industrial slums, but also of Shakespeare and Mozart, also of scientific knowledge and logical thinking. It was paradise, but it wouldn't do, it wouldn't do. They sailed on."[62] It is, above all, John Steinbeck who has correctly summarized the human condition: "Man might be described fairly adequately, if simply, as a two-legged paradox. He has never become accustomed to the tragic miracle of consciousness. Perhaps, as has been suggested, his species is not set, has not jelled, but is still in a state of becoming, bound by his physical memories to a past of struggle and survival, limited in his futures by the uneasiness of thought and consciousness."[63]

In *The Romance of the Forest* (1791), Ann Radcliffe quipped that "there was neither physician nor apothecary in the village, so that nature was deprived of none of her advantages."[64] The paradox is clear. Humankind has developed a "science" to cure illnesses yet in the process has forgotten the remedies once known intuitively, whereas other animals, who, not having acquired the skill of the science of "medicine," still retain their memory and their natural skill. The panacea for almost all ills was "cupping" (i.e., usually meaning bleeding the patient into a cupping bowl to adjust the humours), a remedy first introduced in pharaonic Egypt and Hippocratic Greece that remained equally inefficacious, and even more widely practised, at the close of the eighteenth century. The poor were in one respect fortunate. They could not afford physicians and thus did not die at their hands.

Indeed, it is estimated that until nigh on the close of the nineteenth century, physicians were more likely to kill than to cure their patients. In this instance at least, those who inhabited the state of nature were possessed of original knowledge, or willing to let "nature" take its course, and lived a less precarious life than those who possessed the benefits of civilization. But, as medical knowledge increased in the later nineteenth century, particularly through an understanding of the role of microbes in disease, the situation was reversed. Medicinal civilization appeared to be finally fruitful, but of course, since most of the successful medicines by the end of the first quarter of the twentieth century were derived from invasive experiments on animals, the conflicts between humans and other species were becoming ever greater. The idea of animals intended for human use was at its zenith.

In *Melmoth the Wanderer* (1824), the late-Romantic Irish novelist Charles Robert Maturin expressed sympathy for the snails that devoured his hero's peaches:

> I deem it a sort of cruelty to have them destroyed ... To confess the truth, I have some scruples with the respect to the liberty we assume in the *unlimited* destruction of these lower orders of existence. I know not upon what principle of reason and justice it is, that mankind have founded the right over the lives of every creature that is placed in a subordinate rank of being to themselves ... I cannot, indeed discover why it should be thought less inhuman to crush to death a harmless insect, whose single offence is that he eat the food which nature has prepared for its substance, than it would be were I to kill any more bulky creature for the same reason.[65]

Maturin had qualms about treating even the smallest of animals as creatures that were legitimate means to solely human ends, however insignificant they might appear in the scale of things. They should not be the objects of ill treatment, he opined, merely because such treatment was conducive to human interests. A century later, as a consequence of the successes of invasive experimentation – on all types of creatures, not merely the physically insignificant – Maturin's argument was less persuasive than it had been in his own time. The paradox of human nature continued. For centuries we would perhaps have been healthier by ignoring medical advice and by both recalling the instinctive medical instructions of our own animal nature and learning from that of our fellow creatures. By the twentieth century, medical knowledge was acquired not by learning from animals but by torturing and killing them. There is always a price for culture and civilization – which so often appear prima facie both uncultured and uncivilized.

III

Karl Marx believed that humans are fulfilled by abandoning their animal nature. With considerable wisdom, Marx acknowledged that the distinctively human characteristic lies in the capacity for creative labour (although one might have thought that the beaver, the ant, and the honeybee were not entirely unworthy of similar mention), through which alone, he added, less wisely, humans can experience the fullness of their potential and achieve satisfaction. Marx's complaint about both the present and all hitherto existing society was that it had never allowed humanity to produce "in a human manner"; thus humankind has never realized its true humanness.[66] Indeed, Marx's recognition that we feel more completely fulfilled through productivity rather than leisure is both insightful and sound, if not completely original. Marx had a precursor in the Desert Father Serapion of Arsinoë[67] and in Benedict of Nursia (d. c. 547), the founder of the Benedictine Rule,[68] who established the principles of the dignity of labour, an idea absent from both classical antiquity and pagan Europe, and of manual work as a path to fulfillment and salvation, a view that received considerable sympathy following the Reformation.

In Marx's view, however, history has forbidden the practice of dignified labour. "The result we arrive at then," Marx argues, "is that man only feels himself freely active in his animal functions of eating, drinking, and procreating, at most in his dwelling and dress, and feels himself an animal in his human functions."[69] Marx expresses his insight in recognizing the exclusive source of satisfaction: creative labour. Yet he demonstrates a mere utopianism in thinking both that humans can escape their animality and that the satisfactions of their animality are not a vital and necessary function of being fully human. If one is to achieve satisfaction in one's exclusively (or preponderantly) human functions, one must be satisfied as the human animal, too. For Marx, it is not merely that the human condition is to be improved, but that the animal condition of human beings is to be overcome. Rightly, he postulates a contradiction between essence and existence, humanity and nature, freedom and necessity, individual and species, and objectification and self-affirmation. Yet, come the revolution and the emancipation of the proletariat, all conflict will end (since class contradiction is the sole basis of conflict) and all contradictions will be transcended. And since class conflict has been the sole driving force of history, history itself will end. Humans will have ceased to be animals in any meaningful manner.[70]

All in all, Thoreau and Steinbeck have a more profound understanding of the human condition, of what it is to be human, of the ineradicable animality within us, than does Marx. Despite Marx's attempt to avoid reductionism, he presents an essentially one-dimensional and utopian understanding of humankind. The essential difference between Marx, Thoreau, and Steinbeck is that the first regards the contemporary human condition as a momentary standard waiting to be transcended, whereas Thoreau understands its very essence to lie in its tensions between the peaceful, the aggressive, and the cultural. The tensions cannot be transcended, but they may perhaps be synthesized – or, better said, balanced – more satisfactorily. Steinbeck perceives our historical "becoming" not as a mere social product of the history of class conflict, as Marx and Engels asserted in their most famous phrase: "The history of all hitherto existing society is the history of the class struggle."[71] For Steinbeck, the struggle is a more permanent inner conflict within each individual between the biological self and the cultural self. Alternatively, as Zola wrote, it is the conflict between "the brute and the angel" out of which emerges "the eternal struggle between body and soul" that is the source of our "morality"[72] – or, as Mark Twain wisely recognized, the ensuing morality is "the quality *which enables [man] to do wrong* ... It seems a tacit confession that heaven is for the Higher Animals alone."[73] Human immorality precludes human immortality. The animal lack of the potential for a self-conscious morality ensures immortality. For Thoreau, Steinbeck, Zola, and Twain, there is no solution to the human predicament, no transcendence, but an enlightened awareness of the tensions to be borne and occasionally eased. Tellingly, and paradoxically, Marx, the avowed proponent of industrial and technological progress that redounds to the advantage of the proletariat, describes his well-rounded person, the consequence of the elimination of one-dimensionality from human experience, entirely in preindustrial terms. He is one who "hunts in the morning, fishes in the afternoon, rears cattle in the evening, and philosophizes after dinner, as he has a mind."[74] Never does he manipulate a steam engine, tend a sewage system, or construct a macadamized road. And never does a "she" receive any consideration other than sympathy for her misuse – but without a consideration for her equal capacity to produce creatively. It is one thing to hunt, and fish, and tend cattle, and think. It is another thing to be a mechanical engineer at dawn, an environmental engineer at high noon, and a civil engineer at dusk. And it as an even greater improbability that one would be a cardiologist in the morning, a high-court judge in the afternoon, and a physicist in the evening, as one could not likely possess the requisite knowledge and skill to do more than one of them with even the barest modicum of success. One wonders whether the part-time philosophizing would ever include the contemplation of the interests of the hunted, the fished, or the herded and, if so, what effect this would have on the life of the well-rounded person.

Perhaps the effect would be little, for Marx recognizes "species being" in the human alone. In the *Grundrisse,* Marx wrote of the "great civilising influence of capital," which would eliminate the "deification of nature." His claim was that with the development of industrial civilization, "nature becomes for the first time simply an object for mankind, purely a matter of utility."[75] Animals, therefore, can never be ends in themselves. Nor can the reality of the benefits of specialization be measured against generalization. Specialization loses by default. And so do the animals – they, as a part of nature, become a matter of utility.

By contrast, Percy Bysshe Shelley uses the example of the Argali sheep, which, in domestication, have become corrupted from their magnificent natural selves into a mere shadow of their true nature. Such animals "are called into existence by human artifice *that they may drag out a short and miserable existence of slavery and disease, that their bodies may be mutilated, their social feelings outraged.*"[76] Likewise, Howard Williams wrote in the later years of the nineteenth century that "the natural form and organisation of the original types, the parent stocks of the domesticated Ox, Sheep, Swine, [are] now very remote from the native grandeur and rigour of the Bison, the Mouflon, and the wild Boar."[77] In the early twentieth century, the socialist-feminist author of *Women and Socialism,* Isabella Ford, opined that "in order to obtain a race of docile, brainless creatures, whose flesh and skins we can use with impunity, we have for ages past exterminated all those who showed signs of too much insubordination and independence of mind."[78] Just as, for Marx, humans are corrupted from their essential selves by the oppressions of economic history, for Shelley, Williams, and Ford, so too are domesticated animals. In fact, the rapid pace of modernity itself – whether seen in the rationalization of agriculture and the "creation" of new breeds of domesticated animals, in the seemingly inexorable introduction of labour-minimizing machinery to the industrial process, or in the rendering of traditional and widely cherished skills irrelevant to the new technology – created, in many, a Romantic Luddite reaction against the city, science, and the newfound form of wealth itself. This reaction was marked by a renewed delight in "Nature," which could mean different things to different people, but as Peter Bowler has explained, in "its most extreme form, this delight in Nature could sustain an almost romantic, Arcadian view of Nature as a source of peace and beauty in our lives. As the effects of industrialization began to make themselves apparent, more and more people would begin to long for a past world in which humankind was supposed to have been more in tune with Nature."[79] For many, this "Nature" was whatever had existed in perpetuity, not what new industrial or agricultural knowledge had fabricated.

Not so for Marx. But at least he was politically wise enough to ignore the problem of necessary yet alienating occupations – since he clearly could have had no acceptable answer – unlike the "Utopian Socialist"[80] Charles Fourier, who said, deplorably, yet in recognition of the problem Marx studiously ignored, that the "undesirable jobs would be given to children, for they were often able to make a game out of what appeared to be work to adults."[81] One may doubt that the children chimney sweeps were impressed. They generally died as youths and, if not, invariably were deformed by their labours. Of course, it would be unwarranted to imagine Marx and Fourier alone guilty of this failure to recognize the complexities of human life and of offering utopian solutions – that is, of failing to recognize that certain important occupations would always be unpalatable. Robert Southey, for example, while still dreaming in his radical youth of the "pantisocracy" that he and Coleridge would build on the Susquehanna, imagines idyllically the time to be spent discussing metaphysics while cutting down trees, criticizing poetry while hunting buffalo, and writing sonnets while following the plough.[82] If such was all there was to life, one wonders how there might have been enough capital for future generations to enjoy the same life. Wherefrom the capital to build schools; the division of labour with its surplus to train, employ, and pay teachers; and the ingenuity to produce paper and print books from which to learn, and ink with which to write – let alone to produce the leisured class who wrote the books? The intellectual life depends on surplus value provided from the excesses of a sophisticated economy.

In 1809, once the dream was long gone, Coleridge wrote in *The Friend*: "What I dared not expect from constitutions of Governments and whole Nations I hoped from Religion and a small group of chosen Individuals, and formed a plan, as harmless as it was extravagant, of trying the experiment of human Perfectibility on the banks of the Susquehanna; where our little Society ... was to have combined the innocence of the patriarchal Age with the knowledge and general refinements of European culture."[83] Such conceptions were not unusual, and a number went further than Southey in actually founding their bucolic utopias, the most famous of which was Brook Farm, a Fourierist experiment instituted by the Transcendentalists in 1841. Despite the dreams, such experiments wound up inexorably as nightmares, usually within a handful of years. The primary exception to longevity was the Oneida community, a religious group in central New York State, who believed in communal property and what they called "complex marriages," or what their adversaries named "free love." It lasted from 1848 to 1879; but it, too, suffered in the end from an excess of mutual incompatibilities over the proclamation of love. As usual, *eros* was stronger than *agape,* selfish individualism stronger than altruistic communitarianism. There were many "return to nature" schemes, most of which appeared to founder because they recognized some simple, uniform, and malleable human nature in principle, which could be found nowhere in practice but was consistent with the idea of the perfectibility of human nature. And, again, there were paradoxical versions of this essential humanity. Some, such as Marx, deemed the perfection to lie in that which made us unlike the other animals, a consequence of our progress away from the simplicities of our original selves, while others, such as Thoreau and Steinbeck, held to a vision of humanity in which they recognized the animality of the human condition as a necessary and irremovable – although a far from complete – part. For them, it was a question not of returning to nature but of recognizing that in many respects we had never left it.

IV

There is confusion not only among the various back-to-nature themes, but often within them. And, usually, they lack Thoreau's conscious awareness of the ambivalent nature of human subconsciousness. Explaining how she came to write her 1934 book *Primitivism and the Idea of Progress in English Popular Literature of the Eighteenth Century,* Lois Whitney tells us she was casually reading some novels of the period and was "astonished at the curious mixture of ideas" she "met there – theories of the superiority of primitive man" competing with theories of progress "all huddled together – sometimes two antagonistic points of view in the same sentence." "The primitivistic ideology bade men look for their model of excellence to the first stages of society before man had been corrupted by civilization, the idea of progress represented a point of view that looked forward to a possible perfection in the future. The primitivistic teaching, again, extolled simplicity, the faith in progress found its ideal in an increasing complexity. The former system of thought, finally, taught an ethics based on the natural affections; the latter system was based on an intellectual foundation."[84] Customarily, the rejection of "modernity" arises from the desire to relinquish what are seen as the detrimental aspects of what is pejoratively termed "bourgeois" civilization while maintaining what are seen as its benefits, without recognizing that the disadvantages and the advantages are inextricably linked, that, as often as not, to relinquish the former is to relinquish the latter, and that to maintain the latter is to maintain the former. The vicissitudes of the laws of history do not allow humankind to pick and choose among the inextricably linked.

Sometimes Arcadia represents not a step back in time but a reluctance to take the step forward to modernity that others have taken. Thus, as Paul Johnson remarked, by the 1920s "many, perhaps most, Americans thought of their country, almost wistfully, as the last Arcadia, an innocent and quasi-Utopian refuge from the cumulative follies and wickedness of the corrupt world beyond her ocean-girded shores. But how to preserve Arcadia?"[85] In reality, only in formaldehyde. Just as Lois Whitney had found in eighteenth-century popular novels, so too America saw itself incongruously both as an Arcadia withstanding the pressures of ungodly innovations and, at the same time, as the epitome of progress and radicalism. Its competing heroes were Natty Bumpo and Henry Ford. Andrew Jackson had been, in the same breath, the quintessential frontiersman and the harbinger of capitalism, as lawyer, land speculator, and politician. When Theodore Roosevelt led the Rough Riders up San Juan Hill, Arcadia was only a memory. But it was a potent memory. And an almost perennial one. No other country could embrace the technological age with such enthusiasm and optimism while embracing concomitantly as its primary literary gods the primarily Arcadian figures of Thoreau, John Steinbeck, Gertrude Stein, and Ernest Hemingway, with even Jack London and John Burroughs as lesser luminaries. Even Karl Marx, in his striving after the inevitability of progress, concludes with a return-to-nature image, or at least with a retreat from technological complexity, where one is fulfilled as a fully rounded hunter, herdsman, and critic.

V

Insatiable as a consequence of the contradictions of their acquired nature, humans seek constant remedies for the ills of their condition. For most, the answers lie either in progress toward some intellectually conceived utopia or in some piecemeal practicalities that constantly serve to improve upon the status quo. For many, however, what they have will not do; what they see of the potential for progress is that it offers only more of the same and yet worse. For them the answer must lie in returning to something that we are deemed to have once possessed in a long-forgotten past before we encountered the features of modernity that are deemed inimical to being truly human. All return-to-nature themes are predicated on these premises. Indeed, the very origins of the Judeo-Christian religion constitute a Golden Age myth. Before there is an original sin, before Eve tempts Adam to eat from the fruit of the tree of knowledge, enabling them to become as gods, knowing good from evil, the Garden of Eden is an innocent,

vegetarian paradise in which there is neither mortality nor suffering, nei-
ther intellectual reason nor the infliction of harm. But when Eve fell, in
John Milton's version: "Earth felt the wound, and Nature from her seat /
Sighing through all her works gave signs of woe, / That all was lost."[86]
When Adam, too, succumbed: "Earth trembled from her entrails, as again
/ In pangs, and Nature gave a second groan."[87] The Garden path led out
to the necessity of "woe" and "pangs" – "all was lost"; thereafter, conflict
must reign. Nature sighed.

In the Schocken Bible, a Jewish text from southern Germany (c. 1300),
the vegetarian Golden Age origins of both humans and animals are even
more explicitly expressed than in the "authorized versions":

> God said:
> Here, I give you
> all plants that bear seeds that are upon the face of all the earth,
> and all trees in which there is tree fruit that bears seeds,
> for you shall they be for eating;
> and also for all the living things of the earth, for all the fowl of
>   the heavens, for all that crawls upon the earth in which
>   there is living being –
> all green plants for eating.
> It was so.
> Now God saw all that he had made,
> and here: it was exceedingly good![88]

Indeed, a convincing case can be made that the consumption of flesh
after the Flood was a right granted in an emergency as a consequence of an
environmental catastrophe that had removed the possibility of living year-
round from a plant-based diet. And that, consequently, when the possibility of reverting to a nonflesh diet occurred, when plant food became once
more a viable dietary option, the appropriate step in line with the dictates
of Genesis was to revert to a plant-based regimen. Certainly, when, in the
Schocken Bible, God says that "it was exceedingly good!" he appears to be
referring not merely to what he had created, but to the conditions that
pertained thereto – one of the most significant of which, this rendering of
Genesis would suggest, was the vegetarian lifestyle not only of humans,
but of all of animated nature, an interpretation not at all inconsistent with
the more customary versions of Genesis, in which it is, however, less pronounced. Moreover, even when – whether temporarily or not – meat consumption was allowed, the traditional Genesis 9:4 required that "you must
not eat flesh with life, that is to say blood, in it." Blood was seen as the life

principle, the spark of life, and was not to be consumed in recognition that meat eating was a special dispensation and that the lives of those who were eaten were also of value. In fact, this Jewish prescription continued for several centuries of the Christian era, according to Eusebius's *Ecclesiastical History*.[89] Moreover, at least until the Council of Gangra in the fourth century, vegetarianism was very widely practised in Christianity, diminishing only slowly thereafter – although more rapidly once Augustine had defamed the practice in the fifth century. The idea of enjoying the simplicity and justice of Adam and Eve's peaceable kingdom was very persuasive to many early Christian groups.

Certainly, in many of the world's religions, we encounter a not dissimilar version of "the state of nature." For example, in the Cheyenne creation myth, as we have already noted,[90] we read of the time immediately following creation when: "Every animal, big and small, every bird, big and small, every fish, and every insect could talk to the people and understand each other, for they had a common language and lived in friendship. The people went naked and fed on honey and wild fruits; they were never hungry ... During the days they talked with the other animals, for they were all friends."[91] Soon, as in Genesis, the deity granted the people the right to kill and eat their fellow animals. Likewise, the myths of the Bassari of West Africa teach that the deity Unumbotte gave the people "seeds of all kinds, and said: 'Go plant these'" so that people might live from their fruit. In remarkable similarity to Genesis, "Snake" tempts "Man and his wife" to eat forbidden fruit instead.[92] They become aware of their differences from the animals, develop a language separate from that of the other creatures (that is, their interests diverge), and ultimately become meat eaters. Similarly, the Makritare of the Orinoco believed that there was an initial vegetarian period: "Then Mantuwa, the Jaguar, approached and took a bite of the serpent flesh. That was the first eating of meat. When the others saw the red blood flow, they all pressed in for a mouthful."[93] It is important to recognize not only that according to the myth all animals, including humans, refrained from the consumption of flesh to this point, but also that it was the behaviour of the Jaguar that broke the natural law – humans were followers of the seductive animals. In the case of the Bassari, it was the temptations of the snake that drove humans away from their natural symbiosis with the other animals. Either the gods or the animal masters must grant the concession for flesh eating to be legitimized. Indeed, at the close of the eighteenth century, we find the Anglican priest and Fellow of Trinity College, Cambridge, Thomas Young acknowledging that there could be no good ground for carnivorous behaviour if God had not granted the privilege. Thus, Young remarks that: "After the flood, God by a particular grant, gave permission to Noah and his descendants, to take away the lives of animals for the purposes of food. Now I think it evidently appears from the grant itself, independent of all other arguments, that without it mankind would not have had a right to kill animals for food. For if the right could have been derived from any other source, that grant would have been unnecessary; in which case we cannot conceive that God would in so express, and particular a manner have conferred it."[94] In like vein, the political philosopher and preeminent Anglican theologian William Paley observed a few years earlier: "It seems to me that it would be difficult to defend this right [to eat flesh] by any arguments which the light and order of nature afford; and that we are beholden to it for the permission recorded in scripture, Gen, IX, 1, 2, 3."[95] Clearly, for these divines, humans were by nature vegetarians. They had become meat eaters by being granted a licence to contravene their original nature, just as had the Makritare – the source of one grant being God, and of the other the Jaguar. The implication for the status of animals lies primarily in the fact that, in many cultures, including the Judeo-Christian culture, there is an assumption that in "the state of nature," the interests of human and nonhuman animals were in accord but diverged as humans became creatures of culture.

For many, the state of nature was a condition of mutual sympathy among the species. It was a time when humans and other species were predominantly alike, such that their interests did not diverge. "The Happy Beast" and "The Happy Human" lived symbiotically side by side.

VI

Until the very onset of the Renaissance, Western thought was predicated primarily on the view, adhered to as much by secular as by religious sources, that original sin had removed us from Paradise, to which, because of our fall from grace, because of our essentially sinful nature, we had forfeited the right ever to return in life. History could be none other than a journey of regress.[96] However, the Renaissance brought a new confidence in the capacity for progress, and to such a degree that by the time of the Enlightenment, progress was a predominant theme – notably, although far from exclusively, in the writings of Jacques Turgot and the marquis de Condorcet.[97] Moreover, the idea of perfectibility was on many minds – for example, those of Christian Wolff, Immanuel Kant, William Godwin, and Percy Bysshe Shelley. One often reads that nothing more than constant progress was meant by the quixotic term. Yet this is only a half truth, for in the first edition of *An Enquiry concerning Political Justice* (1793),

William Godwin wrote that soon there would be "no war, no crime, no administration of justice as it is called, no government."[98] Nothing could be more utopian. Indeed, such ideas continued among the harbingers of both anarchism and socialism, and they were not without relevance to those of a general liberal bent. And whereas a more sober reality had set in for Godwin by the third edition of the *Enquiry* – a sobriety from which Marx later failed to benefit – throughout his short life Percy Shelley remained deeply disappointed, and even surprised, that Godwin's forecast had proved inaccurate. For many of those, however, who found something lacking in Enlightenment values – those whom we might dub the Rousseauians – the answer lay in the very opposite of progress, in the re-creation of an Eden or an Arcadia, although via classical Greek rather than Judeo-Christian thought. We encounter such a primitivism in the late-sixteenth-century *Arcadia* of Sir Philip Sidney, and it is alluded to in Shakespeare's *The Tempest* (Act 2, Scene 1; c. 1610). On some readings at least, it is central to Gerrard Winstanley's communism in *The Law of Freedom* (1652). Nonetheless, if primitivism is the antithesis of progressive perfectibility, it remained every jot as much a visionary ideal.

Today, the return-to-nature theme is associated most commonly with the author of the *Discourse on the Arts and Sciences* (1749) and the *Discourse on the Origin and Foundations of Inequality among Men* (1751), although there were in fact many more who subscribed to the doctrine in the late seventeenth and the eighteenth centuries – for example, in Britain, Thomas Tryon in *The Countryman's Companion* (1683), the diarist John Evelyn in *Acetaria: A Discourse of Sallets* (1699), James Thomson in *Seasons* (1726-30), and Alexander Pope in the *Dunciad* (1728) and in *An Essay on Man* (1734)[99] – and there are good grounds for claiming that Rousseau's commitment to the doctrine was a great deal less emphatic than is customarily assumed.[100] Nonetheless, if Rousseau is far less of an exemplar than he is commonly read to be, it was he above all others who influenced vegetarians at the turn of the nineteenth century to find the impetus in his writings to research the history of the doctrine and the advice it has to offer. Thus, for example, soldier and adventurer John Oswald in *The Cry of Nature* (1791),[101] printer and publisher George Nicholson in *On the Primeval Diet of Man* (1801), medieval scholar and antiquarian Joseph Ritson in *An Essay on Abstinence from Animal Food as a Moral Duty* (1802),[102] and West Indian plantation owner and graduate of Christ Church, Oxford, John Frank Newton in *The Return to Nature* (1812)[103] all trod the path of Rousseau on their journey. The Rousseauian theme was still popular in 1856, when Gustave Flaubert referred in *Madame Bovary* to "the cradle of human society": "the savage ages when men lived off acorns in the depths of the forest. Then they had cast off their animal skins, garbed themselves in cloth, dug the ground, and planted the vine. Was this an advance? Didn't their discovery entail more disadvantages than benefits?"[104] A variant of the theme can be found in the glorification of the peasants in Nikolai Gogol's *Dead Souls* (1842):

> "They don't seem to realize that it is because of these luxuries that they them-
> selves have become trash and not men ... Why, thank God, there's still at
> least one healthy class of society left which knows nothing of these sophis-
> ticated fads! We ought to thank God for that. Yes, the man who tills the
> land is more worthy of respect than any" ...
>
> "So you think, sir, that agriculture is the most profitable occupation?"
> asked Chichikov.
>
> "'It's the most righteous, which of course is not the same thing as the
> most profitable.
>
> "'In the sweat of thy face shalt thou eat bread,' it is written. The experi-
> ence of ages has shown that a man who works on the land is purer, nobler,
> higher, and more moral."[105]

Tolstoy held to a similar view, extolling the "peasant thinkers" as the truly wise: "Down with intelligence! Long live simplicity."[106] For Tolstoy, "corruption" grew out of the conditions of the age. To cast it off, one must cast off "civilization."[107] Whereas Marx believed that history had overtaken the peasant class and that *nolens volens* they would be swallowed up by the industrial and urban proletariat, Gogol and Tolstoy regarded this class as our saviour from the infamies of modernity. For Gogol and Tolstoy, a healthy society will resist the diversions of novelty by holding on to and, to the extent that they are already being superseded, reviving the glories of the traditional rural way of life. But we should not be persuaded that the view was rampant, even when it was employed. Sometimes it was offered as a caricature, as in Charles Dickens's *Dombey and Son* when Mrs. Skew-ton wants "Nature everywhere" but not without fine porcelain tea cups, her Arcadia being no wilder than a Swiss farm with cows: "I assure you, Mr Dombey, Nature intended me for an Arcadian. I am thrown away in society. Cows are my passion. What I have ever sighed for, has been to retreat to a Swiss farm, and live entirely surrounded by cows – and china ... I want Nature everywhere. It would be so extremely charming ... [Music has] so much heart in it – undeveloped recollections of a previous state of existence – and all that – which is so truly charming."[108] Dickens is, of course, parodying the Romantic movement, of which he was himself a legatee, but he repeats the sentiments as his own, with a touch of irony to be sure, in a letter he wrote a few days after composing Mrs. Skewton's speech: "All the Savage (I am sure in some former state of existence I was a slap-up Chief: a little Buffalo or a Great Bear or something of that sort) stirs within me, and compels me to go and look out for cottages on banks of Thameses."[109] If nothing else, Dickens's words reflect the pervasiveness of the theme and suggest that there is a little bit of Arcadia in the most urban of us – indeed, in the most Edenic.

More directly antithetical was the firmly *philosophe* Mary Wollstonecraft's remark against Rousseau:

His arguments in favour of a state of nature are plausible, but unsound ... Had mankind remained for ever in the brutal state of nature, which even his magic pen cannot paint as a state in which a single virtue took root, it would have been clear, though not to the sensitive unreflecting wanderer [Rousseau had written the anti-rationalist *Dreams of a Solitary Walker* in 1775], that man was born to run the circle of life and death, and adorn God's garden for some purpose which could not easily be reconciled with his attributes ... Rousseau exerts himself to prove that all was right originally; a crowd of authors that all is now right:[110] and I, that all will be right.[111]

Yet even the modest feminism that Wollstonecraft was promoting, and for which she thought that progressivism was necessary, could be dressed in Edenic language. For instance, the feminist persona in Henry James's *The Bostonians* says:

Good gentlemen all, if I could make you believe how much brighter and fairer and sweeter the garden of life would be for you, if you would only let us help you to keep it in order! You would like so much better to walk there, and you would find grass and trees and flowers that would make you think you were in Eden. That is what I should like to press home to you, personally, individually – to give him the vision of the world as it hangs perpetually before me, redeemed, transfigured, by a new moral tone. There would be generosity, tenderness, sympathy, where there is now only brute force and sordid rivalry. But you do strike me as stupid even about your own welfare![112]

Indeed, it is on this very basis that many feminists make their claim for a greater sensibility toward animals. It is women, it is claimed, who embody "generosity, tenderness, sympathy," while males represent "brute force and sordid rivalry."[113] It is they who represent Eden against the predominantly male theme of Arcadia. Still, the feminist choice of the epithet for their adversaries of "male chauvinist pig" would suggest that their animal sympathy is not what they would make it.

The myth of the "noble savage" was not new, especially on the European continent, even at the turn of the eighteenth century; nor was it previously unassailed. We find it expressed in Peter Martyr's *De Orbe Novo* (c. 1493), in Antonio Pigafetta's *The First Voyage around the World* (1519-22), in Fernández de Oviedo's *La Historia general y national de las Indias* (1535), in Bartolomé de las Casas's *Brief Report on the Destruction of the Indians* (1542), in Girolamo Benzoni's *Historia del Mondo Nuovo* (1565), in *La Piazza* (of about the same date and attributed to Ortensio Landi),[114] in Montaigne's essay "Of Cannibals" (c. 1580), in Gabriel Sagard Théodat's *Great Journey to the Land of the Hurons* (1632), in Gabriel Foigny's *Les Aventures de Jacques Sadeur et le Voyage de la Terre Australe* (1676), in Denis Vairasse d'Alais's *L'Histoire des Sévérambes, Peuples qui Habitent une Partie du Troisième Continent Communement Appelé la Terre Australe* (1677-79), and in the *Mémoires* of the baron de Lahontan (1703). By the time of Rousseau, it was old hat – although he gave it a vibrancy that it had previously lacked. Even Montaigne's essay is notoriously inconsistent and contradictory.

Rousseau was now using the theme to attack the prevailing views of the Enlightenment, which set such store in progress, reason, refinement, and science. Nor was the wisest representative of the Enlightenment slow to respond to Rousseau's onslaught. For example, in *Candide* (1759) Voltaire writes:

> "Do you think," said Candide, "that men have always massacred each other the way they do now? that they've always been liars, cheats, traitors, ingrates, brigands? that they've always been feeble, fickle, envious, gluttonous, drunken, avaricious, ambitious, bloodthirsty, slanderous, debauched, fanatical, hypocritical, and stupid?"
>
> "Do you think," said Martin, "that hawks have always eaten pigeons when they find them?"
>
> "Yes, no doubt," said Candide.
>
> "Well, then," said Martin, "if hawks have always had the same character, why do you expect men to have changed theirs?"[115]

In *L'Ingénu* (1767) Voltaire took the idea of the "noble savage" further to task. He was well aware that the by now customary idealization of the Amerindian as the embodiment of dignified humanity in its original and untainted form was a myth, albeit a popular one. Yet the idea of the "noble savage" as a child of nature – uncorrupted by the artificiality of society, by political and religious authority, and by traditional conventions – was appealing as a stick with which to cudgel both those who praised the past and, contrariwise, those who announced the superiority of the new. It was a tool of convenience for all who bore a grudge. Nowhere is this clearer than in the fact that Rousseau's first essay won the Dijon prize. It was almost certainly read by the Jesuit examiners not only as a denunciation of the Enlightenment, but consequently as an argument in favour of traditional norms – a common enemy implies a common cause – which, of course, was not at all what Rousseau had in mind.

There is always a congenial twist in Voltaire's tales, and the twist in *L'Ingénu* is that the purported Huron turns out to be a Frenchman, albeit a Frenchman who has been fortunate enough to escape the warping of traditional French culture and education. He possesses the natural, noble instincts of the savage. Yet, on being educated to Enlightenment ideals, he becomes a hero to the cause, demonstrating the benefits of Enlightenment humanity and civilization, a happy blend of reason and sensibility. "I would almost be tempted to believe in metamorphosis," he said, "for I have changed from a brute to a man."[116] Whereas Rousseau sees that there is much to be commended in our animal nature, Voltaire wishes to escape the "brute." Yet there is absolutely no reason to conclude that Voltaire was less sympathetic to the animals than was Rousseau. Voltaire's denunciation of Cartesianism and vivisection is without parallel. He describes the accomplishments of birds in nest building and learning, notes the memory, knowledge, and emotions of dogs, and refers to "the compassion we owe all animals." People, he tells us, "must never have observed animals if they cannot recognize their needs expressed in differing tones, their suffering, joy, fear, love, anger, and all their differing sentiments."[117] In *Micromegas* (1752) he expressed his recognition of the difference and fundamental equality of all living beings: "all thinking beings are different, and yet all resemble each other fundamentally in possessing the natural gift of thought and having desires."[118] And in *Zadig* (1748) he cited with evident approval "Zoroaster's great precept: 'When thou eatest, givest also unto the dogs, even should they bite thee.'"[119] Wollstonecraft, too, wrote an educational book for children, in which the first three chapters were devoted to preventing cruelty to animals.[120] If many of those who espoused the return-to-nature theme demonstrated a considerable sympathy toward the animal realm, so too did many of those who rejected it. We cannot take either approach as a better indicator of sensibilities to animals. If the back-to-nature theme encourages us to recognize our similarities with, and relationship to, other species, the universalism of civil society encourages us to recognize individual animals as ends in themselves.

There was a further use of the "noble savage" theme that was the very reverse of that intended by those who introduced and first employed it: the ignoble, and insincere, use of the "noble savage" to justify the most heinous behaviour in one's society. Thus, for example, in the novels of the marquis de Sade and in the comte de Gernande's *Justine, ou Les Malheurs de la Vertu* (1791), appeals are made to the "virtues" of other societies to justify the most libertine and cruel behaviour – for example, the purported lack of sexual restraint among Aboriginals, or the tolerance of cruelty to wives among the Turks, or the perfidy of the Indians and the Asiatics.[121] If these societies were as noble as so many suggested, then surely the behaviour in which they were said to be engaged could be used to justify behaviour that was otherwise condemned by one's own societal standards. The irony was that while in France the Turk and the East Indian could be used as the standard of perfidy and cruelty, the former was customarily used in British literature until the eighteenth century as the standard of humanitarian attitudes to animals, and from the later eighteenth century the East Indian acquired the role. In both the excessive condemnation and excessive adulation, it is probably far more the convenience of having a standard to employ than the real behaviour of any society that is at issue. And the same may be said of our contemporary adulation of the Aboriginal and denigration of the manners of civilization.

If far from unopposed, back-to-nature theorizing proved a popular intellectual pastime in the mid to late eighteenth century – sometimes as little more than a *jeu d'esprit*. James Boswell caught a cold on primitivism. But, then, as Arthur Lovejoy put it, a trifle caustically, "there were few of the intellectual diseases epidemic in his day which he did not catch."[122] Perhaps attempting to do little more than provoke copy for the soon-to-be-famous biography, successfully as it turned out – for he must surely have been able to surmise the potential tenor of Dr. Johnson's response in advance – Boswell raised the topic at dinner:

On the 30th of September [1769] we dined together at the Mitre. I attempted to argue for the superiour happiness of the savage life, upon the usual fanciful topicks. JOHNSON. "Sir, there can be nothing more false. The savages have no bodily advantages beyond those of civilized men. They have not better health; and as to care or mental uneasiness, they are not above it, but below it, like bears. No, Sir; you are not to talk such paradox: let me hear no more on't. It cannot entertain, far less can it instruct. Lord Monboddo, one of your Scotch judges, talked a great deal of such nonsense. I suffered *him;* but I will not suffer *you.*" BOSWELL. "But, Sir, does not Rousseau talk such nonsense?" JOHNSON. "True, Sir, but Rousseau *knows* he is talking nonsense, and laughs at the world for staring at him ... Monboddo does *not* know he is talking nonsense."[123]

Today, Rousseau remains renowned; Monboddo is a nigh forgotten figure. But he was not so in the heart of the eighteenth century. What is of interest here is that to the extent that either was a primitivist, the stages to which each wishes to return are not at all the same. Rousseau wants to return as far as possible to the hunting and pastoral stage of societal development. Most important, however, he knows that we cannot unlearn what we have learned. Accordingly, in reality, our task is to impose on our current cultural conditions the principles of self-sovereignty, equality, and liberty, which he deems the primary but now lost characteristics of early society, to the extent that their achievement is feasible at all. Indeed, as the great German legal historian Otto Gierke remarked, more and more customary in the eighteenth century was the view that "the individual in the state of nature had been his own sovereign. Men were originally free and equal, and therefore independent and isolated in their relation to one another."[124] Quite by contrast with both Rousseau's and Gierke's understanding of the prevailing ideas of the Golden Age, Monboddo desired a return to the ideals of classical Greece and Rome. Those who inhabit the primitive human state, he proclaims, are despicable creatures, not even truly human. Thus "there have been in the world, and are still, herds of men (for they do not deserve the name of nations) living in a state entirely brutish, and, indeed, in some respects, more wild than that of certain brutes, as they have neither government nor arts."[125] Monboddo glories in Latin literary culture – "Horace is my notion of man in his natural and original state"[126] – whereas, by contrast, Rousseau declares: "I venture to affirm that the state of reflection is contrary to nature and that the man who meditates is a depraved animal."[127] The one espouses the heights of the intellectual mind, and the other the feelings of the untutored soul. In fact, Monboddo reflects a common intellectual concern of the turn of the eighteenth century that manifested itself in a reaction against the excesses of the Gothic in both literature and architecture in favour of a return to classical simplicity. Drawing on Dryden and Locke, Joseph Addison wrote in the *Spectator* against poets who sought to display their wit in overdrawn metaphors: "Poets who want this strength of genius to give that majestic simplicity to nature, which we so much admire in the works of the ancients, are forced to hunt after foreign ornaments, and not let any piece of wit of what kind soever escape them. I look on these writers as Goths in Poetry, who like those in architecture, not being able to come up to the beautiful simplicity of the old Greeks and Romans, have endeavoured to supply its place with all the extravagances of an irregular fancy."[128] The "return" of such critics was more a matter of style than of harmony with the natural, except insofar as the simple was deemed a characteristic of the natural, a return to which was known as "Nature's simple plan."

It is, then, difficult to penetrate Johnson's objection to the primitivism of Monboddo, unless it is in Monboddo's concurrence with Rousseau on our consanguinity with the apes. Despite the later rhetoric, this was no original Darwinian discovery. In the medieval era it was often queried whether the apes were simply a different race of humankind. Although Johnson railed against the abject inhumanity of animal experimenters – "a race of men who have practised tortures without pity, and related them without shame, and are yet suffered to erect their heads among human beings"[129] – he had not the mettle to accommodate, at least not to revel in, any very close biological association with the animal realm, despite his famous indulgence of his cat, Hodge. The back-to-nature ideal was not restricted to a few vegetarians who saw humans as vegetable, nut, and fruit eaters in origin but, to Johnson's dismay, was espoused by many dissenting Christians, especially by deists. It was their view that the loving, peaceful, compassionate, and altruistic Christianity of its origins had been corrupted by the priestly class, a view exemplified in Joseph Priestley's *History of the Corruption of Christianity* (1785). Johnson was, of course, very far from having any truck with dissenters, even with non-Tory Anglicans, as indicated in his famous barb about James Granger, vicar of Shiplake in Oxfordshire and author inter alia of *An Apology for the Brute Creation, or Abuse of Animals Censored* (1772): "The dog is a whig. I do not like much to see a whig in any dress, but I hate to see a whig in a parson's gown." Even though one had to await the Victorian era to hear that the Anglican Church was the Tory Party at prayer, Dr. Johnson had already embodied the principle. Still, even the redoubtable Johnson cannot be absolved from inconsistency on the issue. In 1750 we find him praising the altruism of the Golden Age, and in 1753 he is castigating the uncivilized life of the primitive savage.[130] And yet again, in a more youthful wisdom, we find him telling us in 1735, in appraising an account of seventeenth-century travels in North Africa, that a traveller to any part of the globe "will discover, which will always be discovered by a diligent and impartial observer, that wherever human nature is to be found, there is a mixture of vice and virtue, a contest of passion and reason."[131]

What should be clear is that there were two distinctive approaches to the back-to-nature question, each of which could redound to the animals' benefit: we should return to nature in order to repossess the advantages of our original animal condition, thereby becoming once again like the other animals, and accordingly treat them well because we are akin; or we should revel in our progress, which has made us so different from the other species, but treat them well because it is one of the injunctions of our moral nature, in which we are so different from other species, that we should do so.

VII

In 1832 the political scientist, cabinet minister, Oxford don, and editor of the *Edinburgh Review,* George Cornewall Lewis, satirized the very idea of "state of nature" arguments.

> The term, *a state of nature,* has been employed to designate a supposed state of primitive simplicity, before the introduction of the arts of civilization, and the establishment of government and laws. The phrase itself and the theory connected with it have, in this country, been diffused chiefly by the writings and authority of Locke ... As, however, Locke's account is somewhat diffuse and indistinct, I shall prefer giving Pope's description, in his Essay on Man, of the state of nature, and the change from that state to civilization and government, as being shorter and more explicit ... ,

> "Nor think in nature's state they blindly trod;
> The state of nature was the reign of God:
> Self-love and social at her birth began,
> Union the bond of all things and of man.
> Pride then was not; nor arts, that pride to aid;
> Man walked with beast,[132] joint tenant of the shade:
> The same his table and the same his bed;
> No murder cloth'd him and no murder fed.

> *   *   *

> Ah, how unlike the man of times to come,
> Of half that live the butcher and the tomb;
> Who, foe to nature, hears the general groan,
> Murders their species, and betrays his own."[133]

The poet then describes man
    "From Nature rising slowly to Art,"
as addressed by the voice of nature, which enjoins him to take instruction from the lower animals; for example, to learn the art of building from the

bee; the art of ploughing from the mole; of sailing from the nautilus, &c.
Moreover, to imitate forms of government from the same original: a republic from the ants; a monarchy from the bees.

> "Great Nature spoke: observant man obey'd,
> Cities were built, societies were made."

The result of this account seems to be, that in the state of nature God
ruled the world; that is, God alone ruled it, – there being no human rulers.
Benevolence and self-love existed; but notwithstanding the existence of self-
love,[134] all men live in concord, and the feeling of pride was unknown.
There were no arts or government: men lived with the beasts,[135] and sub-
sisted exclusively on vegetable food. In the state of nature men killed neither
beasts nor men. After some time, mankind learnt, by observing some of the
lower animals, to imitate their ways; and having thus invented the arts of
social life, upon the same model they formed societies under an established
government.

Such is an outline of this puerile theory of the progress of society: untenable from its self-contradictions, even as a hypothesis, and distinctly refuted
by facts: a theory which could only have arisen from the distempered imagination of some day-dreamer, and could only have been tolerated by a blind
ignorance or wilful neglect of all history. Pictures of this description may
delight the mind, when presented to it in an avowedly poetical and fabulous
shape, as in the Greek legends of the golden age; but when introduced into
a didactic poem, or a philosophical system of government, they shock the
reason without amusing the fancy.[136]

Yet, valid as Lewis's critique is, it largely misses the point. Indeed, as history, the "state of nature" viewpoint is wildly inaccurate. But as Daniel
Dombrowski has written, although "'once upon a time' stories of a contract between man and animal are merely stories, so are the 'once upon a
time' stories between man and man. In that this condition has not both-
ered the history of social contract theory from Plato to Kant to Rawls,
it should not bother us. That is, these stories of an ancient vegetarian
past, even if not true, offer insights into the beliefs of the people who told
them."[137]

Probably no one was more aware of the ahistorical nature of the theory
than Pope and Locke. The diarist John Evelyn in *Acetaria* (1699) is among
those who appear to take the historical record seriously, referring to "the
sacred records of elder times [which] seem to infer [a vegetarian period]
before there were any flesh shambles in the world."[138] It is one thing to

suggest a prehistorical vegetarian, or at least an altruistic, stage of human history, another to imagine it verified by "sacred records," even if only by implication.

One assumes that Locke and Pope were not describing what they thought of as in detail a historical state but were making observations relevant to human behaviour and ideals. Whether humans were in fact vegetarians in origin, as Pope supposes, is possible but unconfirmed, and most anthropologists seem decidedly dubious. Still, the debate is far from over. What is certain is that humans ate relatively little meat – in ancient Greece, Rome, and the lands of the Bible probably only on the occasion of sacred festivals – and they were probably far more grateful for their food than we are accustomed to be today. Certainly, for both Pope and Locke, we did not respect animals as we should.[139] Even if Pope perhaps exaggerates when he tells us how we learn from the other animals, as Job had announced in the Hebrew Bible,[140] he is indicating that their accomplishments are not negligible and that they are entitled to our esteem and our benevolence. Moreover, many great writers have suggested, from Democritus, the writers of the Bible, and the authors of the bestiaries on, that animals do have certain important lessons to impart. Thus in *Fathers and Sons* (1861), Ivan Turgenev writes:

"Good-bye, brother!" [Bazarov] said to Arkady when he had got into the light cart, and pointing to a pair of jackdaws sitting side by side on the stable roof, he added, "That's for you; follow that example."
"What does that mean?" asked Arkady.
"What? Are you so weak in natural history, or have you forgotten that the jackdaw is a most respectable family bird? An example to you!"[141]

Turgenev had an admirable precursor. In discussing filial devotion, the first-century scholar Philo-Judaeus suggested young storks as a model to follow.[142]

Aldous Huxley has a decided preference for our listening to the lessons from the animals rather than preaching to them, however well intentioned the preacher. Thus in *Island* (1962) – the *Times* rightly called it "One of the truly great philosophical novels" – Dr. MacPhail observes that "'birds don't understand pep talks. Not even Saint Francis's. Just imagine,' he went on, 'preaching sermons to perfectly good thrushes and goldfinches and chiffchaffs! What presumption! Why couldn't he have kept his mouth shut and left the birds to preach to *him*?'"[143] For many, the lessons of history support the idea of "the happy beast," in at least some respects a being superior to the human, an appropriate epithet, even if the idea of a Golden Age is not borne out by recorded history.

VIII

What should be clear is that what are called alternatively, and with nuances of meaning, the themes of return to nature, primitivism, Golden Age, and retrospectivism all contain significantly differing concerns and points of reference. Some see the ideal in some distant past, at or very near human origins, some in a relatively recent past, and yet others somewhere or other along the historical continuum. Indeed, in the Victorian era many looked back nostalgically to some of the conditions of the Middle Ages. In *A Dream of Order: The Medieval Ideal in Nineteenth Century Literature,*[144] Alice Chandler points to Scott, Cobbett, the Lake Poets, Carlyle, Disraeli, Ruskin, and Morris, among others, as those who find there a now-lost chivalry, a simple honesty, a concern for the worker's skill and wellbeing, a bucolic and healthy simplicity – "the essential passions of the heart," wrote William Wordsworth, "find a better soil" among the humble and rustic. The medieval connection to "nature" is exemplified in George Eliot's *The Mill on the Floss,* where "the town of Saint Ogg's – that venerable town with the red-fluted roofs and the broad warehouse gables" "is one of those old, old towns which impress one as a continuation and outgrowth of nature, as much as the nests of the bowerbirds or the winding galleries of the white ants: a town which carries the traces of its long growth and history like a millennial tree ... It is a town 'familiar with forgotten years'... it is all so old that we look with loving pardon at its inconsistencies."[145] If Saint Ogg's is not nature itself, it embodies the idea of an unbroken historical continuity stretching back in perpetuity to the very steps of national origins.

Yet this is only one side of the story. In a discussion of "the use of the Middle Ages as an ideal," Chandler observed that the Anglican cleric Charles Kingsley's *Hereward the Wake* (1866) "fuses hard primitivism with muscular Christianity."[146] This "muscular Christianity" – or "muscular Anglicanism" at any rate – could, however, prove more compelling than any primitivism if the convenience of the argument warranted. Thus at a time when the Puseyites were turning a face of the Anglican Church – and some whole bodies – toward Rome, Kingsley was happy to employ another supposed medievalist in the cause, one who later refused to accompany a Roman cardinal to the Home Office in support of proposed anti-vivisection legislation because he was "the chief representative of Beelzebub in England":[147]

"Man's scientific conquest of nature must be one phase of his kingdom on earth ... And by this test alone I will try all theories, and dogmas, and spiritualities whatsoever – Are they in accordance with the laws of nature? And therefore when your party compare sneeringly Roman Sanctity and English

Civilisation, I say, 'Take you the Sanctity and give me the Civilisation!' The
one may be a dream, for it is unnatural; the other cannot be, for it is natural; and not an evil in it at which you sneer but is discovered, day by day, to
be owing to some infringement of the laws of nature. When *we* 'draw bills
on nature,' as Carlyle says, 'she honours them,' – our ships do sail; our mills
do work; our doctors do cure; our soldiers do fight. And she does not honour yours; for your Jesuits have, by their own confession, to lie, to swindle,
to get even man to accept theirs for them. So give me the political econo-
mist, the sanitary reformer, the engineer; and take your saints and virgins,
relics and miracles. The spinning-jenny and the railroad, Cunard's liners and
the electric telegraph, are to me, if not to you, signs that we are, on some
points at least in harmony with the universe; that there is a mighty spirit
working among us, who cannot be your anarchic and destroying Devil, and
therefore may be the Ordering and Creating God."[148]

If the eternal truths were at stake, the stylistic convenience of primitivism
could be readily discarded, and technological modernity could even be
dressed in naturalistic garb!

Ralph Waldo Emerson went back much further than the medieval era
but not all the way in human history. He said he agreed with the painter
Thomas Cole that "in history the great moment is when the savage is just
ceasing to be a savage ... not yet passed over into the Corinthian civility ...
Everything good in nature and in the world is in that moment of transition."[149] Edward Carpenter's delightfully titled *Civilisation: Its Cause and
Cure* (1880) is very clear on what is wrong, not so transparent on what it
might be replaced with – other than that it occurred before our present
civilization and has been ruined by it. Even those who look to something
close to what they see as human origins possess competing principles – the
Edenists and the Arcadians – their differences further compounded by
whether they glory in the educational foundations of culture or reject it as
the very source of the problem. What the various authors held in common
was the belief that at some point in history, whether almost six thousand
years ago or just a few hundred, humanity had deviated from its appointed
and appropriate path and that the superior conditions reigning in whatever era was deemed the Golden Age should be reinstated. (The maxi-
mum was "six thousand years" since, from the early seventeenth to the
nineteenth century, the prevailing belief, based on Archbishop Ussher of
Armagh's Biblical exegesis, was that creation occurred in 4004 BC.) Well
into the twentieth century, many editions of the King James Bible still contained Ussher's datings in marginal notes, and prosecution witnesses relied
on Ussher's calculations to denounce not just the theory of evolution, but

almost all modern science in the infamous "Creationism versus Evolution" Scopes trial in Tennessee in 1925.

IX

An initial inclination might be to acknowledge the reign of confusion and contradiction and to determine that there is nothing profitable to pursue. All that the back-to-nature thinkers share in common is a belief that, in the words of John Frank Newton, "It is not man we have before us, but the wreck of man."[150] However, we cannot doubt that many of those who looked for their ideals in the past found something of value there for understanding the human-animal relationship. Rousseau, for example, found that however much the compassion of our early character had been destroyed by the development of the luxuries and refinements of a rationalist Enlightenment, there remained beneath the surface the remnants of this compassion, which bore comparison with the natural altruism of the animal world. Thus, as we saw earlier,[151] Rousseau tells us in the *Discourse on Inequality* that "we observe every day the repugnance of horses to tread a living body under foot."[152] In Rousseau's view, "compassion is a natural sentiment,"[153] derived from autochthonous nature in the case of both humans and animals, and it is more readily extinguished in the human than in the animal. The consequence is that "as long as [man] does not resist the inner compulsion of compassion, he will never do harm to another man, or even to another sentient being, except in those legitimate cases where, since his own preservation is involved, he is obliged to give preference to himself."[154] Moreover, in explaining the origins of compassion within an individual breast in his educational treatise *Emile,* Rousseau allows for a genuine empathy for the animals' suffering as well as for that of the human. (Eighteenth-century *pitié,* translated as "pity," it should be noted, often corresponds more closely to our current sense of "compassion."):

> Emile ... will begin to have gut reactions at the sounds of complaints and cries, the sight of blood flowing will cause him an ineffable distress before he knows whence comes this new movement within him ...
>
> Thus is born pity, the first sentiment that touches the human heart, according to the order of nature. To become sensitive and pitying, the child must know that there are beings like him who suffer what he has suffered, who feel the pains he has felt, and there are others whom he ought to conceive of as being able to feel them too. In fact, how do we let ourselves be moved by pity if not by transporting ourselves outside of ourselves and identifying

with the suffering animal, by leaving, as it were, our own being to take on its being. It is not in ourselves, it is in him that we suffer.[155]

If we revert, to the degree feasible, to the conditions of that earlier stage of history in which compassion flourished, we will be able to re-create not only a better world for humans to live in, but a happier and more humane relationship with the animals. For the Rousseauian Joseph Ritson, life in a re-created state of nature would allow humans to divest themselves of the worst characteristics encouraged by civilization – violence and bellicosity: "Man, in a state of nature, would, at least, be as harmless as an ourang-utang."[156]

When, at the turn of the nineteenth century, advocates of vegetarianism sought the origins of a natural humanity, which regarded their fellow animals with compassion, they looked beyond Rousseau to the real or imagined practices of India, to the "Otaheite" (the Hawaiian islanders) and the inhabitants of the "Atlantic islands," and to the writers of classical Greece and Rome for their philosophical evidence, citing Plato, Ovid, Virgil, Porphyry, and even the early medieval Boethius, among others, in their cause. (They could have added Aratus, Tibullus, pseudo-Seneca, and Tacitus to their list of sources.) None was more thorough in his investigation than George Nicholson in *On the Primeval Diet of Man* (1801), although he borrowed much of his evidence – some acknowledged, some not – from John Oswald's *Cry of Nature* (1791). With perhaps a little more credulity than reflection, he cites Porphyry to tell us that the "ancient Greeks lived entirely on the fruits of the earth," as did "the ancient Syrians," and that by "the laws of Triptolemus, the Athenians were strictly commanded to abstain from all living creatures."[157] From Aelian he derives the information that "the ancient Arcadians lived on acorns, the Argives on pears, the Athenians on figs."[158] Via the writings of Diodorus Siculus, we are informed how the fleshless diet of Pythagoras's followers made them "very strong and valorous."[159] From Gellius and Macrobius, we learn that the "Romans were so fully persuaded of the superior effects of a vegetable diet, that besides the private examples of many of their great men they publicly countenanced this mode of diet in their laws concerning food."[160] From here Nicholson goes on to quote a remarkable array of authorities to demonstrate the superiority of a vegetarian diet, the effective use of reason in animals, the abysmal conduct of humankind to animals, the cruelties of sports involving animals, and the similarities of human and animal slavery. In short, the purpose of the return-to-nature theme is to demonstrate that by deviating from our purportedly original and natural diet, we have lost the respect and consideration for the natural realm that is the animals' due not only because *we* owe it as an aspect of *our* moral obligation, but because *they* are entitled to it on account of *their* well-developed senses, their capacities to communicate, their affection, their passions, their "extraordinary exertion of reason," and their learning abilities.[161] The underlying message is that we misjudge animals when we deem them inferior.

Lest we judge Nicholson and those he cites too harshly for their gullibility about the vegetarian conditions of life only a couple of millennia ago, let us, again, not forget that it is possible to revere the vision without according that vision an empirical reality. Thus, in her early poem *Edwin and Eltrude: A Legendary Tale* (1782), Helen Maria Williams, later translator of Bernardin de Saint-Pierre's *Paul et Virginie,* wrote of the "soothing dream / Of golden ages past."[162] The dream was soothing, not historical reality – but nonetheless valuable. In Dostoevsky's *The Devils* (1871-72), Stavrogin has a "Golden Age" vision of a primeval earthly paradise of happiness and innocence, inspired by Claude Lorraine's painting *Acis and Galatea:* "A feeling of happiness, hitherto unknown to me, pierced my heart till it ached." "Here was the cradle of European civilization, here were the first scenes from mythology, man's paradise on earth. Here a beautiful race of men had lived. They rose and went to sleep happy and innocent; the woods were filled with their joyous songs, the great overflow of their untapped energies passed into love and unsophisticated gaiety. The sun shed its rays on these islands and that sea." However, Stavrogin is not persuaded of its reality. It stands instead as the symbol for which humanity, at least while humane, has striven throughout its history. Stavrogin continues: "A wonderful dream, a sublime illusion! The most incredible dream that has ever been dreamed, but to which all mankind has devoted all its powers during the whole of its existence, for which it has died on the cross and for which its prophets have been killed, without which nations will not live and cannot even die."[163] Illusion, indeed; incredible, indeed. But it is the incredible illusion that impels us. And for Dostoevsky, the animals are a part of that primeval innocence, exemplified most completely in *The Brothers Karamazov:*

> For each blade of grass, each little bug, ant, golden bee, knows its way amazingly; being without reason, they witness to the divine mystery, they ceaselessly enact it ... all things are good and splendid, because all is truth ... there is no sin upon them, for all is perfect, everything except man is sinless, and Christ is with them even before us ... Love the animals.
>
> God gave them the rudiments of thought and an untroubled joy. Do not trouble it, do not torment them, do not take their joy from them, do not go

against God's purpose. Man, do not exalt yourself above the animals: they are sinless.[164]

Whether nonhuman animals lived the superior life or not, for Dostoevsky they were the morally superior beings. And their superiority lay in their not having been corrupted by civilization. They have remained true to their original nature.

X

We find the theriophilic message repeated in the early twentieth century, although it is usually intended less to credit the beast's capacities than to suggest by contrast the wretched misery allotted to the human. For example, in *The Everlasting Mercy* (1911), the poet laureate John Masefield says: "If this life's all, the beasts are better." June Dwyer relates how Masefield's verse "does not reflect a lack of appreciation for life's complexities; it expresses a will for a world that is simple and easier to live in than the one he saw."[165] The implication is that the animal's life is better because its needs are simpler and can be met. Those of the human are more complex and cannot be met. Human justice is both a moral necessity and a practical impossibility. Mikhail Sholokhov in *Quiet Flows the Don* (1929) suggests through Uriupin that there is a hierarchy of animals in which humankind is at the bottom, at least in certain circumstances: "To kill your enemy in battle is a holy work. For every man you kill God will wipe out one of your sins, just as he does for killing a serpent. You mustn't kill an animal unless it's necessary, but destroy man! He's a heathen, unclean; he poisons the earth; his life is like a toadstool."[166] A half-century earlier, Victor Hugo captured the cogent essence of the idea: "The butterfly is a success, but Man is a failure. God made a mess of that particular animal."[167]

In fact, throughout the recorded history of ideas, many have argued that animals are not only our equals but in some cases our superiors and that they live a superior life. For them, "the happy beast" is a historical reality. Of course, far more common is the view that the human reigns supreme. Thus in *Memorabilia* (c. 360 BC), Xenophon has his friend Socrates proclaim humankind corporeally "unique in possessing erect posture, hands, speech, sexual appetite 'unbroken to old age'; he is psychically unique in his knowledge of the gods, ability to anticipate and therefore provide against hunger and thirst, cold and heat, and in his ability to learn."[168]

Over two millennia later, in *The Descent of Man*, Charles Darwin repeated

Xenophon's claim about humankind that "all others have yielded before him. He manifestly owes this immense superiority to his intellectual faculties, to his social habits, which lead him to aid and defend his fellows, and to his corporeal structure. The supreme importance of these characters has been proved by the final arbitrament of the battle for life. Through his powers of intellect, articulate language has been evolved; and on this his wonderful advancement has mainly depended."[169]

The following year, in 1872, Nicholas Strakhov was making similar, and indeed customary, points in *The World as Totality*, arguing that humankind is the centre of creation and the most highly developed product of nature. Outraged at this typical example of human hubris, Tolstoy wrote to Strakhov in 1872, telling him that "the zoological perfection of man, on which you lay such stress, is extremely relative, for the very reason that man himself is the judge of it. The housefly is just as much the centre of creation and the most highly developed product of nature."[170] Tolstoy's target was Darwin as much as Strakhov. Indeed, it is because of the importance of reason, speech, and opposable thumbs to humankind that we designate our superiority in these terms. The German chemist Georg Christoph Lichtenberg wrote, in about 1768, that "the most accomplished monkey cannot draw a monkey, this too only man can do; just as it is also only man who regards the ability to do this as a distinct merit ... Just as foolish as it must look to a crab when it sees a man walk forward."[171] As we suggested earlier, the relevant category of acclamation for the bat or dolphin would be echolocation, for the honeyguide navigation and interspecies co-operation, for the wolf loyalty, for the bowerbird decoration, for the assassin bug camouflage, and for the salamander limb regeneration – little of which lies within the equal capacity of the human. Even in those matters in which humans are seen to excel, their abilities are not necessarily exclusive, for, as Joe Ackerley interprets in *My Dog Tulip* (1956): "Dogs read the world through their noses and write their history in urine."[172] This is a bit of a rationalization perhaps but a telling analogy, nonetheless.

The notion of human superiority was given a ribald twist in George Sand's *Indiana* (1831), where Sir Ralph Brown instructs us that "'what in my opinion, constitutes the main superiority of men over animals is that they understand where the remedy lies for all their ills. The remedy is suicide.'"[173] If suicide as a remedy for one's own ills is a greater rarity among nonhuman species, in certain of those species self-sacrifice for the communal interest is not. On these terms, even by human criteria, the superiority of nonhuman species at least merits consideration.

The earliest of the pre-Socratics to anticipate the later view that animals lived a better life than humans was Democritus, a fifth-century-BC atomist whose works are long lost but of which there are fragments as well as commentaries by later writers, who still had access to Democritus's writings. He remarked: "[How much wiser the animals are than man]; they, when they have need of anything, know how much they need; but when [man] needs something, he does not know how much of it he needs."[174] And Plutarch added: "Perhaps we are foolish to admire animals for their learning, although Democritus asserts that we are their pupils in all the most important things – of the spider in weaving and healing, of the swallow in building, of the song birds (the swan and the nightingale) in singing."[175]

There are, to be sure, many other reports of Democritus's statements, which, at the very least, add another dimension to his philosophy – for example, the "gods, both in the past and now, give men all things except those which are bad and harmful and useless."[176] In essence, however, if the Democritan fragments that remain are authentic, this most prolific of the pre-Socratics found humanity to be possessed of the capacity for both good and evil, whereas the animals are more constant in their behaviour. Democritus's contemporary, the dramatist Aristophanes, indicated that the more radical Sophists certainly held to the view of a Hobbesian human-animal identity of character, if not entirely to the notion of an animal superiority. In Aristophanes's *Clouds*, Strepsiades strikes his father and justifies his act with the remark: "Look at cocks and other such animals – they punish their fathers; and how do they differ from us – except that they don't make Acts of Parliament."[177] Aristophanes's own view was that the "naturalism" of these Sophists would "turn men into the likeness of savages or even of animals"[178] – for him a less than pleasing prospect. Nonetheless, theriophily is not absent from his play *The Birds*. Pisthetairos is disillusioned with the warlike failures of Athenian humanity and enlists the help of the birds to found "Cloudcuckooland," a utopian commonwealth in the sky, where, a bird tells us, "men of every nation / Confer on us, I understand, / Ecstatic approbation."[179]

The fourth-century-BC Greek Cynic philosopher Diogenes took the idea of animal superiority further than most others, before or since, at least if the account by Dio Chrysostom (d. after AD 112) is to be believed. Chrysostom was a Greek Sophist who resided in Rome, when not under banishment, and who shared with Plutarch in the revival of Greek literature. While there is no hard evidence to support Chrysostom's account, it conforms to the view held of the Cynics generally that they believed humankind to be an unhappy species because of its desires for luxuries and because of the softness that had developed in the human character as a consequence. By contrast, animals remained true to their original natures, becoming neither effete nor weakly. They were "happy beasts."

[Diogenes] used to say that men because of their softness live more wretched lives than the beasts. For these have water for their drink and grass for their food; most of them go naked the year round; they never enter a house, and make no use of fire. Yet, unless someone kills them, they live out their full span that nature has allotted to each of them; and all alike go through life strong and healthy, with no need of doctors or of drugs. Men, however, though they are exceedingly fond of life and invent so many devices for postponing death, for the most part do not in fact even reach old age; and their life is passed under a burden of ailments so numerous that it is not easy to name them all. Nor do the drugs which the earth produces suffice them; they must resort to knife and cautery as well ...

To the objection that it is impossible for man to live like the animals, by reason of the tenderness of his flesh and because he is naked, being unprotected by fur, as most animals are, or by feathers, or by a tough hide, [Diogenes] would reply that the reason for men's extreme tenderness lies in their manner of life – in the fact that they avoid both sunlight and cold. It is not the bareness of their bodies that causes the trouble. And he would point to the frogs and many other animals much more delicate and less protected than man, some of which nevertheless not only withstand cold air but are even able to live through the winter in the coldest water. He pointed out also that in man himself the eyes and face need no covering; and that in general no animal is born in a place in which it cannot live. If this were not so, how could the first human beings to be born have survived, without fire or houses or clothes or any food except what grew wild?[180]

Yet Ernest Barker considers Diogenes a mere moderate among his associates. Discussing the views of Antisthenes, a contemporary of Plato and founder of the Cynics, Barker tells us:

Apparently, he held that the wise man would not live in a State according to its enacted laws, but would live by the law of virtue, which is universal; while he believed that the nearer man approached to "the nature of the animals" (a subject on which he also wrote), the better it would be for human life. As it is used by Antisthenes, the parallel of animal life serves to point to the cry – Back to Nature: abandon cities, laws, and artificial institutions for all that is simple and primitive. It is the cry of the Radical Sophists: it is the cry of Rousseau in his youth. When we come to Diogenes, the greatest of the Cynics, we find a greater moderation and a different atmosphere.[181]

A later passage evokes not only images of Diogenes's taking up his abode in a tub and, upon seeing a peasant lad drink from his hands, discarding his last utensil, a cup, but also images of the nineteenth-century French social theorist of anarchism, Pierre Proudhon, who declared that "property is theft."[182] We read from Diogenes: "Do you not see these beasts and birds? how much more free from trouble they live than men, and how much more happily, too; and how much healthier and stronger they are, and how each of them lives as long as is possible for it. Yet they have neither hands nor human intelligence; but they enjoy one supreme good which overbalances all their disadvantages: they possess no property."[183] So why has our civilization chosen, generally speaking, to ignore the idea of "the happy beast"? As Chateaubriand wisely wrote: "no one is interested in beings who are perfectly happy."[184] No one wants to be outdone. We want to be protective heroes. As Machiavelli and Hobbes understood, we seek "glory."

If Seneca, whose views are notably ambiguous, did not believe that we had much to learn from the animals in terms of skills, at least he thought that the example of Diogenes could instruct us in the value of living in accord with their simplicity, even though it was an unachievable ideal:

> How, I ask, can you consistently admire both Diogenes and Daedalus? Which of these two seems to you a wise man – the one who devised the saw, or the one who, on seeing the boy drink water from the hollow of his hand, forthwith took his cup from his wallet and broke it ... The things that are indispensable require no elaborate pains for their acquisition; it is only luxuries that call for labor. Follow nature, and you will need no skilled craftsmen ... Nature was not so hostile to man, that, when she gave all the other animals an easy role in life, she made it impossible for him alone to live without these artifices.[185]

Still, if culture had made the simple life an impossibility for humankind, Seneca understood that the simple life had much *in principle* to commend it. The animals are not troubled by remorse and premonitions as are humans. "And so foresight, the noblest blessing of the human race becomes perverted. Beasts avoid the dangers which they see, and when they have escaped them are free from care; but we men must torment ourselves over that which is to come as well as over that which is past."[186] Human superior abilities diminish human happiness. Animal inferior abilities increase their happiness.

It would seem likely that many of those who ascribe superiority to the animals do so primarily in order to condemn their fellow humans, a practice that has certainly not abated in contemporary animal-rights literature. Thus when Jonathan Swift in *The Beasts' Confession* (1738) claimed to do "great honour to his own species almost equalling them to certain brutes," we can make a confident guess that he is talking partly tongue in cheek to lampoon humanity and perhaps *secondarily* to celebrate the animals. So too perhaps when Voltaire tells us in the *Philosophical Dictionary* (1764) that bees are superior to humans in that the secretions of the former are benevolent and those of the latter noxious. By contrast, Antisthenes appears to believe that life according to nature as the animals live it is a truly superior life. And Diogenes would appear, on the basis of the way he chose to live his own life in austerity and simplicity, to believe that, as a rule, the animals lived superior and more fulfilling lives by the absence of property. The institution of private property was for Rousseau, too, that which prevented humans from fulfilling themselves according to their nature: "The first person who, having fenced off a plot of ground, took it into his head to say *this is mine* and found people simple enough to believe him; was the true founder of civil society."[187] And for Rousseau, civil society was what brought about the division of labour, which proved the foundation of the loss of the altruistic and compassionate character that was endemic to the primitive and the animal.

Philemon (361-262 BC), a Greek poet of the New Comedy, shared the characteristic of his companions of demeaning humans through comparison with the animals – reason and vainglory made humans less happy creatures than their animal relatives:

> Much the most wretched of all animals is man, should anyone examine into his way of life. For he lives a life over-careful in all things and he is usually wholly at a loss about things and always toiling. And to all other beasts earth has freely given their daily bread, providing them herself, not taking ... but for men she grudgingly pays back the seed, as though returning the capital alone, and finding in drought or frost a pretext, withholds the interest. And perhaps because we alone give her trouble and turn her topsy-turvy, she takes this vengeance upon us.[188]

Of course, while this compares humans and animals to the advantage of the animals, it is clear that the animals are only a trope. Philemon is concerned not with the animals but with the human as the victim of Nature's injustice. It is a form of self-indulgent victimhood, from which our own era suffers in as great an abundance as did the era of the New Comedy. We cannot bear the consequences of treating ourselves as responsible for our own ills. Let us then blame them on others.

Philemon continues in much the same vein: "Oh thrice blessed and thrice happy in all things are the beasts who do not reason! None of them are ever put to the necessity of arguing, nor is any other such evil thing imposed upon them, but whatever nature each one bears, this he has as his law. But we men live an unenviable life. We are slaves to opinions in our law-making, slaves to our ancestors, slaves to our descendants. It is not possible to avoid evil, but we always find some excuse for not avoiding it."[189] Reason, of which humankind is so proud, and which is often referred to as that which ennobles humans over all other beings, is now seen as the source of our failures, for humans lack the certainty accorded other animals. Exemplified here again is the "woe is me" syndrome. Moreover, our individuality, which Saint Augustine and Thomas Hobbes later deemed a primary human characteristic that dignified us above "the lower animals," is used to belabour our problems: "Why in the world did Prometheus, who they say made us and all the other living beings, give to the beasts each a single nature according to its kind? All lions are brave; all hares, again, are timid. One fox is not a dissembler by nature and another sincere, but if one should collect thousands of foxes, he would see one nature in each and every one of them and one way of life. But as great as is the number of human individuals and of their bodies, so great is the spectacle of their ways of life."[190] Of course, Philemon's natural history leaves a great deal to be desired. Other species possess significant individuality, too, as all guardians of companion animals will know – to imagine ourselves "owners" rather than guardians is a delusion. But there is more reason for denying individual characteristics to a particular species than for failing to get to know them as the animals that they are. This is the case because, *from our perspective,* there is a certain commonality to each species that hides its individuality from us – at least when we are not thinking of those companions that we regard as members of our own families – precisely because we see them in contrast to ourselves as a species. And, no doubt, they too in judging us will assume a greater species consistency than exists in reality. Each species sees, at least at first blush, from the perspective of its own species character and even then only from the individuality of the particular member of the species.

When we turn to Menander, student of Theophrastus and the foremost poet of the New Comedy, we encounter a view that, while concerned primarily to chastise humanity for its capacity for self-destruction, also expresses a sincere respect for the ass – a respect not infrequently seen in later literature, notably from the pens of Henry Fielding, comte de Buffon, Samuel Taylor Coleridge, Laurence Sterne, and Robert Louis Stevenson:[191] "All the animals are most blessed and have much more intelligence than man. Look first at this ass. He is admittedly born under an unlucky star. No evil comes to him through himself, but whatsoever evils Nature has given him, these he has. But we, aside from the necessary evils, invent others by ourselves. We are pained if someone sneezes; if someone speaks ill, we are angry; if someone has a dream, we are frightened; if an owl hoots, we are terrified. Struggles, opinions, contests, laws, all these evils are added to those in nature."[192] What is especially surprising about Menander's remarks is that the animals are said to have much more intelligence than humankind. Perhaps, as James E. Gill has remarked: "Such passages ... are difficult to assess in the absence of a thorough knowledge of their contexts for, especially in comedy, such complaints can rarely be understood to represent in any strict way the author's philosophical views. It is probably safe, however, to regard them as standard comic comparisons given currency by contemporary philosophical speculation."[193] Even if the view expressed is perhaps not that of Menander himself, it appears to have been fairly widely held, if not perhaps in the extreme form that Menander gives it.

As discussed earlier,[194] there are a number of subtle differences between, for example, reason, intelligence, and rational intuition in Greek thought. But it is rare for a higher degree of any of them to be ascribed to the animals over the humans. Almost certainly, Menander is intentionally exaggerating the theriophilic point of view for dramatic effect. Most of those who ascribe intelligence to the animals are content to show either that it is quite adequate for the animals' needs or that it is quite similar to, but not superior to, human reasoning – although it might sometimes be offered as a capacity that, when coupled with a finesse of sense, is beyond all human capacity. Plutarch, for example, tells us:

> Even to this day the Thracians, whenever they propose crossing a frozen river, make use of a fox as an indicator of the solidity of the ice; if she perceives by the sound that the stream is running close underneath, judging that the frozen part has no great depth, but is only thin and insecure she stands stock still and, if she is permitted, returns to the shore; but if she is reassured by the absence of noise, she crosses over. And let us not declare this a nicety of perception unaided by reason; it is, rather, a syllogistic conclusion developed from the evidence of perception. "What makes noise must be in motion; what is in motion is not frozen; what is not frozen is liquid; what is liquid gives way."[195]

The third-century Greek Pyrrhonist Sextus Empiricus also related the use of animal intelligence that, when coupled with a superior sense of smell, produced a result beyond human capacity:

> According to [the Stoic philosopher] Chrysippus, who shows special interest in irrational animals, the dog even shares in the far famed "Dialectic." This person [i.e., Chrysippus], at any rate, declares that the dog makes use

of the fifth complex indemonstrable syllogism when, on arriving at a spot where three ways meet, after smelling at the two roads by which the quarry did not pass, he rushes off at once by the third without stopping to smell. For, says the old writer [Chrysippus], the dog implicitly reasons thus: "The creature went by this road, or by that, or by the other; but it did not go by this road or by that; therefore it went by the other."[196]

The use here by Sextus Empiricus of the phrase "irrational animals" is most instructive when the very purpose of the exercise was to demonstrate animal rationality. One finds an even more extraordinary and equally instructive usage in the title given to Plutarch's "On the Use of Reason by 'Irrational' Animals." The application of the term "irrational" to animals was a concession to contemporary linguistic convention, not a reflection of the writers' views on animal reason. When one encounters the phrase in the classical period, there is no guarantee that the writer denies a significant capacity for reasoning to nonhumans.

As Lovejoy and Boas made clear, following in large part the evidence of S.O. Dickerman, many of the ancients acknowledged a physical superiority of the animals over the humans. For example, Philo-Judaeus recognized the superiority of the bull in power, of the hare and the hunting dog in speed, of the eagle and the hawk in vision, and of the canine in sense of smell. Cicero allowed that if the hedonism of the Epicureans were valid, the beasts would exceed humans in happiness – but he certainly denied the premise.[197] For Cicero, as for Xenophon and Aristotle, animals exist for the sake of humankind. Again, however, it would be unwarranted to ignore the complexities of Cicero's views as well as those of Aristotle. In a letter to Marcus Marius in August 55 BC, Cicero abominated the mistreatment of animals in the Circus Maximus and sympathized with the "impulse of compassion" shown by the commoners toward the animals, reflecting a feeling that the animals "had something human about them."[198] And Aristotle commented favourably on the nest-building abilities of the swallows and their tender care for their offspring, on the organization of the cranes for their defence, on the capacity for self-medication of the goats, and on the singing instruction of the nightingale mother to her scions[199] – points repeated frequently in the classical literature thereafter.

By contrast with Philo-Judaeus, Diodorus Siculus was not convinced that animal superiority was restricted to the senses – or, at least, so he wrote. Diodorus reported a conversation between various philosophers, conducted at the behest of Croesus, king of Lydia. In response to the question from Croesus concerning who were the most courageous of beings, Anacharsis replied: "the wildest of the beasts. For they alone die willingly for freedom."

Who, then, is the most just? Again, Anacharsis gave the same reply, saying "they alone live in accordance with nature, not law. Nature [is] the work of God, but law a convention of man, and it is more just to employ the devices of God than those of man." Finally, he is asked whether the beasts are also the wisest. Indeed so, since "to reverence the truth of nature above the conventions of law was the most characteristic possession of wisdom."[200] It seems certain, however, that Diodorus is far more concerned with providing a clever argument in refutation of customary human conceits about the value of their cultural acquisitions than with portraying "the wildest of the beasts" as in reality the superior beings. Nonetheless, the fact that in *De ira* Seneca goes to some pains to refute the belief that humans have much to learn from animals indicates the degree to which the belief must have been held, at least in the late classical period: "He errs who adduces [the animals] as an example for man, for in them impulse takes the place of reason. In man reason takes the place of impulse." Seneca goes on to elaborate the quite antithetical impulses of different species, concluding "why should you propose to man such infelicitous examples [as anger, timidity, flight], when you have the world and God, whom man, alone of all the animals understands, so that he alone may imitate him?"[201]

While Ovid is also decidedly unwilling to countenance ideas of animal superiority, he is convinced that animals find happiness more readily not on account of *their* superior abilities but because of what is presented as the narrow-mindedness of human social conventions. However, Ovid prefaces an illlustrative legend by stating that the "story I am going to tell is a horrible one: I beg that daughters and fathers should hold themselves aloof ... and refuse to believe this part of my tale." It is the story of Myrrha, Cinyras's daughter, falling in love with her father:

> The girl realized what was happening to her, and fought against her horrible desire. "What am I thinking of, what am I scheming?" she said to herself. "I pray to the gods ... to prevent me from my crime. But is it? No fault can be found with this kind of love on the grounds that such affection is unnatural, for other animals mate without any discrimination; there is no shame for a heifer in having her father mount her, a horse takes his own daughter to wife, goats mate with the she-goats they have sired, and birds conceive from one who was himself their father. Happy creatures, who are permitted such conduct! Human interference has imposed spiteful laws, so that jealous regulations forbid what nature itself allows."[202]

Ovid presents a picture of Myrrha horrified at her own impulses and disputing with herself whether what she desires is a crime or subject to justification by nature. Clearly, to the extent that the use of animal analogy is not here a mere rationalization, Myrrha's argument involves the notion that natural law applies alike to humans and animals and that the evidence for one may thus be justifiably applied to the other. Indeed, the premise of Myrrha's argument appears to have been taken up later in the foundations of Roman law. The third-century Syrian and premier Roman jurist Ulpian, primary author of the *Digest*, which formed the basis of the *Corpus Juris Civilis*, which in turn laid much of the foundation for Roman law, argued that the *ius animalium* (natural justice for animals) was a part of the *ius naturale* (natural law, or natural justice, in general). The formulation provides the very opening section of the *Digest* itself:

> Private law is threefold; it can be gathered from the precepts of nature, or from those of the nations, or from those of the city. Natural law is that which nature has taught all animals; this law indeed is not peculiar to the human race, but belongs to all animals ... From this law springs the union of male and female ... the procreation of children and their education ... the law of nations is that which mankind observes. It is easy to understand that this law should differ from the natural, inasmuch as the latter belongs to all animals, while the former is peculiar to men.[203]

Moreover, the *ius gentium* (the prevailing law found by the Romans as they ventured into alien territories) "differs from the natural [*ius naturale*] because it is common only between men, while [*ius naturale*] is common among all living creatures."[204] Indeed, the *ius gentium* countenanced slavery, a practice quite incompatible with the *ius naturale*. If, as is customary, we take natural law as the highest of the categories of law and thus as that to which substantive laws must conform – as Paul Sigmund explained, "a conflict between natural law and an existing legal provision was grounds for annulment of the positive law"[205] – then, in terms of Ulpian's formulation, Myrrha's argument has merit, at least if we accept her evidence as both valid in the instances of the animals she names and valid for animals in general, not necessarily a wise supposition. Whether it truly has merit is of less importance than the fact that formulations in these terms, both in the telling of myths and in the formulation of legal principles, suggests a sense of a closer human-animal proximity than we customarily imagine. There is seen to be a common justice that applies to humans and animals alike, that applies to the legal system dominating the classical world and greatly influencing its successors, and that is seen to arise from the common laws of nature.

Pliny the Elder's encyclopedic account of the state of knowledge in the first century provides another instance of the victimhood syndrome. Yet he gives the game away by heading his section on "Man" in the *Natural History* with the caption: "Man is the highest species in the order of creation." Despite human heights, however, Pliny bemoans the fact that:

4. All other animals know their own natures: some use speed, others swift flight, and yet others swimming. Man, however, knows nothing unless by learning – neither how to speak nor how to walk nor how to eat; in a word the only thing he knows instinctively is how to weep. And so there have been many people who judged that it would have been better not to have been born, or to have died as soon as possible.

5. Man alone of living creatures has been given grief, on him alone has luxury been bestowed in countless forms and through every limb – and likewise ambition, greed and a boundless desire for living, superstition, anxiety about burial and even about what there will be after his life ends. No creature's life is more fragile; none has a greater lust for everything; none, a more confused sense of fear or a fierce anger. To sum up, other creatures spend their days among their own kind. We see them gather together and take their stand against other species; fierce lions do not fight among themselves, serpents do not bite serpents – not even sea-monsters and fish act cruelly, except against different species. But man, I swear, experiences most ills at the hands of his fellow men.[206]

Humankind may be "the highest species in the order of creation," but there can be little doubt that, for Pliny, the other animals are happier. Indeed, Pliny acknowledged that other species know what is healthful for them while humans do not.[207] If Pliny is right about the happiness of the beasts, we must abandon the notion that humanity is the highest species or reject the common belief that happiness is the appropriate goal.

The same may be said for the argument of the Christian-Socialist cleric Charles Kingsley in *Yeast* (1851). Humans live as good a life as the animals, better than most, unless one belongs to the lower human orders. "The finest of us are animals after all, and live by eating and sleeping: and taken as animals, not so badly off either – unless we happen to be Dorsetshire labourers – or Spitalfield weavers – or colliery children – or marching soldiers – or, I am afraid, one half of English souls this day."[208] For Kingsley, the animals live a better life than the oppressed working classes. If relieved of their oppression, however, the workers would achieve their rightful place among the highest species. The most ambitious classical argument for animal superiority and the one most frequently borrowed from in the Renaissance, and later, was Plutarch's essay "On the Use of Reason by 'Irrational'

Animals," based on a loose adaptation of the tenth book of Homer's *Odyssey*.
Some Greek warriors have been captured by the enchantress Circe and trans-
formed into swine. Odysseus (Ulysses) tries to persuade their spokespig,
Gryllus ("grunter"), to have them converted back to humans again.

GRYLLUS [to Odysseus]: ... what you're scared of is people changing from
  a worse state to a better one ... you're trying to persuade us, whose lives
  are filled with countless advantages ... to sail away with you after having
  changed back to human beings, when there is no creature worse off than
  a human being ... Let's start with the virtues, which you humans obviously pride yourselves on: you consider yourselves superior to animals in
  morality, intelligence, courage and all the other virtues. Now, you're really
  clever, Odysseus, so answer this question of mine. You see, I once over-
  heard you telling Circe about the Cyclope's land and how it is never
  ploughed and never sown at all, but is such good and innately generous
  land that it spontaneously yields produce of every kind. So my question
  is: which do you prefer, the Cyclope's land or harsh Ithaca, which sup-
  ports only goats and which grudgingly gives its farmers a return on their
  considerable efforts and hard labour, a return that is small in quantity,
  poor in quality, and of low value? ...
ODYSSEUS: ... Although my own native land occupies a higher place in my
  heart and my affection, my acclaim and admiration go to the Cyclope's
  land.
GRYLLUS: So we'll say that this is how it is: the most intelligent man in
  the world expects to value and appreciate different things from those that
  he is drawn to and loves. And I assume your answer is also relevant to the
  mind, given that the same goes for it as the land: the better mind is that
  which produces its spontaneous crop, so to speak, of virtue without involving hard work.[209]
ODYSSEUS: I grant you that assumption as well.
GRYLLUS: Well, you're now admitting that the animal mind is better
  equipped by nature for the production of virtue, and is more perfect. I
  mean without being instructed or schooled – without being sown or
  ploughed, as it were – it naturally produces and grows whatever kind of
  virtue is appropriate to a given creature.
ODYSSEUS: Virtues in animals, Gryllus?
GRYLLUS: Yes, virtues of every kind, more than you'd find in the greatest
  human sage. Let's start with courage if you like, which is a particular
  source of pride to you ... when animals fight with one another or with
  you humans, they do not employ tricks and stratagems; they rely in their
  battles on blatant, bare bravery backed up by genuine prowess ... You

don't find animals begging or pleading for mercy or admitting defeat.
Cowardice never led a lion to become enslaved to another lion, or a horse
to another horse ... if you think that you humans are superior to animals
where courage is concerned, then why do your poets use as epithets for
the best fighters "wolf-spirited," "lion-hearted" and "of boar-like" bravery?
Why do none of them describe a lion as "human-hearted" or a boar as "of
man-like bravery"? ... animals draw on pure passion in their fights, whereas
for you humans it is diluted with rationality ...

Animals ... have minds which are completely inaccessible to and un-
approachable by alien feelings and lifestyles ... They keep their distance
from elegance and extravagance in their way of life, and assiduously protect their self-restraint from their better self-government by restricting
the number of the desires which live with them and by forbidding entry
to extraneous ones ... [A]nimals' intelligence refuses to accommodate any
expertise which has no point or purpose; and it doesn't regard the necessary skills as matters to be imported from others, or as things to be taught
for money ... but rather it generates the necessary skills on the spot, out
of itself, as if they were native and natural products ...

You see, if you reply with the truth – that nature is their teacher – then
you're attributing animals' intelligence to the source of all authority and
skill. And if you want to deny the name of reason or intelligence to what
animals have, then it is time for you to try to find a more attractive and
distinguished name for it, since there is no doubt that the faculty it constitutes is both practically better and more impressive than human intelligence. Ignorance and lack of information play no part in it; rather it is
a self-taught and self-sufficient faculty. And this is not a sign of weakness,
but is because of the strength and perfection of its natural virtue that it
does without any educational contribution that external agents might
make of its intelligence ...

In case you're not convinced we [animals] can learn skills, then listen
while I demonstrate that we even teach them. For instance, hen partridges
teach their chicks to avoid danger by concealing themselves – to lie on
their backs and hold a clump of earth in front of themselves with their
claws. And consider storks: you can see them on the roofs, the mature
ones in attendance instructing those which are tentatively learning to
fly. Nightingales give singing lessons to their chicks, and those which
are caught while they are still young and are brought up in captivity by
humans don't sing very well, as they've been separated from their teacher
too soon ...

ODYSSEUS: ... Do you claim that even sheep and donkeys are rational?

GRYLLUS: My dear Odysseus, these are precisely the ones to provide very strong evidence of the fact that animals' nature is not lacking in reason and understanding. A given tree is not more or less mindless than any other tree; they are all equally unconscious, because none of them has a mind. Analogously, a given creature would not give the impression of being intellectually less alert and quick than any other creature, if they did not all have reasoning and understanding, but in varying degrees.[210]

Perhaps Plutarch is not convinced of the absolute and overall superiority of the nonhuman species – although if we take him entirely at his word, he is – but beyond doubt, he believes that the animals have the advantage in certain respects. Their intelligence meets their needs better than human intelligence meets human needs, and their attributes fulfill their requirements as the animals that they are better than human attributes help humans to satisfy their natures.

There are occasions when the Bible is no less restrained than Plutarch about superior animal knowledge. For example, in Job 12:7-9 we are told: "You have only to ask the cattle, for them to instruct you, and the birds of the sky for them to inform you. The creeping things of the earth will give you lessons, and the fish of the sea provide you with an explanation: there is not one such creature but will know that the hand of God has arranged things like this." Further in Proverbs 6:6-8 we are instructed: "Idler, go to the ant; / ponder her ways and grow wise: / no one gives her orders, no overseer, no master, / yet all through the summer she gets her food ready, / and gathers her supplies at harvest time." And in Jeremiah 8:7 we read: "Even the stork in the sky / knows the appropriate season; / turtledove, swallow and crane / observe their time of migration. / But my people do not know / Yahweh's laws" (*The New Jerusalem Bible* translation).

Famously, in *Utilitarianism* (1863) John Stuart Mill proclaimed, as we saw earlier, that "It is better to be a human being dissatisfied than a pig satisfied."[211] One wonders whether Mill had read Plutarch. After all, Mill merely asserts his case rather than arguing it. He assumes that his audience will concur. Not only Plutarch, but also many of the classics, would have rejected his view. Again, Mill insists that the intellectual pleasures are superior in kind. "Of two pleasures," he writes, "if there be one to which all or almost all who have experience of both give a decided preference ... that is the more desirable pleasure ... Now it is an unquestionable fact that those who are equally acquainted with, and equally capable of appreciating and enjoying both, do give a most marked preference to the manner of existence which employs their higher faculties."[212]

"Unquestionable fact" or not, the Greek philosopher Cyrenaios attempted to refute such claims by valuing bodily pleasure above intellectual

pleasure because it is more intense and powerful. Moreover, Ovid and Diodorus Siculus would have looked at the animals in their pleasures and then
looked askance at John Stuart Mill. And Menander, Democritus, and Plutarch would have wondered, even if Mill were right, whether the animals
got a greater quantum of pleasure out of their mental activity than did
humans. Tolstoy came very much to the same conclusion – the human pursuit of pleasure was in fact alienating rather than fulfilling. Thus in *The Law
of Love and the Law of Violence* (1908), Tolstoy tells us that humans

> are only differentiated from animals by the fact that since time immemorial
> animals have kept the same stomachs, claws and fangs, while humanity moves,
> ever more rapidly, from dirt roads to railroads, from horses to steam, from
> spoken sermons and letters to the printing press, telegraph and telephone;
> from sailing boats to ocean liners; from side arms to gunpowder, cannons,
> machine guns, bombs and bomber planes. And life with telegraphs, tele-
> phones, electricity, bombs and aeroplanes, and with enmity between all
> peoples, who are guided not by some unifying spiritual principles but by
> alienating animal instincts, and who use intellectual faculties for their own
> pleasure, is becoming more and more futile and calamitous.[213]

To be sure, Tolstoy constantly advises us of human rational superiority.
For example, in *A Confession and Other Religious Writings* he writes:

> The difference between man and the animals is that an animal's cognitive
> faculties are restricted to what we call instinct, while man's basic cognitive
> faculty is his reason. The bee gathering his food can have no doubt as to
> whether he is doing something that is good or bad. But a man reaping the
> harvest, or gathering fruit, cannot help wondering whether he is undermin-
> ing the growth of future crops, or whether he is depriving his neighbours of
> food. Nor can he help thinking about what will happen to the children he
> is feeding, and much else besides ... He cannot be satisfied by the same
> things that guide an animal's behaviour.[214]

As we saw earlier,[215] Twain recognized the same distinction, but despite
the great superiority of human reason, humans were the inferior beings
because they chose the immoral rather than the moral. For Tolstoy, too –
with the exception of the human as a religious being[216] – the human's superior intellectual abilities, which are used for the pursuit of pleasure, and
ultimately of war, denigrate rather than elevate the human. As he rumi-
nated in *Resurrection* (1898),

the jackdaws, the sparrows and the pigeons were cheerfully getting their nests ready for the spring, and the flies, warmed by the sunshine, buzzed gaily along the walls. All were happy – plants, birds, insects and children. But grown-up people – adult men and women – never left off cheating and tormenting themselves and one another. It was not this spring morning which they considered sacred and important, not the beauty of God's world, given to all creatures to enjoy – a beauty which inclines the heart to peace, to harmony and to love. No, what they considered sacred and important were their own devices for wielding power over each other ...

[T]he grace and gladness of spring had been given to every animal and human creature.[217]

And the animals had respected the gift, whereas the adult humans had cast it aside. It is difficult, on the basis of such reasoning, to see how humans can be entitled to superior consideration. Any right to which superior reason might have given entitlement is removed by the manner in which this reason is employed. Aristotle said that the human "when perfected, is the best of animals; but if he be isolated from law and justice he is the worst of all" (*Politics* I.2.15). Along with Twain, Tolstoy, Diogenes, Philemon, Pliny, and Plutarch, Aristotle would surely, on this basis, have been compelled to conclude that humans lacked the willingness to pursue a justice that might place them at the apex of the animal kingdom. They must thus belong to the lowest part of animality.

XI

Generally speaking, the medieval era, at least on the religious and philosophical surface, was decidedly less kind to animal capacities than had been the classical age. Thus writes George Boas:

In Origen, we find a defense of man's technological achievements which is based ... on a familiar pagan *topos,* the hostility of Nature to man. Celsus, the arch-enemy of the Alexandrine father, had apparently been playing upon this theme, the most popular expression of which became the Latin slogan, *natura non mater sed noverca* [nature is not the mother, it is the stepmother], a slogan which was supposed to prove the superiority of the brutes over men. Origen, instead of justifying man's position as a necessary consequence of the Fall, argues that God did indeed create him in want, but at the same time endowed him with the power to remedy his deficiencies. Necessity, he maintains, is the mother of invention, and the brutes' lack of need is a defect, not an advantage. For feeling no lack they invent nothing.[218]

The idea of nature as the cruel stepmother has a venerable history, and not merely in *Cinderella*. We find it in Pliny when he tells us of the cruelties inflicted upon humankind by nature: "She is rather a cruel stepmother."[219] Almost a millennium and a half later, Leonardo da Vinci employs the same trope: "O Indifferent nature, whereof art thou so partial, being to some of thy children a tender and benignant mother, and to others a most cruel and pitiless stepmother?"[220] The difference is that, for Leonardo, nature has turned her wrath undeservedly on the patient and servile ass. In *Contra Celsum,* Origen remarked that:

> the irrational have their food all ready for them, but they have not even an impulse towards the arts. And they have also a natural covering. For they have hair or feathers, or scales, or shells. And let this be our defence against these allegations of Celsus, "We, laboring and perishing in our labor, are fed in scarcity and with toil. But for [the animals] all things are grown unsown and unploughed" ... We at any rate, though much weaker than the animals in body, and smaller than some to a great degree, are stronger than the beasts by our understanding ... The Creator has made them all slaves of the rational animal and of his natural understanding.[221]

It is notable that in Celsus's argument, the animals have remained in the Golden Age of the state of nature, whereas humans have begun the slippery slide away from their peaceful Edenic nature and toward chaotic progress via lusty Arcadia.

If the notion of the animal as superior seems to diminish in medieval theology, we can still find inklings of it (and sometimes more than inklings) in late-classical and medieval legends insofar as humans are constantly having to rely on animals' abilities to help them achieve their ends. Thus in the Welsh *Mabinogion,* King Arthur's courtiers have to rely on a host of animals to help them discover the whereabouts of the kidnapped Mabon and then to transport them to the place of captivity; and in the same myth, the lion and the knight engage in reciprocally supportive behaviour, each being superior in his own way. It is a story that began in Rome with Androcles facing a lion in the Roman amphitheatre and that is repeated in Germany and Italy with Saint Ambrose and then in France with Yvain. In Russia the lion was replaced by a bear, and Saint Ambrose by Saint Sergey. In the second-century apocryphal *Acts of Paul,* it is Saint Paul who faces a lion in the Colosseum. The lion, recognizing Saint Paul as the one who baptized him and brought him to God, consequently refuses to harm him. In the seventh century John Moschus told the story of Abba Gerasimos, who removed a reed stuck in the paw of a lion on the banks of the

River Jordan. Thereafter, they became firm friends and undertook several adventures together. The lion's magnanimity is central to each of these stories. Indeed, for all intents and purposes they are the same story set in different contexts. And they all emphasize a human-animal reciprocity.

In *Gudrun* a swan helps the eponymous heroine to claim her lover. Lohengrin's search for the Holy Grail is also aided by a swan. In *The Ettin Langshanks,* Dietrich has a noble horse save him from death. In the *Ràgnar Lodbrok Saga,* magpies come to the aid of Krake. In the medieval version of *The Legend of Alexander the Great,* the steed Bucephalus, praised for having saved the emperor from many a peril and for having "stood him in stead, in battles full and hard," has even "won him worship in war."[222] While there can be little doubt that humans are viewed as the primary purpose of life, we find also both reciprocity with the animals and an acknowledgment of their superior abilities in certain respects. Certainly, in the Carolingian legend of Rinaldo and Bayard, the horse is more faithful than the knight.[223] In the *Mabinogion* the lion is ultimately more valiant than Owain, although Owain's valour is necessary to save the lion from the serpent.[224]

The Vision of William concerning Piers the Ploughman is a fourteenth-century allegorical, alliterative, unrhymed poem that acknowledges a certain animal superiority.

Then Nature approached me, calling me by name; and he bade me take heed, and gather wisdom from all the wonders of the world. And I dreamt that he led me out on to a mountain called Middle-Earth, so that I might learn from all kinds of creatures to love my Creator. And I saw the sun and the sea, and the sandy shores, and the places where birds and beasts go forth with their mates – wild snakes in the woods, and wonderful birds, whose feathers were decked with many colours. And I could see man and his mate, in poverty, and in plenty, in peace and war; and I saw how men lived in happiness and misery both at once ...

And I perceived how surely Reason followed all the beasts, in their eating and drinking, and engendering of their kinds ...

And I beheld the birds in the bushes building their nests, which no man with all his wits, could ever make. And I marvelled to know who taught the magpie to place the sticks on which to lay her eggs and breed her young; for no craftsman could ever make such a nest hold together, and it would be a wonderful mason who could construct a mould for it.

And yet I wondered still more at other birds – how they concealed their eggs secretly on moors and marshlands, so that men could never find them; and how they hid them more carefully still when they went away, for fear

of the birds of prey and of wild beasts ... And when I saw all these things
I wondered what master they had, and who had taught them to rear their
houses so high in the trees, where neither man nor beasts could reach their
young.[225]

Under the guise of reason and nature, nonhuman animals were the happier creatures.

Even when we turn to the pre-Augustinian theologians, we find a measure of recognition of animal abilities. Thus, as we saw earlier,[226] the fourth-
century Arnobius emphasized human-animal similarities, concluding that
"even in those things [that the animals] make with beaks and claws, we see
that there are many appearances of reason and wisdom which we men are
unable to copy, however much we ponder them, although we have hands
to serve us dextrously in every kind of work."[227] And his contemporary
Lactantius acknowledged an exclusively human religious sense, but other
characteristics were held largely in common: "Can anyone deny that there
is reason in them when often they delude man himself?"[228] Nonetheless,
the notion of the animal as a superior being was largely lost among the
writing clergy. But not completely. Bernard of Siena, whose mind was of
the Renaissance if his dates – 1380-1444 – were not, was an exception. The
Franciscan saint – perhaps more a composer of sermons than a true theologian – was convinced of the moral superiority of the animals: "Look at
the pigs who have so much compassion for each other that when one of
them squeals, the other will run to help ... and you children who steal the
baby swallows. What do other swallows do? They all gather round together to try to help the fledglings ... man is more evil than the birds."[229]
Walter Map, Archdeacon of Oxford at the end of the twelfth century –
perhaps more a writer of profane history than a theologian – was also an
exception. He wrote in *De nugis curialium* of the degeneracy of humanity
as exemplified in courtly life: "Now how comes it that we men have
degenerated from our original beauty, strength and force, while other living creatures in no way go astray from the grace first given to them? The
creatures of earth, sea and air – everything except man – rejoices in the life
and powers with which they were created. They, it seems, have not fallen
out of their maker's favour."[230] Both saw something superior about animals
and their lives. But if theriophily was in retreat in general, it would not
be so for long, for the Renaissance saw both a revival of human hubris
and, at the same time, a self-questioning of human conceits. History as an
ongoing battle of competing ideas continued but with even greater conflicts of ideas than before – if for no other reason than the Renaissance's
accentuation of individuality.

5

Theriophily Redivivus

I

AVIEW PREVAILS IN THE LITERATURE ON animal ethics that the age of "humanism" was decidedly detrimental to animal interests. Thus, for example, in *Animal Liberation*, Peter Singer writes: "The central feature of Renaissance humanism is its insistence on the value and dignity of human beings, and on the central place of human beings in the universe. 'Man is the measure of all things,' a phrase revived in Renaissance times from the ancient Greeks is the theme of the period ... the Renaissance humanists emphasized the uniqueness of human beings, their free will, their potential, and their dignity; and they contrasted all this with the limited nature of the 'lower animals.'"[1] One would have thought that the discoveries by Copernicus, Galileo, and Keppler, perhaps the defining moments of the new era, which removed the earth from the centre of the universe around which all else revolved, would have cast some doubt on the idea of the centrality of human affairs. The Renaissance introduced the idea that became the governing conception of the Enlightenment: all nature, humanity included, was continuous, simple, and uniform. In fact, nothing could have been more significant in warning humanity against a belief in its primal significance. And, indeed, this is precisely why some declared the findings heretical. And others founded in opposition the Accademia Nazionale dei Lincei (National Academy of the Lynx), the Italian forerunner of all national scientific societies, to promote objective knowledge and freedom of thought. Moreover, if the Reformation was more liberal, enlightened, and individualistic in allowing personal interpretations of the scriptures without a priestly intermediary, it was also more conservative in emphasizing the wretchedness of humanity against the optimistic Thomist view of human nature and faith in reason. Indeed, if there was not too much in which Luther and Calvin concurred, they were in unison

against the optimism of Aquinas. They applauded together the Augustin-
ian declaration of human depravity, skepticism toward reason, and reliance
on faith.

Ignoring the complexities of competing tendencies in human thought,
Richard Ryder claims that "the growing anthropocentrism of the Renaissance heralded several centuries of outstanding cruelty."[2] The generally
astute Mary Midgley suggests that the "humanist [attitude] ... makes it
hard to see that man might have any significant link with other species."[3]
Yet this is to miss the psychological importance of the humanist revival.
What humanism overcame was the truly ecclesiastical medieval attitude of
what the Thomist scholar Jacques Maritain described as "keeping men from
thinking about themselves." In the second half of the fourteenth century,
Francis Petrarch announced: "I am an individual and would like to be
wholly and completely an individual; I wish to remain true to myself, so
far as I can."[4] This is indeed the reveille of the Renaissance, although to be
accurate, as Michael Seidlmayer has demonstrated convincingly, the clarion was heard a century earlier at the time of the Hohenstaufen – "the
most creative and, intellectually, the most important of the Middle Ages
... one of the summits in the development of European culture. It was *then*
and not at the time of the Renaissance, that what Burkhardt called 'the
discovery of the world and man within it' really began."[5]

Petrarch's statement may be interpreted as egotism. Yet decidedly it is
not. It is the announcement of an awakening from the slumber of seeing
everything about the individual in terms of relationships – to family, for
example. The story of Romeo and Juliet is a striking instance of the conflict arising from the new awareness of the importance of the self rather
than of one's family. Knowing one's place, especially as to class, had long
been the hallmark of one's identity – through traditional kinship associations rather than choice and reason. Church and state, for example, were
primary but frequently conflicting obligations to which one's self must be
inescapably subordinated. In 1555 at the Augsburg Peace of Religions, the
principle was instituted of the monarch of each participant state determining the religious affiliation of his subjects: *cuius regio, eius religio* – both
nationality and denomination were determined for one. To be sure, it was
a compromise curiosity in the absence of a practicable principle but nonetheless a worthy testament to the collectivist ethic of a dying world.

Petrarch's words are a prerequisite for being able to recognize individuals as ends in themselves – which is in turn a prerequisite for seeing animals
as ends in themselves and not merely as means. It is only when we can look
to ourselves and say "I" that we can look to the animals and acknowledge
their right to say "I," too, which was precisely how Leibniz understood the

matter. If the Renaissance, and its aftermath, witnessed a deplorable conception of the human as the only truly worthy animal, as with the Cartesians, it also experienced a conflicting and growing compassion and respect for our fellow creatures among the large body of anti-Cartesians. If the Renaissance was a spur to the most heinous vivisection, it was also a spur to its equally vehement denunciation. Only a proportion of this denunciation was based on a conception of the animal as a superior species, but much was based on a conception of the animal as most seriously worthy of ethical consideration.

The very frequency of the mention of René Descartes in books pertaining to the history of animal ethics is reflective of the great influence that the frequently dubbed "father of modern philosophy" is believed to have wielded. To be sure, his rationalist skepticism was of the most profound importance for his era and continued thereafter. He also made advances in optics, algebra, physiology, and law. Yet, as we have already observed,[6] the influence of his notion of the animal as a machine, as an automaton, is greatly exaggerated. Perhaps the notion's primary influence was to provide easy sport for those who wanted to denounce his views as entirely contrary to logic and the experiences of common sense. Even in France, where the doctrine was rather more persuasive than elsewhere, more stood up to denounce it than to embrace it.

Despite this, we are left with a decidedly contrary impression in the animal-welfare and animal-rights literature. For instance, Jim Mason writes that "More than any other thinker, Descartes detached humanity from the natural world and set it up as the ruling class, aloof from and absolutely unrelated to its underlings. From him we get the thinking that prevails in the modern era."[7] Roderick Frazier Nash tells us that "Descartes provided a general philosophy of the irrelevance of ethics to the human-nature relationship." He is so convinced of the pervasiveness of this view in the Old World that he tells us, offering us a complete contrast, that "at the height of Descartes's influence in Europe, the New Englanders endorsed the idea that animals were not unfeeling machines."[8] He is referring to the proclamation of "The Body of Liberties" of the Massachusetts Bay Colony in 1641, which required that "no man shall exercise any Tirrany or Crueltie toward any bruite Creature which are usuallie kept for man's use." "Perhaps," he adds, "the task of creating a new society in a wilderness made the [American] Puritans more mindful of comprehensive ethical principles derived from a state of nature."[9] The fact that "The Body of Liberties" was authored by a recently immigrated English Puritan, Nathaniel Ward, might have given cause for further reflection. But so might Ward's borrowing of the phrase "the cruelty used to the beasts" from a justification for a piece of Irish animal-welfare legislation in 1635.[10] And so might the outlawing of public animal-baiting during the English Commonwealth. A host of British Puritan statements of the time could be offered as parallels to the sentiments behind "The Body of Liberties." One 1625 example from William Hinde, perennial curate of Bunbury in Cheshire, will suffice: "I think it utterly unlawful for any man to take pleasure in the pain and torture of any creature."[11] Nash's assumption of the ubiquity of Cartesianism has misled him into imagining, without due warrant, a uniqueness in the American experience.

In a discussion of the role of Christian philosophy in general and of René Descartes in particular, James Serpell referred to the "Cartesian idea that [animals] were incapable of suffering." Such ideas, he adds, "provided human beings with a licence to kill; a permit to use or abuse other life-forms with total impunity."[12] He quotes Keith Thomas in support of his thesis: "In drawing a firm line between man and beast, the main purpose of early modern theorists was to justify hunting, domestication, meat-eating, vivisection." A more accurate portrait would have been sketched if he had quoted further from Thomas: "though Descartes's work was disseminated in England, the country threw up only half a dozen or so explicit defenders of the Cartesian position ... the theologian Henry More was more representative of English opinion when he bluntly told Descartes in 1648 that he thought this a 'murderous' doctrine. Most later English intellectuals felt with Locke and Ray that the whole idea of beast-machine was 'against all evidence of sense and reason' and 'contrary to the commonsense of mankind.'"[13] Moreover, as we saw to a degree in the last chapter and shall see further in this, the reception of Descartes in his homeland, if more favourable than in Britain, produced a similar array of denunciations.

II

The most consummate Renaissance man – indeed, the exemplar for whom the term was coined – was Leonardo da Vinci. Not only was he one of the world's greatest artists – by whatever criteria one might wish to employ – but he also contributed to the development of physics and biology, to the understanding of the potential for human flight and submarine travel, and to the comprehension of the relationship between time, space, sound, music, painting, sculpture, and architecture. He was, moreover, no less emblematic of the new humanism than Boccaccio, Petrarch, or Sir Thomas More. Yet he does not elevate humans over the animals, as one might imagine in light of the manner in which the term "humanism" has been portrayed in the animal-rights literature. Indeed, a significant measure of theriophily can be found in his writings insofar as he regarded human behaviour as sometimes being lower than that of the beasts:

> Animals will be seen on the earth who will always be fighting against each other with the greatest loss and frequent deaths on each side. And there will be no end to their malice; by their strong limbs we shall see a great portion of the trees of the vast forest laid low throughout the universe; and when they are filled with their food, the satisfaction of their desires will be to deal death and grief and labour and fear and fright to every living thing; and from their immoderate pride they will desire to rise towards heaven, but the excessive weight of their limbs will keep them down. Nothing will remain on earth, or under the earth, or in the waters, which will not be persecuted, disturbed and spoiled, and those of one country removed to another. And their bodies will become the tomb and means of transit of all living bodies they have killed.[14]

Of course, the animal that the vegetarian Leonardo is describing is the human animal – the oppressor and murderer of other species and devastator of the land, the marine, and the subterranean environments – through decimation of the forests, largely for ship building, through fishing, and through mining, respectively – a remarkable prescience! Certainly, the "central feature" of Leonardo's thoughts is not the "insistence on the value and dignity of human beings." To be sure, in *Oration on the Dignity of Man*, Giovanni, Conte Pico della Mirandola, wrote that "nothing in the world can be found that is more worthy of admiration than man"[15] – as Peter Singer points out. However, Singer does not point out that Mirandola's *bête noire*, the Catholic Church, denounced him scathingly for his presumption. Nor does he mention that on occasion Mirandola was known to contradict himself and belittle the presumptions – "vanities" he calls them – of science and philosophy in favour of traditional simplicity.[16] More typical of the era than Mirandola, and decidedly Catholic, were the words of the early-Renaissance German monk Thomas à Kempis in *The Imitation of Christ:* "While man extends his power in every direction, he does not always succeed in subjecting it to his own welfare. Striving to penetrate farther into the deeper recesses of his own mind, he frequently appears more unsure of himself. Gradually and more precisely he lays bare the laws of society, only to be paralysed by uncertainty about the direction to give it."[17] Nor was Kempis without recognition of animal significance, for he wrote that "if thy heart were right, then every creature would be for you a mirror of life and a book of holy teachings."[18] Of course, Kempis was not a humanist figure as such, but his beliefs and values nonetheless pervaded what is known as the Renaissance era rather more convincingly than did those of Mirandola. And, indeed, as precept rather than practice, so did those of Saint Philip Romolo Neri (1515-95), known as the "Apostle of Rome," who stressed prayer over reason, and whose single daily meal was "bread and water, to which a few herbs were sometimes added."[19] Moreover, it would be remiss to ignore the denunciation and prohibition of bullfighting on the grounds of its cruelty by Pope Pius V in 1567, even though the proscription was generally ignored by the Spaniards against whom it was directed.

There was as much self-questioning as conceit in the humanist era. Not only does Singer not recognize the ambivalence of the Renaissance period, one side of which is represented by the extremely influential Kempis and Philip as well as by the great Protestant reformers from Luther to Melancthon, but he offers Mirandola's work as a *typical* example of "the self-indulgent essays" penned by "the Renaissance writers."[20] He treats Leonardo as an aberration. Kempis and Philip he ignores. The thinkers of the Reformation he fails to recognize. The greatest minds of the Renaissance period receive short shrift in favour of a thinker of lesser rank and less influence. Of course, Singer is right to mention this hubris as an aspect of the Renaissance, but he is quite wrong to paint a picture of the Renaissance almost entirely in these colours. Again, a picture-in-the-round is required. Indeed, as George Boas showed almost three-quarters of a century ago, the Renaissance abounded in theriophily, a viewpoint that requires a different perspective on the Renaissance than that offered to us in much of the animal-oriented literature.[21]

III

As with a fair portion of the theriophilic material in general, it is difficult to discern just how seriously the sixteenth- and seventeenth-century examples are to be taken, although it should be noted that it is almost as difficult to discern just how seriously the "noble savage" and Edenic literature is to be taken in its suggestion that the primitive way of life is superior to that of Western civilization. What is clear, however, is that both forms of argument indicate that there is something about the primitive, or "animal," life that has been lost in the "progress" undertaken by the "advanced" societies. As we have seen,[22] following the "discovery" of the Americas, there was a fair industry in books proclaiming the greater virtue of the "noble savage," who also lived a happier life because it was lived in greater conformity with the laws of nature. Theriophily is simply a step further back in the same argument, suggesting that if primitive humans have escaped the perils of civilization and on that account live more fulfilled lives, then such should be true for the animals a fortiori.

Pierre Bayle's famous essay "Rorarius," in his *Historical and Critical Dictionary* (1697), tells us that Rorarius Hieronymus was papal nuncio at the court of King Ferdinand of Hungary and that he was the author of "a book that deserves to be read."[23] It would appear, on the author's own testimony, that he was engaged in "a discussion with a learned man" about the relative merits of the emperors Charles V, Otto the Great, and Frederick Barbarossa. "It took no more than this," Bayle remarks, "to make Rorarius conclude that beasts are more rational than men, and immediately he began composing a treatise on the subject."[24] The treatise appears to have been first published in parts in 1544 and 1547, then largely forgotten (although Montaigne certainly had a copy, of which he made use in the 1570s). A new edition appeared in Paris in 1645, according to Bayle, followed by its being "reprinted in Holland more than once." Boas mentions a 1648 Paris edition, which he believed to be the first since 1547, and then tells us that it was reprinted in Paris in 1654, 1666, and 1728 – no mean testament to at least a measure of popularity. Although Rorarius was Italian by birth, it was in France, secondarily in the Netherlands – a very common place of publication in the sixteenth and seventeenth centuries, especially when censorship was feared – that his work was most readily received. Indeed, it was to a Frenchman that it was dedicated.

In Rorarius's dedicatory epistle to the Bishop of Arras, dated 1 March 1547, he states: "I had written two little books wherein I showed that brutes frequently make a better use of their reason than men; and I was prompted to do this in order to repress the impudence or, rather, madness of some persons, whose eyes are not strong enough to behold the glories of Charles V, the greatest of all emperors."[25] As Bayle points out, the reader of the dedication "will find the author strongly prejudiced in favor of Charles V, and a great flatterer." Indeed, it is not unreasonable to conclude that Rorarius wrote the work as much to demean his intellectual adversaries and to ingratiate himself with the current emperor's allies as to elevate the animals. Just as the primitivists were often more concerned to denigrate the braggarts of civilization than to describe the real culture of Aboriginal society (and they still are), so too some of those who wrote of the superior reason of the animals had only a modicum of interest in the real intellectual capacity of animals. Nonetheless, if either the Aboriginals or the animals are to serve their rhetorical function, it must be possible to portray them convincingly as in some manner superior to those who are to be censured for their conceits. If the animals had not been presented as in some manner possessing a type of reason that was better measured to fulfill them as the kind of beings that they are, the argument would have been dismissed out of hand. In fact, as we shall see, the timbre of the times made it quite plausible to suggest that the manner in which human reason failed to fulfill people as the kinds of beings that they are reflected poorly on humans compared with animals. The very fact that the argument could be persuasive suggests a measure of respect for animal life that is denied by those who project Renaissance humanism as essentially anti-animal. Indeed, when Bayle tells us at the close of the seventeenth century that "for a long time people have maintained that beasts do have a rational soul"[26] – and this in the century of Descartes, Malebranche, and Bossuet – it ought to prompt us to reconsider the prevailing judgment with regard to this era in the vast majority of animal-rights literature.

Rorarius is, by almost any standard, wildly anthropomorphic in his analysis. For example, he deems wolves and lions more intimidated, more fearful of similar punishment, if they encounter one or two of their conspecifics nailed to a gibbet at the crossroads than is a human thief by being branded with a hot iron or by the loss of his ears.[27] Yet such anthropomorphism, if quite unjustified ethologically, reflects a very real belief that humans and other animals share a great deal in common in their motives and consequent behaviour. Animals are not essentially different from humans but behave according to the same psychological laws as humans. Rorarius does not appear to have believed that animal reason was more capable of abstract tasks than human reason. Far from it. It was simply that animal reason functioned more efficiently and honourably in assisting animals to reach the goals appropriate to their psychological and moral nature than did human reason function in assisting human beings to reach the goals that are appropriate to their psychological and moral nature. Thus he tells us that the superior reason of the brutes permits them to fear death less than do humans; it restrains their anger, their appetites, so that a lion will show more restraint toward a woman than a man, and will desist entirely from venting his wrath on a child, unless quite ravenous. Certainly, this extends the application of "reason" beyond its normal limits, at least as we employ the term today. It nonetheless suggests an animal superiority. Humans, he argues, are far more cruel than the animals. He lists a litany of the most heinous human abuses – including one in which a nobleman ordered his servant boy burnt alive because he did not bring him his candle at the appointed time.[28] By comparison, lions are truly generous in their behaviour. Indeed, by contrast with humans, Rorarius continues, animals possess in greater strength the virtues of moderation, of chastity, as elephants do not commit adultery, and of piety, as animals are more reverent than humans. Animals can be seen to comprehend humans; thus they must be able to understand themselves, the implication being, for Rorarius, that they possess language.[29] Rorarius's list of animal characteristics bears some similarity to that of the French army surgeon Ambroise Paré published around the same time. Paré claims that animals possess the virtues of "magnanimity, prudence, fortitude, clemency, docility, love, carefulness, providence; yea knowledge, memory, &c." Moreover, "they have taught man many things." And "the hare is eminent for memory, the dog for apprehension and fidelity, the serpent for wisdom, the fox for subtlety, the dove for charity and innocence, the elephant for docility, modesty and gratitude. Pliny said [the elephant] comes near the understanding of man[30] ... the ape is eminent for imitation and understanding, the turtle for love, the crocodile for deceit, the lamb for patience, the wasp for anger &c." Paré, however, is a little more circumspect than Rorarius, telling us that animals possess all the human qualities "both good and bad," and all the human faculties, only in some of which the animals excel, although "for his five senses [man] is by them excelled."[31] Of course, it is not necessary to demonstrate that animals were regarded as superior in order to show that their status in the Renaissance was greater than that customarily ascribed to them.

The notion of human-animal psychological similarities also follows from the Renaissance practice, begun apparently in the late-medieval era, of charging animals with criminal offences, arraigning them before courts of law, usually ecclesiastical courts, trying them, and, if duly convicted, inflicting the punishment of the courts upon them.[32] Formal recordings have been kept of 189 of these trials.[33] In fact, although this practice had dwindled by the eighteenth century, its last recorded use was not until the early twentieth century in Switzerland. Beetles, asses, cockchafers, mice, turtledoves, wolves, horses, rats, and especially pigs, since they roamed freely and had abundant opportunity to commit felonies, were burned at the stake, hanged, and, since the source of their criminality was often infestation of their soul by devils, exorcised. Yet if we find that preposterous today, we should also recognize that the practice arose out of a conception of animals as responsible and answerable for their actions in the same manner that humans are responsible and answerable for their actions. Nor should we imagine animals different from humans on account of the devils by whom they had been visited. Exorcism of infested humans was even more common.

A famous case at Autun in Renaissance France in 1522 demonstrated not only the assumption of animals' responsibility for their actions but the seriousness and solemnity with which court, laity, and legal profession treated the prosecution of animals:

The rats of the diocese were ordered brought before the court of the bishop's vicar, accused of having wantonly destroyed the barley harvest. They were summoned in due form by an official to appear on a certain day. Counsel for the defense was Bartholomew Chassenée, afterward president of the Parlement de Provence and a great figure in French legal history.

When his clients failed to appear, he pleaded that the summons was too individual and that not a few but all the rats of the diocese should be informed. The curates of every parish were then ordered to notify all rats residing in their vicinity, but again Chassenée's clients failed to appear. This time he argued that, since the rodent population was widely scattered, they should be given time to prepare for such a migration. Again the judge found the plea reasonable and postponed the hearing. On the next date set, the defense sprang its big surprise. The rats could not appear, because, although the summons implied the full protection to the parties both going and coming, they were afraid of the plaintiff's cats. Therefore the defense asked the judge to bond each complaining peasant for the good conduct of his cat while the rats were before the court. In much alarm over the difficulty of restraining their pets, the plaintiffs promptly demurred, and proceedings were adjourned *sine die*.[34]

Nor should this ascription of moral responsibility to animals surprise us. It was enjoined, in an era in which Biblical pronouncements were accepted as literal truth, by Genesis 9:5, where we are told, "I shall demand account of your life-blood, too. I shall demand it of every animal, and of man." If humankind is responsible for its criminal acts, so too are animals.

While the criminal prosecution of animals can scarcely be said to have been to the practical advantage of the animals, it reflects nonetheless a certain type of respect for them that diminished as a more accurate understanding of the nature of animals and their motives developed. Fortunately, at the same time the criminal prosecutions also diminished.

Among the theriophilic pioneers of zoology mentioned by Boas in *The Happy Beast in French Thought of the Seventeenth Century* is one Pierre Gilles, who wrote a commentary on Aelian, *De vi et natura animalium* (On the life and nature of animals), which was published at Lyons in 1533. Like Rorarius, and on similar grounds, he maintains the superiority of animal reason and insists that animals are able to speak. It is we, with a few notable exceptions, who are lacking in not being able to understand them, even though at least some of them are able to understand us. Gilles records the reasoning of dogs and the nightingale instructing her young in the art of song. He mentions animals' memory, their care for their young, and their moral excellence as well as their ability to learn from humans. Some men are fiercer and more savage than the animals, and we have learned much of value from the animals. He goes on to list the various virtues of a large number of them. As Boas observes of Gilles's approach, "its point of view is at least clear; it is that of an admirer of the beasts who finds them superior to men."[35] After mentioning a few other contributors, Boas concludes that "it is therefore certain that theriophily was common enough in France in the sixteenth century and was part of a general primitivistic current of thought."[36]

For later theriophilist zoologists it is less likely to be the superiority of the animals' reason that is mentioned than their greater proximity to nature, which affords them a greater likelihood of being fulfilled as the animals that they are. Nonetheless, in *De animalibus insectis,* published at Bologna in 1602, the respected Ulysse Aldrovandi allots seven pages to the rationality of the ants. He describes in detail their piety and the organization of their cities. In 1615 Jeremia Wilde published a whole volume on ants (*De Formica*)[37] in which he discusses their minds, their ethical attributes, their religion, their government, and their peaceful and warlike qualities. As Boas says, he "puts them, to be sure, in the class of irrational animals, but the absence of this faculty does not prevent their being almost divine."[38] Wilde indicates that nature wished to express *perfectionem omnem* (all perfection) in this animal. Given the classical proclivity to equate the approximation to divinity with the capacity for philosophical reason, Wilde's judgment comes as a surprise, especially when he accords the ants memory and imagination but not ideas. On the other hand, if we recall Plato's categorization of the ants and the bees as the animals into whose souls the best of humans below the level of the philosophers will enter in metempsychosis,[39] perhaps a different category of divinity might be imagined – a moral rather than an intellectual category. Certainly, Wilde writes of the ants' frugality, vigour, temperance, communitarianism, and understanding of each other. They live a life not of refinement and luxury but one close to nature, wherein they are superior to the humans who engage in the excesses of civilization.

In much of the theriophilic literature, we find the same kind of antiintellectualism that appears in the primitivist literature, as we noticed in the writing of Rousseau.[40] But it is also to be found in such disparate writers as Thomas Hobbes, who emphasized the destructive ends to which reason can be put, such as lying and corruption, hence the prime necessity of an undisputed political order. And it is likewise found in Edmund Burke, who emphasized the practical limitations to reason and charged that those who rely on it at the expense of prudence, experience, traditional wisdom, and political compromise destroy the very foundations of order and liberty on which their societies are based. Anti-intellectualism is levelled against those who unwisely imagine that their unaided reason will be able to design effective and just civil societies and will permit them to live a sound and fulfilling life. In the Renaissance we find the Italian Ortensio Landi's *Paradossi* – one of the most widely read books of the time, at least in its French translation (*Paradoxes,* c. 1553), but with frequent new editions under different titles – suggesting that "it is better to be ignorant than learned (*sçavant*)" and that "the poor rustic (*villageois*) is more at ease in life than the freeman (*citoyen*)."[41] All the heresies, he writes, come from the educated. Landi praises those who have the wisdom to keep the learned from public office since they "fear that these lettered people, through their great learning ... will disturb the peace and sound order of the common-wealth."[42] It is unlikely that Burke ever read Landi, but they offer similar messages. Lilio Gregorio Geraldi makes very similar points on the futility of learning in his popular *Progymnasma* (1552). While we have every reason to believe that these authors were only half serious – after all, paradoxes and what we might term intellectual gymnastics, as the term *Progymnasma* suggests, were the theme – they helped to prepare an attitude both prim-itivistic and theriophilic. Yet, as Boas has indicated, the issue of superior animal reason was no simple matter for the theriophilists:

> The question of human misery and animal happiness was complicated by the question of evaluating reason. If one decided that the beasts could not reason but that man could, then in order to maintain one's theriophily, one had to deprecate reason and theriophily became linked with anti-intellectualism. If on the other hand one did not wish to deprecate reason, and yet did not wish to abandon theriophily, one had to prove that either the beasts reasoned as well as human beings or better. Intellectualistic theriophily and anti-intellectualistic theriophily were not kept neatly apart. Both strains were sounded by the same men and sometimes in the same work.[43]

As we saw earlier,[44] in her *Primitivism and the Idea of Progress in English Popular Literature of the Eighteenth Century,* Lois Whitney was astonished to find in some novels of the period a "curious mixture of ideas ... theories of the superiority of primitive man" competing with theories of progress "all huddled together – sometimes two antagonistic points of view in the same sentence." Clearly, here as there, we encounter a rationalization that was taking place in the authors' efforts to bring conceited civilized humans down to size or to elevate denigrated "savages," or animals, to their right-ful place of respect.

These influential works by Landi and Geraldi were immediately followed by those of Petrus Boaystuau (Pierre Boaistuau), who wrote a tragedy of Romeo and Juliet long before Shakespeare and who penned his *Theatrum Mundi* (c. 1566) in two parts, one to demonstrate the misery of humankind and the other its excellence – the first part being almost four times the length of the second! Boaystuau repeats the famous adage of Pliny that nature is more a stepmother than a mother to humankind, since the animals arrive in this world all clothed – "some have feathers, some body hair, some leather, some scales, and some wool."[45] Only humans are born naked. The beasts cure their illnesses with the aid of nature, not with the aid of physicians, at whose mercy we are and "whose labours we are required to engage very expensively, and who are very often the cause of our deaths, since the majority of their laxative medicines are nothing other than veritable hammers with which to kill men."[46] Boaystuau continues at great length to describe the greater merits of the animal over the human. He writes of "Nature's greater kindness to the beasts" and of their cleverness. Boas sums up Boaystuau's profoundly theriophilic views:

> If we wish to contemplate and admire human wisdom and prudence, we have only to turn to the tiny beasts we daily crush beneath our feet, "who in such things surpass mankind." They have natural virtues; we acquired. They know one another; they seek the useful and avoid the harmful; they can easily deceive men by their ruses. They have both our melancholy and our gay arts. Think of the marvels of the social insects: the ants who build three caverns in their ant-hills, one for parliaments, assemblies, councils, and assizes, one for a storehouse, one for a cemetery. They are more moral than we: they are not murderous; they respect their parents; they sacrifice themselves for their children. Not only are their individuals more virtuous than we, but so are their states. One might think Nature had given them to us as a lesson in good living.[47]

Nor was the lesson of the last sentence lost on posterity. In *Les Misérables* (1862), Victor Hugo repeated the adage with greater eloquence, and a little less theriophily: "Animals are ... the portrayal of our virtues and vices made manifest to our eyes, the visible reflection of our souls. God displays them to us to give us food for thought."[48]

By contrast with his image of "happy beasts," Boaystuau gives us one of human misery that persists throughout life in "family, commerce, government, the church, the army." None is free from its ravages. Whereas "the beasts fight only from hunger or to protect their young" and only with the arms provided by nature, we war with artificial contrivances against our fellow humans – foreigners, barbarians, and Christians alike. For Boaystuau,

Boas instructs us, "Man ... is surrounded by causes of misery; diversity of religious belief, war, pest, famine, disease, poisons, the elements, hostile beasts, avarice, ambition, pride, envy, pomp, superstition, love, old age, and death."[49] Unlike for the animals, misery is his lot. To be sure, he possesses certain virtues, but, unlike for the animals, these virtues are far outweighed by those things that detract from his wellbeing.

Giovannbattista Gelli's remarkable *Circe* (published in Italian in 1549 and in French in 1550) is an embellishment of the famous Gryllus episode written by Plutarch.[50] In Plutarch's version, which is loosely based on Book 10 of Homer's *Odyssey*, Circe has enchanted a number of Greek warriors and changed them into pigs. Ulysses tries to persuade Gryllus that he and his fellow swine should desire a reconversion to their human state, which Gryllus gracefully declines, arguing that the life of pigs – and, indeed, of other animals – is preferable to that of life as a human being. In Gelli's adaptation of the story, Circe, who has this time turned those she has enchanted into various animal forms rather than into pigs alone, grants the animals the right to resume their human shape if they should so desire. Ulysses attempts to persuade each of the animals in turn but succeeds only with the elephant.

The Oyster has an innate dread of his enemies, which shows that he stands higher than man in the natural order of things. The Mole points out to Ulysses that man alone must work for a living. He suffers drought and famine. Human life "is a continual battle against hostile forces, which is reason enough for you to enter it weeping – which none of the animals do."[51] The Snake demonstrates that humans are dependent on artificial means to maintain their health, of which the beasts have no need. The Hare informs Ulysses that humans must either govern or be governed – both of which conditions are full of misery. Further, the Hare opines, neither riches nor servitude exist among animals, and the animals are very much the better for it. In response to Ulysses's claim that man could avoid these problems by controlling his desires, the Hare retorts that "Nature should not have given him desires, so difficult to master." The Roebuck enumerates four human failings: the anxieties produced by a private-property system, the uncertainties of economic competition, which is a consequence of such a system, man's suspicion of his compatriots, and his fear of the law. The feminist Doe points out the losses she would incur in returning to a human state as a slave to her husband. She desires equality, but male-dominated human society will not grant it. The Lion recounts the evils of ambition, envy, and anger, from which animals do not suffer. The Horse emphasizes the peacefulness of his life and the "perfection" that is fitting to his species and his nature. The Dog asserts animal superiority through the very lack of reason. Animals thereby live by the superior rules of Nature rather than by the inferior rules of Art. As evidenced by the lives of ants, bees, and spiders, they have great prudence, demonstrating the superiority of instinct over reason. The Calf acknowledges that each of the animal species has its peculiar vice: the gluttony of the pig, the cruelty of the tiger. The human, however, has all the vices of all the animals collectively. Moreover, animals pursue natural justice by the very nature of their being. If humans know justice, they do not practise it, for they disobey the laws of nature. In turn, Oyster, Mole, Snake, Hare, Roebuck, Lion, Horse, Dog, and Calf turn down the offer of a return to human life. Life as the animal that they are is preferable to that of the human being.

However, when Ulysses turns to the Elephant, who was a philosopher before his enchantment by Circe, he finds a willing listener. The life of reason is the life to which he will choose to return. Although the book ends with a "Hymn to Reason," it is not certain whether, for Gelli, the philosopher is, à la Plato, the highest form of human being or, à la Burke, the least wise of humans for his reliance on intellect alone. Certainly, whichever way the reader may interpret the tale, there can be no doubt that *Circe,* which "had a great vogue and was often imitated," Boas tells us,[52] succeeded in persuading the readers that there was much eminently worthwhile about the life of the animal. In at least some respects, animal life was superior. And, in part, the superiority arose out of a lack of need for reason, for, as Tolstoy explained over three centuries later, "As soon as a man applies his intelligence and only his intelligence to any object at all, he unfailingly destroys the object."[53] When humans think reason their most important attribute, they become mere one-dimensional beings.

In Germany in the first half of the sixteenth century, we find one of the most famous men of the time, Martin Luther, expressing the view that there was something more admirable in the life of the birds than in that of humans. On seeing a bird wing to roost, he remarked:

That little fellow has had his supper, and he is now going to sleep quite secure and content.

Like David, he abides under the shadow of the Almighty, and lets God take care. How happy are the little creatures singing so sweetly, and hopping from branch to branch. We might well take off our hats to them and say, my dear Herr Doctor, we could not have learned thy art of trustfulness. Thou sleepest all night without care, in thy little nest; thou risest joyful in the morning, and praisest God, and then seekest thy daily food. Why cannot I, poor fool that I am, live like these little saints in the fulness of content and trust.[54]

Theriophily, even if intended to be taken only half seriously, turns up in the most surprising of places!

IV

The most famous theriophilist since antiquity was Michel Eyquem, seigneur de Montaigne. It is not that his theriophily was more emphatic than that of others. Indeed, it was, like that of most others, somewhat restrained. Rather, it was Montaigne's reputation as an essayist – he was both founder of the form of writing and coiner of the term – that led to the currency of his ideas not only in the France of his own era but thereafter throughout the Western world. Plutarch excluded, prior to the twentieth century, Montaigne's elevation of the status of animals was the most widely read among the well-educated.

Each of the above-mentioned books on the animal cause was in Montaigne's library, and there is every indication that he had read them, and in some cases he referred to them explicitly. He was, moreover, not just a theriophilist, but a primitivist, at least after a fashion, and there are indications that he was influenced by Peter Martyr and Girolamo Benzoni, and perhaps, if to a lesser degree, by Gonzalo Fernández de Oviedo, Pietro Bembo, and Francisco López de Gomara.[55] Such was the breadth of the primitivist message.

The very nature of his primitivism offers insights into how his theriophily might be interpreted. Montaigne's idealization of the American Natives and criticism of his compatriots was undeniably quixotic. In his essay "Of Cannibals" (1580), he wrote of the "naturalness so pure and simple" of the Aboriginal peoples of the Americas, "whose countries never saw one palsied, bleary-eyed, toothless, or bent with age ... [their] whole day is spent in dancing." Without even a hint of a sense of his own ingenuousness, he insists that "what we actually see in these nations surpasses not only all the pictures in which poets have idealized the golden age and all their inventions in imagining a happy state of man, but also the conceptions and very desires of philosophy."[56]

Yet the ideal is difficult to classify. It is certainly not Edenic. But it is difficult to comprehend even as Arcadian. He acknowledges both the cannibalism and the broader carnivorousness of the Aboriginals, whereas, in the classical rendering of the Golden Age, a principled vegetarianism was the rule. The purpose of killing and eating fellow humans arose not out of any dietary necessity but, Montaigne tells us, as a ritualization of "ultimate revenge." In the traditional version of the Golden Age, humankind gathered the existing abundance of food, which they ate raw. Now, for sustenance, the Natives hunted with bows, killed animals, and cooked. Likewise, in the utopian past, humankind lived at peace both with each other and with the animals. Now the Natives went to war with their neighbours. Indeed, Montaigne posits "resolution in war" as one of their primary virtues. In the primordial age of gold, humankind was reported to be compassionate. Now, if a soothsayer "fails to prophesy correctly, and if things turn out otherwise than he has predicted, he is cut into a thousand pieces if they catch him."[57] Clearly, the behaviour does not fit the prior conception of a people which "surpasses ... the golden age ... and very desires of philosophy." Even an Arcadia is unlikely to be recognized in the seeking of "ultimate revenge" through cannibalism, in "resolution in war" as a primary virtue, or in the murder of faulty prophesiers. Montaigne's completed picture does not seem to conform to what he calls "naturalness so pure and simple." Indeed, the very idea is denied by his statement that: "I am not sorry that we notice the barbarous horrors of [their] acts, but I am heartily sorry that, judging their faults rightly, we should be so blind to our own."[58] The real purpose of Montaigne's essay, and one that has lessons for our reading of his theriophily, is here clarified. It is not to praise Aboriginals because they are wholly admirable, but rather to employ them as a convenient means of criticizing the excesses and injustices of one's own society – its greed, its luxury, and its conceit. Further, the Aboriginals are not only mistreated by their Western conquerors, but also classified unjustifiably as inferior beings for whom such treatment may be deemed acceptable. In fact, even if their society is not a utopia, it nonetheless has lessons to impart to the West, just as the West has lessons to impart to Aboriginal societies. The Aboriginals are entitled to respect and just treatment, both practically and categorically.

We encounter in "Of Cruelty" and "Apology for Raymond Sebond" – Montaigne's foremost contributions to theriophily, the second being the primary – the same kinds of hyperbole and rationalization that we witnessed in "Of Cannibals" and in some of the books that influenced him. Thus bees have the best form of government:

> Is there a society regulated with more order, diversified into more charges and functions, and more consistently maintained than that of the honey-bees? Can we imagine so orderly an arrangement of actions and occupations as this to be conducted without reason and foresight?

> Some, by these signs and instances inclined,
> Have said that bees share in the divine mind
> And the ethereal spirit.
> – Virgil[59]

The web-weaving of spiders reflects, for Montaigne, the depth of their deliberation and thought: "Why does the spider thicken her web in one place and slacken it in another, use this sort of knot, now that one, unless she has the power of reflection, and thought, and inference?"[60] The nest building of swallows proves that they are expert architects, masons, and meteorologists:

> Do the swallows that we see on the return of spring ferreting in all the corners of our houses search without judgment, and choose without discrimination, out of a thousand places, the one which is most suitable for them to dwell in? And in that beautiful and admirable texture of their buildings, can birds use a square rather than a round figure, an obtuse rather than a right angle, without knowing the properties and their effects? Do they take now water, now clay, without judging that hardness is softened by moistening? Do they floor their palace with moss or with down, without foreseeing that the tender limbs of their little ones will lie softer and more comfortable on it? Do they shelter themselves from the rainy wind and face their dwelling toward the orient without knowing the different conditions of those winds and considering that one is more salutary to them than the other? ...
>
> We recognize easily enough in most of their works, how much superiority the animals have over us and how feeble is our skill to imitate them.[61]

There is a tension between Montaigne's claim of rationality for the animals and his insistence that their instinctive behaviour moves them closer to the divine than if they acted from free will. The overall effect of the essays is to demand greater respect for the animals as the beings that they are and to demonstrate that they possess at the very least a greater degree of similarity to humans than was often accorded, even if it was not a decided rational, emotional, and constructive superiority.

If we bear in mind that these are exaggerations, or at least an interpretation that would commonly be regarded as exaggerated, that they arise from a careful, if not entirely successful, study of the real lives of the animals, that they indicate quite remarkable animal talents that we are apt to overlook – and should one want to argue that they are no more than the products of evolution, so too are most human talents – and that the error lies in ascribing too much of the behaviour to reflection and deliberation, we can readily recognize that Montaigne's theriophily should require only a modicum of refinement in light of more recent ethological study to provide a solid basis for an animal ethic. Indeed, when the physiological exaggerations are removed, Montaigne's words sound quite restrained on occasion. He held in "On Cruelty" that there is "a kind of respect and a

general duty of humanity which tieth us ... unto brute beasts that have life
and sense, but even to trees and plants ... unto men we owe justice, and to
all other creatures ... grace and benignity ... there is a certain commerce
and mutual obligation between them and us."[62] At this juncture Montaigne
appears to be advising us, in line with a tradition of Stoic thought, that we
owe other creatures a significant obligation, but it is less than the justice
we owe to our fellow humans.

As in "Of Cannibals," Montaigne intends in "Apology for Raymond
Sebond" to prick the human bubble: "Presumption is our natural and original malady. The most vulnerable and frail of all creatures is man, and at
the same time the most arrogant ... It is by the vanity ... of this imagination that he equals himself to God, attributes to himself divine characteristics, picks himself out and separates himself from the horde of other
creatures, carves out their share to his fellows and companions." To the
animals, "he distributes ... such portions of faculties as he sees fit. How
does he know by the force of his intelligence, the secret internal stirrings
of animals? By what comparison between them and us does he infer the
stupidity that he attributes to them?"[63] Having earlier informed us of the
superiorities of the animals, he now turns to stressing equality:

> We are neither above nor below the rest; all that is under heaven, says the
> sage, incurs the same law and the same fortune.
>
> All things are bound by their own chains of fate. (Lucretius)[64]
>
> There are certain affectionate leanings which sometimes arise in us without
> the advice of our reason, which come from an unpremeditated accident that
> others call sympathy: the animals are as capable of it as we are. We see horses
> forming a certain familiarity with one another, until we have trouble mak-
> ing them live or travel separately; we see them apply their affection to a cer-
> tain color in their companions, as we might to a certain type of face, and
> when they encounter this color, approach it immediately with joy and dem-
> onstrations of good will, and they take a dislike and hatred of some other
> color. Animals, like us, exercise choice in their amours and make a certain
> selection among females. They are not exempt from jealousies, or from ex-
> treme and irreconcilable envy.[65]

Yet, even if we find that Montaigne is coming closer to an egalitarian
argument, we still encounter such theriophilic views as: "As for friendship,
theirs is without comparison more alive and more constant than that of
men."[66] "Animals are much more self-controlled than we are."[67] "As for

domestic management, not only do they surpass us in their foresight in piling up and saving for the time to come, but they also possess many parts of the knowledge necessary for it."[68] "As for fidelity, there is no animal in the world as treacherous as man."[69] In the final analysis, animals possess all the better characteristics of humanity, but usually in greater abundance, and if they share any of humanity's malevolent characteristics, they possess them in far less degree. It is difficult, ultimately, to find much restraint in Montaigne's theriophily – even if we can be confident that it is written at least as much to denigrate human hubris as to elevate the status of animals.

V

Montaigne's essays provoked a considerable response, sometimes favourable, even doting, sometimes captious, and sometimes caustically critical – which is precisely what one would expect if one understands the history of ideas in Western society as an ongoing unresolved debate, with differing views and emphases in the ascendant at different times, and none at all in the ascendant at some times. And, certainly, as the premier literary figure of his age, Montaigne could not fail to excite, and, as often as not, he was read as though he had invented the whole idea of animal superiority, or at least revived an issue dead since Plutarch – when in fact, as we have seen, he had a number of sixteenth-century precursors, some of whom were themselves widely read. Not surprisingly, Montaigne's disciples, as disciples are wont, preferred to award the title to the master alone.

The earliest of these disciples is Montaigne's friend Estienne Pasquier, who despite acknowledging his essay on the subject a paradox, nonetheless opens his piece in *Les Lestres d'Estienne Pasquier* (1586): "Do not imagine that I am not in earnest; for I consider myself to belong to that group of people who thinks that nature has been too indulgent to the other animals, as compared to us."[70] He follows Pliny in telling us that Nature has been a stepmother to us and a mother to the other species – who arrive fully shod and clothed, are able by nature to construct their own homes, and find their own food. And if they do not possess Reason, that is to their advantage since it was Adam's acquisition of Reason that had human beings cast out of Eden. Again the paradox: having denied the value of Reason, Pasquier asks how we know that the animals do not possess it. They show, he says, all the evidence of rationality. He then parades the usual litany of animal good character – their charity toward the sick and reverence toward the dead. Bees and ants take care of their less fortunate brethren. Pelicans display parental devotion, and storks filial piety. Lions are magnanimous –

sufficiently proverbial that Henry Fielding described the lion as a "magnanimous beast" in *Joseph Andrews* (1742)[71] – and beasts are capable of shame, the example being that elephants copulate only in the dark. Animals are prudent, show friendship, and know natural medicine – and do not have to rely on drugs as humans do: "we are obliged to admit that it is excellent and beneficial medicine to take no medicine at all."[72] There is no greater contrast than animals' obedience to the universal laws of nature, while human laws change from country to country and epoch to epoch. And if animals lack speech, it is to their benefit, for speech is the source of discord: "What bolsters heresies, what encourages law-suits, what makes a man an adulterer with his neighbour's wife, if it is not the capacity for speech?"[73] While there is no indication that Hobbes ever read Pasquier, there is little doubt that they shared a common recognition that those attributes widely thought to raise humanity above the beasts could be interpreted differently. Even if Pasquier's primary purpose was to wound human pride, his claim to be in earnest about the superiority of animal character should be given due consideration. Certainly those who wrote against the theriophilists thought the claim a sincere one.

Turning to Pierre Charon's *De la Sagesse* (On wisdom; 1601), we can recognize one who is undoubtedly sincere in his belief that there is great neighbourliness (*voisinage*) and kindredness (*cousinage*) between humans and animals, humans being neither entirely above nor entirely below their relatives. Humans are superior in the powers of their soul; they are more subtle, vivacious, and inventive; they have better judgment and more choice; they have speech, hands, and great diversity of limbs, which provides improved bodily service. On the other hand, the animals have better health and vigour, they are moderate in their diet and behaviour, and they are subject to neither shame nor prejudice, superstition nor ambition, avarice nor envy. Each species has some particular field of excellence in which the capacities of humankind are exceeded – for example, the building capacities of the swallows and the spiders, the musical abilities of the nightingale, and the medicinal secrets known by yet others – and Charon quotes Pliny and Plutarch as his authorities. Again, the superior health of the animals became almost proverbial. It was John Frank Newton's view in his *Return to Nature* (1811) that "if men were universally as healthy as these wild animals, they would certainly exceed the age of one hundred and fifty years."[74]

Despite the apparent balance between the superiorities of humankind and those of animalkind, the book concentrates on the virtues of the animals. He uses the examples of the Thracian fox and the dog at the crossroads[75] as well as other tales to show the superior reasoning power of animals, which, Charon says, most people falsely attribute to instinct. Still, this too would be to the credit of the beasts, for it would show that they live by Nature rather than by Art. Charon concludes that animals reason well, if less well than do humans. But then some humans reason better than others, and there are greater differences among humans than beween humans and animals. Mark Twain's words in "Dick Baker's Cat" are representative of the genre inspired by Charon: "[Dick Baker's Cat] had more hard, natchral sense than any man in this camp – 'n' a *power* of dignity – he wouldn't let the Gov'ner of Californy be familiar with him." And as for his powers of reasoning, "Sagacity? It ain't no name for it. 'Twas *Inspiration!*"[76]

A further follower of Montaigne was one François de La Mothe le Vayer – member of the Académie Française and subject of an article in Bayle's *Historical and Critical Dictionary* – who, in his *Politique du Prince* (1654), repeated the classical division of government into monarchy, aristocracy, and democracy and then determined that the animals possess the same divisions, the bees being governed along monarchical lines, the cranes as an aristocracy, and the ants in democratic manner. In his *Petits Traités ou Lettres*,[77] he argued that wild animals are antagonistic to humankind only because we make them so and that if we treated them well, they would be friendly to us. While we should be grateful for the status that God has accorded us, we should still think on the possibility that "the cats convince themselves perhaps that rats and mice exist only for them to grow fat on"[78] – a clear warning to those who imagine that animals are intended for human use.

In the same period, we encounter more philosophically sophisticated defences of animal reasoning in, for example, the writings of the court physician Marin Cureau de la Chambre. In *Traité de la Connoisance des Animaux, où tout ce qui a esté dict Pour et Contre le Raisonnemment des Bestes est examiné* (Treatise on animal cognition, where all that has been said for and against animal reason is examined; 1671), he argues, against those who would determine all animal behaviour to be instinctive, that to insist upon such a determination would be to explain the known by the unknown. Moreover, if instinct were understood, "it would be found to be no different from, no more remote than, Reason."[79] Nonetheless, Cureau's interests are in the philosophical minutiae and complexities of epistemology. There is little evidence that he saw himself as an advocate for the animals, although his was a determined denunciation of what he saw as the irrational objection to animal reason among the Cartesians. To be sure, he writes of the animals' marvellous industry, ingenious foresight, social organization, prudence, gratitude, and generosity,[80] but much of his work involves fine distinctions between imagination, reason, experience, instinct, innate and acquired ideas, and the like. Nonetheless, he offers a rather different message in *Discours de l'Amitié et de la Haine qui se trouvent entre les Animaux* (Discourse on friendship and hatred to be found among the animals; 1677). Here Cureau writes of the aesthetic sense of the animals, as later did Charles Darwin, although for Darwin the sense is far more restricted. Despite stating that "the nests of humming-birds and the playing-passages of bower-birds are tastefully ornamented with gaily-coloured objects [which] shows that they must receive some kind of pleasure from the sight of such things," Darwin concludes that for "the great majority of animals ... the taste for the beautiful is confined, so far as we can judge, to the attractions of the opposite sex."[81] Not so for Cureau, who believes that certain species are attracted to certain colours and that the "nightingale must love sound for its own sake, otherwise it would not diversify its notes as it does" – and, of course, there is aesthetic sexual attraction, too. Moreover, the animals are capable of friendship not merely for the pleasures it brings (*amitié délectable*) but as a mark of authentic goodwill (*amitié honnête*). And their hatred, where it is present, is a rational hatred toward those who might do them harm. Still, Cureau decides that animals are intended by God for human use and that they lack immortal souls – all of which goes to demonstrate how complex was the status of animals in sixteenth- and seventeenth-century France.

A decidedly less learned, but nonetheless prolific, contributor to the debate was René des Cériziers, who acknowledged animal rationality, although of a decidedly inferior kind. In *La Morale* (1643), he opined that the animals possess "a light shade of our enlightenment and a weak imitation of our reason."[82] Nonetheless, he repeated the traditional view initially ascribed to Democritus and later addressed by Plutarch that we derive many of our arts and skills from the animals – they are our instructors, not we theirs. In addition to the more common examples – such as our learning societal organization from the ants and bees, military drill from the cranes, and weaving from the spiders – he also announced that we derive the art of bloodletting from the sea-horse, although I have been unable to discern on what empirical ground he made the claim. Others have suggested, perhaps ingenuously, that we learned the art from the bat, and some, perhaps a touch more sensibly, that the source of our knowledge was itchy hippopotami scratching on trees until they bled and other animals scratching themselves to the same end, which appeared to provide some relief. Moreover, menstruation was a natural occurrence, following which there appeared to have been some improvement in demeanour. Whatever the source of the practice of bloodletting might have been, some, both in antiquity and the Renaissance, claimed it a wise course of action learned by humans from the animals.

The practice of human bloodletting seems to go back at least as far as the early Egyptians, and it was certainly practised in Greece long before Hippocrates. Nonetheless, it was Hippocrates of Cos (460-370 BC) who provided the medicinal justification for the usage. He argued that when a person was healthy, all body systems were in balance. Ill health resulted when they were out of balance. Letting blood would restore the balance. Both Galen (AD 130-210), court physician to the Roman emperor and an early vivisectionist, and Marcus Aurelius elaborated on the Hippocratic view, asserting that the basis of the balance, or harmony, to be achieved was the four humours ("cardinal humours," they were sometimes later called), or bodily fluids: blood, phlegm, yellow bile, and black bile. Each of these "humours" was associated with a specific psychological characteristic that affected the health of the body. Indeed, these humours were generally deemed to determine a person's physical and mental qualities, as reflected in the ascendance of a given humour in a given individual. Blood was indicative of a sanguine personality, involving laughter, a love of music, and a passionate disposition. Phlegm represented the lethargic and dull personality. Yellow bile epitomized a person of a choleric disposition, quick to anger. Finally, black bile represented melancholia, the personality of the depressed. It was the purpose of the physician to restore harmony among these humours, when they were imbalanced, by the employment of emetics (medicines to induce vomiting), cathartics (medicines to release pent-up emotions), purgatives (laxatives), and, the prime surgical remedy, bloodletting, or phlebotomy – from *phleb* (vein) plus *tomia* (cutting). The purpose of bleeding was to reduce excess circulation, slow the pulse, and diminish irritation, all deemed to be the cause of inflammation and subsequent psychological discomfort – physical and psychological characteristics were deemed to be irremediably interrelated. Bloodletting remained a common treatment for fevers, inflammations, a variety of disease conditions, and, quite improbably, but absolutely certainly, haemorrhage – until the work of Louis Pasteur in the 1860s demonstrated the role of microbes in disease. Even then, the practice retained a certain vogue until the early twentieth century and could still be found generally recommended in the 1923 edition of Sir William Osler's *Principles and Practices of Medicine*.

What is of significance for the status of animals is not only that the practice of bloodletting was deemed to be learned from the animals, but that it, along with the medicinal "balancing" recipes, was practised on animals, too. Indeed, it was a common practice with horses in classical Greece and continued until the collapse of the Roman Empire, beginning in the fifth century AD.[83] Horses were the most prized of animals in antiquity, and we find praise lavished on them for their psychological and physical

characteristics in, for example, Homer's *The Odyssey* and *The Iliad* (c. eighth
century BC). If bleeding was appropriate for humans, why not for the most
esteemed of animals? If the human emotions needed to be counterbal-
anced, why not those of the "goodly steeds ... the lightest of foot and chief
in strength ... leaping high and speedily accomplishing the way," as *The
Odyssey* has it?

With the revival of classical learning, sensibilities, and aspirations in the
Renaissance, we find the practice readily restored. In the years around 1550,
a time when records were still poorly kept, and much that was kept later
lost, we still find the reporting of 126 German instances of animal bloodletting.[84] In Giovanni Battista Ferraro's 1560 book on breeding, managing,
and treating horses,[85] not only is there a section on the use of medicines
to restore the "humours," but the last section discusses animal surgery,
concentrating on phlebotomy, and the final page is a fine folding wood-
cut – metal engraving was a later invention – of a "bloodletting horse." In
the eighteenth century the rights to perform equine throat and teeth clean-
ing, as well as the sole privilege of bloodletting, were closely guarded by
the Leipzig farrier guild.[86] Moreover, despite Cartesianism, these veterinary
practices continued for as long as similar practices were common in human
medicine. The logical implication is that veterinary practitioners – who
were, in effect, until the early nineteenth century and the origins of a veterinary profession, primarily farriers – assumed, first, that human and animal anatomies were homologous, a fact demonstrated beyond reasonable
doubt by Giovanni Borelli in 1680 when he showed in *De moto animalium* (On the movement of animals) that the same laws as those determining human movements governed the wings of birds, the fins of fishes, and
the legs of insects; and second, that animals, especially horses and cattle,
for it is they on whom the bloodletting was primarily practised, possessed
the same physical and psychological characteristics as humans. Of course,
the farriers did not assume that humans and animals shared identical
abilities and demeanour. But if the ultimate purpose was to balance the
humours – the competing jovial, slovenly, ill-tempered, and melancholic
dispositions – then these must have been regarded as present in nonhuman
animals, too. A basic, if not complete, human-animal identity was their
perception.

Descartes's reception on the issue of animals as automata was, as we have
noticed, more favourable in France than elsewhere, which is precisely why,
perhaps, France also witnessed such a plethora of denunciations of Cartesianism – an instance of the Newtonian law of physics that for every action
there is an equal and opposite reaction. Of course, there are no such *laws*
in history, but there are certainly such propensities. Still, there are some

who in their reaction preferred to ignore the arguments of those against whom they were reacting. For instance, in *La Fine Philosophie accomodé à l'Intelligence des Dames* (Tasteful philosophy adapted for the intelligence of ladies; 1660), René Bary trots out the customary paradox that learning and reason will not provide virtue and that the beasts engage in learning and reason and possess virtue. The beasts have practical knowledge: "Birds in flight will never swoop on their prey when it is too far away."[87] They possess theoretical knowledge (or knowledge of universals, that quality customarily denied the animals): "The image, for example, that the sheep has of the wolf is not the image of this particular wolf or that particular wolf ... since there is some common characteristic possessed by all wolves ... the author of nature has etched in her memory the image of this commonality."[88] They have memory, the power of recall, the capacity to compare ideas – all the customary human faculties indeed. To be sure, in the final analysis, a little Cartesian influence creeps in: "Supernatural happiness is a great boon, and the understanding of animals is circumscribed; felicity is a purely spiritual bounty, and the understanding of animals is an as if it were corporeal faculty."[89] Of course, the immediate question arises of what is meant by "as if it were corporeal." It is rather more than Descartes allows to the beasts but is in the end not much less discouraging: "Purely spiritual good is incorruptible, and faculties which are as if they were corporeal are perishable."[90] "As if it were corporeal" amounts to pretty much the same as "corporeal."

There can be little doubt that the abundant theriophily in Renaissance France was rather less extravagant in its claims after Descartes, although it certainly continued to exist throughout the eighteenth century. Moreover, it was a movement of sufficient strength that one cannot accord the Cartesians even a temporary victory against it. It should be recalled that the Cartesian critique of theriophily denied not only intelligence to the beasts, but, if not entirely consistently, all sensation. Moreover, it denied the animals not only an immortal soul, but even a sensitive soul. Although France was the country in which Cartesianism had more success than any other, there was a healthy denunciation of it there, too. If a less than earnest willingness to ascribe immortal souls to animals was present among the French, there was certainly an ascription of a far higher status to animals than is customarily acknowledged. It was only the likes of Descartes himself, the Jesuit François Garasse, Pierre Chanet, Nicolas Malebranche, Jacques-Benigne Bossuet, and the seminarians of Port Royal who maintained the thesis of *bruta non sentiunt*. Even those who gave the animals short shrift, the Schoolmen and the church itself, found Cartesianism untenable. Moreover, there were those who were willing to go a long way in their elevation of the animals. Nowhere was that more notable than among the satirists and the poets, who enjoyed a suitable art form through which to deny humans their hubris, and, above all, to cock a snook at the far-famed Descartes.

In 1672 the famous author of letters addressed to her daughter, Mme de Sévigné,[91] wrote sardonically of Descartes's famous "beast-machines": "Machines which love, machines which choose one fellow over another, machines which are jealous, machines which are afraid! Surely, surely, you are making fun of us; not even Descartes could have aspired to get us to believe that."[92] And Bernard Fontenelle provided what was the snappiest rebuttal of Cartesianism, excluding the already noted quip of Lord Bolingbroke that the plain man would persist in believing that there was a difference between the town bull and the parish clock:[93] "You say that the animals are both machines and watches, don't you? But if you put one male dog machine in close proximity with a female dog machine, a third little machine may be the consequence. In their place, you may put two watches in close proximity with each other for the whole of their lifetime without their ever producing a third watch. Now, according to our philosophy, all those things that have the capacity to render three out of two possess a greater nobility which elevates them above the machine."[94]

In *Histoire des Oiseaux* (c. 1662), Cyrano de Bergerac represents the birds as having taken command. They have charged Cyrano with crimes against birddom and hold court:

> It would be absurd to believe that an entirely nude animal, which nature herself did not take the trouble to furnish with the necessities of life in bringing him into the light of day, was capable of reason like [the birds]. Further, they add, [perhaps this could be so] if he were a little closer to our shape, instead of quite unlike and quite frightful, a bald creature in fact, a plucked bird, a chimera composed of all kinds of natures, who is afraid of everything – this, I say, is man: so stupid and conceited as to convince himself that we were made for him; man who, with his perspicacious soul, cannot distinguish between sugar and arsenic, and who swallows hemlock which, in his fine judgment, he has mistaken for parsley; man who maintains that one only reasons in accord with the senses, but who has himself the weakest, most sluggish, most unsound set of senses of all the creatures; man, in fact, who, to complete the picture, nature has made like a monster, but who has nonetheless infused him with the ambition to dominate all the animals, is hereby sentenced to death.[95]

In order to avoid his fate, Cyrano claims to be a monkey rather than a man, but he fails all the tests of monkeyhood set for him and is duly charged with being a man, for which the incontrovertible evidence is:

First, in that he lies so shamelessly in maintaining that he is not one; second, in that he laughs like a madman; third, in that he cries like an idiot; fourth, in that he blows his nose like a peasant; fifth, in that he is plucked like a mangy fellow; sixth, in that he carries his tail before him; seventh, in that he always has a quantity of tiny square pieces of grit in his mouth that he lacks the courage either to spit out or to swallow; eighth, and finally, in that each morning he raises his eyes, his nose, and his broad beak, clasps his open hands together, palm against palm, pointed towards heaven, making them joined as one, as though he were bored at having two separate ones, he snaps his legs in half, falling on to his shanks, I have taken notice that then, as he drones magical words, his broken legs reattach themselves, and he rises as merry as before.[96]

Derisory taunts at religious ritual are frequently encountered among the members of the theriophilic school.

In Cyrano's view, at least as expressed by the birds, equality is the first law of the commonwealth, and this fact humans disregard by persecuting the birds, which they hunt, kill, and eat. In a nice gibe at the Cartesian experimenters, he says that humans believe God to have given the birds entrails only to show humans the future – but, after all, humans are not to blame for their absurdities since they lack the requisite reason to understand them. Still, they possess the faculty of will, which they use to attack the birds without cause, and they corrupt such birds as hawks and falcons to betray their own kind. Cyrano is accordingly condemned to be eaten by flies, but he escapes this destiny when it is shown that he was once kind to a parrot and that he had the wisdom to acknowledge that birds are capable of reason. The birds are more generous than humans.

Certainly, Cyrano is less restrained than most theriophilists. Still, we can find the same respect for animals, if not quite the same grandiloquence, in Molière's *L'Avare* (The miser; 1668). Instead of treating humans as necessarily careless of animal interests, Molière chooses to depict both sensibility and cruelty:

MAÎTRE JACQUES [*the coachman and factotum*]: ... you were saying ... ?

HARPAGON [*the miser*]: That my carriage needs to be cleaned and my horses got ready to drive to the fair.

MAÎTRE JACQUES: Your horses, sir? My word, they're in no condition at all to walk. I won't tell you they're down on their litter, the poor beasts don't have any, and it would be no way to talk; but you make them observe such austere fasts that they are nothing any more but ideas or phantoms or shadows of horses.

HARPAGON: They're sick indeed. They don't do anything.

MAÎTRE JACQUES: And because they don't do anything, sir, don't they
need to eat anything? It would be much better for them, poor creatures,
to work a lot and eat likewise. It breaks my heart to see them so emaci-
ated; for the fact is I have such a tender feeling for my horses that when
I see them suffer, it seems to be happening to me. Every day I take things
out of my own mouth for them; and, sir, it's a sign of too harsh a nature
if a man has no pity on his neighbor.[97]

There is both compassion for, and empathy with, the horses. And insofar
as they suffer, and can be seen to suffer, Molière is offering us an implicit
denunciation of Descartes, one that he clearly expects his audience to share.

The "Eighth Satire" of Nicolas Boileau (1663) begins with the proclamation that of all animals man is the most stupid. To be sure, he is lord of
creation, and endowed with reason, but quite lacking in wisdom, which
requires an equanimity untroubled by desire. And of that humankind is incapable. We are treated to the by now customary animadversions – humans
are inconstant, ambitious, lustful, avaricious, and hateful. Wolves do not
entrap their fellow wolves; tigers do not divide their territories into competitive camps; bears do not go to war with bears; vultures do not attack
their compatriots. Only a human deems it a matter of honour to slash his
fellow human's throat. Humans have reason, but it turns them to folly. Animals lack reason, but they are not so foolish as to bow down before idols.
Asses with the light of nature fare better than humans with reason; bears
are not troubled by superstition or fancies. As the ass says in conclusion:

> Seeing these madmen on all sides, doctor,
> It may be said in all goodwill, without any jealousy ...
> Upon my word, man is just an animal, no more than we.[98]

Likewise, in her famous poem *Les Moutons* (The sheep), Antoinette
Deshoulières writes of their being without cares, as much loved as loving,
unencumbered with useless desires, following nature without ambitions,
reputation, shares, deceptions:

> We, however, have the distinguishing factor of reason
>   And you are ignorant of its use.
> Innocent animals, you have nothing to be jealous of.[99]

Finally, let us mention Jean de La Fontaine, who took Descartes to task
for his view that "the animal is a machine." He is aghast that, according

to Descartes, "everything is done without choice, and by springs; / Neither sentience, nor soul; everything is material."[100] Indeed, if this were true of the animals, why not equally so of the humans? Must we not grant the animals at least one distinction from the plants? Do they not use their reason as do humans? Does not the aged and exhausted stag compel a younger stag to take his place? Does not the mother partridge lead the hunters away from her brood by distracting their attention? Do not the beavers employ a division of labour in building their dams? Do not Polish foxes possess the military art against their foes?[101] Still, in the final analysis, La Fontaine, destructive as he is of Cartesianism, nonetheless acknowledges a superior human soul. The animal soul is like that of a child – which is, of course, quite enough to confound the Cartesians, who deny them any kind of soul, but not enough to satisfy those like Pierre Bayle who see an essential similarity in all souls. Again, one is perplexed by the fact that even some of the French theriophilists do not deign to grant the animals an immortal soul.

VI

Theriophily flourished outside France and Italy, too, but it was generally later in origin, although, as we saw in Chapter 4, William Langland in *Piers the Ploughman* and Walter Map in *De nugis curialium* contributed to the genre in the Britain of the late Middle Ages.[102] If the eighteenth-century poetry and prose of Jonathan Swift, William Shenstone, John Hildrop, and Anne Finch, Countess of Winchilsea, fit nicely into the theriophilic theme, in the sixteenth and seventeenth centuries the British concern was more with compassion, and the prevention of cruelty, toward animals than with theriophily. Still, there are several examples of theriophily worthy of mention. For instance, in 1664 Margaret Cavendish, the Duchess of Newcastle, reflected on whether other species possessed forms of knowledge unattained by humankind:

> For what man knows whether fish do not know more of the nature of water, and ebbing and flowing and the saltness of the sea? Or whether birds do not know more of the nature and degrees of air, or the causes of tempests? Or whether worms do not know more of the nature of the earth and how plants are produced? Or bees of the several sorts of juices and flowers than men? ... Man may have one kind of knowledge ... and creatures another way, and yet other creatures' manner or way may be [as] intelligible and instructive to each other as Man's.[103]

It was, however, in her poetry that her sensibilities flourished. She repeated the above prose ideas in verse in "A Discourse of Beasts," "Of Fishes," and "Of Birds." In "A Moral Discourse between Man and Beast," she gave the advantage in different respects to each. Despite the overall balance, however, "several *Creatures* by several Sense, / Have better far (than *Man*) *Intelligence*."[104] In "Of the Ant" she praises their communal ownership, their industrious endeavours, and their superiority over humans in lacking humans' worst passions:

> They live as the *Lacedaemonians* did,
> All is in *common, nothing* is forbid.
> No *Private Feast,* but altogether meet,
> *Wholesome,* though *Plain,* in *Public* do they eat.
> They have no *Envy,* all *Ambitions* down,
> There is no *Superiority,* or *Clown.*
> No *Stately Palaces* for *Pride* to dwell,
> Their *House* is *Common,* called the *Ants Hill.*
> All help to *build,* and keep it in *repair,*
> No 'special *work-men,* all *Labourers* they are.
> No *Markets* keep, no *Meat* they have to sell,
> For what each one doth eat, all *welcome* is, and well.
> No *Jealousy,* each takes his *Neighbours Wife,*
> Without *Offence,* which never breedeth *Strife.*
> Nor fight they *Duels,* nor do give the *Lie,*
> Their greatest *Honour* is to live, not die.
> For they, to keep in *life,* through *Dangers* run,
> To get *Provisions* in 'gainst *Winter* comes.
> But many lose their *Life,* as *Chance* doth fall,
> None is perpetual, *Death* devours all.[105]

Of course, this is all utopian panacea. Few lived in any more *"Stately Palaces"* than did the Duchess of Newcastle herself; and she didn't bequeath her privileges either to the plebeians or to the beasts.

Shakespeare exhibits a great deal of compassion and respect for animals, but only once, through the voice of Isabella in *Measure for Measure* (Act 3, Scene 1; c. 1605) does he approach the theriophily theme:

> The sense of death is most in apprehension,
> And the poor beetle that we tread upon,
> In corporal sufferance finds a pang as great
> As when a giant dies.

The sectarian Thomas Edwards wrote of an essential human-animal equality, if not of an animal superiority, in his *Gangraena* (1646): "God loves the creatures that creep on the ground as well as the best saints; and there is no difference between the flesh of a man and the flesh of a toad."[106] And the radical Ranter Jacob Bottomley wrote mid-century, with a pantheistic touch: "I see God in all creatures, man and beast, fish and fowl and every green thing from the highest cedar to the ivy on the wall; and that God is the life and being of them all."[107] And the phrase "my fellow creature," referring at least sometimes to "fellow animals," almost becomes a commonplace. It can be found, inter alia, in the writings of the Dissenter Edmund Bury and the Presbyterian minister John Flavel.[108] None of these, however, possessed the customary irreverence and religious skepticism that we frequently encounter among the French and Italian theriophilists. Nonetheless, all the elements of the tradition are as prominent in John Wilmot's "Satyr" (1675) as they are in the texts of any French or Italian writer:

> Were I (who to my cost already am
> One of those strange prodigious Creatures *Man*)
> A Spirit free to choose for my own share,
> What Case of Flesh, and Blood, I pleas'd to weare,
> I'd be a *Dog,* a *Monkey,* or a *Bear,*
> Or any thing but that vain *Animal,*
> Who is so proud of being rational.
> The senses are too gross, and he'll contrive
> A Sixth, to contradict the other Five;
>
> And before certain instinct, will preferr
> *Reason,* which Fifty times for one does err ...
> For all his Pride and his Philosophy,
> 'Tis evident, *Beasts* are in their degree
> As wise at least, and better far than he.
> Those *Creatures,* are the wisest who attain,
> By surest means, the ends at which they aim ...
>
> Whose Principles, most gen'rous are and just,
> And to whose *Moralls,* you would sooner trust.
> Be judge your self, I'le bring it to the test,
> Which is the basest *Creature Man,* or *Beast?*
> *Birds,* feed on *Birds, Beasts,* on each other prey,
> But Savage *Man* alone, does *Man* betray:
> Prest by necessity, they Kill for Food;
> *Man* undoes *Man,* to do himself no good.[109]

If theriophily was less frequent in the Britain of the seventeenth century, it came into its own in the more cynical eighteenth century – the French and Italians learned their cynicism earlier. There is no better capture of the early-eighteenth-century British wit than Jonathan Swift's *The Beasts' Confession to the Priest, on Observing how most Men mistake their own Talents, Written in the Year 1732*, in which he employs goats, apes, wolves, pigs, and asses to symbolize the worst aspects of humanity. In the "Advertisement" to the *Confession* he advises that "the following Poem is grounded upon the universal Folly in Mankind of mistaking their Talents; by which the Author doth a great Honour to his own Species, almost equalling them with certain Brutes." The opening and closing lines of the poem will reflect the tenor of Swift's mind:

> When Beasts could speak (the Learned say
> They still can do so every Day)
> It seems they had Religion then,
> As much as now we find in Men ...
>
> It happen'd when a Plague broke out,
> (Which therefore made them more devout)
> The King of Brutes (to make it plain,
> Of Quadrupeds I only mean)
> By Proclamation gave Command,
> That ev'ry Subject in the Land
> Should to the Priest confess their Sins

In turn, each of the animals denies its principal vice. The wolf admits to fasting too strenuously, the ass to being too jovial, the pig to being too refined in its eating habits, the ape to being too stoic, and the goat to being too chaste. Next, in turn, we meet an equally improbable honourable lawyer, a cringing place-seeker, a fawning chaplain, a humanitarian physician, an honest politician, and a cardsharp who complains that he always loses.

> *Esop;*
> I would accuse him to his Face
> For libelling the *Four-foot* Race.
> Creatures of every Kind but ours
> Well comprehend their nat'ral Power;
> While We, whom *Reason* ought to sway,
> Mistake our Talents eve'ry Day:
> The Ass was never known so stupid

To act the Part of *Tray,* or *Cupid;*
Nor leaps upon his Master's Lap,
There to be stroak'd and fed with Pap;
As *Esop* would the World perswade;
He better understands his Trade:
Nor comes when'er his Lady whistles;
But carries Loads, and feeds on Thistles;
Our Author's Meaning, I presume, is
A Creature *bipes et implumis,*[110]
Wherein the Moralist design'd
A Compliment of Human-Kind:
For here he owns, that now and then
Beasts may *degen'rate* into Men.[111]

Neither Leonardo nor Montaigne nor Cyrano could have been any more mordant. Or perplexing.

There is no doubt that Swift gives a clear indication both in the title and in the opening of the advertisement that his primary purpose is to prick the human bubble. Yet, as he continues, the idea that the brutes are truly superior gains ground. Likewise, the opening lines of the poem itself refer indirectly to human religious insincerity, while the closing lines suggest a very real comparative failing on the part of humans to match the self-understanding of other species. The animals know their own true natures, while humans imagine themselves to be something other than they are. *The Beasts' Confession* is in the same genre as Swift's "A Voyage to the Country of the Houyhnhnms" of *Gulliver's Travels* (1726), a land where the horses (Houyhnhnms) govern and where the humans (Yahoos) are treated as an inferior species, of value only to the extent that they are instrumental to the purposes of horses. Soame Jenyns followed the same line in 1757, but in deadly earnest rather than in satire. God, he wrote, "has given many advantages to Brutes, which Man cannot attain to with all his superiority ... we are not in the state of existence as our ignorant ambition may desire."[112]

In Alexander Pope's *An Essay on Man* (1734), humans learn their arts from the animals, which in the manner of Montaigne are not one whit the less than the lessons of antiquity, and human reason and justice are outshone by instinct:

Thus then to man the voice of Nature spake:
"Go, from the creatures thy instructions take:
Learn from the birds what food thy thickets yield;
Learn from the beasts the physic of the field

Thy arts of building from the bee receive;
Learn of the mole to plough, the worm to weave;
Learn of the little nautilus to sail,
Spread the thin oar, and catch the driving gale.
Here too all forms of social union find,
And hence let reason, late, instruct mankind:
Here subterranean works and cities see;
There towns aerial on the waving tree.
Learn each small people's genius, policies,
The ants' republic, and the realm of bees;
How those in common with all their wealth bestow,
And anarchy without confusion know;
And these forever, though a monarch reign,
Their separate cells and properties maintain.
Mark what unvaried laws preserve each state,
Laws wise as Nature, and as fix'd as fate.
In vain thy reason finer webs shall draw.
Entangle Justice in her net of law,
And right, too rigid, harden into wrong;
Still for the strong too weak, the weak too strong.
Yet go! and thus o'er all the creatures sway,
Thus let the wiser make the rest obey:
And for those arts mere instinct could afford,
Be crown'd as monarchs, or as gods adored.[113]

John Hildrop rounds out the theriophilic theme of the first half of the eighteenth century – which was thereafter to become a commonplace – in his *Free Thoughts upon the Brute Creation* (1742):

Shew me one species of animal more ridiculous, more contemptible, more detestable, than are to be found among the silly, the vicious, the wicked part of mankind. Are apes and monkeys more ridiculous or mischievous creatures than some who are to be found in the most polite assemblies? Is a *poor dog* with four legs, who acts agreeably to his nature, half so despicable as a *sad dog* with two, who with high pretension to reason, virtue, and honour, is every day guilty of crimes for which his brother dog would be doomed to hanging? Is a swine that wallows in the mire half as contemptible as a drunkard or a temperance? What is the rage of tigers, the fierceness of lions, the cruelty of wolves and bears, the treachery of cats and monkeys, when compared with the cruelty, the treachery, the barbarity of mankind? The wolf and the tiger, that worry a few innocent sheep, purely to satisfy hunger, are harmless animals when opposed to the rage and fury of conquerors, the barbarity and cruelty of tyrants and oppressors, who uninjured, unprovoked, lay whole countries to waste, turn the most beautiful cities into ruins, and sweep the face of the earth before them like an inundation or devouring fire.[114]

Hildrop's message differs somewhat from a number of others in that there is no adulation of the animal character – indeed, different species are described as fierce, cruel, and treacherous. There is instead a conviction that however harmful animals may be, humans are decidedly worse. A scatological touch in similar vein was added in the Victorian lyric poet Walter Savage Landor's "A Quarrelsome Bishop":

> To hide her ordure, claws the cat;
> You claw, but not to cover that.
> Be decenter, and learn at least
> One lesson from the cleanlier beast.[115]

We can also find something of the theriophilic message in the *Spiritual Canticle* (1584) by the Spanish Discalced Carmelite Saint John of the Cross, who wrote of the animals' "natural gifts and graces," which God gave them "to make them finished and perfect ... He left them clothed with beauty; communicating to them supernatural being ... all clothed with beauty and dignity ... abundance of graces and virtues and beauty wherewith God endowed them."[116] However, while this may contain some of the content of French and Italian theriophily, especially in its recognition of a beautiful and dignified clothed animal in contrast to a nude human, it has none of its savour. Irreverence and a decidedly secular flavour run through French and Italian theriophily, almost unsullied with piety. It is the philosophy of the *libertins,* not of the devout. But then the same was largely true of the classical tradition of Democritus, Pliny, and Plutarch, which it revived.

While, as noted, after Hildrop theriophily became common, there are still unusual and surprising instances that bring a slight astonishment to the reader. Thus, for example, we find Granville Sharp, the Anglican political radical and opponent of the slave trade, contending in his 1777 *A Tract on the Law of Nature* that animals are superior to humans because "they never commit suicide." He continues:

> They are never known to violate that universal principle, self-love, except it be for a *reasonable cause,* that they risk their own lives in defence of their young, to preserve their species, or through gratitude, as dogs will defend their masters, which surely is no depravity!

To what extraordinary cause then shall we attribute this very singular superiority of brutes, in a circumstance so necessary to happiness? The cause is obvious: Brutes have never been subjected to *spiritual delusions,* or been actuated by *infernal spirits,* since the time that the serpent deceived our first parents![117]

An anonymous pamphlet of 1823, *Cursory Remarks on the Evil Tendency of Unrestrained Cruelty, particularly on that Practiced in Smithfield Market,* expressed a more common opinion. While humans were creatures of reason and animals creatures of instinct, humans misused their attributes and were superior only in their depravity. The practices witnessed at Smithfield Market in London exemplified the "malignant moral distemper" that contrasted humans to those on whom the cruelties were performed. It was, so it was argued in the pamphlet, the duty of Christian philanthropists to take a stand against these crimes.[118]

In 1838 we find the Reverend William Drummond asking of man:

Can he match the elephant in strength, the horse in fleetness, the lynx or the eagle in vision, the spider in delicacy of touch, or the hound in scent? Can he elaborate any article like the honey of the bee, or concoct a poison like the rattle-snake's? Even in intelligence he is excelled by them, for they make use of that portion of understanding which God has given them, as God designed, and "know and reason not contemptibly;" but he abuses his intellect and the gifts of God, and reasons, if reasoning it may be called, in a way which even they, if they could speak, would call contemptible.[119]

In the extremely influential *The System of Positive Philosophy* (1851), the reputed founder of the discipline of sociology, Auguste Comte, declared that the "happiness of living for others is not entirely monopolised by Man. Many animals possess it likewise, and indeed give evidence of a higher degree of sympathy than our own."[120] It was dogs to whom he first and foremost referred. Writing in the *North American Review* in 1891, the English Romantic novelist Ouida (Louise de la Ramée) remarked that "those who are great or eminent in any way find the world full of parasites, toadies, liars, fawners, hypocrites; the incorruptible candour, loyalty and honour of the dog are to such like water in a barren place to a thirsty traveller."[121] In 1903 Elijah Buckner observed of "the faculty of *love,* which is the highest attribute of the soul, and the strongest tie that binds together all living creatures, [that] we are bound to acknowledge that it is manifested in a greater degree in lower animals than in man." He arrives at this conclusion by quoting Jesus' words "Greater love hath no man than this, that a man lay down his life for his friends" and by then insisting that "lower animals, especially dogs, have repeatedly sacrificed their lives for their friends, no one will pretend to deny."[122] He next regales us with several canine accounts to justify his point, concluding that "such self-sacrifice can only be compared to the rarest and noblest acts of man."[123] They are commonplace in the world of dogs. Moreover, "man is the only animal on earth which kills for fun, and in this respect is the meanest creature God ever made."[124]

Against those who deem reason the all-important factor of human superiority, Buckner follows the naturalist Louis Agassiz in elevating instinct. In making his case for animal immortality, Buckner announces: "Instinct is the highest attribute God possesses, and a man who would deny that it is immortal, could with the same consistency, deny that the omniscience, omnipotence, or any other attribute of God is immortal. Instinct is so closely interwoven with all the phenomena we see and know around us, that if it is not an immortal attribute of God, we know nothing that is. It is knowledge from God, directly imparted to animals, not once, but continually." Buckner does not refrain from adding the theriophilic rider: "animals are endowed with instinct superior to man."[125]

Whether theriophily is used to puncture human pretensions or to elevate the real world of animal life, it cannot be denied as an intrinsic aspect of human consciousness, present in varying degrees throughout human history. To imagine that, following the classical era, respect for animals does not begin until Bentham or Darwin is to ignore a great deal of the history of ideas.

6

Symbiosis

ANIMALS AS MEANS AND AS ENDS

I

I F SOME CULTURES ARE "AT ONE WITH nature," Westerners, in Woody Allen's famous phrase, are "at two." And of course, there is a convincing ring of truth about the quip. Indeed, it has been a common critique of Western society. For example, Francis Bacon is customarily assigned the dubious compliment of being the one, as William Leiss wrote in *The Domination of Nature,* "to formulate the conception of human mastery over nature much more clearly than had been done and to assign it a prominent role in men's concerns."[1] And Jim Mason added: "Secularist and materialist as Bacon was, he gave up none of the human supremacy and dominionism expressed in Genesis. If anything his human chauvinism was even more swaggering ... Such a supremacist view gave humans virtual ownership of nature with a secular kind of right to do anything they pleased with it."[2] Clearly, an understanding of the nature to be controlled and an explanation of the purpose of that control are demanded.

The words of Leonardo da Vinci, whom Peter Singer regards as an exception to Renaissance dominionism, calling him "one of the first genuine dissenters,"[3] but who is in fact in many respects a representative figure – indeed, the very epitome of Renaissance man – should be an appropriate starting point. "O indifferent nature, whereof art thou so partial," Leonardo wrote in following Pliny on nature as a cruel stepmother rather than a mother. But unlike Pliny, as we have observed, he regarded nature as having been a cruel stepmother to the nonhuman animals as well as to the human.[4] Nature is no ally. Moreover,

> the rat was being besieged in its dwelling by the weasel which with continual vigilance was awaiting its destruction, and through a tiny chink was considering its great danger. Meanwhile the cat came and suddenly seized hold of

the weasel and forthwith devoured it. Whereupon the rat, profoundly grateful to its deity, having offered up some of its hazel-nuts in sacrifice to Jove,
came out of its hole in order to repossess itself of the lately lost liberty, and
was instantly deprived of this and of life itself by the cruel claws and teeth
of the cat.[5]

At the very outset of the Renaissance, Leonardo was expressing a view that,
rather than being a dissenting opinion, pervaded the Renaissance. Nature
is the embodiment of cruelty. It is neither an ideal nor a salvation but a cruel
force that must be dominated and suppressed if justice is to be achieved.
Leonardo glories in the beauties of nature and captures them magnificently
in paint. But he is repelled by nature's indifference to justice. It is not
merely human cruelty to animals that repels him – although this most certainly does repel him, as his *Notebooks* demonstrate – but also "the cruel
claws" of animal reality, even if they are ordained by nature.

Even the commonly condemned Francis Bacon was, at least on occasion,
aware of the limits to the domination of nature, insisting that if humans
do not possess the habit and nature of goodness, then "man is a busy, mischievous thing." Nor is the control of nature to be undertaken for the sake
of humans alone, for the "inclination of goodness is imprinted deeply in
the nature of man; insomuch that, if it not issue towards men, it will take
unto other living creatures"[6] – the first being humans, to be sure, but
nature is to be controlled for the sake of other animals, too. Moreover,
"there is implanted in man by nature a noble and excellent spirit of compassion that extends itself to the brutes which by the divine ordinance are
subject to his command. This compassion therefore has a certain analogy
with that of a prince toward his subject."[7] Animals, Bacon believes, are subject to human authority, as Genesis indicates – he is not as much of a secularist as Mason imagines – but nonetheless it is an authority over animals
that corresponds to the obligation implied in the authority of a prince
over his subjects. If nature is to be controlled for human ends, this control must also include the interests of the animals. In such a view – and,
as we shall see, it is not an uncommon view – there is a clear distinction
between nature herself and the animals of which she is in part constituted.

It is a matter of some moment that those who are most outspoken in
the condemnation of nature are precisely those who either have or are
regarded to have the strongest concern for animals and their interests. For
example, the great Romantic novelist Victor Hugo, who was, inter alia,
honorary president of the Societé Française contre la Vivisection, wrote
somewhat opaquely in *Les Misérables* that nature is "pitiless ... There are
times when nature seems hostile ... 'After all, what is a cat?' he demanded,

'it's a correction.' Having created the mouse God said to himself, 'That was silly of me!' and he created the cat. The cat is the *erratum* of the mouse."[8] Charles Darwin was, somewhat by contrast, immediately translucent. Writing to J.D. Hooker in 1856, he observed: "What a book a Devil's Chaplain might write on the clumsy, wasteful, blundering low & horridly cruel works of nature."[9] The anti-vivisectionist dramatist George Bernard Shaw, who pronounced himself "a vegetarian on humanitarian and mystical grounds" and who had "never killed a fly or a mouse vindictively or without remorse," could also welcome the change "from the brutalizing torpor of nature's tyranny over Man into the order and alertness of Man's organized dominion over nature."[10]

In *Theory of the Future of Man* (1854), the founder of Positivism, Auguste Comte, wrote of the benefits to be derived from the control of nature. "The better use will have as consequences, respect for our voluntary [animal] assistants and increased use of the blind forces of nature ... From the animals as from men, we should demand not merely automatic service, as opposed alike to economy and morality ... As salutary for the instinct as for the heart, the discipline of synthesis will make us shrink from substituting animals for men [in vivisection], the priesthood of Sociocracy more even than the priesthood of Theocracy being disposed to insist upon the constant respect of our [animal] auxiliaries."[11] For Comte, not only could nature be controlled for the benefit of humankind and the "higher" of the "lower animals," but the animals could come to cooperate willingly with humankind.

For Thomas Hardy, who was an early convert to Charles Darwin's theory of evolution and what he saw as its ethical implications for the benefit of the animals, nature is that which has to be subdued if humans and animals are to be treated justly. In *Two on a Tower* (1883), Hardy referred to, and commented on, "nature's cruel laws." By the time of *Jude the Obscure* (1895), the theme was recurrent. Hardy mentions: "the perception of the flaw in the terrestrial theme by which what was good for God's birds was bad for God's gardener ... Nature's logic was too horrid for [Jude] to care for. That mercy towards one set of creatures was cruelty towards another sickened his sense of harmony ... O why should nature's law be mutual butchery! ... Cruelty is the law pervading all nature and society; and we can't get out of it if we would."[12]

In a letter to the *Academy and Literature* (May 1902), he attacked the "sophistry" of Maurice Maeterlinck's *Apology for Nature*:

> In your review of M. Maeterlinck's book you quote with seeming approval his vindication of Nature's ways, which is, (as I understand it) to the effect

that, though she does not appear to be just from our point of view, she may practise a scheme of morality unknown to us, in which she is just. Now, admit but the bare possibility of such a hidden morality, and she would go out of court without the slightest stain on her character, so certain should we feel that indifference to morality was beneath her greatness.

Far be it from my wish to distrust any comforting fantasy, if it can be barely tenable. But alas, no profound reflection can be needed to detect the sophistry in M. Maeterlinck's argument, and to see that the original difficulty recognized by thinkers like Schopenhauer, Hartmann, Haeckel, etc., and by most of the persons called pessimists, remains unsurmounted.

Pain has been, and pain is: no new sort of morals in Nature can remove pain from the past and make it pleasure for those who are its infallible estimators, the bearers thereof. And no justice, however slight, can be atoned for by her future generosity, however ample, so long as we consider Nature to be, or to stand for, unlimited power. The exoneration of an omnipotent Mother by her retrospective justice becomes an absurdity when we ask, what made the foregone injustice necessary to her Omnipotence?

So you cannot, I fear, save her good name except by assuming one of two things: that she is blind and not a judge of her actions, or that she is an automaton, and unable to control them ...

... to model our conduct on Nature's apparent conduct, as Nietzsche would have taught, can only bring disaster to humanity.[13]

In a 1906 letter to Frederic Harrison, he indicated, like Leonardo, that the animals themselves were a part of the problem: "In regard of [blood] Sport, for instance, will ever the great body of human beings ... ever see its immorality? Worse than that, supposing they do, when will the still more numerous terrestrial animals – our kin, having the same ancestry – learn to be merciful? The fact is that when you get to the bottom of things you find no bed-rock of righteousness to rest on – nature is *un*moral – & our puny efforts are those of a people who try to keep their leaky house dry by wiping the waterdrops from the ceiling."[14] Whereas Shaw rejoices in the future control of nature that will bring justice for humans and animals, Hardy recognizes that there are no easy solutions, probably no solutions at all, at best "wiping the waterdrops from the ceiling." In 1910 he wrote to Sidney Trist, editor of the *Animals' Guardian,* despairing at his "own conclusion" of "the difficulty of carrying out to its logical extreme the principle of equal justice to all the animal kingdom. Whatever humanity may try to do, there remains the stumbling-block that nature herself is absolutely indifferent to justice, & how to instruct nature is rather a large problem."[15] For Hardy, the failure of Francis Bacon and those who tried to control nature was not their sense of justice – indeed, this was the only way justice could be achieved. No, his despair was that, regardless of how much we tried to manipulate nature for the ends of justice, we were almost certain to fail. Nature herself would thwart all our efforts.

Undoubtedly, humanity has failed miserably in its attempts to control nature. In its naive efforts it has devastated the interests of the environment itself, of animals, and even of humans. But let us not imagine that justice would somehow have been achieved if we had left "nature" well enough alone. All the injustices of which Leonardo, Victor Hugo, Charles Darwin, Thomas Hardy, and countless others complained would have remained, and in certain respects been exacerbated, if the Francis Bacons of this world had not attempted their ultimately impossible and counterproductive task. Neither should we ever imagine that the control of nature was undertaken without all concern for the constituent elements of nature nor that it always failed their interests. Oddly, it is now that we are beginning to recognize our failings in our historical attempts to control nature that we are slowly coming to understand what a wise control of nature would involve – at the very least, insofar as we recognize the need to undo a great deal of the harm we have committed. Certainly, we should not forget that the attempt to rescue animals from species extinction is just as much an attempt to control nature as those human actions that brought them to the present predicament.

II

The most significant question to be asked is: What is the proper human relationship to nature, and how may it be determined? For many, the ideal appears to lie in some kind of symbiosis, which, scientifically, is an interaction between two different organisms living in close physical association, especially one in which the symbionts benefit mutually from the arrangement. Such a form of symbiosis is known as mutualism. In the biological sciences, the term is often used to depict an obligative relationship in which neither species can live without the other (for example, protozoans in the guts of termites digest the wood ingested by the termites). There are, however, other forms, such as commensalism, in which one species (the commensal) obtains nutrients, shelter, support, or locomotion from the host species, which is substantially unaffected (for example, remoras are able to obtain both locomotion and food from sharks). The boundaries of the concept are nonetheless fluid – indeed, unclear. As John Steinbeck wrote in *The Log from the Sea of Cortez* (in effect, as the representative of the marine biologist Ed Ricketts):

It would seem that the commensal idea is a very elastic thing and can be extended to include more than host and guest; that certain kinds of animals are often found together for a number of reasons. One, because they do not eat one another; two, because these different species thrive best under identical conditions of wave-shock and bottom; three, because they take the same kinds of food, or different aspects of the same kind of food; four, because in some cases the armor or weapons of some are protection to the other (for instance, the sharp fins of an urchin may protect a tide-pool johnny from a larger preying fish); five, because some actual commensal activities may truly occur. Thus the commensalitie may be very loose or very tight and some associations may partake of a very real thigmotropism.[16]

("Thigmotropism" is a tropism – the involuntary organization of an organism – in which contact with a solid surface is the orienting factor.) A third form of symbiosis is parasitism, the best-known example of which is the practice of cuckoos and cowbirds engaging in brood parasitism. They lay eggs in the nests of other bird species and have their young raised by the foster parents, whose impulse is to feed the alien young, even though it is often at the expense of their own offspring. Clearly, the scientific use of the term "symbiosis" covers a wide range of behavioural circumstances, none of which correspond to what is meant by those whose ideal is a symbiotic life.

What is customarily meant by those who write of symbiosis as an appropriate political goal is a mutually beneficial arrangement covering the lives of many or all species in which all live together to a common advantage. Yet the examples given hint at the nature of the problem for those who want to make some kind of mutually benevolent relationship in nature into a practical ideal. Naturally, there is nothing we can, or should, do to change the behaviour of cuckoos and cowbirds. Nor can we, or should we, change the commensal behaviour of the remoras. Nor can we turn tigers into pussy-cats – and, even if we could, they would be the torturers and killers of birds and rodents. George Orwell was a profound observer of nature and felt an authentic respect for its animals. Yet his friend Kay Welton "discovered that he knew a great deal about birds and was 'passionately fond' of them ... she later learned that he was equally fond of cats, but that he could 'never square the fact that cats killed birds.'"[17] Mutualistic symbiosis flies in the face of ethological reality. At best, we can be realistically concerned only with how humans may alter their behaviour to live with greater respect for other species than they now do. It would be inappropriate even to suggest the desire for a harmony, for such harmony would depend on our having the right and capacity to adjust the behaviour of others toward us as well as ours toward them.

In fact, "harmonization" has sometimes been the aim. Thus the Zoological Society of London, the founding organization of Regent's Park Zoo, which was the first modern public zoo – although designed in imitation of the private Jardin des Plantes in Paris – declared in 1826 as one of its principal purposes "the introducing and domesticating of new Breeds or Varieties of Animals ... likely to be useful in Common Life." At least one contemporary, the eminent literary critic, poet, and essayist Leigh Hunt, was not convinced of its wisdom. Indeed, he criticized the keeping of all wild animals because their incarceration would only lead to lingering deaths. The Zoological Society's only domesticating success was with the golden hamster – at least, if we ignore its *ultimately* great success in domesticating the public to its aims after 1840 (for several years entry was limited to subscribers). It numbered both Charles Dickens and Charles Darwin among its frequent and enraptured visitors. Yet the idea of domestication was not based entirely on the desire to use animals for human ends. There was also a sense that if we treated animals well, fed them appropriate food designed to limit aggression, and trained them in appropriate altruistic behaviour, we could manage to secure an accommodation of their interests to ours, with the result that animals and humans could live in greater harmony. In *The Ethics of Diet* (1883), Howard Williams reported that "there are well-authenticated instances, even in our own times, of true *carnivora* that have been fed, for longer or shorter periods, upon the non-flesh diet."[18] Williams appears to believe that *carnivora* may be made less harmful, and not themselves be harmed, by a change in their diet. Today many pet-owning vegetarians share the same view. In *My Father's Island: A Galapagos Quest,*[19] Johanna Angermeyer tells how in mid-twentieth-century Ecuador the belief still reigned that if you fed an ocelot a vegetarian regimen, its nature would change in accord with the gentleness of the diet. Most striking, however, are the writings of Auguste Comte, one of the greatest influences on the thought of the nineteenth century. In his *Theory of the Great Being* (1854), he wrote of the possibility of interspecies altruism and of what, in effect, amounted to mutualistic symbiosis between humans and certain other species that he regarded as potentially altruistic:

Our conception of the constitution of the Great Being remains defective unless we associate with man all the animal races which are capable of adopting the common motto of all the higher natures: *Live for Others*. Without the animals, the Positive Synthesis could but imperfectly form the permanent alliance of all voluntary agents to modify the external conditions of our life so far as they are modifiable ... Political action recognizing [the benevolent instincts] as supreme, is enabled to carry out the largest plans, by bringing

all our practical remedies to bear on the direct improvement of man's condition upon earth, in concert with the animal races, which, as sympathetic,
are justly associated with Humanity.[20]

The only animals that Comte mentions explicitly are the horse and the
dog, but there is no indication that he believed mutualistic symbiosis with
humans to be restricted to these domesticated species. Clearly, the kind of
thinking engaged in by the founders of the London Zoological Society res-
onated even more sonorously with Comte.

Some, however, suggest that nature itself is a form of symbiotic mutualism in which the overall benefit is obtained not by achieving the ends of
individuals but by means of a harmony among species in which nature,
red in tooth and claw, behaves in such a way as to maintain a healthy bio-
diversity – a sustainable development, as the trite but harmful phrase goes.
Already in 1889, the naturalist Bradford Torrey observed that:

> The whole earth is one field of war. Every creature's place upon it is coveted
> by some other creature ... The import of this apparent wastefulness and cru-
> elty of Nature, her seeming indifference to the welfare of the individual, is
> a question on which it is not pleasant, and, as I think, not profitable to dwell.
> We see but parts of her ways, and it must be unsafe to criticise the working
> of a single wheel here or there, when we have absolutely no means of know-
> ing how each fits into the grand design, and, for that matter, can only guess
> at the grand design itself.[21]

For Torrey, while the human individuals are entitled to be treated as ends
in themselves, nature forbids such treatment for wild animals. Even if
Torrey exaggerates the Hobbesian notion of nature – "a war of all against
all" – his message about our ignorance cannot be gainsaid.

Ralph H. Lutts, professor of environmental studies and museum direc-
tor for the Audubon Society, took up Torrey's theme a century later, commenting that some environmentalists and animal-welfare advocates have
"difficulty dealing with change":

> Their efforts to preserve, conserve, and protect wildlife are often efforts to
> maintain the status quo without regard for the ecological context. However,
> they inevitably run up against questions of what is not worth protecting;
> when are preservation efforts working against normal ecological change; and
> when do efforts to prevent the death of animals represent a rejection of death
> as a necessary component of living systems? Throughout there is also the
> question of how we are to think of and value wild animals, if not as pets or

furry little people ... When is it appropriate to intervene, and when is it not?
... Whatever the answer, it does not lie alone in a responsibility to protect
the lives of individual wild animals ... [Unwisely] some people are morally
committed to undo the "wrongs" against wildlife that are committed both
by humans and by nature itself.[22]

Clearly, what Lutts is indicating is that the moral rules we apply to
humans, and even to pets, are not the rules to be applied to wildlife. Wild
animals are not, and should not be regarded as, ends in themselves. They are
to be viewed as members of a species. It is the species, not the individual,
that is important – but even then not primarily as an entity in itself but to
the extent that it contributes to the wellbeing of the overall environment.
Ecological health is the end in itself. If this proposition is to be maintained,
we must ask of Lutts the very question with which we began the first chapter. If there are no disjunctions in evolution, a theory that Lutts himself
makes a cornerstone of his treatise, how is it possible that humans can
have earned themselves the right to be treated in a different way from the
rest of creation? What is it that sets humanity apart? Lutts's answer is that
"Humans are [a] rarity in nature. We produce relatively few young that have
a prolonged childhood, and we give them tender, loving care. Family members, friends, and associates are important to us as individuals."[23] If it is the
importance of individuals qua individuals to a species' mentality that constitutes their entitlement to be treated as individual ends, as entities who
matter in and for themselves, then surely Lutts would have to acknowledge whales, dolphins, chimpanzees, gorillas, orangutans, wolves, and horses
(and, of course, a great deal more since all individual animals of any complexity pursue pleasure, avoid pain, and, other things being equal, seek to
avoid death) as entitled to the same *kind* of rights as humans, which involves
the fulfillment of their species-specific needs as the beings that they are.

Nonetheless, Lutts has an important point to make. The health of the
overall environment is vital to the interests of all species and all individuals.
Our treatment of other animals must be consistent with ecological interests. And where the individual may be seen as unimportant to the species
and the environment, less stress needs to be laid on its rights as an individual. But what Lutts does not seem willing to countenance is that, since
there are no discontinuities in evolution, the treatment of humans must,
at least prima facie, meet the same criteria – and perhaps do so even more
readily, for nothing could be more profitable to environmental health than
the culling of a billion humans.

Certainly, ignoring the human population problem, where animals demonstrate the capacity for self-sacrifice, as with prairie dogs, the interests of

their community as a whole may be deemed more important than the rights of the constituent parts. If human individuals are to be treated as ends, then, unless some sounder criteria can be offered other than those promoted by Lutts, so too must much wildlife. Certainly, animal ethics is far more complicated than those who would think in terms of abstract rights, or susceptibility to pain, or environmental harmony as the sole relevant criteria would have us believe. And with such creatures as prairie dogs, willing to sacrifice themselves for the common good, there may be an added criterion whereby they may be entitled to a greater consideration than that to which Benthamite pleasure-pursuing humans are entitled. Or at least, we should be willing to countenance the probability that no ethical theory currently in play is adequate to decide the human-to-animal relationship respecting the rights and wellbeing of each.

One of the more common approaches to human superiority and the proclamation of greater human rights is offered in *The Animal Contract* by former curator of the London Zoo and former Fellow of Wolfson College, Cambridge, Desmond Morris. He describes how for a million years our ancestors hunted prey and avoided predators. "They killed and ate only what they needed to survive, and they destroyed only those life forms that threatened their well-being. Their response to other animals was very different from ours. In many ways they considered them their equals or superiors. Many animals had faster legs, a better sense of smell, stronger teeth and more acute hearing. Our ancestors were right to respect them."[24] Morris goes on to explain how we used our brains rather than our brawn, how the hunting lifestyle shaped our personalities, making us more cooperative than our fellow animals, and how becoming bipedal freed our hands to perform myriad tasks. We bonded in pairs, developed strong territoriality, the division of labour, speech, and religion. "All of this," we are told, "resulted from one simple switch in feeding from fruit-gathering to hunting. It was a major transformation that set us on our pathway to global success."[25] The natural response is to note that, despite Morris's glorification of the human, we are less cooperative than some species, certainly than the honeyguides, the prairie dogs, and the wolves, and are decidedly more competitive than any other species. Surely, many species have a greater sense of territoriality than humans. And if our division of labour is more complex than that of other species, it is no more of a reality than hunting lionesses, the macho behaviour of silverback gorillas, or the many species that set sentinels without specific regard to gender. He goes on to castigate vegetarianism – "for the growing child a meatless diet would in most instances be disastrous and would result in a rapid decline in the local population."[26]

Go tell that to the population of India, which grew from 238 million in 1901 to a billion by the end of the century – and not by a greater increase in the flesh-eating than in the vegetarian population! The overall effect of Morris's approach is to provide a sense of human superiority in which humans should treat animals respectfully, but as our inferiors – as, in essence, intended for human use. We are superior as a consequence of our accomplishments and have a corresponding right to use animals for our ends. Morris might have been enlightened by a reading of the classical and Renaissance theriophilists!

Perhaps the most sophisticated version of symbiosis is offered by John Steinbeck, again speaking on behalf of, or at least with the advice of, Ed Ricketts, but not without his own inimical communitarian orientation:

> Our own interest lay in relationships of animal to animal. If one observes in this relational sense, it seems apparent that species are only commas in a sentence, that each species is at once the point on the base of a pyramid, that all life is relational to the point where an Einsteinian relativity seems to emerge. And then not only the meaning but the feeling about species grows misty. One merges into another, groups melt into ecological groups until the time when what we know as life meets and enters what we think of as non-life: barnacle and rock, rock and earth, earth and tree, tree and rain and air. And the units nestle into the whole and are inseparable from it. Then one can come back to the microscope and the tide pool and the aquarium. But the little animals are found to be changed, no longer set apart and alone. And it is a strange thing that most of the feeling we call religious, most of the mystical outcrying which is one of the most prized and used and desired reactions of our species, is really the understanding and the attempt to say man is related to the whole thing, related inextricably to all reality, known and unknowable. This is a simple thing to say, but the profound feeling of it made a Jesus, a Saint Augustine, a Saint Francis, a Roger Bacon, a Charles Darwin and an Einstein. Each of them in his own tempo and with his own voice discovered and reaffirmed with astonishment the knowledge that all things are one thing and that one thing is all things – plankton, a shimmering phosphorescence on the sea and the spinning planets and an expanding universe, all bound together by the elastic string of time. It is advisable to look from the tide pool to the stars and then back to the tide pool again.[27]

For Steinbeck and Ricketts, neither individuals nor species can be the ultimate ends in themselves; indeed, in a significant manner they do not really exist as organisms but are parts of the greater organism of life itself, which constitutes the true end of all action. As Steinbeck wrote in an un-published manuscript of 1934, "Argument of Phalanx": "Man is a unit of the greater beasts, the phalanx. The phalanx has pains, desires, hungers and strivings as different from those of the unit man as man's are different from those of his cells." If man is an end in himself, so too is the species, the phalanx – "unit of the greater beasts" – and, to borrow a term from Ralph Waldo Emerson, which Steinbeck himself liked to use, the "over-soul" itself. For Steinbeck, when species interests are in conflict, it is the "over-soul" that prevails. Yet Steinbeck proclaimed for humans alone all the benefits of individuality and self-responsibility in addition. In the end, he retreats into a Taoist quietism: "Criticize nothing, evaluate nothing. Just let the Thing come thundering in – accept and enjoy. It will be chaos for a while but gradually order will appear and an order you did not know."[28] Of course, such an attitude debars one from moral judgments on *logical* grounds – but not, it would appear, *in fact.*

In his obituary "About Ed Ricketts," which he appended to *The Log from the Sea of Cortez,* John Steinbeck described his closest friend and mentor, whose views on nature largely informed his own:

> His scientific interest was essentially ecological and holistic. His mind always tried to enlarge the smallest picture. I remember him saying, "You know, at first view you would think the rattlesnake and the kangaroo rat were the greatest of enemies since the snake hunts and feeds on the rat. But in a larger sense they must be the best of friends. The rat feeds the snake and the snake selects out the slow and weak and generally thins the rat people so that both species can survive. It is quite possible that neither species would exist with-out the other." He was pleased with commensal animals, particularly with groups of several species contributing to the survival of all. He seemed as pleased with such things as though they had been created for him.[29]

Life itself before the phalanx, the phalanx before the species, and the species before the individual – this is indeed ecological and holistic think-ing. The relevant question is whether human individuals are included in the practical considerations. Are humans to be culled when evidence dem-onstrates the damage they do? Will humans be deprived of their habitats when rabbits are at stake? Are individuals, whether human, hare, leopard, or loon to be denied their individuality and not to be treated as ends in themselves? These are paradoxes to which no adequate answers are ever offered. Steinbeck does not seem to think that the questions need an answer, for somehow a symbiosis is achieved:

To each group, of course, there must be waste – the dead fish to man, the broken pieces to gulls, the bones to some, and the scales to others – but to the whole, there is no waste. The great organism, Life, takes it all and uses it all. The large picture is always clear and the smaller can be clear – the picture of eater and eaten. And the large equilibrium of life of a given animal is postulated on the presence of abundant larvae of just such forms as itself for food. Nothing is wasted; "no star is lost."[30]

In fact, Steinbeck's vision, important as a correction to our customary individualistic optimism as it is, is one-sided. This is not a world of eat and be eaten alone. It is also a world of affection, courage, sacrifice, fidelity, altruism, and skill, of jealousy, hatred, cowardice, disloyalty, selfishness, and incompetence – among many species, not just among humans. Moreover, an earthquake can have such a destructive effect that both the predator and the prey are engulfed. Death takes it all and (mis)uses it all.

In many respects the idea of symbiosis is indeed akin to that of the "over-soul," the universal soul, or the universal spirit. The prey does not die. Its soul, its quintessence, is released from the body, and returns to the collective spirit, from which it is reborn again in the body of another animal, going forth to be killed again as a part of the eternal cycle of life and death. Steinbeck's symbiosis works in the same way and can come to be readily employed to justify hunting and specimen collecting, both important parts of Steinbeck's own life. The idea functions as a justification for killing since killing is not really killing. Those who care for animals in and for themselves, both as individuals and as species, must ultimately be repulsed by such a moral system – as, indeed, must those who care for humans, if humans are not to be excluded from the holistic ethic. Nonetheless, there are enough partial truths in Steinbeck and, to a lesser degree, in Torrey, Lutts, and Morris that they must not be ignored. There is a contradiction, or a series of contradictions, between the interests of each animal as an end in itself, as a member of a particular species, as a member of a particular species in conflict with the interests of other species, and even just as a part of Life itself (or, now, more commonly via James Lovelock, of Gaia itself).[31] That these interests are in conflict is undeniable. And that ethical decision making always involves giving one priority over another is equally undeniable. What this means is that ethical decision making is not merely a matter of having the right moral system – such as utilitarianism, or mutualistic symbiosis, or one or another of the systems of virtue – but must in some manner involve some kind of balance among competing principles achieved by the exercise of considered judgment in each and every practical situation.

III

There is a pervasive belief that the idea of animals as intended for human use has dominated Western thought throughout its history. Thus, for example, we read in Lynn White Jr.'s influential paper on "The Historical Roots of Our Ecologic Crisis" that "by destroying pagan animism, Christianity made it possible to exploit nature in a mood of indifference to the feelings of natural objects ... Christianity not only established a dualism of man and nature but also insisted that it is God's will that man exploit nature for his proper ends."[32] And James Serpell adds that "according to this tradition, the Earth and the animal and plant species were created specifically to serve the interests of humanity."[33] Peter Singer relates how "Aristotle holds that animals exist to serve the purposes of human beings" and that "it was the views of Aristotle rather than Pythagoras that were to become part of the later Western tradition."[34] In fact, the exhortation did not begin with Aristotle, for what Richard Ryder coined "speciesism" was already present in the fifth century BC with Xenophon in his *Memorabilia* (4.2.9-12) when he stated that "the beasts are born and bred for man's sake."

We have already dealt with the interpretation of the Biblical exhortation to dominion.[35] But it remains important to investigate the claim that in the Western tradition animals have been regarded as intended for human use. Again, there is no doubt that such a view has been continually present, but the story, as in all the history of animal ethics, is a far more involved tale than that which is usually related to us. We should also ask ourselves, since Xenophon felt compelled to make his provocative statement, whether there was an animal-sympathetic tradition that he was opposing. After all, we don't normally feel compelled to proclaim a specific assertion unless its antithesis has worthy adherents. While we can find both Cicero and Marcus Aurelius – indeed, a number of the Stoics – following Xenophon, we can also find the proposition most rigorously denounced in the writings of Porphyry:

> If God fashioned animals for the use of men, in what [manner] do we use flies, lice, bats, beetles, scorpions, and vipers? ... And if our opponents should admit that all things are not generated for us, and with a view to our advantage, in addition to the distinction which they make being very confused and obscure, we shall not avoid acting unjustly, in attacking and noxiously using those animals that were not produced for our sake ... I omit to mention, that if we define, by utility, things which pertain to us, we shall not be prevented from admitting that we were generated for the sake of the most destructive animals, such as crocodiles ... and dragons.[36]

It is precisely this point of view that we find repeated by Thomas Hobbes in the seventeenth century: "I pray, when a lion eats a man and a man eats an ox, why is the ox made more for the man than the man for the lion?"[37] In fact, the new science, so admired by Hobbes, had already by the end of the twelfth century encouraged a denunciation of the anthropocentric view. Thus Moses Maimonides writes in *The Guide to the Perplexed:* "And if the earth is thus no bigger than a point relative to the sphere of the fixed stars, what must be the ratio of the human species to the created universe as a whole? And how then can any of us think that these things exist for his sake, and that they are meant to serve his uses?"[38] In the seventeenth century, we can find Margaret Cavendish, the Duchess of Newcastle, railing against the Stoic view in *The Hunting of the Hare:*

> Yet man doth think himself so gentle, mild,
> When he of creatures is most cruel wild.
> And is so proud, thinks only he shall live,
> That God a god-like nature did him give.
> And that all creatures for his sake alone,
> Was made for him to tyrannize upon.[39]

Across the channel, the philosopher and theologian Pierre Charon (1541-1603), close friend of the redoubtable Montaigne and author of *On Wisdom* (1601) and *Three Truths* (1594), had already advised his audience that "Man scruples not to say that he enjoyeth the heavens and the elements, as if all had been made, and still move only for him. In this sense, a gosling may say as much, and perhaps with more truth and justness."[40] Just over a century later, Alexander Pope employed the same popular example in *An Essay on Man* (3.1.45-46): "While man exclaims, 'See all things for my use!' / 'See man for mine!' replies a pampered goose."

In fact, at the close of the seventeenth century, we encounter an acknowledgment not only that regarding animals as intended solely for human use is untenable, but also that it has long been rejected by all those who have a modicum of sophistication. Thus writes the naturalist and taxonomist John Ray in *Wisdom of God Manifested in the Works of Creation* (1691):

> It is a generally received Opinion that this visible world was created for Man; that Man is the end of Creation, as if there were no end of any Creature but some way or other serviceable to man. This opinion is as old as *Tully* [i.e., Marcus Tullius Cicero] ... But though this be vulgarly received, yet Wise Men nowadays think otherwise. Dr. More [the Cambridge Platonist] affirms "The Creatures are made to enjoy themselves as well as to serve us, and that

it's a gross piece of Ignorance and Rusticity to think otherwise." And in another place, "This comes only out of Pride and Ignorance or a haughty Presumption, because we are encouraged to believe, that in some sence, all things are made for Man, therefore to think they are not all made for themselves. But he that pronounceth this, is ignorant of the Nature of Man, and the Knowledge of Things."

I believe there are many Species in Nature, which were never yet taken notice of by Man, and consequently of no Use to him, which yet we are not to think were Created in vain; but it's likely (as the Doctor [More] saith) to partake of the overflowing Goodness of the Creator, and enjoy their own beings.[41]

For More and Ray, only the ignorant and the rustic, the proud and the presumptuous, would imagine animals made solely for human ends. If the animals are also there "to serve us," such a recognition is necessary if one is to be both honest and carnivorous. Indeed, all meat-eating societies, all societies that put the ox to the plough or the water buffalo to the dray cart, or that wear eagle feathers and bison robes, or that ride horses and milk cows, believe that animals exist at least in part for human use.

Oddly, in that he certainly used animals in the most heinous manner for human ends in his vivisection practices, René Descartes, too, acknowledged that: "It is not at all probable that all things have been made for us in such a way that God had no other end in view in making them ... We cannot doubt there is an infinity of things which exist now in the world, or which formerly existed and have now ceased to be, which have never been seen by any man or been of use to any."[42] He wrote further in a 1645 letter to Princess Elizabeth: "For if a man imagine that beyond the heavens there exist nothing but imaginary spaces, and that all the heavens are made solely for the sake of the earth, and the earth for the benefit of man, the result is that he comes to think that this earth is our principal dwelling-place and this life the best that is attainable by us; and also that, instead of recognizing the perfections which we really possess, he attributes to other creatures imperfections which do not belong to them, in order to raise himself above them."[43]

"What is man," asked Blaise Pascal "in the midst of infinity?"[44] Robert Boyle, too, in seeing the countless stars through a telescope, was compelled to reject the idea that everything was created for human benefits. Even those who were the greatest misusers of animals discarded the view of what Peter Singer called "the central place of human beings in the universe."[45] Boyle was quite capable of his invasive methods because he chose, as later did Charles Darwin, to place the importance of knowledge over the importance of the pain and suffering of the animals. Nor were these exceptions in rejecting the view that everything existed for the sake of humans. Writing of the eighteenth century, Arthur Lovejoy tells us that:

> It was implied by the principle of plenitude that every link in the Chain of Being exists, not merely and not primarily for the benefit of any other link, but for its own sake, or more precisely, for the sake of the completeness of the series of forms, the realization of which was the chief object of God in creating the world. We have already seen that, though essences were conceived to be unequal in dignity, they all had an equal claim to existence, within the limits of rational possibility; and therefore the true *raison d'être* of one species of being was never to be sought in its utility to any other.[46]

This is not to suggest that the idea of animals for human use disappeared. It is to indicate that the course of the history of ideas is played out on a stage of competing conceptions without a script and by actors who do not know the outcome of the drama. Indeed, the curtain will drop one night on a scenario that will be quashed at the following performance.

Despite the animadversions of More and Ray about the ignorance, rusticity, and presumption of those who fail to recognize that animals exist for their own sakes, we can still count their contemporary, the illustrious archbishop of Cambrai, François Fénelon (1651-1715), as one of the many still following the adage of Xenophon, Aristotle, the Stoics, Saint Augustine, and Saint Thomas in telling us that "not only the plants but animals are made for our use." And what of wild animals that prey on humans? "If all the countries were peopled and made subject to law and order as they should be, there would be no animals that would attack man."[47] And why has God permitted the existence of such animals? "One finds ferocious animals only in the remote forests, where men of a bold, fighting disposition may go to engage in a game which resembles warfare, so that there will never be any necessity to have a real war between nations."[48] Fénelon employs the particular twist of the great-chain theory that saw progress through the links in terms of increasing perfection. Thus the gift of speech made humans "more perfect" than other animals. More entertainingly, and no less paradoxically, Fénelon added: "it happens equally often that I am more perfect when I remain silent than when I talk."[49] It is indeed reflective of the paradoxes of the human mind that while Fénelon revelled in the idea of progressive perfection, his *Télémaque* (1699) belongs to the genre of return-to-nature literature.

The language of discourse of the seventeenth century was still pre-Kantian and thus archaic to modern ears. Kant revolutionized the language of ethics in the eighteenth century with the development of the categorical imperative, according to which each person is entitled to be treated as an end in himself or herself and never as a mere means. Notoriously, in Kant's view: "Animals are not self-conscious and are there merely as a means to an end. The end is man." Although Kant was himself not without animal sympathies,[50] it should scarcely surprise us that he met with a spirited disapproval from those who thought in terms of an unbroken human-animal continuity. Thus, for example, Arthur Schopenhauer observed indignantly that "general morality [is] outraged by the proposition [of Kant] that beings devoid of reason (hence animals) are *things* and therefore should be treated merely as *means* that are not at the same time an *end.*"[51] It should be noted, however, that for Schopenhauer with regard to humans and animals, as for Kant with regard to humans, the categorical imperative did not deny that beings were, or should be, means but only that they should not be solely so. It was, however, Johann Wolfgang Goethe who accepted the Kantian philosophy and language – as would most people in this increasingly individualistic era – but who turned the imperative in order to accommodate the interest of animals. Convinced as he was, as a result of his zoological research, that human-animal parallels were complete and that animals possessed consciousness, he concluded that:

> Each animal is an end in itself, it emerges fully formed
> From nature's loins, and produces perfect offspring.
> All its limbs are formed in accord with eternal laws,
> And the most unusual form preserves the secret of the primal pattern.[52]

All animals are perfect in themselves as the beings that they are, and they are entitled to be treated as ends in themselves. Contemporaneously, Samuel Taylor Coleridge announced a combination of the individualist and communitarian conceptions of animals as ends: "Nature has her proper interest; & he will know what it is, who believes and feels, that everything has a life of its own, & that we are all *one Life.*"[53] We can find the combination of the individualistic and communitarian messages also in William Blake. In *Visions of the Daughters of Albion,* he allows for the sacredness of each individual life: "And trees & birds & beasts & men, behold their eternal joy! / Arise, you little glancing wings, and sing your infant joy! / Arise, and drink your bliss, for every thing that lives is holy."[54] In *The Book of Thel,* he adds the communitarian aspect: "every thing that lives / Lives not alone, nor for itself."[55] The most significant part of the Blake message is that animals and humans alike are a part of the same fellowship of life. If animals are not entirely ends in themselves, neither are humans.

Despite the greater spiritual depths of such authors, many lesser lights continued to express the idea of animals as ends in themselves in more traditional terms and in more traditional antipathy to those who imagined themselves a separate and special part of nature. For instance, the parliamentarian and commissioner of the Board of Trade Soame Jenyns wrote in 1757 in theriophilic manner that God "has given many advantages to Brutes, which Man cannot attain ... Is not the justice of God as much concerned to preserve the happiness of the meanest insect which he has called into being, as of the greatest Man that ever lives? Are not all creatures we see made subservient to each others uses?"[56]

In 1802 the antiquarian Joseph Ritson opined:

The *sheep* is not so much "design'd" for the *man*, as the *man* is for the *tyger;* this animal being naturally carnivorous, which man is not; but *nature* and *justice,* or *humanity,* are not, allways, one and the same thing ...

If god made *man*, or there be any *intention* in *nature,* the life of the *louse,* which is as natural to him as his frame of body, is equally sacred and inviolable with his own ...

[T]here is neither evidence nor probability, that any one animal is "intended" for the "sustenance" of another, more especially by the privation of life. The lamb is no more "intended" to be devoured by the wolf, than the man by the tyger or other beast of prey ... such reasoning is perfectly ridiculous![57]

Robert Southey mocked those who would treat the pig as a mere instrument of human ends, noting that the pig did not share the opinion. In "The Pig, a Colloquial Poem," the narrator tells of "pig-perfection" and notes that the pig itself is:

> a democratic beast,
> [Who] knows that his unmerciful drivers seek
> Their profit, and not his. He hath not learnt
> That pigs were made for man, – born to be brawn'd
> And baconized; that he must please to give
> Just what his gracious masters please to take.[58]

In "The Dancing Bear," Southey lampooned those parliamentarians who held to the antiquated doctrine:

> [W]e are told all things were made for man;
> And I'll be sworn there's not a fellow here
> Who would not swear 'twere hanging blasphemy

To doubt that truth. Therefore, as thou wert born,
Bruin, for Man, and Man makes nothing of thee
In any other way – most logically
It follows thou wert born to make him sport;
That that great snout of thine was form'd on purpose
To hold a ring; and that thy fat was given thee
For an approved pomanium! To demur
Were heresy. And politicians say
(Wise men who in the scale of reason give
No foolish feelings weight) that thou art here
Far happier than thy brother bears who roam
O'er trackless snow for food ... Besides
'Tis wholesome for thy morals to be brought
From savage climes into a civilized state.

Into the decencies of Christendom.
Bear, Bear! It passes in the Parliament
For excellent logic, this!

Of course, we are not intended to take Southey too seriously in his denial of a parliamentary animal sympathy. There were a significant number of parliamentarians who held the same view as Southey. If some parliamentarians derided animal protective legislation, we should not forget that early legislative defeats were often narrow, and opponents sometimes thought of themselves as the proper representatives of the oppressed humans, the lower classes whose pleasures, rather than those of the wealthy, were likely to be diminished by the proposed legislative measures.[59]

The incomparable Lord Erskine, who led the parliamentary fight for animal protection, argued: "For every animal which comes into contact with Man, and whose powers, and qualities, and instincts, are obviously constructed for his use, Nature has taken the same care to provide, and as carefully, and as bountifully as for Man himself, organs and feelings for its own enjoyment and happiness."[60] Reflecting the by now more rigorous scientific interest in nature, William Kirby and William Spence declared with consummate caustic wit in their groundbreaking *An Introduction to Entomology* (1815): "The *gastrophilus equi* can subsist no where but in the stomach of the horse or ass; which animals therefore, this insect might boast, with some show of reason, to have been created for its use rather than for ours, being to us useful only, but to it indispensable."[61] In Anne Brontë's 1847 novel, *Agnes Grey,* the eponymous heroine responded laconically and sardonically to the suggestion that "the creatures were all created for our convenience" with "I thought the doctrine admitted some doubt."[62] George Sand in *Indiana* (1831) was less restrained: "Yours is the god of men, the king, the founder, and protector of your race [males]; mine is the God of the universe, the creator, the support, and hope of all creatures. Yours has made everything for you alone; mine has made all species for each other."[63]

The Irish clergyman William Drummond, writing in *The Rights of Animals* (1838), derided the once customary Aristotelian view:

> Aristotle held an opinion which egregious pride and arrogance may still defend – that as nature has made nothing imperfectly or in vain, she must of necessity have made all things for man – a conclusion which the premises by no means warrant; as if man were the sole object of creation, and that the thousand and ten thousand species of nature could not exist happily, many of them might say much more happily without him ... We should indeed hold an overweening idea of our own importance, were we to imagine that the magnificent frame of nature was formed solely on our account, and not for myriads of other animated beings.[64]

It was now the norm to acknowledge the idea of animals as means alone to be a barbaric relic of a long-derided human hubris. Indeed, Drummond was capable of even greater rhetorical inquiry: "On what principle can it be believed that the allwise Creator formed such an infinite variety of creatures, only to furnish subjects to gratify the cruel and destructive propensities of man? Is there no distinction to be made betwixt a privilege and an absolute independent right?"[65] His answer was more than rhetoric: "Beasts, birds, fishes, insects, as well as men, were formed to taste the pleasures of existence; and they, as well as men, are furnished with innumerable sources of enjoyment, passions, appetites, affections, feelings, solitary or social, conjugal, parental, and to a certain extent intellectual and moral, all of which have their proper objects of gratification."[66] Remarking in *The Veterinarian* in 1839 on misinterpretations of Holy Writ, William Karkeek observed that:

> One of the most preposterous and absurd conclusions that have been arrived at in this manner is, – THAT ALL THE INFERIOR ANIMALS WERE CREATED FOR THE SOLE USE OF MAN. "This monstrous faith of many made for me," says [Robert] Southey [in *The Doctor*, vol. 4, 1834], "seems rather unreasonable. – Made for thy use, tyrant that thou art, and weak as thou art tyrannical! ... Made for thy use, indeed, when so many seem to have been made for thy punishment and humiliation."[67]

Writing a year later in the same journal, W.C. Spooner seems to take the matter a little further, arguing that animals, unlike humans, have no future existence. Consequently, "born for the present, for the present only are they adapted."[68] Thus they are even more immediately and materially ends in themselves than are humans. In what respects may Charles Darwin be suggested to have exceeded such sentiments?

Jules Michelet believes not merely that the birds possess eternal souls, but that "to reveal the bird as a *soul*, [is] to show that it is a *person*."[69] In the United States John Muir was asking: "What good are rattlesnakes for? ... good for themselves, and we need not begrudge them their share of life."[70] And John Howard Moore insisted in stentorian voice: "*All* beings are ends; no creatures are *means*."[71] Such representative declarations should suffice to demonstrate that the history of ideas pertaining to animals has been a debate rather than a declaration of an orthodoxy. Moreover, the issue is not merely between those who see animals as intended for human use and those who think in terms of animals as ends in themselves. There is also a third category, exemplified by Coleridge's conception of *"one Life,"* Blake's suggestion that "every thing that lives / Lives not alone, nor for itself," Jenyns's idea of creatures that are "subservient to each others uses," and Sand's notion that God "has made all species for each other." For such thinkers, there is a notion of the appropriate relationship between humans and animals as one that is at least quasi mutually symbiotic, satisfying both each individual as an end and the relationship between individual animals while also allowing for the fulfillment of each species as an end in itself.

Until the symbiotic prophecies of Isaiah 11 come to pass, such must remain an idyllic yearning. Nonetheless, for it to be retained as an unachievable ideal in which its spirit is approximated to the degree possible is only a desirable improbability. If, however, such steps are taken further, as with Torrey, Lutts, and especially Steinbeck, however well intentioned these steps may be, they lead ultimately to a corporatism for animals in which they are always means: to their own species, to other species, or to Life itself. If one is to countenance such a view, one must either apply it to humans equally or find some manner of determining a special status for humanity. And, to date, all arguments from Xenophon on that have claimed a special status for humans have been countered with well-considered claims that there is nothing unique in the human condition or at least nothing of sufficient exclusivity to allow the human to be classified entirely separately. This is not to argue that it is necessarily impossible to discover such a principle but to indicate that any justification for treating different animals in a different manner requires an investigation into their species-specific needs. And if the needs of humans differ from those of other species, then to this extent other species require treatment by humans of a special and distinctive kind in order that their needs be met. Yet it is impossible to follow such a line of reasoning if we accept the implications of ecological holism. On such terms, individual animals cannot be ends in themselves. Nor, if we are to remain consistent, can individual humans.

7

Evolution, Chain, and Categorical Imperative

I

IN THE OPENING PAGES OF THIS BOOK, we noted the encomium
accorded Charles Darwin and his theory of evolution as well as the
"revolution" that "evolution" is commonly said to have spawned in the
understanding of the human-animal relationship, particularly with regard
to ideas of kinship. It was suggested there that the honours accorded Darwin were far in excess of the originality of his contribution. The claims
usually made for Darwin, which were significant, are not primarily on
account of his sympathy for animals but largely because human-animal
continuity was so emphatically demonstrated by his theories. The idea has
long struck me as exaggerated since the notion of human-animal continuity not only has a lengthy history – and is the very mainstay of, for example, Plutarch, Lucretius, Porphyry, Leibniz, and Bayle – but also does not
in fact seem to have been advanced any further by Darwin's novelties,
despite the originality of natural selection as an explanation for the method
of evolutionary adaptation. I do not here wish to show that there has been
a continuous debate about, and considerable recognition of, animal worthiness in the Western tradition, although that is certainly true, but that
the *scientific* and *philosophical* contributions of others prior to Darwin did
a great deal to stress ideas of continuity and kinship, even in the works of
those who were not especially remarkable in their practical sympathies for
animals.

The failure to recognize the significance of Johann Wolfgang Goethe to
the history of animal ethics represents one of the lacunae in the literature.
In the history of ideas concerning the status of nonhuman animals in ethical debate, one looks in vain for anything more than a passing reference
to the work of Goethe (1749-1832).[1] Occasionally, one may read that he held
to an embryonic theory of evolution or that Charles Darwin acknowledged

him to be an adamant prior proponent of evolutionary principles.[2] Or we may encounter the view expressed by Marian Scholtmeijer that "in Goethe's *Elective Affinities [Wahlverwandtschaften]* and *Faust,* the very few places in which animals are mentioned at all occasion expressions of disgust."[3] And while such luminaries as Voltaire, Bentham, Schopenhauer, and John Stuart Mill[4] are accorded pride of place in the postclassical literature that touches on the history of animal ethics, Goethe is customarily ignored, trivialized, or condemned. Yet in reality, Goethe's contribution to the understanding of the relationship of humans to animals is profound, sympathetic, and most importantly, in his application of Kant's categorical imperative to animals, groundbreakingly provocative.

Why, then, has Goethe received in some instances a bad press and in others no press when, I am suggesting, his contribution to the debate is both original and stimulating? In the primary study of the implications of evolutionary theory for human responsibilities to animals, James Rachels's *Created from Animals: The Moral Implications of Darwinism,*[5] there is no mention of Goethe (or Maupertuis, or Herder, or Buffon, or Robinet) and little mention of any earlier evolutionary thinker. Yet Darwin's precursors were well recognized. For example, in 1903 Elijah Buckner wrote that "this theory has been addressed by several learned naturalists of Europe, but Charles Darwin is the modern champion of the theory that man and lower animals are descended from 'only four or five progenitors.'"[6] To be sure, a few paragraphs are devoted by Rachels to Erasmus Darwin and to Jean-Baptiste Lamarck,[7] but since they erred about the method of evolution, their significance to the *Moral Implications* of evolutionary theory is dismissed out of hand. Yet, surely, it is not natural selection – Darwin's original contribution to evolution – that has moral implications but evolution itself, the fact of being *Created from Animals,* and there were many before Charles Darwin who argued the case for our animal descent. Indeed, in "An Historical Sketch on the Progress of Opinion on the Origin of Species previously to the Publication of this Work," which Darwin appended to later editions of *The Origin of Species,* he mentioned some thirty forerunners, among whom, citing Geoffroy Saint-Hilaire as his source, he noted that "there is no doubt that Goethe was an extreme partisan of ... [evolutionary] views."[8] Had Rachels not restricted his attention to Charles Darwin, the idea of being "created from animals" might have been seen to have implications for a much lengthier history of the perception of the human-animal kin relationship.

In the sentence following her debunking of Goethe's animal orientations – which makes me immediately think contrastingly of the role of ravens and deer in *Faust, Part Two* – Scholtmeijer adds that "in its religious and supernatural mode, the Romantic imagination was inclined to jump from the individual person to the panorama of nature, overleaping the animal in the process."[9] Insightful, and indeed accurate, as is Scholtmeijer's observation for some of the Romantics some of the time, there are numerous exceptions: Blake with his "Robin Red breast in a Cage" and "Tyger, Tyger, burning bright," even if we don't know quite what to make of the feline beast; Wordsworth with his "Hart-Leap Well," "The White Doe of Rylstone," "Music," and "Foxey"; Southey with the parliamentarily misjudged dancing bear and Phillis; and Coleridge with the young ass tethered near his mother and the well-loved "man and bird and beast" of "The Rime of the Ancient Mariner."[10] Nonetheless, Goethe's naturalist sentimentalism in *The Sorrows of Young Werther* fits Scholtmeijer's categorization well, and on at least some occasions, Ann Radcliffe and Percy Bysshe Shelley fall into the same excess. And the exclusion of animals from nature is exemplified in some of Goethe's 1770s poetry, as with: "Wie herrlich leuchtet / Mir die Natur! / Wie glänzt die Sonne! / Wie lacht die Flur!" (How wonderfully Nature lights the path for me! How the sun gleams! How the fields laugh!) Nonetheless, to classify Goethe as a Romantic is misleading. Certainly, he is one on occasion, but his inclinations varied from Sturm und Drang to Realism to Classicism. "It is probably best to speak of the 'styles' of Goethe," suggests Stuart Atkins, "since we cannot reconcile the enormous differences" between the various stages of his writings.[11] As a philosopher, he is a thoroughgoing Idealist and almost as much Platonic as Kantian. To understand Goethe as a Romantic, especially one of the traditional "religious mode," is to miss the complexity, the eclecticism, and the variety of Goethe – in the final analysis, he is simply Goethean, which includes a healthy dose of Kantianism, whose anthropocentrism he nonetheless, as we shall see, turns upside down.

There is another aspect to Goethe's work that should not be overlooked. Of course, his accomplishments – and they were legion – arose primarily from the multifaceted originality of his incomparable mind, but they were also enhanced by his employment at the court of Weimar. The "spirit of Weimar," as Ute Lischke has remarked, was "cosmopolitan, tolerant, humane"[12] – a far cry from contemporary culture not just elsewhere in Germany, but throughout most of Continental Europe. Goethe was inspired by the liberality of Weimar on his arrival – he soon came to champion it – and he ultimately symbolized it not only in his lifetime, but henceforth. To be sure, as a Weimar bureaucrat, he sullied himself by procuring recruits for the Prussian army, and his administration of mines does not appear to have been undertaken in an especially humanitarian spirit. But the intellectual atmosphere, the music, the theatre, even the masked balls, reflected the unrivalled joy of Weimar society; and, if some found Weimar's own fashion magazine and the production of plaster busts to decorate bourgeois homes rather pretentious, they were nonetheless a part of the liberalization of culture beyond traditional Teutonic boundaries – women too were "encouraged to cultivate their minds vigorously and were given the opportunity to do so."[13] Goethe's literary and scientific profundity came from his inner being, but his residence in a vibrant and pioneering Weimar was no hindrance to his intellectual achievements and to the development of his humanitarian consciousness, of which his recognition of the significance of the human-animal relationship is an important part.

II

The once standard English-language biography, J.G. Robertson's *The Life and Work of Goethe* (1932),[14] set the tone in trivializing Goethe's role in evolutionary theory. Robertson insists that "we are hardly justified in drawing the conclusion that Goethe was a Darwin before Darwin."[15] And, of course, he is right – few would have thought to claim otherwise.[16] Certainly, Goethe had no conception of evolution by natural selection – although one might suggest that he was closer to it than those who regarded the role of will as a primary causal factor in evolution. The effect of Robertson's statement, however sound it may be, is to minimize the importance of Goethe's recognition – he was not alone – that all animals, including human animals, are related both anatomically and historically and that there is a single template from which all animals are derived. If Goethe's theorizing does not allow him to see *how* animal descent occurs, he is every bit as much aware as Darwin of the moral implications of the idea of continuity and kinship implicit in all evolutionary theory. What Goethe and Darwin share in common is an understanding of the role of scientific discovery in our understanding of the human-animal relationship.

Although Goethe's sensibilities toward animals have gone largely unrecognized and unexplored, there are notable partial exceptions. For example, in his renowned *The Great Chain of Being*, Arthur O. Lovejoy cites Germany's national bard, writing to Frau von Stein in 1786: "Had I time in the short span of a single life, I would devote myself to extending it to all the kingdoms of Nature – to her entire domain."[17] Of course, the skeptics will feel entitled to interpret this, and not entirely without justification, as an expression of his scientific interests and awe for the abstractions of nature rather than as any indication of his recognition of the moral significance of nature's constituent parts. However, as we shall see, for Goethe,

the conceptions are not incompatible, just as they were not unrelated for Leibniz or Charles Darwin. More emphatically, Nicholas Boyle, in the first volume of his monumental biography of Goethe, writes of "the view Goethe seems to have held from 1783 onwards that the supreme religious issue is not the relation between men and gods ... but between men and animals."[18] Nothing Darwinian can be imagined to have exceeded this conviction! William Rounseville Alger wrote in 1860 of Goethe sympathizing "with all lower forms of life."[19] And Sir Arthur Helps observed in 1872 Goethe's acknowledgment of a fraternal relationship to the animal realm.[20] Even in such instances, however, the story is broached but not told. Lovejoy offers us only parenthetical, if illuminating, remarks on Goethe. Boyle's task is to display the whole range of Goethe's polymath wisdom in the context of the ideas and politics of his era – as dramatist, poet, novelist, scientist, administrator, diplomat, and statesman. It is not his intent to provide an integrated analysis of Goethe's thoughts on animals. Alger refers to Goethe in passing in alluding to the third-century story of "The Seven Sleepers of Ephesus" that Goethe had retold. And Helps noted his companionship with the animal realm only in citation.

There have been several illuminating studies of Goethe's scientific research, most of which concentrate on his investigations into optics, botany, and geology rather than on his anatomical research, although the last certainly has not gone unmentioned.[21] There is none, however, at least to the best of my knowledge, that relates his anatomical studies to the moral status of animals or to the ethical implications of his scientific and philosophical conclusions. Those who have analyzed Goethe's anatomical studies have been concerned with elucidating his status as a scientist and his role in the development of scientific thought. They have not been intent on fleshing out an understanding of the ethical aspects of the human-animal relationship.

Goethe's contribution to the understanding of the human-animal relationship lies in three distinct, yet related, areas. The first is evolutionary theory, with regard both to homologies and to descent, and the implications of recognizing the essential – if, for Goethe, incomplete – human-animal similarities. In this respect, Goethe belongs to a small but significant group of late-eighteenth-century scholars who came independently and almost simultaneously to explicitly evolutionary conclusions – which prompted Charles Darwin to ponder the "singular instance of the manner in which similar views arise at about the same time ... Goethe in Germany, Dr [Erasmus] Darwin in England, and [Etienne] Geoffroy Saint-Hilaire in France, came to the same conclusions on the origin of species in the years 1794-5."[22] And Darwin might have added Herder to the list. What distinguished

Goethe's contribution from those of Erasmus Darwin and Geoffroy Saint-Hilaire (but not from that of his Weimar colleague Johann Gottfried Herder) was his explicit rejection, as we have noted, of the notion that "will" could be a primary causal factor in evolutionary change. In this respect, his theory shared rather more with that of Charles Darwin than did those of most of his contemporaries, including Jean-Baptiste Lamarck, who developed the first detailed evolutionary theory in 1801, elaborating it further in 1809 and 1815. The second area influenced by Goethe is the age-old idea of the *scala naturae* (chain of being), or *scala creatorum* (chain of creatures) as it was alternatively known. As I emphasized in the first chapter, the customary dichotomy between evolutionary and chain ideas cannot be maintained, at least insofar as almost all evolutionary thinkers have adopted an underlying notion of the great chain as the criterion for our ethical relationship to animals, even if it is not strictly compatible with evolutionary theory as science, as many insist.[23] Goethe followed such thinkers as Joseph Addison and John Locke in allotting the animals a respected place in the scale, but in contrast to Addison and Locke, who thought of each step in the chain as a step closer to perfection, Goethe followed and elaborated the conceptions of Leibniz in arguing that each animal is a perfection in itself as the animal that it is and with the needs that it has. If a case can be made that, strictly speaking, chain and evolution are not rigorously compatible, both Goethe and Charles Darwin conceptualized the status of animals as though they were. The third area to which Goethe contributed, as we noticed in passing in the previous chapter, is that of the Kantian categorical imperative. Whereas in the Kantian view each person is to be treated as an end and never as a mere means, thus restricting application of the categorical imperative to humans alone, Goethe, quite in contrast to Kant, extended this to the animal realm.

Themes – rather than theories – of evolution have a lengthy history, at least in a primitive and inchoate form. Indeed, they have been so prevalent that, in the revised edition of his 1894 classic *From the Greeks to Darwin: The Development of the Evolution Idea through Twenty-Four Centuries,*[24] Henry Fairfield Osborn enumerated more than fifty significant individual contributors to the concept prior to Charles Darwin. And Osborn doesn't mention such early intimations of evolution as that issuing from the Sumerian schools of around the third millennium BC, which are reported to have taught that originally people walked with all limbs on the ground and ate herbs with their mouths like sheep – a vague portent of the idea that humans are descended from vegetarian apes. Thales of Miletus (c. 600 BC), customarily acknowledged, via Aristotle's identification, as the first of the pre-Socratic philosophers, held to the view that "nothing comes into being out of nothing" and looked to the waters of Mother Sea as the matter from which all things arose and out of which they exist. To interpret this as evolutionary thought might be deemed an unjustified intellectual leap if Thales's student Anaximander had not understood the master's teachings in this vein. Whatever Thales's views, it is clear that Anaximander himself espoused evolutionary principles. In the third century, Hippolytus of Rome reported him to have maintained the view that "humans originally resembled another type of animal, namely fish" (*Refutation of all Heresies,* 1.6.6). More emphatically, in the first century Plutarch records Anaximander's view "that originally humans were born from animals of a different kind, because the other animals can soon look after themselves while humans alone require a long period of nursing; that is why if they had been like this originally they would not have survived" (*Miscellanies,* fragment 179.2, as reported in Eusebius, *Preparations for the Gospel,* 1.7.16). The details may be dubious. But the fact of an evolutionary interpretation is not. Moreover, Plutarch claims that Anaximander drew moral conclusions from this theorizing: "so Anaximander, having declared that fish are at once the fathers and mothers of men, urges us not to eat them" (*Table Talk,* 730DF).[25] Indeed, this underscores the crux of my case that continuity and kinship alone, not kinship via natural selection, are the basis of the moral implications of evolutionary thought – and, in fact, no less the basis of nonevolutionary thought, although there are bases other than kinship for having obligations toward animals. Moreover, it should be clear that the moral implications drawn by Anaximander are far in advance of those drawn by Charles Darwin and his fellow early theorists of evolution by natural selection – not one of whom among the prominent, so far as I can discover, at least before J. Howard Moore some half-century later, drew moral conclusions from it about the inappropriateness of consuming one's kin. Both in the pre-Socratic period and, more resoundingly, in the Stoic period, consanguinity was deemed a primary basis of moral obligation. Throughout the whole classical period, those who emphasized our similarities to other species claimed that we shared *communitas* with them and hence had moral obligations to them. Those who stressed the differences denied *communitas* and correspondingly rejected the moral obligations. The contrasting emphases retained their force at least until the beginning of the Enlightenment, with scholars such as the naturalist John Ray delighting in the human-animal similarities, while the eminent natural-law theorist Samuel Pufendorf could reach no other conclusion than that "the Creator established no common right between man and brutes, [and] that no injury is done brutes if they are hurt by man, since God himself made such a state to exist between man and brutes,"[26] although humans, Pufendorf adds, may damage their own character by treating animals with gratuitous cruelty. Moreover, it would be a generality that applies, by and large, throughout human intellectual history, quite independently of evolutionary theory – the more we acknowledge the proximity of other animal beings to ourselves, the more likely we are to acknowledge them as entitled to moral consideration; the further away we see them from our own natures, the less likely we are to grant them ethical status, although some, such as Scully and Linzey, would predicate their obligation to animals precisely on the grounds of the *fact* of their inferiority to us.[27] Certainly, even if we concentrate our thoughts on kinship, neither evolution by natural selection nor even evolution in general is necessary for a recognition of human-animal proximity.

The Pythagorean Empedocles (c. 450 BC), renowned for his criticism of animal sacrifice on ethical grounds and for his vociferous espousal of an ethical vegetarianism, seems to subscribe to, or at the very least to hint at, a form of evolutionary thought in his reported statements that "the parts of animals are mostly formed by chance" (Aristotle, *Physics,* 196a.24) and "that some hybrids were generated differently in the blending of their forms but connected by the unity of their bodies (Aelian, *The Nature of Animals,* 17.29).[28] Certainly, Osborn was convinced. "Empedocles of Agrigentium," he tells us, "took a great stride beyond his predecessors and may justly be called the father of the evolution idea."[29] Even if one is not as readily convinced as Osborn, one can scarcely fail to notice the emphasis on homologies – on the essential unity of bodies. Moreover, as a Pythagorean, Empedocles acknowledged the doctrine of reincarnation that regarded human souls as migrating into animal souls after death, and vice versa, thus emphasizing a direct spiritual relationship among all species, which encouraged our obligation not to eat our relatives.

Charles Darwin himself acknowledged the significance of pre-Socratic evolutionary ideas but passed them over without comment. On the other hand, he readily recognized the import of Aristotle among the many forerunners he mentions, some of whom are in turn passed over by Osborn. Darwin writes of the significance of Aristotle's understanding of "the merely accidental in nature" and of "adaptation." Moreover, according to Darwin, he had argued that "whatever things were not thus constituted [i.e., appropriate to the adaptive needs of the animal] perished, and still perish."[30] In fact, Darwin has Aristotle hopelessly wrong, for here Aristotle was criticizing others who held to such a version of evolution – which only encourages us to recognize the pervasiveness of the idea. In fact, Aristotle held to a different evolutionary theory – a complete ascending gradation, or as Osborn has it, "a progressive development corresponding with the progressive development of the soul. Nature, [Aristotle] says, proceeds constantly

by the aid of gradual transitions from the most imperfect to the most perfect."[31] For Aristotle, evolution and the chain are in accord. Moreover, "the
merely accidental in nature" has no role. While Aristotle, who subscribes
to a form of evolutionary theory, also stresses the human-animal differences
and the rights of humans over animals, Plutarch, Lucretius, and Porphyry,
among several more in the classical era – for example, Theophrastus, Philo
of Alexandria, Ovid, Seneca, and Sextus Empiricus – make no reference to
evolutionary ideas but emphasize the biological and psychological homologies among species as well as the consequent sense of kinship. Plutarch
and Porphyry, in particular, draw extensive moral implications from such
kinship, and if Lucretius does not seem so direct, Montaigne manages to
draw the implications from Lucretius's work for him, quoting him to demonstrate that animals have a limited language and well-developed sentience.
Montaigne offers such phrases from Lucretius as "all things are bound by
their own chains of fate" and "all things go their own way, nor forget / Distinctions by the law of nature set." These sentiments are cited in further-
ance of Montaigne's own claims that "we are neither above nor below the
rest: all that is under heaven ... incurs the same law and the same fortune"
and that while "there are orders and degrees ... it is under the aspect of one
and the same nature."[32] It would be inappropriate to regard classical evolutionary theories as refined advances in scientific knowledge; rather, they
are more like inspired guesses. But they indicate clearly that several early
philosophers and embryonic scientists found the ideas of homology, descent,
and modification (adaptation) useful for understanding humanity's biological and historical relationships to other species. And in many instances,
they drew moral conclusions from their empirical findings.

III

Now, this brief sketch of classical evolutionary themes would be of little
relevance to Goethe's thoughts on evolution if the Germanic bard had
not been steeped in classical thought. The fact that he was is indicated
very clearly in *Faust, Part Two,* on which he had worked intermittently for
decades and which he finished shortly before his death in 1832 – although
as early as 1806 he had declared: "The whole is forthcoming, not yet all
written but composed."[33] In this work, he expressed evolutionary ideas
through the characters of Thales and Proteus, the sea god whom Goethe
also employs elsewhere to suggest the prophesier of the future, the germ
from which the future will be constructed:

Proteus: The waves give life more growth and ease;
Come now to the eternal seas
With Dolphin-Proteus.
          (*He transforms himself.*)
               Forth we ride!
Mount my back, and Joys attend you,
Thus I bear you and befriend you:
Let the ocean be your bride.
Thales: Submit to a request so winning,
And start to be at the beginning.
Accept swift working of the plan:
Then, following eternal norms,
You move through multitudinous forms,
To reach at last the state of man.
       (*Homonculus mounts upon the Proteus dolphin*).[34]

(Homonculus, or homunculus, is an as yet undeveloped humanoid being.) Like Thales, Goethe postulates the development of humankind from life in water. Like Aeschylus in *Prometheus Bound* and Aristotle in the *Nicomachean Ethics*, Goethe sees evolution as an ascending gradation. But he goes further than Aeschylus and Aristotle in understanding that common origins had implications for friendship, which suggests the obligations of interspecies community, whereas the idea of friendship developed by Aristotle is for (free, male, Greek) humans alone.[35] Osborn points out that Aeschylus in *Prometheus Bound* – "All arts among the human race are from Prometheus," Aeschylus tells us – revealed "the biology and anthropology of his day, the spirit of Empedocles and of Anaxagoras in setting forth the principle of the moral, social, intellectual and spiritual evolution of man." Anaxagoras believed that all species existed in germ or seed form throughout history and were realized as Nature purposed. Osborn continues: "Aeschylus thus ranks as the first poet of Evolution, to be followed by Lucretius, and by Erasmus Darwin, Goethe, and Tennyson."[36]

In fact, Goethe was a prime mover in the revival of classical thought and in the understanding of its relevance to contemporary philosophy. To be sure, classical learning increasingly imbued the eighteenth-century European mind in general – indeed, increasingly since the Hohenstaufen era – and was championed by the Renaissance. But Goethe gleaned more from it than did most, in part through his productive sojourn in Italy. Yet the *Italienreise* of 1786-88 and 1790 was already predicated on his admiration for classical antiquity. He was fully aware, as indicated by his use of Thales as the character to introduce evolutionary ideas, that the very earliest aspects of the classical tradition offered a preliminary account of descent and modification and of their irreversible laws, without introducing the animal's will as the agent of change. Evolution occurs "following eternal norms," not from spirited animal intervention. Moreover, as Elizabeth Rotter indicated in her *Goethes Urphänomen und die Platonische Idee* (1913), Goethe was an emphatic adherent of Plato's Pythagorean thought system. His notions of the *Urpflanze* (the prototypical plant) and the *Urbild* (the prototypical form of the animal) can be understood only in the context of Plato's theory of Forms: the conception that there is an idea, a form, an original more "real" than its empirical manifestations that underlies each immanent existence and constitutes its essence. To express the matter in more contemporary Kantian terms, Goethe was seeking the *noumena* (things as they are in themselves – i.e., *Dinge an sich*) behind the *phenomena* (things as they seem to us).[37] Or perhaps in Goethe's case, there is an element of Aristotelianism in which the Form is evinced in its natural development, as when an acorn reaches its fulfillment as a full-grown oak. The *Urbild* and the *Urpflanze* are, for Goethe, the templates – the kernels within which the secret of the transformation is held – from which all animal and vegetable forms are derived. However, as Stuart Atkins has rightly explained, Goethe "soon abandoned ... his early idea that there might be an actual prototypical plant ('Urpflanze') with the basic features of all plants and restricted that term and its analogies to mean more or less what today are called models."[38] Indeed. But models, ideational archetypes, are precisely what Plato means by Forms. To reject the idea that there is an original historical plant that contains the germs of all plants is not to reject the ideational Form of all plants. In fact, as Goethe wrote in an unmailed letter of 1816, "I was at that time [before the Italian journey] seeking the *Urpflanze* [the original plant], unconscious that I was seeking the idea, the conception."[39] Thus, it might be suggested, in line with classical conceptions, that when he continued to write of the "metamorphose" of plants and animals, his conception contained both the idea of "metamorphosis" (i.e., development from a template) and "metamorphology" (the essence that lay behind the structure). Still, by consensus, the latter diminished in Goethe's mind from the mid-1780s.[40]

IV

Of course, Goethe's ideas exist in the context not only of classical thought, but also of the eighteenth-century age of Enlightenment, which was healthily tinged with Romanticism by the close of the century. Until the eighteenth century – and, for many, far later – Christian thought in general insisted on the fixity of species and the literal interpretation of the account of creation in Genesis 1, in which the creatures of the sea and air were created "in their own species" (verse 21) on the fifth day, and creatures of the land "each in its own species" (verse 25) on the sixth day. Clearly, serious naturalistic inquiry was discouraged by such a doctrine, especially since its denial might have resulted in prosecution, more often in social and intellectual persecution. Nonetheless, it would be remiss to ignore Saint Augustine's fifth-century comment – the germ of which appears to have been derived from Anaxagoras and was later borrowed by Robinet – that while all species had been created by God, some were but seeds that appeared at a later stage of history[41] – which, while not itself evolutionary, was not the least inimical to such conceptions. It was not in fact until toward the close of the sixteenth century, marked by the theological conservatism of the Spanish Jesuit Francisco Suarez, that the tide of thought generally sympathetic to evolution was stemmed, the idea of special creation becoming more or less orthodoxy until the later nineteenth century – although we might note that there was a quite significant number of the heterodox around.

What hindered a convincing victory for the anti-evolutionists was Francis Bacon's contemporary doctrine of the "Two Books," these being the "Word of God" and the "Works of God" – which encouraged philosophers to go beyond the scriptures in order to find God's message in the scientific investigation of nature – that is, in the "Works of God." So central was this doctrine to making reliance on the theologically questionable discoveries of scientific investigation acceptable – or, at least, less dangerous – that as late as the mid-nineteenth century Charles Darwin would choose Bacon's definition of the "Two Books" to face the title page of *The Origin of Species*. Moreover, in general, the Reformation, despite its frequent and unabashed intolerances, diminished reliance on clerical authority – although we have to look to Locke at the end of the seventeenth century for an adoption of tolerance in principle, and even then only of tolerance for fellow Protestants! It was Burke in the later eighteenth century who first prominently proposed political rights for Catholics, Jews, Muslims, and even pagans. And the Renaissance not only helped to infuse a spirit of independent intellectual adventure, but also reinvigorated classical ideals and ideas that had previously been valued only to the degree they were useful to more-or-less orthodox Christianity. In this vein, Saint Augustine had co-opted Plato into the authoritarian orthodoxy of the church, and Saint Thomas Aquinas had borrowed widely from Aristotle – "the Philosopher," as Thomas invariably called him in order to add a more liberal dimension. Yet, if the intent was to adapt Aristotelianism to the service of Christianity, the effect was perhaps equally that of opening late-medieval society to some of the less immediately threatening of classical Greek intellectual ideas. Aquinas laid emphasis on faith's compatibility with reason, the value of the intellect as a primary source of knowledge, the legitimacy of secular authority *and* its capacity to improve life on earth, which was openly acknowledged as worth living for its own sake. As he wrote in the *Summa Theologica:* "Man is by nature a social animal. Hence in a state of innocence (if there had been no Fall) men would have lived in society. But a common social life of many individuals could not exist unless there were someone in control to attend to the common good."[42] Indeed, and this was a portent of the Renaissance itself, each individual counted as a worthwhile end in the mundane world – the very basis of the later Kantian categorical imperative. There is a humanitarian radicalism in Aquinas that is all too rarely recognized. Of course, it is not present in Augustine, whose era would have been too early for humanist thought. In particular, Augustine's authoritarian political position was rendered necessary by the defeats suffered by the Roman Empire, which the pagans laid unfairly at the feet of the "effete" Christians, and also by the fact that Rome suffered from recurrent political crises, civil war, invasions by the "barbarians," inflation, and cessation of trade from the middle decades of the third century on. The defeats did not arise as a consequence of the acceptance of Christianity and were endemic by the time of Augustine, but they were blamed on Christianity nonetheless.

Bishop Marbod of Rennes (d. 1123) and John of Salisbury (d. 1180) offered even earlier contributions than did Aquinas to this elevation of personal identity and the need to eliminate individual alienation, which, they recognized, suffused the feudal order. People knew themselves only as parts of a greater whole, not as individuals.[43] Marbod argued in contrast that a man should "stand there in himself." The alternative was "to let him be drawn out of his own being" and consequently to let him "lose cognizance of his own self." Salisbury announced that "the greatest danger is the estrangement (*alienatio*) of man from himself, whereby he loses all 'dignity of his nature and his person' – in other words his personal identity. A man then plunges into a 'land of oblivion' (*terra oblivionis*), in which he is unaware of himself, and finally succumbs to simple 'human degeneration' (*degeneratio hominis*)."[44] This view, which rang the death knell of the feudal order and which is more commonly associated with Petrarch, may be summed up in the contention of Aquinas two centuries earlier that there is not only the "good" of the *civitas,* but also a "human good which does not consist in a community but pertains to each individual as a self."[45]

It was not necessary to be related to another to have an obligation to another – an idea that could be adapted freely to animals from its application to humans. Even though we had to wait more than a half-millennium for Kant to develop the categorical imperative, its substance can be seen here already adumbrated. This welcoming of selfhood, and the accomplishments dependent thereon, set the tone and provided the context for the intellectual adventures of the subsequent eras, including the eighteenth century. And these adventures embraced the study of nature, which was now less restrained than previously owing to the sense of a *bonum universale*, which in turn hindered not only capitalism and usury but also intellectual endeavour and the spirit of discovery, as well as submission to what was acceptable to religious, and even to secular, *ministerium*. Although authority was still secure, at least it was no longer sacrosanct.

It should not occasion surprise, then, that the Enlightenment, building upon both the late-medieval era and the Renaissance, witnessed a revival of independent thought around a host of issues, not least that of the relationship of humans to other species. If Aquinas recognized Aristotle as "the Philosopher," whose thoughts were held to embody much of Christianity without being Christian, one would also have to recognize Aristotle as the first of ethologists, the scientific investigator who had spent four years on the island of Lesbos researching the empirical reality of marine animals. It is thus in the Aristotelian spirit that we witness not only the art of Leonardo da Vinci and Albrecht Dürer, based on serious (although usually noninvasive) scientific research on human and animal anatomy, but also the science of Carl Linnaeus, founder of the first convincingly systematic taxonomy of plants and animals in 1735. The Greeks had distinguished between land and sea animals, thus placing squid and porpoises along with cod and mackerel. Until the eighteenth century, it was common to categorize flying birds and flying fish together because they were airborne or to yoke reptilian turtles and mammalian armadillos because of their protective coverings. By employing systematic anatomical comparison, Linnaeus categorized and devised scientific names for some 4,200 species of animals and 7,700 species of plants and provided the foundation for all modern taxonomy. We should not, of course, imagine that Linnaeus solved all problems associated with animal categorization. Buffon thought the whole idea flawed since nature "amazes more by her exceptions than her laws," and Oliver Goldsmith declared animal classification "but a very trifling part of its history."[46] Goethe wrote that "genus can change to species, species to variety, and in other conditions varieties can change *ad infinitum* ... and yet those farthest separated from each other have a pronounced relationship."[47] Moreover, as Darwin observed over a century later, "I was much struck how entirely vague is the distinction between species and varieties ... I look at the term species, as one arbitrarily given for the sake of convenience to a set of individuals closely resembling each other, and that it does not essentially differ from the term variety, which is given to less distinct and more fluctuating forms."[48] It is not a point Linnaeus would have taken amiss, for we find him seriously doubting the immutability of species and writing to the German botanist Johann Gmelin in 1747 "demanding that you show me a generic character ... by which to distinguish between Man and Ape. I myself most assuredly know of none. I wish someone would indicate one to me. But, if I had called man an ape, or vice versa, I would have fallen under the ban of the ecclesiastics. It may be as a naturalist I ought to have done so."[49] By acknowledging what different animals shared with each other and what distinguished them from each other, Linnaeus was able to produce a taxonomy that acknowledged the commonality of all animal life while recognizing their genus- and species-specific characteristics. Yet, as his letter indicates, he found it impossible to find a generic characteristic to separate humans from apes – although, of course, the different "apes" (i.e., human, chimpanzee, gorilla, and orangutan) were non-interbreeding species. Nothing could have been more significant in diminishing the logical and classificatory grounds for the hubris of humans who imagined themselves as essentially superior beings, of a different order from the rest of creation. From Linnaeus on, it was increasingly more difficult to conceive of humans as not being animal in the same way that other animals were animal. Still, in the order of Primates, Linnaeus made a clear distinction between "homo" and "simia" – literally making flat-nosedness the primary separating characteristic! Today, *Homo sapiens* is still assigned a separate family in the Primate order, whereas chimpanzees, gorillas, and orangutans have been categorized together in the Pongid family since the 1950s – "Pongid" having no other claim to biological distinction than that it is derived from the Congolese *mpongo,* a term used for apes. The rationalizing circularity could not be more evident! Writing in 1871, Charles Darwin wisely remarked that "if man had not been his own classifier he would never have thought of founding a separate order for his own reception."[50] Yet it is noticeable that in scientific classification today, Darwin has not been fully heeded. Humans are still given their own exclusive family. Indeed, in a 1997 comparison of the results of ethological research on the sense of self in monkeys, apes, and humans, D. Hart and M.P. Karmel concluded that there is a qualitative difference between the monkeys and the apes, whereas the difference between the great apes and humans is "largely quantitative."[51] If zoological evidence alone were more significant than vanity, the classification of humans today would be less flattering to them. Indeed, a good case can be made that the classification of apes as somehow superior to their fellow animals – many animal-rights advocates demand special rights for apes through the Great Ape Project – is itself a reflection of the perceived human superiority. Apes are accorded preferential consideration precisely because they are so like humans! Still, it is perhaps the most appropriate way to convince humans of their lack of uniqueness – at least a lack of uniqueness of a different order from the uniqueness of every other species.

The French naturalist comte de Buffon could write in *Histoire naturelle* (1749-67) that "this orang-outang is a very singular brute, which man cannot look upon without contemplating himself, and being convinced that his external form is not the most essential part of his nature."[52] Like Linnaeus, Buffon hints clearly at a common ancestry among animals – he even wrote of the development and "degeneration" of species – but could never bring himself to accept evolutionary ideas unequivocally. He vacillated frequently. Indeed, in his "Historical Sketch" of the evolutionary idea, Charles Darwin described Buffon as "the first author who in modern times has treated [mutability of species] in a scientific spirit," although since his "opinions fluctuated at different periods, and as he does not enter on the causes of the transformation of species,"[53] Darwin chose to ignore his contribution. Yet, as I have already suggested, it is the fact rather than "the causes of transformation of species" that is relevant to the status of animals. There is no doubt that Buffon understood that humans find a reflection of themselves in the great apes. Moreover, even if he constantly equivocated, his description of the pig could scarcely be seen as anything other than distinctly evolutionary, especially with regard to the idea of vestiges, which played such a significant role in later evolutionary theory: "The pig does not appear to have been formed upon an original, special, and perfect plan, since it is a compound of other animals; it has evidently useless parts, or rather parts of which it cannot make any use – toes, all the bones of which are perfectly formed, and which, nevertheless, are of no service to it. Nature is far from subjecting herself to final causes in the formation of her creatures."[54]

Around the same period (1745 and 1751), we find the French mathematician Pierre Maupertuis writing of the chance combinations of elementary particles, of the elimination of species by the accidents of nature, and of the derivation of all present species from a small number, perhaps a single pair, of original ancestors.[55] A similar thesis was propounded by the popular French Encyclopedist Denis Diderot (1749 and 1754) – higher animals may all be descended from "one primeval animal"[56] – to which the fellow *philosophe* Jean-Baptiste Robinet added in *De la Nature* (1761-68): "I doubt not that there was a time when there were not yet either minerals or any of the beings that we call animals; that is to say, a time when all these individuals existed only in germ, and not one of them had come to birth ... At least it appears certain that Nature has never been, is not, and never will be stationary, or in a state of permanence; its form is necessarily transitory ... Nature is always at work, always in travail, in the sense that she is always fashioning new developments, new generations."[57] In *Système de la Nature* (1770), the materialist and atheist Baron d'Holbach queried "whether Nature be not now assembling in her vast laboratories the elements fitted to give rise to whole new generations, that will have nothing in common with the species at present existing. What absurdity, then, would there be in supposing that man, the horse, the fish, the bird will be no more? Are these animals so indispensable to Nature that without them she cannot continue her eternal course? Does not all change around us? ... Nature contains no constant forms."[58] Contemporaneously with d'Holbach's publications on dying and regenerating Nature, Charles Bonnet developed a complex progressive theory of an infinitely self-differentiating Nature in *Palingénésie philosophique* (1770) – "Whether it can be called a form of 'evolutionism' is a question of terminology" was Arthur Lovejoy's cautious conclusion[59] – and Montesquieu hinted about the multiplication of species after flying lemurs had been discovered in Java.[60] Although Lovejoy was a little cautious about Bonnet, he was rather less so about others, dating the birth of "modern evolutionism" to 1754-55, with Maupertuis's dismissal of the idea of predilineation, Buffon's recognition of homologies of structure in the most diverse of animals, and Diderot's conception of descent from a common prototype in *Pensées sur l'Interprétation de la Nature* (1756).[61] All these, and several more, including the redoubtable German philosopher and mathematician Gottfried Wilhelm Leibniz, set the intellectual stage for the more explicit and self-conscious evolutionary theories developed in the 1790s. In Leibniz's case not only do we find his philosophical recognition that each animal capable of conceiving of itself as an "I" is entitled to be treated as an "I," but we also encounter Kant's recognition of his animal sensibilities: "Leibniz used a tiny worm for the purposes of observation, and then carefully replaced it with its leaf on the tree so that it should not come to any harm through any act of his. He would have been sorry – a natural feeling for a humane man – to destroy such a creature for no reason."[62] Neither evolution by natural selection nor evolution itself was at all necessary for sensibility toward animals – but merely a recognition of the inner impulses of our natural compassion.

V

Of the three naturalists whom Charles Darwin recognized as his eminent evolutionary precursors of the 1790s, Goethe is customarily treated as the least significant. For example, in Henry Milner's *The Encyclopedia of Evolution* (1990), Etienne Geoffroy Saint-Hilaire and Erasmus Darwin are allotted their own sections, while Goethe is relegated to parenthetic commentary along with Lorenz Oken under "Naturphilosophie." Which of the three ought to be accorded temporal priority – or whether Goethe's Weimar colleague Johann Gottfried Herder trumped all of them – is an unanswerable and unprofitable question, for each developed his theory out of prior studies that already contained the germs of their evolutionary ideas. What is perhaps clear is that Goethe was less systematic in his theorizing and provided less detail than either Geoffroy Sainte-Hilaire or Erasmus Darwin.

The details of Geoffroy Saint-Hilaire's theories emerged only in 1828, although he had indicated quite clearly already in 1795 that different species are various degenerations of the same type and that animal structures undergo changes and transformations over time. It was not, however, until 1828 that he stated explicitly that the same forms have not been perpetuated since the origin of all things and that the conditions of life, or the *"monde ambiant,"* were the agents of change. Like Thales and Anaximander (and, by dramatic implication, Goethe), he postulated the evolution of animals from sea life. While Geoffroy Saint-Hilaire is not, in his own writings, at all precise on the effects of "will" on change, he nonetheless associated himself directly with Jean-Baptiste Lamarck's early-nineteenth-century theorizing in their public and bitter contest with Baron (Georges) Cuvier, stalwart defender of the fixity of species, who assigned humankind to a separate order of Bimana, distinct from all other species. It was Lamarck's – and consequently Saint-Hilaire's – misfortune to develop a theory that served later to condemn Lamarck. He argued, incorrectly, that if, for example, a relatively short-necked giraffe continually strove to reach the higher leaves of a tree, it would eventually stretch its own neck and that its offspring would be likely to inherit a rather longer neck. Lamarck did not consider this conception of the "inheritance of acquired characteristics" an original discovery, nor even an especially important part of his theory, but through Trofim Lysenko's anti-genetic theories of adaptation developed in the 1930s, which received official support of the Soviet Central Committee of the USSR in 1948, the term "Lamarckian" – applied by Lysenko to his theory – came to be associated indelibly with the inheritance of acquired characteristics. Lamarck became the unwitting proponent of an underpinning of Communist ideology rather than an exponent of a scientific theory.

In fact, many more than Lamarck and Geoffroy Saint-Hilaire in the eighteenth and nineteenth centuries considered "will" and "striving" important aspects of inheritable biological change. Indeed, until the early twentieth century and the general recognition, then acceptance, of Gregor Mendel's work on pea genes as demonstrating the method of inheritance, there was no adequate alternative theory to "Lamarckism." In fact, Charles Darwin himself subscribed at least in part to the theory of the inheritance of acquired characteristics, while noting that there was also contradictory evidence. What is significant here is that as a result of his anatomical research, Goethe rejected the theory out of hand, insisting that neither humans nor other animals could alter the course of evolutionary history. Even if Goethe had no conception of the manner of evolution – natural selection, as understood by both Charles Darwin and Alfred Russel Wallace and as first publicly communicated in 1858 – he understood better than Geoffroy Saint-Hilaire, Lamarck, Erasmus Darwin, and even the early natural-selection theorists themselves what was *not* the agent of change.

There was one fundamental matter of primary importance in which Goethe and Geoffroy Saint-Hilaire were in complete agreement. As expressed in the latter's words in *Philosophie anatomique* (1818) "There is, philosophically speaking, only a single animal."[63] This corresponds precisely to Goethe's idea of the *Urbild* of animal life. It was not, however, at all an entirely new conception, nor was it necessarily inimitable to materialistic conceptions of evolutionary change, as is frequently argued.[64] The history and implications of the concept are laid out by the realist novelist Honoré de Balzac in his 1845 preface to the series *La Comédie humaine* (fourteen years before *The Origin of Species,* we might remind ourselves):

> It is a mistake to suppose that the controversy which in these latter days has arisen between Cuvier and Geoffroy Saint-Hilaire rests upon a scientific innovation. Synthetic unity filled, under various definitions, the greatest minds of the two preceding centuries. In reading the strange books of those mystical writers who drew science into their conceptions of the infinite, – such as Swedenborg, Saint-Martin, and others; also the writings of the great naturalists, Leibnitz, Buffon, Charles Bonnet, etc., – we find in the monads of Leibnitz, in the organic molecules of Buffon, in the vegetative force of Needham, in the encasement germs of Charles Bonnet, who was bold enough to write in 1760, "animal life vegetates like plant life," – we find, I say, the rudiments of that strong law of self-preservation upon which rests the theory of synthetic unity. *There is but one animal. The Creator used one and the same principle for all organized being. An animal is an essence which takes external form from the centres or conditions in which it comes to its development. All*

zoological species grow out of these differences. The announcement and pursuit of this theory, keeping it as he did in harmony with preconceived ideas of the Divine power, will be the lasting glory of Geoffroy Sainte-Hilaire, the conqueror of Cuvier in this particular branch of science, – a fact recognized by the great Goethe in the last words which came from his pen.[65]

What Balzac is rightly indicating is that there is no insurmountable contradiction between the idealist conception (say Platonic) and the materialist conception (say that of d'Holbach) with regard to the question of "synthetic unity." If there is ultimately an idea, or Form (or spirit or soul) that underlies each material existence, this will not prevent us from understanding the material laws of nature. For both the idealist and the materialist, there is but one animal – all are structured along the same principles, whether these principles are derived from God, Nature, or biological accident, whether they are underlain by an idea or are no more than the products of material history. There is still one model of all animality.

As we have noted, in the late seventeenth century, the Italian physicist and mathematician Giovanni Borelli demonstrated in *On the Movement of Animals* how the same laws governed the wings of birds, the fins of fishes, and the legs of insects. In the second volume of this work, he applied the same physical laws to the movements of muscles, the circulation of blood, and the process of respiration. It is in this sense that, for both Geoffroy Saint-Hilaire and Goethe, "there is but one animal" – that is, all animals share the same basic principles and operate under the same basic laws. Indeed, the very classification of "animal" relates to their common evolutionary inheritance. Whether ultimate reality consists in Platonic Forms, Democritan or Epicurean atoms, Leibnizian monads, or Buffon's organic molecules is irrelevant to understanding the essential "synthetic unity," as Geoffroy Saint-Hilaire termed it, or the *Urbild* of the animal, as Goethe has it. Certainly, when James Joyce in *Ulysses* announces in distinctly Platonic form (if of a medieval bent, filtered through Duns Scotus and William of Ockham) that "horseness is the whatness of allhorse," he does not imagine such an intrinsically idealist expression to be inconsistent with materialist evolutionary theory. And while unitary conceptions of animals do not imply evolution, they are nonetheless a prerequisite for the understanding of evolution. Moreover, the same moral principles appear to derive from the "synthetic unity," the *Urbild*, evolution itself, and evolution by natural selection. None appears to offer a firmer ethical foundation than any other.

Of the three 1790s evolutionists whom Charles Darwin acknowledged in his "Historical Sketch," it is undoubtedly his grandfather, Erasmus Darwin, who contributed most to the theory, if only in that he asked most of the relevant questions pertaining to the development of an appropriate theory. In *The Botanic Garden* (1790), he developed Linnaeus's taxonomy in poetic form, offering slightly veiled hints about plant evolution. But it was in *Zoonomia, or the Laws of Organic Life* (1794-96) that he asked explicitly whether all living beings came from a single ancestor, much as Diderot had done, and how one species could develop from another. He hypothesized that overpopulation increased competition and that competition was an agent of evolutionary change. In essence, this, too, was the underpinning of Charles Darwin's *The Origin of Species*, the third chapter of which is entitled "The Struggle for Existence." Charles says of it that it "is the doctrine of Malthus, applied to the whole animal and vegetable kingdoms."[66] Erasmus also argued that humans were related to apes and monkeys and that sexual selection was a factor in evolution – a matter of sufficient moment to Charles Darwin's findings that his 1871 work was entitled not just *The Descent of Man* – the title by which it is customarily known – but *The Descent of Man, and Selection in Relation to Sex*. Indeed, parts two and three of this work, which constitute over two-thirds of the whole, are devoted to sexual selection. Yet Erasmus Darwin's work was marred far more emphatically than that of Geoffroy Saint-Hilaire, and significantly more so than that of Lamarck, by the frequent claims that it was the "urges" of plants and animals, their "lust, hunger, and anger," that developed their inheritable forms. Lysenko would have erred less if he had termed his theory "Erasmus Darwinism" rather than "Lamarckism."

Normally, scientists are inclined not to wear their loves on their lapels. But scientists who are also poets may be excused a modicum of exception. While respect for other species runs through Erasmus Darwin's studies, there is only an occasional hint of compassion. He is far more concerned with understanding the animal world than with paying it homage. Nonetheless, we can find him admonishing us that human wisdom is not so different from the instinctive wisdom of the wasp, bee, or spider, which links "the reasoning reptile to mankind."[67] And in *The Temple of Nature* (vol. 4, 427-28), humankind is told that it "Should eye with tenderness, all living forms, / His brother emmets, and his sister worms." But it is when he moved away from animal studies themselves and strayed into educational writing that he was decidedly less reluctant. In *A Plan for the Conduct of Female Education in Boarding Schools,* he advises his readers: "A sympathy with the pains and pleasures of others is the foundation of all our social virtues. 'Do as you would be done by' is a precept which descended from heaven ... This compassion, or sympathy with the pains of others, ought also to extend to the brute creation, as far as our necessities will admit ... To destroy even insects wantonly shows an unreflecting mind, or a depraved heart."[68] There is no more explicit expression of compassion in the writings of Erasmus's more celebrated grandson.

VI

Goethe's evolutionary ideas were inspired by his anatomical research on the intermaxillary bone of the human jaw. For those – and there were many – who wished to trumpet a human exclusivity, the apparent absence of an intermaxillary bone, the *os maxillare,* in the human skull was interpreted to show that, while still animal, humankind had been made by God in a different manner from any other species and with a capacity unmatched by any other species. It was believed by the most celebrated anatomists of the time, Blumenbach, Camper, and Sömmering, for example, that humans lacked the contentious bone and that there was thus something lacking in the proclaimed human-animal continuity of such thinkers as Locke and Leibniz. It was a precariously slim argument anyway – almost comical, in fact – on which to base an *essential* difference between humans and other species, especially apes. But every straw must be grasped when so much is at stake – humans alone had been made in the image of God! For Goethe, his discovery of the intermaxillary bone removed the last hindrance to seeing humanity as a "living concept" that constituted a step in a sequence of varying animal forms, thus confirming Leibniz's principle of continuity. "The harmony of the whole," Goethe wrote in a letter to Hans Knebel, "makes every creature what it is, and man is man by the form and nature of his upper jaw, as well as the form and nature of the last phalanx of his little toe. Then again every creature is but a tone, a modulation of a great harmony, which must be studied as a whole and in all its grandeur; otherwise each individual part is but a lifeless letter. This little work [on comparative anatomy] is written from this point of view and that is really the interest that lies concealed in it."[69] The point of view to which Goethe refers is identical to that of Leibniz when he seeks to show how a being can have both an individual identity of its own and yet also belong to a larger order of which it is an intrinsic part, both acting upon and being acted upon. Thus, in this way, an individual, whether human or nonhuman animal, is an entity in him- or herself, a member of a species, and a part of nature as a whole, without losing any of the individuality. Both animal and nature are organisms. Individualist and communitarian ethical theories remain in constant tension – more or less true or false not with respect to the matter in and of itself but with respect to the individualist or communitarian context in which the question is raised.

As a consequence of his 1784 anatomical studies, Goethe concluded aright that humans did indeed possess the *os maxillare,* although to have been more accurate, he would have written of the *vestiges* of the intermaxillary bone, in the same manner that humans have vestiges of a tail and of gills – Thales and Anaximander were right about the fishy origins of humans. But Goethe had no adequate evolutionary concepts nor language thereby derived in which to think and write of vestiges, as did Charles Darwin a century and a half later.

Persuaded by Leibniz's principle of continuity, to which we shall return, Goethe had proclaimed already in his 1783 poem "Das Göttliche" (Divinity) that it was only an ethical sense, which he perceived as derived from culture and art, rather than anything anatomical, that separated humans from other animals.

> For that alone
> Distinguishes him
> From all beings
> That we know.[70]

To be sure, humans were elevated above other beings but not on account of any inherent bodily differences. Now Goethe's 1784 anatomical discovery confirmed it.

Armed with his anatomical conclusions, in the 1790s Goethe extended his thoughts toward evolution, noting that: "Nature can attain to whatever she sets out to achieve only by means of a gradual succession. She makes no leaps. She could not, for instance, make a horse, if it had not been preceded by all the other animals, as a kind of ladder by which she ascended to the structure of the horse."[71] The ancient Latin saying *Natura non facit saltum* (Nature makes no leaps) – is often incorrectly attributed to Charles Darwin and is certainly seen as the underpinning of what is termed Darwinian gradualism. The phrase was in fact already being used by Linnaeus early in the eighteenth century and was made much of by Jean-Baptiste Robinet in *De la Nature* (1761).[72] It was now being applied – at least in German translation – to evolutionary notions by Goethe a century and a half before Darwin appropriated the phrase. Goethe is arguing that Nature herself cannot skip the steps made necessary by her own laws. She, too, is curtailed by necessity. This inability of Mother Nature is repeated in "The Metamorphosis of Animals," probably composed around 1800:

All its limbs are formed in accord with eternal laws,
And the most unusual form preserves the secret of the primal pattern
                ... the eternal mother is quite
Unable to horn lions, though she were to muster all her might,
For she has insufficient power to plant both rows of teeth and to promote
antlers and horns.

Moreover, even human will is unable to effect new forms. Even "the more
noble creatures" are "bound by the holy limits of life's generation. No god
extends these boundaries, nature respects them, / For only thus limited
was perfection ever possible."[73] Nature's laws are inviolable. Neither the
will of the gods, nor of humans, nor of other animals is able to transcend
them. The limbs of a given animal are always compatible with each other.
"Monsters" – such as the centaur or chimera – *must* belong to the realm of
mythology alone. Perfection is possible only in the fulfillment of the animal as the member of the species to which it belongs.

Goethe and Herder had discussed these ideas together at considerable
length from 1783 to 1784. And we find Herder expressing very similar conceptions in his *Ideen zur Philosophie der Geschichte der Menschheit* (Ideas for
a philosophy of human history; 1784-91), where the thoughts are more
clearly expressed, or perhaps more accurately, are more capable of ready ren-
dition into comprehensible English:

The human species has been praised for possessing in the most perfected
form all the powers and capabilities of every other species. This is patently
untrue. Not only is the assertion incapable of empirical proof; it is also logically insupportable, for it is self-contradictory. Clearly, if it were true, one
power would cancel out the other and man would be the most wretched of
creatures. For how could man at one and the same time bloom like a flower,
feel his way like the spider, suck like the butterfly, and also possess the muscular strength of the lion, the trunk of the elephant and the skill of the beaver?
Does he possess, nay does he comprehend, a single one of those powers,
with that intensity, with which the animal enjoys and exercises it? ...

No creature, that we know of, has departed from its original organization
or has developed in opposition to it. It can operate only with the powers
inherent in its organization, and nature knew how to derive sufficient means
to confine all living beings to the sphere allotted to them.[74]

Goethe's (and Herder's) idea of a limited perfection of each species
in light of its own self is that the animal is perfect as the being that it is
with the purposes and needs that it has. Its form would not be perfect if its

purposes and needs were any other than what they are. Perfection is not absolute but relative to each animal's intrinsic nature. Goethe continues this approach in contemplating the hollow spaces in the human skull: "In this case the question Why? would not lead us very far, whereas the question How? teaches me that these cavities are the remnants of an animal skull, which are found larger in proportion in rudimentary organisations, but in man, in spite of high development, have not been entirely lost."[75] Likewise, "creatures gradually evolv[e] as plants and animals out of a relation in which it is scarcely possible to draw a separating line between them and develop toward perfection in two opposite directions, so that in the end the plant culminates in a tree, enduring and stationary, while the animal reaches its highest degree of locomotion and freedom in its crowning representative, man."[76] Here Goethe expresses ideas of the great chain and evolution – even if in the form of a ladder – in a single breath. More exclusively evolutionary are the lines from *Eins und Alles* (One and all):

> To metamorphose the creation,
> Lest rest become complete stagnation,
> Eternal, living motion works ...
> This endless force, itself exerting,
> Creating forms and these converting,
> Doth only seem at times to rest.[77]

It is improbable that Goethe acquired a very clear conception of descent until very near the end of his life when he reviewed the controversy between Cuvier and Geoffroy Saint-Hilaire, coming down uncompromisingly on the side of the latter. Yet if the controversy clarified the idea of evolution for him, he knew that this was where he had always been, even if his evolutionary expressions were sometimes opaque. "This event is for me," he said, "one of altogether incredible value, and I have a right to rejoice that I have finally lived to witness the general victory of a cause to which I have devoted my whole life, and which is preeminently my cause."[78] In describing his conversations with Herder around the time that Herder wrote *Ideen zur Philosophie der Geschichte der Menschheit* (which was, by consensus, a product in part of Goethe's mind), Goethe tells us: "Our daily conversation was occupied with the very beginnings of the water-earth and the organic creatures that have been developing upon it since the earliest times. The very beginning and the ceaseless continuation through development were always talked about."[79] And, indeed, Goethe even managed to bring the Malthusian element into his theory, of which Charles Darwin later made so much. Prometheus, benefactor of humankind in providing fire and the arts, is speaking:

The lot vouchsafed to man is that bestowed on beasts,
Upon whose archetype [*Urbild*] I have myself improved:
It is that one opposes the other, all alone,
Or else in troops, and foe press foe with grinding hate,
Till stronger over weaker brutal triumph gain.[80]

Even if not always clearly expressed, the germ of a thoroughgoing evolutionary theory is present in Goethe's thought, exemplified perhaps most clearly when he wrote of the fossil of an extinct species of gigantic ox: "In any case this ancient creature may be considered a widely distributed extinct parent stock of which the common ox and the zebu may be looked upon as descendants."[81]

What is of significance for the status of animals is that in *Faust, Part One* (c. 1800), in the section on "Wood and Cave," Goethe allows us an insight into the respect he accords other species and into the fraternal and friendly nature he envisages of the human-animal relationship. It is Faust who is speaking:

Sublime spirit, you gave me, gave me all
I asked for ...
You gave me the grandeur of Nature as
My kingdom, the power to experience her, to
Enjoy her. You did not permit me merely
A cold astonished visit, but granted me
A look into her deep breasts,
As into the bosom of a friend.
You lead the ranks of living beings
Before me, and teach me to know my brothers
Of the still bush, the air and water.[82]

Moreover, this look into the lives of his mammalian, avian, and piscine "brothers" allows him an awareness of the "deep mysterious wonders" of his own self. Like Buffon,[83] he understood that to experience those who are like ourselves is to gain an insight into ourselves. Further, to the extent that he acknowledges the immortality of the soul at all, he finds a common destiny for man and beast. Thus he closes his narration of the traditional legend of "The Seven Sleepers of Ephesus" (1809) with the words:

But no more to king or people,
Did the chosen reappear.
For the Seven, who long had tarried –

Nay, but they were eight in number,
For the faithful dog was with them –
Thenceforth from the world were sundered.
The most blessed Angel Gabriel,
By the will of God Almighty,
Walling up the cave for ever,
Led them into Paradise.[84]

VII

There is much to be said for the argument of H.B. Nisbet that Lucretius, author of *De rerum natura (On the Nature of the Universe)*, had a particular appeal to the eighteenth century on account of the "immense popularity of didactic poetry" in the period, his "uncompromising intellectualism," his "preoccupation with nature to the detriment of theology," and his belief that knowledge alone, especially scientific knowledge, was "the path to human salvation." Moreover, so Nisbet persuades us, Goethe's naturalism in "The Metamorphosis of Animals" follows Lucretius in its depiction of nature as a mother-goddess, in its repudiation of teleology – the animals' purpose is a natural consequence of its shape rather than of something imposed upon it from outside – in its testament to the constancy and necessity of law, and in its affirmation of the law of compensation, or the law regarding "the correlation of parts."[85] And, we might add, the piece quoted above from Herder shows a similar indebtedness to Lucretius. Moreover, just as Goethe allowed his recognition of human-animal homologies to influence his view of the status of animals, so too does Lucretius. In perhaps the most famous, certainly the most moving, passage of *De rerum natura*, Lucretius encourages us to acknowledge the compassion of the cow as no different from that of a human mother:

> Outside some stately shrine of the gods incense is smouldering on the altar. Beside it a slaughtered calf falls to the ground. From its breast it breathes out a hot stream of blood. But the bereaved dam, roaming through green glades, scans the ground for the twin-pitted instrument of familiar feet. Her eyes roll this way in search of the missing young one. She pauses and fills the leafy thickets with her plaints. Time and again she returns to the byre, sore at heart with yearning for her calf. Succulent osiers and herbage fresh with dew and her favourite streams flowing level with their banks – all these are powerless to console her and to banish her burden of distress. The sight of other calves in the lush pastures is powerless to distract her mind or relieve her distress.[86]

While it may be true that Goethe's animal sensibilities are never expressed with quite the level of compassion demonstrated by Lucretius, we might note that no more than a handful of others ever reach this plateau – perhaps just Plutarch, Porphyry, Leonardo, and Schopenhauer, to snatch at the most prominent. What needs to be asked is whether Goethe takes a philosophical position that might be seen to advance that of Lucretius. Was Goethe's "Metamorphosis of Animals" entirely Lucretian, as Nisbet argues, or something more?

VIII

Although Goethe was influenced by Lucretius, as he was by the classical tradition in general, he was also a follower of Leibniz (primarily with regard to the principle of continuity), of Spinoza (primarily with regard to the unity and divinity of Nature), and especially of Kant (not least with regard to the categorical imperative). Indeed, Goethe had begun to study Kant's work in earnest in the early months of 1789. So convinced did he soon become of the validity of the Kantian schema that, in the words of Nicholas Boyle, "he struggled to find an expression for his anatomical ideas that could withstand Kantian criticism."[87] Writing in 1804, Goethe observed that "in considering the events of modern times, we may appropriately remark ... that no man of learning has with impunity rejected, opposed, or spurned the great philosophical movement which Kant began."[88] According to Kant, as explained by Nicholas Boyle, Reason "envisages a world in which we would have direct knowledge of all things as they are in themselves and of their relation to appearances: we could see ourselves as the monadic substance underlying, and giving unity to, all our contingent experiences; we could see the world as a fully harmonious organic whole in which none the less we acted freely as independent agents; we could have a clear vision of the Supreme Being and his perfections, from which all less perfect beings derive."[89] In reality, however, Kant informs us, we can know only imperfect things, and then obscurely. We are simply quite incapable of representing the ultimate reality of things, any attempt at which will result in error and contradiction. We cannot know the existence or nature of things. However, as in almost all relativist arguments, there is the positing of an objectivity toward which we may move, if never reach, and the further away from which we move, the less understanding we will have. At least in principle, if never completely in fact, for Kant, Reason posits the reality of each individual being yet at the same time understands this reality as an essential part of a greater whole, each individual being possessing degrees of perfection as each moves further away from the Supreme Being. For Kant, even if we cannot experience ultimate reality, in both philosophy and science, we should proceed *as if* we could – a phrase constantly repeated by Kant. We must thus seek to systematize Nature in our minds. In reality, there are frequent and occasionally significant breaks in the scale of beings. Yet, for Kant, the Leibnizian model of a continuous, uninterrupted chain is the model to conceptualize and employ in naturalist inquiry in an attempt to reduce the irregularities and gaps in the perfect order that Leibniz believes the chain as such and the beings of the chain to possess in themselves. Goethe's contributions to the study of plant and animal anatomy, and even to evolution itself, can be understood precisely in these Leibniz-Kantian terms, which, of course, in turn have classical origins – not least in Plotinus, who both conceptualized the One and helped to systematize the chain of realities.

It is, however, in the development of the categorical imperative – simply put, the notion that an individual person is to be considered primarily as an end rather than solely as a means – that Kant contributes most to ethics. Indeed, the Kantian categorical imperative is viewed as a major turning point in ethical theory. To be sure, his formulations are refinements, a rendering in more precise terms, of the implications of much of Western culture for over two thousand years. Thus, in the Hebrew Bible in Leviticus 19:18, we encounter the idea that "you will love your neighbour as yourself." In the New Testament we read in Matthew 7:12: "So always treat others as you would like them to treat you." Nor is this a New Testament innovation, for the verse continues: "that is the Law and the Prophets." Matthew 19:19 repeats Leviticus: "You shall love your neighbour as yourself." And as we have already noted, Aquinas claimed that there is "a human good ... which pertains to each individual as a self."[90] Moreover, a good case can be made that the very message of the Renaissance, exemplified by Petrarch, was the recognition of the worthiness of each individual as a self.

Kant formulated the categorical imperative in slightly different ways in *The Foundations of the Metaphysics of Morals* (1785) and *The Critique of Practical Reason* (1788). The most famous and most frequently cited formulations are to be found in the first of these books: "Act only on that maxim by which you can at the same time will that it should become a general [universal] law," and "act in such a manner that the maxims of your behaviour should become through your will a general [universal] law of nature." It is, however, with the formulation of the notion of the end in itself that Kant announces what, for most people, is the most significant expression of the imperative. This is formulated as: "Act in such a manner that you always treat humanity, both in your own person and that of any

other, always as an end, and never solely as a means."[91] There is a variant
of this sentiment – far less frequently cited but no less prominent in Kant's
formulation – that is of some significant import to animal ethics: "Humankind and, in general, each rational being exists as an end in itself, not merely
as a means for the use of this or that person's arbitrary will."[92]

Normally, we are told that Kant indicated that animals lacked self-
consciousness and consequently lacked the capacity to be ends in themselves and that they were therefore means to human ends, although as a
matter of human character we were obligated to treat other species with
moral consideration. And no doubt this is, in general, the correct reading
of Kant. Nonetheless, in this last quoted formulation of the imperative,
Kant states unequivocally that the principle applies beyond humans to
"each rational being," and if, as is implied, the category of rational beings
is not exhausted by the category of human beings, then what can these
other rational beings be but at least some of the more complex animals
(unless Kant intends it to apply to the angels he refers to in the Great
Chain of Being, which is improbable since they are not subject to being used
by humans)? Thus, if the angels are not meant, these animal beings, too,
for Kant, at least in this expression of the categorical imperative, appear to
be ends in themselves, which may never be used solely as a means, in the
same manner that humans may never be used solely as a means.

There would thus appear to be some ambivalence in Kant's conclusions.
There is, however, less ambivalence about Goethe's use of the principle to
apply to animals, although it could be argued that he is making a scientific rather than a moral point. But in reply one could respond that it is
a scientific point with moral implications: animals were not created for
human use. In "The Metamorphosis of Animals," Goethe announces that
"each animal is an end in itself" (*Zweck sein selbst ist jegliches Tier*). Goethe
is borrowing the very words employed by Kant in the formulation of the
categorical imperative and applying them to animals. "Humankind and in
general each rational being exists as an end in itself" – that is, as *Zweck
an sich selbst.*" Goethe's statement is of major significance in the development of animal ethics. Kant had, by consensus, revolutionized the formulation of the principles of ethics by his pronouncement of the categorical
imperative. Now, Goethe was ensuring that the pronouncement would not
be limited to human ethics, as Kant, at least most of the time, had insisted
it be restricted.

In describing Goethe's philosophy in general, Albert Bielschowsky tells us:

If the value of each being is to be derived from that being itself, then every
creature must have its purpose in itself, and cannot be explained by external

purposes; much less by subordination to the purposes of man, – who, in spite of Copernicus, still considered himself the centre of the universe. This teleological way of thinking, however, still held sway over the investigators of nature and prevented the scientific comprehension of organic nature and the progress of investigation. In his energetic rejection of teleology our poet stood almost alone. His philosophical teacher [Kant] had, with his usual acumen, long ago discovered the anthropomorphism of final purposes and had declared that "all final causes are human interventions." In this particular Goethe followed him unconditionally.[93]

Yet, even though Kant had rejected anthropomorphism, he had espoused anthropocentrism. On this particular point, Goethe rejected his philosophical teacher unconditionally. Animals were ends in themselves; they were not instruments of human purpose. Anthropocentrism, too, must be discarded. And, contrary to much unwarranted speculation on the novel implications of Charles Darwin's findings, this occurred long before the pronouncement of evolution by natural selection. Goethe was giving Kantian expression to the rejection of anthropocentrism found, for example, clearly stated in Lucretius and more clearly developed in Leibniz. Even Kantian language could be used to deny Kant's Aristotelian and Thomist presumptions that animals were intended for human use.

Now, it would be misleading to leave the impression that Kant thought that animal use was at the absolute discretion of humans. To be sure, he tells us that "so far as animals are concerned, we have no direct duties. Animals are not self-conscious, and are there merely as a means to an end. That end is man. We can ask, 'Why do animals exist?' But to ask 'Why does man exist?' is a meaningless question." Despite his philosophical acceptance – at least most of the time – of the proposition that animals were for human use, his advocacy of appropriate behaviour toward animals did not fall far short of the recommendations of those who did consider them ends in themselves. Such are the incongruous perils of philosophy, or more definitively, of abstract language, which controls its employer far more than the employer controls the language! Thus Kant continues:

Animal nature has analogies to human nature, and by doing our duties to animals in respect of manifestations which correspond to manifestations of human nature, we indirectly do our duty toward humanity. Thus, if a dog has served his master long and faithfully, his service, on the analogy of human service, deserves reward, and when the dog has grown too old to serve, his master ought to keep him till he dies. Such action helps to support us in our duties toward human beings, where they are bounden duties. If then any acts of animals are analogous to human acts and spring from the same principles we have duties toward the animals because thus we cultivate the corresponding duties toward human beings. If a man shoots his dog because the animal is no longer capable of service, he does not fail in his duty to the dog, for the dog cannot judge, but his act is inhuman and damages in himself that humanity which it is his duty to show towards mankind. If he is not to stifle his human feelings, he must practise kindness toward animals, for he who is cruel to animals becomes sharp in his dealings with men. We can judge the heart of a man by his treatment of animals. Hogarth depicts this in engravings [of *The Four Stages of Cruelty*]. He shows how cruelty grows and develops. He shows the child's cruelty to animals, pinching the tail of a dog or cat; he then depicts the grown man in his cart running over a child; and lastly, the culmination of cruelty in murder. He thus brings home to us in a terrible fashion the rewards of cruelty, and this should be an impressive lesson to children. The more we come in contact with animals and observe their behavior, the more we love them, for we see how great is their care for their young. It is then difficult to be cruel in thoughts even to a wolf ... A master who turns out his ass or his dog because the animal can no longer earn its keep manifests a small mind.[94]

The difference between Kant and Goethe is that the latter believes that we have a responsibility toward animals because *they* matter, because *they* are ends in themselves. Cruelty is an abomination because of the harm it does to the animals, not merely because it degrades the humans who commit the cruel acts. Still, there is no doubt that Kant and Goethe could have rejoiced together in the view that "we can judge the heart of a man by his treatment of animals."

IX

Arthur O. Lovejoy informs us that "among the great philosophic systems of the seventeenth century, it is in that of Leibniz that the conception of the Chain of Being is most conspicuous, most determinative, and most pervasive. The essential characteristics of the universe are for him plenitude, continuity, and linear gradation. The chain consists of the totality of monads, ranging in hierarchical sequence from God to the lowest grade of sentient life, no two alike, but each differing from those just below and just above it in the scale by the least possible difference."[95] In Leibniz's conception of the chain:

All the different classes of beings which taken together make up the universe are, in the ideas of God who knows distinctly their essential gradations, only so many ordinates of a single curve so closely united that it would be impossible to place others between any two of them, since that would imply disorder and imperfection. Thus men are linked with the animals, these with the plants and these with the fossils, which in turn merge with those bodies which our senses and our imagination represent to us as absolutely inanimate. And since the law of continuity requires that when the essential attributes of one being approximate those of another all the properties of the one must likewise gradually approximate those of the other, it is necessary that all the orders of natural beings form but a single chain, in which the various classes, like so many rings, are so closely linked one to another that it is impossible for the senses or the imagination to determine precisely the point at which one ends and the next begins – all the species which, so to say, lie near to or upon the borderlands being equivocal, and endowed with characters which might equally well be assigned to either of the neighboring species.[96]

Clearly, Leibniz is emphasizing the similarities to each other of all the beings close to each other in the chain. Thus humans share much in common with mammals, especially apes. And, given that when "the essential attributes of one being approximate those of another all the properties of the one must likewise gradually approximate those of the other," for Leibniz, as for Linnaeus, the distinction between human and chimpanzee, between dog and wolf, between domestic cat and lynx, must be very minor, indeed.

Oddly, this is not how the idea of the chain of being is usually interpreted in the literature that touches upon the history of animal ethics. For example, under "chain of being" in the index to Marian Scholtmeijer's *Animal Victims in Modern Fiction,* the reader is invited to see "dominance, human; hierarchy; supremacy, human."[97] Richard Milner tells us that the scale treated the inferior realm as "base."[98] David Maybury-Lewis describes the idea as one in which "all living forms were part of a series that ranged from the simplest organisms to the most complex and most perfect, which was man. Humankind was therefore central to God's plan for this world, just as this world was considered to be the centre of the universe ... Whether man's dominion was guaranteed by the Bible or science, the result was the same – the natural world was his to exploit."[99]

We have already taken a look at the generality of the claim that humans were deemed entitled to exploit the world and that animals were intended for human use, both of which were found to be, at best, one-sided views of the historically continuous debate. Similarly, if we investigate the reality of the idea of the Great Chain of Being, we find that it is sometimes used to elevate humans above other species and to accord them the right to use animals for their own ends, although almost never arbitrarily. Certainly, some, such as François Fénelon, thought of humans as more perfect, but he did not believe that this relieved him of all obligation toward his fellow creatures. Others used the idea of the chain to stress human-animal similarities. For instance, in the third century Plotinus writes:

> We must conclude that the universal order is for ever something of ... justice and wonderful wisdom ... from what we see in the All, how this order extends to everything, even to the smallest, and the art is wonderful which appears, not only in the divine beings but also in the things which one might have supposed providence would have despised [for] their smallness, for example, the workmanship which produces wonders in rich variety in ordinary animals ... It would not have been right for all things to have been cut off from each other, but they had to be made like each other, in some way at least.[100]

In the fifth century, Macrobius claimed all beings from the highest to the lowest to be "mutually linked together and without a break."[101] In the sixteenth century, Cardinal Saint Robert Bellarmine remarked: "God willed that man should in some measure know him through his creatures, and because no single creature could fitly represent the infinite perfection of the Creator, he multiplied creatures and bestowed on each a certain degree of goodness and perfection, that from these we might form some idea of the goodness and perfection of the Creator, who, in one most perfect essence, contains infinite perfections."[102] Saint John of the Cross added contemporaneously that animals are "finished and perfect."[103] For John Locke, "the several *Species* are linked together, and differ but in almost insensible degrees."[104] In *An Essay on Man* (1734), Alexander Pope is emphatic:

> Vast chain of being! which from God began,
> Natures ethereal, human, angel, man,
> Beast, bird, fish, insect, what no eye can see,
> No glass can reach; from Infinite to thee,
> From thee to Nothing ...
> Connects each being, greatest with the least;
> Made Beast in aid of Man, and Man of Beast;
> All serv'd, all'd serving! Nothing stands alone;
> The Chain holds on, and where it ends unknown.[105]

We even find the principle expounded in children's literature. In Sarah Trimmer's *Fabulous Histories, Designed for the Instruction of Children, Respecting their Treatment of Animals* (1786), we read that, in addition to humans, God "made all other living creatures likewise; and appointed them in their different ranks in creation, that they might form together a community, receiving and conferring reciprocal benefits."[106] In *The Book of Thel* (1789), William Blake offers an enigmatic account of the great chain, emphasizing the communal spirit as residing in everything from the earthworm and the very clay in which the worm lives all the way to the seraphim. At the turn of the nineteenth century, Priscilla Wakefield was telling us that "the different orders of beings approach each other so closely, and are so curiously united by links, partaking of the nature of those above and those below, that it requires a discerning eye to know what rank to assign them."[107] And the veterinarian William Youatt, repeating Alexander Pope in his *An Essay on Man,* declared "ALL ARE BUT PARTS OF ONE STUPENDOUS WHOLE" as the motto for his 1839 study of human obligations to animals.[108] For each of these, humanity stood in the middle of the continuum (not at the summit, as Maybury-Lewis indicates) as the highest of animals, with the classes of angels and God ranked above them. Equally for each of these, the continuum showed great proximity and considerable similarity among all the animals, and responsibility toward our fellow creatures was a primary implication. If today one is tempted to find the inclusion of the celestial spirits in the chain merely a device designed to make the human position an apparently less exalted and boastful one, we should recall that, in the days before the intrusion of a deistic rationalism into religion, angels were treated with deference and devotion. They play a significant role in Dante's *Divine Comedy* and in Milton's *Paradise Lost.* In the fifth century, Pseudo-Dionysius the Areopagite gave the angels their own codification, from seraphim and cherubim to archangels and angels, thus allotting them a specific place in the chain. If the modern intellectual world cannot take such an arrangement seriously, it should still not forget that the distant past did so in great earnest, with every kind of being having its allotted and, in most instances, respected place.

Goethe may be readily read as fitting neatly into the great-chain form of thought, even if he gives little consideration to the heavenly hosts. His poem "Limits of Humanity" shows humankind "lowly inclining / In infantile awe" beneath the immortals. And the poem concludes:

... races unnumbered
Extend through the ages,

Linked by existence's
Infinite chain.[109]

(By "races," Goethe here means species.) In "Divinity," as we have seen, Goethe emphasizes human and animal similarities, save for the ethical sense derived from culture. In the "Metamorphosis of Animals," he acknowledges the advantages of increasing complexity. Nature's distribution of talents "quietly favoured / The merry efforts of the children of multiple needs." And he mentions "the power of the more noble creatures."[110] Moreover, in his comments on evolution, he frequently refers to humanity as the destiny, the highest stage of animal development: natural history moves "through multitudinous forms / To reach at last the state of man." For Goethe, evolution is the driving force in fulfilling the ends of the Great Chain.

The evolutionary spiral is driven toward humankind as Nature's destiny: "Nature, in order to arrive at man, institutes a long prelude of beings and forms which are, it is true, deficient in a great deal that is essential to man. But in each is manifest a tendency which points to the next form above."[111] For Goethe, each species is perfect as the being that it is with the purposes and needs that it has. Yet succeeding stages of evolution provide different forms of life with higher purposes and needs, for which their own perfection is required. The laws of nature are eternal, producing ever greater structural differentiation and functional specificity, until humans emerge as the most developed of Nature's species. Clearly, there is more Ernst Haeckel[112] than Charles Darwin in such a version of evolution. Nonetheless, as was argued in the first chapter, the moral underpinnings of the evolutionists' arguments, if not evolution itself, point in the same direction. Thus, famously, in 1959 the Darwinian scholars W.M.S. Russell and R.L. Burch introduced the principles that are today the mainstay of animal experimentation ethics: replacement, reduction, and refinement (the 3 Rs).[113] Wherever possible, invasive animal research should be replaced by other forms of study; techniques should meanwhile be developed to permit a reduction in the numbers used and to refine existing methods so that pain and suffering are reduced. Of course, one may adopt such a system, however humane it may be, only if one has decided that it is in principle satisfactory to use animals for human ends. Humans must thus be deemed of a higher order than nonhuman animals, and the much vaunted neutrality of Darwinism is abandoned – if not by the implications of the theory, at least by the moral principles and practice of most of those who are instrumental in maintaining the theory. While there is a scientific gulf between

Goethe and Charles Darwin, the moral underpinnings of their theories are in practice the same. Indeed, notions of differing levels of complexity, differing levels of capacity for pleasure, enjoyment, pain, and suffering, and differing levels of ability to reason, to remember, and to anticipate are factors that have been seen to affect the degree of consideration to be allowed to different species in the classical, medieval, Renaissance, Enlightenment, Victorian, and modern eras alike. Darwinism did not change the relevant factors at all.

There would thus appear to be something suspicious about Edward Clodd's Darwinian claim that "man's belief in his exceptional place in nature is a relic of the anthropocentric theory"[114] if in fact those who concur with Clodd and Darwin still treat humans as essentially superior. Again, there may be satisfactory criteria for treating humans more favourably than other species, and degrees of sentience, anticipation, rationality, and the like may be considered relevant factors. If so, and if traditional evolutionists treat humans more favourably, then it serves to deceive to point out how evolutionary theory leads to different conclusions on the status of animals than those posited by the great-chain theory.

8

Kinship and Evolution

THE DARWINIAN MYTH

I

I<small>T HAS BEEN ONE OF THE RECURRING</small> observations of this book that if there is a consensus on any issue pertaining to the history of animal ethics, it is that Charles Darwin's theory of evolution had a profound, not to say revolutionary, impact on the recognition of our kinship with, and corresponding obligations to, other animals. Nonetheless, I have argued that such a thesis is untenable. It is customarily assumed that in coming to recognize our evolutionary descent from other species, our corresponding kinship with them, and our essential similarities to them, via the findings of Charles Darwin (and, secondarily, Alfred Russel Wallace), we came to acknowledge that other animals are entitled to greatly increased moral consideration. It is, for example, as we saw in the previous chapter,[1] the thesis of James Rachels's *Created from Animals: The Moral Implications of Darwinism;* it is also an assumption or argument to be found in much modern animal-welfare and animal-rights literature, including the well-conceived. For instance, Marian Scholtmeijer's generally admirable study of *Animal Victims in Modern Fiction* opens with the claim that "the Darwinian revolution profoundly altered society's conception of animals."[2] In *Deep Vegetarianism,* Michael Allen Fox refers to "the work of Charles Darwin (1809-82), which breached the species barrier so dramatically."[3] Mary Midgley expounds upon "the Darwinian theory of evolution" that "radically shakes the massive traditional picture of ourselves."[4] The veterinary ethicist Rev. Giles Legood claims that "Darwin blew apart the almost universally accepted theory that human beings were set apart biologically from the rest of the natural order."[5] In decidedly more modest tone, Hilda Kean observes in *Animal Rights: Political and Social Change in Britain since 1800* that "the work of Darwin and his contemporaries was to have a profound influence on cultural and intellectual life – and on the popular perception of animals."[6] This is certainly not one of the more bloated statements about the effects of Darwinism. Yet even here the perspective is misleading if one imagines the statement to mean that the popular perception of animals became profoundly different from what it was before Darwin and that the change, to the extent that there was one, was due primarily to Darwin. One wonders, certainly, whether the qualifying term "and his contemporaries" means only those who agreed with Darwin or whether it includes also those who strenuously opposed him, in which case the proposition is almost trivial. Further, Christine Kenyon-Jones remarks that "Darwin's struggle to de-centre man from the universe included a struggle with human-centred language."[7] She writes of a "general characteristic of approaches in pre-Darwinian imaginative literature"[8] that stands in stark contrast to post-Darwinian literature. Yet no convincing evidence of the post-Darwinian metamorphosis is offered. Nor are we informed, if a significant change took place at a later date, why we should assume with any confidence that it was as a consequence of Darwinism. If it was, we need to be given the evidence, not speculation alone, in refutation of the hypothesis frequently argued in this study that many prior naturalists contributed to the diminution of anthropocentrism at least in no less a degree than Charles Darwin. Included among them are Leibniz, Goethe, Etienne Geoffroy Saint-Hilaire, and Erasmus Darwin.

Others went even further in their proclamation of the evolutionary insurrection. Thus Peter Singer tells us that "intellectually the Darwinian revolution was genuinely revolutionary. Human beings now knew they were not the special creation of God, made in the divine image and set apart from animals; on the contrary human beings came to realize that they were animals themselves."[9] It is difficult to reconcile this claim with Singer's contrasting but equally one-sided pronouncement in the *Oxford Companion to Philosophy* that in "Western ethics non-human animals were until quite recent times accorded a very low moral status."[10] Other than in the writings of Jeremy Bentham, so we are told, it was not until the 1970s that the rights of animals received recognition.[11] Yet, even if we accepted the accuracy of the statement, it would be strange, if Darwinian theory were truly so "genuinely revolutionary," that someone writing over a half-century before *The Origin of Species* could have so readily recognized what Singer perceives to be the novel implications of Darwinism and that for over a century after the book's publication most others failed to recognize its ethical implications. Indeed, one of the most common counterarguments to Darwinism was to be found in the popular argument from design, which was employed, as one of very many examples, in Sydney Dyer's *Home and Abroad, or the Wonders of Familiar Objects,* published in Philadelphia in 1880 and advocated widely among Baptists; it was certainly as sympathetic to animals as anything in Darwin. Moreover, one wonders where Singer could possibly have derived the idea that it was only with Darwin that humans came to recognize themselves as animals! That would be to deny the very lengthy history of observations on homology and evolutionary theory to which I have frequently alluded. An instance could readily be offered of W.C. Spooner, writing in *The Veterinarian* in 1840, where he takes it as a given that "we must suppose that it has been satisfactorily proved that reason is possessed by brutes."[12] To add one more observation, made by no ally of Darwin, the Christian-Socialist cleric Charles Kingsley observed unmistakably in *Yeast* (1851) that "The finest of us are animals after all." Perhaps only a few Cartesians ever denied that humans were animals – but they appeared at the same time to suggest that we were not really people in any meaningful sense either – for although, like animals, we had a body, in essence we were, unlike animals, *res cogitans* (thinking things). It was minds, not people, that thought.

In similar vein to Singer, as we noted in the first chapter, in Michael W. Fox's *Inhumane Society,* we read that "as the concept of human superiority is, as Charles Darwin emphasized, logically and ethically untenable, the only grounds for contending that it is humankind's right to exploit animals are based in custom and utility."[13] One wonders where Darwin is supposed to have emphasized that "the concept of human superiority" is "logically and ethically untenable" – certainly in nothing that I have read by Darwin. And the very reverse is to be found on almost every other page of the first seven chapters of *The Descent of Man.* In light of such emphatic pronouncements from such a variety of sources on the paramount importance of Charles Darwin to the development of animal ethics, my claim of the relative insignificance of evolution by natural selection to the understanding of our moral relationship to other species needs to be carefully articulated and addressed.

The idea that Darwinism revolutionized our attitudes to animals is certainly not restricted to recent literature. For example, the Dorset novelist and poet Thomas Hardy equated evolution with the discoveries of Charles Darwin, announced that "as a young man" he "had been among the first acclaimers of *The Origin of Species,*" and wrote a New York correspondent in 1909: "The discovery of the law of evolution [of Charles Darwin], which revealed that all organic creatures are of one family, shifted the centre of altruism from humanity to the whole conscious world collectively. Therefore, the practice of vivisection, which might have been defended while the belief ruled that men and animals are essentially different, has been left without any logical argument in its favour." In a letter of 10 April 1910 to the secretary of the Humanitarian League, Hardy observed that "the most far reaching consequences of the establishment of the common origin of all species is ethical ... Possibly Darwin himself did not wholly perceive it, though he alluded to it."[14] Would it have surprised the animal advocate and anti-vivisectionist Hardy to realize that the social-critic novelist Charles Dickens[15] and the pre-Darwinian veterinary author William Youatt[16] opposed animal experimentation far more strenuously than did the Darwin who killed pigeons for his research, albeit with a heavy heart, and who opposed vociferously the legislative restrictions to animal experimentation that were suggested and that almost carried in 1876, being replaced at the last moment by a far weaker act supported by the Darwinian physiologists?

Indeed, if Darwinian theory was so significant, and if we accept Hardy's view that its effect on the justification of vivisection was a vital characteristic of the ethical dimensions of the theory, then on what grounds could so many pre-Darwinian and anti-evolutionist figures have opposed vivisection and so many Darwinians supported it? Writing in 1711, the great essayist Joseph Addison, for example, related his despair at "a very barbarous experiment" by one "who was well skilled in dissections" in support of his demonstration of the humane parental love in animals.[17] The early-eighteenth-century poet Alexander Pope opposed the invasive researches of his friend Dr. Stephen Hales, proclaiming that "he commits most of those barbarities with the thought of being of use to man; but how do we know that we have a right to kill creatures we are so little above as dogs for our curiosity, or even for some use to us?"[18] Samuel Johnson, as we noted earlier, castigated vivisectors in 1765 as "a race of men who have practised tortures without pity, and related them without shame, and are yet suffered to erect their heads among human beings."[19] Concluding a condemnation of the late-eighteenth-century vivisection of rabbits by a certain Dr. John Caverhill, the *Monthly Review* expostulated, "surely there are *moral* relations subsisting between man and his fellow-creatures of the brute creation."[20] In 1791 the atheist John Oswald referred to the "Barbarians" who "at the shrine of science, have sacrificed the dearest sentiments of humanity."[21] Famously, the deist Voltaire railed against the "barbarians who seize this dog, such a prodigious friend of mankind, nail him to the table, and dissect him in order to show you his veins."[22] But the devout were no less adamant than the infidels. For instance, the naturalist Charles Bonnet in *Contemplation de la Nature* (1765) tells us that "one cannot read without emotion the story of a bitch, who, while being vivisected, began to lick her pups, as though it were they who needed to be consoled in their torment." The primitivist Bernardin de Saint-Pierre announced in his *Études de la Nature* (1784): "I pay heed to the fact that cruel experiments are performed every day on animals to discover the secret intimations of nature, but they succeed in merely distorting the operation of the animals' spirit ... sending their nerves into convulsions."[23] Abbé Jacques Delille was more direct: "Desist, you barbarians."[24] In 1838 Rev. William Drummond deemed vivisection the "love of science perverted."[25] Already in his 1835 poem, *Pleasures of Benevolence,* he had asked those who study "the mysteries of nature" to "forbear / To purchase knowledge at the price, too dear, / Of violated mercy." And in *The Bird* (1856), Jules Michelet complained of science that it "cannot study unless it kills; the sole use of which it makes of a living miracle is, in the first place, to dissect it. None of us carry into our scientific pursuits that tender reverence for life which nature rewards by unveiling to us her mysteries."[26] None of these fairly representative compassionate pre-Darwinian writers found evolution a prerequisite for the understanding of the horrors of vivisection.

Another opponent of vivisection, the biologist J. Howard Moore, whom we met briefly in the first chapter, argued with the greatest eloquence in both *The Universal Kinship* (1906) and *The New Ethics* (1907) that Darwinism proved the kinship necessary for a meaningful theory of the ethical relationship between humans and animals, implying that no previous analysis had provided such a justification.[27] One wonders whether Moore had pondered the fourth-century words of Saint John Chrysostom: "Surely, we ought to show [other species] great kindness and gentleness for many reasons, but above all because they are of the same origin as ourselves."[28] One wonders equally whether Moore had compared Darwin with Saint Basil of Caesarea, who wrote of "the sense of fellowship with all living things, our brothers the animals to whom thou hast given the earth as their home in common with us."[29] The Romantic poet Samuel Taylor Coleridge, an avowed opponent of Erasmus Darwin's version of evolution, declared in 1802 that "Nature has her proper interest; and he will know what it is, who believes and feels, that everything has a life of its own, that we are all *one Life*."[30] In the delightful poem about the young ass tethered near his mother, Coleridge feels compassion for the "Poor little foal of an oppressed race ... I hail thee *Brother*." He did not need evolution to know that he was related to, had a kinship with, the donkey. All it took was "a fellowship of Woe." Writing in 1903 with a sympathy for Darwin, but in adamant opposition to Darwinism, Elijah Buckner tells us that "the Creator has given us to understand we must acknowledge a kinship to all animals. There is a relationship, by descent, between all species of the animal kingdom, as each species has descended from its own progenitor, and all progenitors have one common Parent or Creator. This is the Bible theory of evolution."[31] Do these sentiments not demonstrate as much a sense of kinship as anything in Darwin? To be sure, not all Darwinians were of one mind. For instance, the Darwinian physician W. Lauder Lindsay proclaimed in 1879 that "Man's claim to preeminence on the ground of the uniqueness of his mental constitution is as absurd and puerile ... as it is fallacious."[32] But then Lauder Lindsay was not only disagreeing with other Darwinians, but also concurring with a point of view held by many for a long period before Darwin. Nonetheless, as was almost universal in the Victorian era, Lindsay likewise wrote of "the lower animals" and even included the phrase in the title of his best-known work.

II

There are three ways in which Darwin's theory may be interpreted to have fundamentally changed our attitude to animals. It may be suggested: 1) that Darwin's own attitudes, and his discussion of our similarities to, and kinship with, other species differed fundamentally from those of previous commentators and that he offered a novel account of the extent of those responsibilities; 2) that the theory aroused public awareness for the first time of our similarities and relationships to other animals and that this awareness encouraged a hitherto absent respect for them; 3) that Darwin and his followers showed themselves more sympathetic than others to animal interests in the issues that arose in the decades following the publication of *The Origin of Species* and *The Descent of Man*. On all three counts Darwinism fails the test. Certainly, Darwin's theory of evolution is of the greatest scientific significance. If he did not discover evolution, he certainly developed clearer conceptions of it. If he did not discover natural selection, as some maintain, he provided both the evidence and the systematization in which it could be made plausible, and even probably verifiable (although a clearer demonstration would have to await the broadcasting of Gregor Mendel's work on pea genes). Yet, from the perspective of animal ethics, Darwin offered few revelations, if any, about the way in which animals were entitled to be treated.

Now it would be churlish to deny that Darwin himself cared deeply for animals, at least when he was not engaged in his pastime of sport hunting – which he gradually abandoned as he matured – or killing animals for his research. Thus in *The Descent of Man* Darwin described sympathy for animals as a "virtue ... one of the noblest with which man is endowed" but unfortunately not one universally practised for: "Sympathy beyond the confines of man, that is, humanity to the lower animals, seems to be one of the latest moral acquisitions."[33] In an article on "Vermin and Traps" in the *Gardener's Chronicle* in 1863, he excoriated those who would set such traps: "If we attempt to realise the suffering of a cat, or other animal when caught, we must fancy what it would be to have a limb crushed during a whole night, between the iron teeth of a trap, and with the agony increased by constant attempts to escape ... where game keepers are not human, or have grown callous to the suffering constantly passing under their eyes, they have been known by an eyewitness to leave the traps unvisited for 24 or even 36 hours."[34] Darwin noted that "in the agony of death a dog has been known to caress his master, and everyone has heard of the dog suffering under vivisection, who licked the hand of the operator; this man, unless the operation was fully justified by an increase of our knowledge, or unless he had a heart of stone, must have felt remorse to the last hour of his life."[35] Here we must begin to recognize the early limits to Darwin's compassion, real as it was. When scientific knowledge is at issue, the suffering of the animal must be subordinated to it. Quite clearly, without diminishing an appropriate recognition of Darwin's animal sympathies, one could enumerate many others from Plutarch, Porphyry, and Montaigne, through Pierre Bayle, Henry Fielding, James Thomson, and Oliver Goldsmith, to Emanuel Swedenborg, Emily Brontë, Arthur Schopenhauer, and the Lake Romantics, and many, many more, who evinced a greater sensibility toward animals than did Charles Darwin.[36]

Certainly, Darwin goes to considerable lengths to describe the similarities between humans and other species and to propose this as a primary indicator of our responsibilities toward other species. Of the greatest significance, however, is the fact that all the categories Darwin employs and all the conclusions Darwin reaches were already used and expounded in numerous pre-Darwinian studies pertaining to animal ethics. In Chapters 3 and 4 of *The Descent of Man*, "Comparison of the Mental Powers of Man and the Lower Animals," Darwin tells us that there his object was "to show that there is no fundamental difference between man and the higher animals in their mental faculties."[37] He concludes an initial discussion with the assertion that "it has, I think, now been shown that man and the higher animals, especially the Primates, have some few instincts in common. All have the same senses, intuitions and sensations – similar passions, affections and emotions, even the more complex ones, such as jealousy, suspicion, emulation, gratitude and magnanimity; they practice deceit and are vengeful; they are sometimes susceptible to ridicule, and even have a sense of humour; they feel wonder and curiosity; they possess the same faculties of imitation, attention, deliberation, choice, memory, imagination, the association of ideas and reason, though in very different degrees."[38] Animals are next said to possess "the social instincts [that] lead an animal to take pleasure in the society of its fellows, to feel a certain amount of sympathy with them, and to perform various services for them."[39] Nonetheless, we are told that "of the differences between man and the lower animals, the moral sense or conscience is the most important."[40] Darwin concludes that "there can be no doubt that the difference between the mind of the lowest man and that of the highest animal is immense [which, we might note in passing, seems to offer a quite different emphasis from the claim "that there is no fundamental difference between man and the higher animals in their mental faculties"] ... Nevertheless the difference in mind between man and the higher animals, great as it is, is certainly one of degree and not of kind."[41] It is precisely because of these similarities that Darwin concludes that "humanity to the lower animals" is "one of the noblest [virtues] with which man is endowed."[42] The relevant question, of course, is whether this analysis differs in any significant manner from pre-Darwinian accounts. Almost all the literature relating to animal ethics has assumed the answer to be in the affirmative.

Yet it is not. Immediately, one notices the similarity to the views of the devout theist and anti-materialist Sir Benjamin C. Brodie, who was president of the Royal Society when *The Origin of Species* was published and who had a few years earlier written in his *Psychological Enquiries* (1855): "The mental principle in animals is of the same essence as that of human beings; so that even in the humblest classes we may trace the rudiments of these faculties to which, in their state of more complete development, we are indebted for the grandest results of human genius. I am inclined to believe that the minds of the inferior animals are essentially of the same nature with that of the human race."[43] Moreover, writing in the same year that Darwin composed, after decades of thought, *The Origin of Species* (1858), the eloquent American theologian William Rounseville Alger observed: "Of late years the tendency has been to assimilate instead of separating man and beast. Touching the outer sphere, we have Oken's homologies of the cranial vertebrae. In regard to the inner sphere, we have a score of treatises, like Vogt's Pictures from Brute-Life, affirming that there is no qualitative, but merely a quantitative distinction between the human soul and the brute soul. Over this point the conflict is still thick and hot."[44]

If we turn to the study of the Nonconformist cleric and veterinarian William Youatt on *The Obligation and Extent of Humanity to Brutes,* published in 1839, twenty years before *The Origin of Species* and thirty-two years before *The Descent of Man,* we will be struck immediately by the degree of similarity between the concepts and categories employed by Youatt and those later employed by Darwin. Moreover, it should be clear that each uses the prevailing categories of ideas of human-animal continuity prevalent in the nineteenth century. There is nothing revolutionary, not even anything slightly novel, about Darwin's analysis and conclusions.

In very similar vein to the approach Darwin would take later, Youatt wrote of the animals' senses, emotions, consciousness, attention, memory, sagacity, docility (i.e., the capacity for learning), association of ideas, imagination, reason, instinct, social affections, moral qualities, friendship, and loyalty – each of which is acknowledged to exist in other species and to differ from human attributes only by degree.[45] Besides his evolution by natural selection, Darwin's concepts of human-animal comparison add nothing of substance to the notions already expressed by Youatt, which are in turn nothing out of the ordinary in the history of thought with regard to animal attributes. To be sure, not all were of one mind; and not all, even among those from whom one might have expected something better, rose to the acknowledgments of either a Youatt or a Darwin. For instance, as Hilda Kean has related, according to the Zoological Society of London in 1831, the "brown bear allegedly possessed a high degree of brute force, intellectual stupidity, and insatiable and gluttonous voracity ... the black ape's expression was 'peculiarly cunning'; the leopard had a moral character of suspicion, presenting an air of 'malignity and wiliness.'"[46] It would appear that Youatt, who was proud to describe himself as "Veterinary Surgeon to the Zoological Society of London,"[47] exercised very little authority over the pronouncements of this society.

With respect to the intellectual qualities, Youatt tells us, in language bearing a striking resemblance to the future expressions of Darwin, that "we are endeavouring to shew that the difference between [humans and nonhuman animals] in one of the most essential of all points, is in degree and not in kind."[48] Later in the same year, in *The Veterinarian,* and still over thirty years before Darwin, we find William Karkeek repeating the phrase: "We are ... endeavouring to shew that the difference between them, is in degree, and not in kind."[49] As an instance of consciousness and attention, Youatt tells us of the "terrier, with every faculty absorbed in his eager watching for prey."[50] As examples of memory, he refers to the training of the horse and the dog.[51] He observes that "connected with memory is association of ideas, i.e., when the occurrence of a certain event brings to our mind a whole train of thoughts: and, one link being obtained, the entire chain of reasoning occurs"[52] – and animals possess this capacity, although it has not been uncommon to deny it to them. Youatt regales us with dog stories to make his point. Imagination, he tells us, "is a faculty of mind by which we recall parts and portions of former impressions, and combine them in different ways forming new images, fanciful, or sublime, or ridiculous."[53] We are offered the animal's capacity for dreaming as an illustration of the use of imagination in animals. Youatt returns to "reason" – "the power by which we deduce one proposition from another, and proceed from premises to consequences."[54] Several dog tales are offered in support of animal rationality. But the most significant is the repetition of an account of wasp reasoning taken from Erasmus Darwin's *Zoonomia, or the Laws of Organic Life* (1794-96).

> Dr. Darwin used to tell the curious story of a wasp. As he was walking one day in his garden he perceived a wasp upon the gravel-walk with a fly almost as large as itself, which it had just caught. Kneeling down in order to observe the manoeuvres of the murderer, he saw him distinctly cut off the head and body; and then, taking up the trunk, to which the wings still remained attached, he attempted to fly away. A breeze of wind, however, acting upon the wings of the fly, turned the wasp around, and impeded or forbad his progress. Upon this he alighted again on the gravel, and, having thus removed the cause of his embarrassment, he flew off with his booty.
>
> Could any process of reasoning be more perfect than this? "I shall not be able to carry away the whole of this fellow now I have got him," – the plunderer may be considered as saying to himself – "therefore I'll leave his head and his tail behind, and take home his chest; but the wind gets hold of his wings, or twirls me around: I must rid myself of these sails, or I shall never reach my nest in safety."
>
> Some reasoning of this kind must have passed through his mind, or his actions are altogether inexplicable. Instinct has, and can have, nothing to do with it.[55]

In discussing instinct,[56] Youatt attempts to show the similarities in human and animal instinct and seeks to demonstrate that much that is so regarded in animals is a product of variability instead. Nest building, for example, differs from species to species and individual to individual – he offers a host of examples – reflecting considerable ingenuity on the part of the bird rather than mere instinct, even though the actions may be prompted by instinct. Voltaire had made the same case in shorter compass three-quarters of a century before Youatt: "How pitiful, how impoverished, to have said that animals are machines lacking knowledge and feeling, which always conduct themselves in the same manner, which learn nothing, improve nothing, etc. What! The bird which builds his nest in a semi-circle when he attaches it to a wall, a quarter circle when built on a corner, and a full circle in a tree, this bird does everything in the same manner?"[57]

Finally, Youatt turns to the moral qualities: "What, of brutes? Ay! and strongly developed, and beautifully displayed, and often putting the biped to shame. I begin with the social affections. They are as necessary in the little republics of the brutes as among any of the associations of men. They are the cement which binds together the different parts of the fabric. 'But,' say some, 'they are mere instincts!' We care not for that. *These instincts or propensities are the foundation of every virtue in the human being;* and in the quadruped they cannot escape our regard and admiration."[58] Youatt finds abundant evidence of parental affection, and of conjugal attachment and fidelity among the brutes, but relatively little filial affection because of their short period of infancy.[59] He finds much of merit in animal capacity for friendship and loyalty.[60] "The disposition and habits of the different domesticated animals – the spirit and courage of the horse – the patient endurance of the ox – the intelligence and fidelity of the dog – these are subjects which will pleasantly and profitably employ our study, and excite our attachment to the animal and our admiration of the skill displayed in the structure of each."[61] However inferior animals may be to humans by degree, they enjoy their own perfection: "*They are all perfect in their way.* Their propensities and their reasoning are precisely what they should be."[62] Here Youatt's conclusion follows that of Leibniz and Goethe – animals are perfectly formed to meet the needs and purposes that they have as the animals that they are. These are the bases on which Youatt insists that our obligations to our fellow creatures are predicated – beyond those of Biblical edicts, which Youatt reads to require human moral responsibility to all other creatures.[63] In *Animal Minds and Human Morals,* Richard Sorabji writes of Darwin as "arguing in *The Descent of Man,* though not with perfect consistency, that there is no psychological attribute of humans which is not also found in animals to some degree, reason included."[64] Sorabji thinks that Darwin is adding something original to the debate. In fact, he is following a long line of predecessors.

Youatt was not being, nor was he claiming to be, especially original. The French army surgeon Ambroise Paré was already writing in the mid-sixteenth century, as we have noted, that "magnanimity, prudence, fortitude, clemency, docility, love, carefulness, providence, yea knowledge, memory & c. is common to all brutes."[65] And in the mid-seventeenth century, the Puritan leveller Richard Overton was citing Paré with admiration and approval. We find extensive listing of similar animal attributes in the writings, for example, of Rorarius, Gilles, Bary, de la Chambre, Bayle, Voltaire, and George Nicholson, with Nicholson citing them from a broad variety of sources. In the early eighteenth century, we encounter the influential Bishop of Durham, Joseph Butler, taking it as common knowledge that animals as well as people share "apprehension, memory, reason ... affection ... enjoyments and sufferings."[66] In the year before the publication of

Youatt's treatise, we find William Drummond in *The Rights of Animals* proclaiming that the "inferior animals have passions, feelings, sensibilities, as well as the lordly man. Many of them have the conjugal, the paternal, and maternal affections strong and indomitable, even in torture and death ... Many of them, it must be admitted, are endowed with properties so closely approximating to those of the creature we call rational, that it is by no means easy to shew an essential difference. It cannot be denied that they possess moral affections, for they have gratitude; nor intellectual faculties, for they have memory; and that each of their powers conducts to the proper end with unerring certainty."[67] Edward Jesses's widely popular *Gleanings in Natural History,* already in its sixth printing by 1845, had chapters on the "Language of Insects," "Gratitude of Animals," and "Love of Animals for Their Young." And three years after Darwin's *The Descent of Man,* we find the Reverend J.G. Wood following a line similar to that of Darwin but completely uninfluenced by evolution. He sets out to show "that the lower animals share with man the attributes of Reason, Language, Memory, a sense of moral responsibility, Unselfishness and Love, all of which belong to the spirit and not to the body."[68] Nor were they doing anything more than following the classical line of discussion. For example, in the third century Porphyry was writing of animal sensation – the capacity to feel pleasure and pain – and of the memory, reason, and language of animals. Moreover, he pointed out that while each other species possessed a common language, humans were inferior in being able to converse only with their fellow Greeks, or fellow Italians, and so forth.[69] Almost two centuries earlier, Plutarch had observed in his essay "On the Use of Reason by 'Irrational' Animals" that other species possess "morality, intelligence, courage, and all the other virtues" and that the faculty of reason they possess "is both practically better and more impressive than human intelligence."[70] Nor were these stalwarts of animal advocacy much of an exception in the argument. For instance, in the third century, Lactantius wrote the first systematic account in Latin of the Christian attitude to life:

> The highest good of man ... is in religion alone. For the other things, even those which are thought proper to man, are found in other animals also. For when they discern and distinguish their own voices among themselves by proper notes, they seem to speak together. Some type of laughing also appears in these, since by a stroking of the ears and a contraction of the mouth and a movement of the eyes into frolicsomeness, they either laugh at men or at each other. What of their mates and their own offspring? Do they not bestow on them something like mutual love and indulgence? They surely have a providence, for they look out for things for themselves for the future and they

lay away foods. Signs of reason are also caught in many. For when they seek
useful things for themselves, beware evil, avoid danger, and prepare hiding
places for themselves opening out to several exits, certainly they manifest an
understanding of something. Can anyone deny there is reason in them when
often they delude man himself?[71]

Some – Pliny the Elder, for example – went further, allotting even a religious sense to the animals, especially to elephants; some two millennia later,
Elijah Buckner hints at something similar for dogs.[72] And in the late Middle Ages, William Langland's *Piers the Ploughman* postulates reason as lying
more with the animals than with humankind. If Charles Darwin may be
said to have overcome some of the excesses of some of his forerunners –
although not by too great a degree, for he was certainly guilty of a similar
surfeit of anthropomorphism – he does so in no greater degree than Youatt.
Indeed, with regard to the status of animals, in what manner can anything
in Darwin's theory of evolution be said to have improved on the 1836
proposition of Ralph Waldo Emerson that "each creature is only a modification of the other; the likeness of them is more than the difference, and
their radical law is one and the same"?[73]

Contrary to what we read in most of the recent literature bearing on the
history of animal ethics, it would be difficult to conceive of any manner
in which Darwin's theory of evolution, certainly not as used by Darwin in
The Descent of Man, brings about any substantial change in the grounds
that entitle nonhuman animals to moral consideration. He is merely extending the form of argument employed in the classical, medieval, and
Renaissance periods a little bit further, although no further – indeed, less
far – than had already been undertaken by some of his Wilhelmian and
Victorian predecessors.

III

One could argue that previous theories of evolution had far less public
impact than did that of Charles Darwin. And this is indeed true, although
perhaps much less significantly than is customarily imagined. Erasmus Darwin's theory, for instance, appeared in a 55-page chapter entitled "Generation" in Volume I of *Zoonomia, or the Laws of Organic Life,* published in
the spring of 1794. Starting on page 482 of a 586-page volume, the chapter received at first very little notice in the book's reviews. And when, a
year later, the "subversiveness" of the chapter finally received indignant
attention, and again in 1798 when evolutionary theory was ridiculed in

the *Anti-Jacobin* by Tory statesman George Canning, there was no T.H. Huxley waiting in the wings to turn a vitriolic disagreement into a *cause célèbre*. Certainly, evolution was not ignored. It was ridiculed by the Tory press. Attached to Canning's political diatribe was the famous cartoon by James Gillivray of 1 August 1798 attacking all purported contemporary subversives from Godwin to Paine to Priestley.[74] Erasmus Darwin was lampooned as an ape carrying a basket of "Jacobin Plants" on his head. Moreover, the book reached the less than sympathetic attention of the Vatican. Erasmus Darwin reported with evident pleasure that the *Zoonomia* as a whole was "honoured by the Pope by being placed on the Index Expurgatorius."[75] And, as Darwin's biographer rightly reported, "his evolutionism was obnoxious in time of war."[76] But, as we noted in the Introduction, in time of war the establishment had rather more to worry about than scientific theories that impinged on religious orthodoxy. Erasmus Darwinism was sneered at and cast adrift rather than rammed and sunk. British sympathy for the French revolutionaries – and Darwin had been an avowed admirer of the American rebels (as had Edmund Burke), if less ostentatiously of the French – engaged more of the attention of the authorities, bringing the suspension of habeas corpus, the carried-out threat of transportation to the prison island of Australia for fourteen years, and even the failed threat of execution. It was not that the Crown did not prosecute but that juries would not convict on capital charges. William Blake was one of those charged for his apparent declaration on behalf of Napoleon, but the evidence was inadequate to convict. And Byron continued to have a bust of Bonaparte on his desk. But his crime was ignored, perhaps on account of his nobility or perhaps because his apparent incestuous relationship with his half-sister Augusta Leigh overshadowed his other sins.

Samuel Taylor Coleridge was one of those most outraged by early evolutionary theory. He wrote to William Wordsworth, censuring him for not denouncing the idea of human-ape similarities in *The Excursion* (1814), even though the pair held similar views: "I understood that you would take the Human race in the concrete, and have exploded the absurd notion of Pope's *Essay on Man*, [Erasmus] Darwin and all the countless Believers – even (strange to say) among Xtians of Man's having progressed from an Ouran Outang state – so contrary to all History, to all Religion, nay to all Possibility."[77] Coleridge's and Wordsworth's mutual rejection of evolution did not prevent their profound sympathy and respect for their fellow creatures. Neither did it result in a major public vituperation of evolution nor in significant expressions of scientific sympathy.

It should certainly not go unnoticed that Coleridge regarded the belief in evolution as widely accepted in the early nineteenth century, even "among

Xtians" – almost a half-century before *The Origin of Species*. Nor, in Coleridge's view, was it a rarity a century and a quarter before the *Origin*, as
exemplified by Alexander Pope, who derived his information from Edward
Tyson's 1699 work hinting at evolution, *Orang-Outang; sive Homo Sylvestris;
or the Anatomy of a Pigmie Compared to that of a Monkey, an Ape, and a Man*.

France was perhaps further advanced in public dissemination of the
idea. Geoffroy Saint-Hilaire, as we have seen, propounded the view in the
earlier nineteenth century that "there is, philosophically speaking, only a
single animal." The prolific French novelist Honoré de Balzac advertised
the evolutionary message to the whole literary world in his popular series
of books *La Comédie humaine*. In the preface that he wrote in 1845 as a
generic preamble to the whole series, which was hence included in each
post-1845 printing of those works published prior to 1845 as well as in all
subsequent novels and their translations – comprising some seventy volumes! – Balzac announced approvingly, and, indeed, as the underlying
theme of his own novels, that all animals, human animals included, were
created on "one and the same principle" and that all animals behaved according to similar natural laws – in conformity "with preconceived ideas of the
Divine power."[78] Balzac, one would imagine, reached a far wider audience
than Charles Darwin and T.H. Huxley, even though he did not instigate
a furore, perhaps because the theory – at least in Balzac's version of a self-
interest paradigm, which he saw as following from evolutionary theory –
was more demeaning of humans than elevating of animals. And since all
this came from "the Divine power," any lack of orthodoxy was limited in
its effect by a more convincing and accommodating religious obeisance –
indeed, by a far more sincere one than that occasioned by the mandatory
nods to the Deity by both Erasmus and Charles.

The primary point, however, is that the ensuing furore after the publication of *The Origin of Species* had very little to do with our appropriate
attitudes to animals – to the extent that there was a furore, for that, too,
is customarily grossly exaggerated,[79] even though the level of interest aroused
in prior discussion was evidenced by the fact that the first printing sold
out on the first day of publication. Rather, the furore revolved around our
descent from animals and the implications that this might have for scripture and revealed religion – and even for the fear that if humans come
to believe that they are just like all the other animals, they will behave
accordingly, and a civilized and law-abiding society will be lost forever! As
far as attitudes to animals were concerned, if evolutionary theory were valid,
we were descended from animals and owed them a responsibility because
they were our close biological kin. If evolutionary theory were invalid,
we were not descended from animals, but we owed them a responsibility

because we were built on the same principles as they – more-or-less axiomatic among the well-educated at least since the studies of Giovanni Borelli in 1680 – and because they, too, were sentient creatures of God and thus related through our common Father. Of course, these were not the only views. There were also animal-inconsiderate views to be found rampant among both the evolutionists and the orthodox. Certainly, the maintenance of evolutionary theory was no sign of a guarantee of sensibility toward animals. Nor was a denial of Darwinism in any manner a hindrance to such sensibilities.

It is a matter of some significance that a number of the most noteworthy post-Darwinian animal advocates spent considerable energy denouncing evolution by natural selection as inimical to appropriate respect for animals. The devout ethical vegetarian and opponent of vivisection Leo Tolstoy – who never used a whip on his mounts: "I talk to my horses; I do not beat then" – warned his son Sergei against "Darwinism," which "won't explain to you the meaning of your life and won't give you guidance for your actions."[80] For Tolstoy, Darwinism destroyed the ideal of love for fellow creatures that abounded in primitive Christianity. The quite secular but equally ethical vegetarian and even more emphatic opponent of vivisection George Bernard Shaw castigated Darwinian "Natural Selection" as "no selection at all, but mere dead accident and chance."[81] Expounding a philosophy that he conceived as an adaptation of Schopenhauer, but which in fact corresponded more closely to the creative-evolution ideas of Henri Bergson, Shaw wrote his French translator, Augustin Hamon: "I am above all things a believer in the power of Will (Volonté). I believe that all evolution has been produced by Will, and the reason you are Hamon the Anarchist, instead of being a blob of protoplasmic slime in a ditch, is that there was at work in the Universe a Will which required brains & hands to do its work & therefore evolved your brain and hands."[82] Shaw did not see Darwinism as an appropriate basis for his own extensive animal sensibilities. His animal ethics were quite secure without it. For Shaw, Darwinism removed individual responsibility for actions and thereby denied all meaningful ethics. In his *Mutual Aid: A Factor of Evolution,* the communitarian anarchist Peter Kropotkin criticized Darwinism as overly reliant on the concepts of competition and struggle and for thus failing to give adequate emphasis to the cooperation and altruism present in both human and nonhuman animals, which were, Kropotkin believed, the very basis of morality. For Kropotkin, Darwinism eradicated compassion as the foundation of justice.[83] Likewise, in *The Study of Animal Life* (1892), the naturalist J. Arthur Thomson argued not only that Darwinism was one-sided in its emphasis on "the struggle of life," but also that there was a significant

danger of drawing erroneous moral conclusions from its apparent implications. For Thomson, Darwinism failed to understand the real, and better,
nature of our animal relations. The physician and anti-materialist Elijah
Buckner, a protagonist of animal futurity, acknowledged in 1903 his admiration for "Darwin's great labors and profound researches into the dark
and hidden mysteries of nature, yet there is no doubt his theory has been
a great hindrance to the cause of humanity, and has worked a great hardship to suffering animal life."[84] Stephen Coleridge, the head of the National
Anti-Vivisection Society, writing in 1913, regarded Darwin's work as demeaning humanity, which was "once the supreme work of God at the
head of His universe," by reducing it "to an accidental development of an
arborial ape."[85] He described the 1876 Cruelty to Animals Act, which had
been reluctantly welcomed by Darwin even though he thought it far stronger
than appropriate, as legislation that "expressly exempted the vivisector from
observing the law of the land, and permitted him to do what cabmen and
costermongers remained punishable for doing. It legalized the torture of
animals, domestic or wild, if the torture were inflicted by a selected class
of persons."[86] Opposition to vivisection did not, for Coleridge, require a
subscription to evolutionary theory. Indeed, evolutionary theory removed
humankind from its special God-given position as the defender of morality and the protector of animals.

IV

Darwin's own animal sensibilities oscillated between a deep caring for his
fellow creatures and a far deeper caring for his own intellectual discipline.
He was, for example, appalled at his own experiments on pigeons. "I love
them to the extent that I cannot bear to skin & skeletonise them," he wrote.
"I have done the black deed and murdered an angelic little Fan-tail Pointer
at 10 days old," he added.[87] Appalled as he was, he continued the skinning,
the skeletonizing, and the "murdering." Animals mattered. Knowledge mattered far more.

He wrote a correspondent in 1871: "You ask my opinion on vivisection.
I quite agree that it is justifiable for real investigation on physiology; but
not for mere damnable and detestable curiosity. It is a subject which makes
me sick with horror, so I will not say another word about it, else I should
not sleep tonight."[88] Perhaps not vivisection for "damnable and detestable
curiosity" – which was already illegal, at least by implication if not in practice, under earlier statutes – but animal experimentation for the sake of
pure knowledge, not merely experimentation intended to have medical

benefits, was quite acceptable to Darwin. More conservatively, the majority of those who wrote in favour of animal experimentation favoured its use only when direct medical benefits were seen as its object. Darwin was not so sympathetic to the interests of the animals.

Before the days of the Great Vivisection Debate of the 1870s, Frances Power Cobbe, journalist, redoubtable Kant scholar, Unitarian preacher, and ardent anti-vivisectionist, found Darwin's "gentleness" toward a pony and "fondness" for his dog "very pleasing traits in his character." Yet "Mr. Darwin eventually became the centre of an adoring *clique* of vivisectors who (as his biography shows) plied him incessantly with encouragement to uphold their practice, till the deplorable spectacle was exhibited of a man who would not allow a fly to bite a pony's neck, standing before all Europe as the advocate of vivisection."[89]

If Charles Darwin was a reluctant supporter of vivisection, he was nonetheless an adamant supporter not only of the practice, but of the practice unrestricted and unregulated. To one of his daughters, he wrote:

I have long thought physiology one of the greatest of sciences, sure sooner, or more probably later, greatly to benefit mankind; but, judging from all other sciences, the benefits will accrue only indirectly in the search for abstract truth. It is certain that physiology can progress only by experiments on living animals. Therefore the proposal to limit research to points of which we can now see the bearings in regard to health, etc., I look at as puerile ... I conclude, if (as is likely) some experiments have been tried too often, or anaesthetics have not been used when they should have been, the cure must be in the improvement of humanitarian feelings.[90]

His exasperation at opposition to his views may be said to have encouraged a lack of humanitarian feelings of his own. When a bill to control vivisection was introduced into Parliament and, according to Darwin's finest biographers, "fell into Cobbe's lap ... Darwin vented his spleen in *The Times*, the old patrician targeting those women who 'from the tenderness of their hearts and ... their profound ignorance' oppose all animal experimentation."[91] On two grounds Darwin's diatribe was quite unjustified. Cobbe was certainly not ignorant. Few knew more about the conditions of animal experimentation than did Cobbe. Nor were the primary abolitionists women; rather, they were men of high professional and social status.

In actuality, evolution was far from bringing about an increase in sensibilities, we find Darwin's position in almost exact correspondence with that of the Anglican cleric and Fellow of Trinity College, Cambridge, Thomas Young, who was writing in 1798. In his *An Essay on Humanity to*

Animals, published over seventy years before *The Descent of Man,* Young discusses some of the vivisection practices and concludes: "Upon the whole, that cruelty does actually take place in Anatomical experiments seems indisputable, but in what particular instances I must leave to the judgment and conscience of Anatomists to determine; taking liberty of repeating to them, that every experiment is cruel, without having for its object the leading to some great and public good."[92] Both the Christian Cambridge don and, over seven decades later, the evolutionist deplored the cruelties of vivisection but found the potential benefits to humans more important than the lives of animals. And each, rather than legislating controls, was willing to leave the matter in the hands of the experimenters themselves. In fact, Darwin's view was certainly less enlightened than that expressed by the Anglican reverend Arthur Broome in his formulation, as the first honorary secretary of the Society for the Prevention of Cruelty to Animals, of the *SPCA Founding Statement* in 1824. Therein, the society condemned "the practice of dissecting animals alive, or lacerating, mutilating, and inflicting torture upon them in various modes, to satisfy an unprofitable curiosity."[93] And this founding statement was a great deal less enlightened than the third earl of Carnarvon's presidential address to the society in 1837. Having denounced the practices of the animal experimenters as an affront to Christian sensibilities, he asked his audience: "Will you be able to restrain your indignation then, when you are calmly told that it is better to leave such matters to the *'discretion'* of individuals? In other cases, the law of outraged morals steps in to protect and avenge; but against these cases, offensive to the light, outraging decency, repugnant to generous sympathy and to the Christian faith, the law deals not its thunders."[94]

To be sure, Darwin supported some controls in the end. When alternative bills were laid before Parliament and it was clear that some legislation would succeed, Darwin opted for the very much weaker one. Probably because of Darwin's reputation as a man of great animal sensibilities whose theories revolutionized our attitudes to animals, his role in the legislative development has sometimes been entirely misconstrued. For instance, Lewis G. Regenstein, writing in 1991 and apparently relying on C.W. Hume's 1957 account, states that "Darwin was himself a defender of animals and helped promote England's 1876 Cruelty to Animals Act, which regulated experiments on animals."[95] In fact, not only was Darwin opposed to the strong bill initially proposed, but he was also disturbed by the compromise bill ultimately enacted. By contrast, many of the explicitly Christian animal advocates of the 1870s and '80s would, as we shall see, find Young's view both naive and inconsiderate of the respect to which animals were entitled. They supported the stronger legislation opposed by Darwin, which, even then, was nowhere near as strong as they would have liked. Indeed, it was Darwin and his followers who maintained the traditional Aristotelian, Thomist, and Kantian view that animals were for human use, and it was the Christian opponents of vivisection who opposed this view of animals, at least with regard to the issue of vivisection.

V

Both major and more modest intellectual figures of the nineteenth century contributed the warning salvos to the Great Vivisection Debate that would erupt in the mid-1870s. The influential veterinarian William Youatt was the author of the standard works on the dog, the horse, cattle, sheep, and the pig. He was editor of *The Veterinarian,* veterinary surgeon to the Zoological Society of London, lecturer on veterinary medicine at University College, London, and honorary veterinary surgeon to the Society for the Prevention of Cruelty to Animals. He opened *The Obligation and Extent of Humanity to Brutes* (1839) with a chapter on "The Obligation of Humanity to Animals as Founded upon the Scriptures" and continued with a discussion of duties of Christians to animals, before turning to "On the Dissection of Living Animals." There he recounts the innumerable cruelties of vivisection, argues for strenuous legislative controls, almost forty years before controls would become a reality – while acknowledging the potential medical benefits of animal experimentation itself, which are, he claims, the only legitimate grounds for experimentation – and cites "the recorded opinions of several of the most eminent surgeons against the dissection of living animals."[96] Youatt's concerns about vivisection, and those of the surgeons he cites, are certainly more emphatic than those of Charles Darwin.

In his 1859 inaugural address as president of the Royal College of Veterinary Surgeons, Charles Spooner, professor at the Royal Veterinary College, remarked that "vivisection for physiological exploration may or may not be justifiable, in rare instances, but, if practised, it always ought to be done under some anaesthetic influence; and the doing of it should be avoided by every conscientious physiologist, whenever possible. The [Royal] Society for the Prevention of Cruelty to Animals [RSPCA] must keep its eyes open to check the tendencies to these horrid practices, which, it is to be feared, are budding forth in this country, and bring public opinion and the law of England to bear if necessary, to root them out (loud applause)."[97] In 1867 all ten members of the Royal College of Veterinary Surgeons' Court of Examiners for Scotland signed a formal protest about the practice.

In 1864 the RSPCA offered a prize of £50 for an essay against vivisection in English and 1,000 French francs for an equivalent work in French. In 1866 the editors of *The Veterinarian* reviewed the prize-winning essays very favourably, adding: "We have already expressed our opinion respecting vivisection, and stated our conviction to be that very rarely is it called for, and that the benefits supposed to result from it, as a general rule, are at best dubious ... We feel assured that we carry with us in these remarks the right feelings of our profession, as a body, since paramount in their duties is the manifesting of kindness and humanity to the brute creation."[98]

In an 1866 article entitled "Inhumane Humanity," Charles Dickens, who appears to have taken little interest in Darwinism[99] – and rather more, but not a great, interest in Christianity, despite his having composed delightful prayers for his children – castigated contemporary practices of vivisection, noting in particular how the humanitarian Oliver Goldsmith would be appalled at current usages, were he still alive, and concluded that "Man may be justified – though I doubt it – in torturing the beasts, that he himself may escape pain; but he certainly has no right to gratify an idle and purposeless curiosity through the practice of cruelty."[100]

Charles Dodgson (i.e., Lewis Carroll of *Alice in Wonderland* fame), Anglican deacon and Oxford logician and mathematician, who was sympathetic, if less than wholeheartedly, to evolutionary theory,[101] concluded his 1875 article on "Some Popular Fallacies about Vivisection" with the claim: "We have now, I think, seen good reason to suspect that the principle of selfishness lies at the root of this accursed practice ... creating a new and more hideous Frankenstein – a soulless being to whom science shall be all in all."[102] Later, the novelist Wilkie Collins, famed author of *The Woman in White* and *The Moonstone,* referred to vivisection in his *Heart and Science* (1883), which he wrote to plead "the cause of the harmless and affectionate beings of God's creation," as "this infernal cruelty." He denounced the motives of the experimenters as "All for Knowledge! all for Knowledge!"[103] Generally speaking, for those who opposed animal experimentation, animals were more important than knowledge gained at the animals' expense; for those who supported animal experimentation, knowledge was the supreme goal, or at least so their accusers claimed – and not without a measure of justification. After all, Darwin confessed to it.

Faced with the reality of a far greater Victorian doubt about the legitimacy of a justification for vivisection, modern commentators have suggested frequently that vivisection became more acceptable later as a consequence of the availability of anaesthesia to reduce pain and suffering. In fact, anaesthesia was readily available at the time of the discussions regarding the passage of the 1876 Act. Indeed, the bill that was finally rejected would have required anaesthesia in all vivisection. The reality is that many of the most prominent Victorians opposed vivisection quite simply on the grounds that it was unjustifiable to use animals in experimentation for human ends. The use of anaesthesia would not eliminate the crime.

While it would be appropriate to quote from such radical animal advocates as Anna Kingsford, Edward Maitland, Annie Besant, Louise Lind-af-Hageby, and Leisa Schartau to demonstrate the depths of the Victorian and Edwardian antipathy to animal experimentation which was not predicated on evolutionary ideas, I shall restrict my evidence to those whom one might count among the elite not because their views are more important but in order to avoid any suspicion that I am taking my evidence from the less prominent and less culturally central figures in society.

With increasing Victorian distaste for what they saw as the barbarism of vivisection, Frances Power Cobbe, former reluctant vivisector George Hoggan, and editor of the *Spectator* Richard Hutton joined forces to found an anti-vivisection society in November 1875. Initially named the Victoria Street Society – from its London location – it later became the National Anti-Vivisection Society. Its immediate membership included the Anglican archbishop Thomson of York; the Roman Catholic cardinal Henry Ernest Manning; John Coleridge (by 1880 Lord Chief Justice Coleridge); Alfred, Lord Tennyson, the poet laureate since 1850; John Ruskin, Slade Professor of Art at Oxford University; and prominent literary figures Robert Browning and Christina Rossetti. Its first president was the most prominent social reformer of the nineteenth century: the seventh earl of Shaftesbury. Each of these original members opposed animal experimentation on the basis of what they regarded as explicitly Christian principles, including Cobbe, who, as a Christian Unitarian, was appalled at the vivisector, "the smooth cool man of science ... [who] stands by that torture trough."[104] Not all were opposed to evolution. Tennyson, for example, a vice president of the society, developed a pre-natural-selection Lamarckian-style theory of evolution in *In Memoriam* (1850) and in *Maud* (1855). And his Swedenborgian Christianity was a little less than orthodox. Nonetheless, for all of them, admittedly more emphatically in some cases than in others, it was Christianity, not evolution, that was at the basis of their animal sensibilities and their opposition to vivisection. Indeed, the subscription to explicitly Christian justifications in both parliamentary debate and secular journals rings strange to modern ears.

In an 1882 article in the *Fortnightly Review*, Lord Chief Justice Coleridge condemned the practice of vivisection and indicated where his upper-chamber parliamentary vote would lie on any future occasion: "What would our Lord have said, what looks would He have bent, upon a [laboratory] chamber filled with 'the unoffending creatures which He loves,' dying under torture deliberately and intentionally inflicted, or kept alive to endure further torment in pursuit of knowledge? ... the mind of Christ must be the guide of life. 'Shouldst thou not have had compassion upon these, even as I had pity on thee?' So He seems to say to me; and I shall act accordingly."[105] Coleridge made it abundantly clear that total abolition of vivisection was his preference – he could never trust the intentions and machinations of the vivisectors – but he was prepared to support legislation for strict controls, provided that they were far less lenient than those contained in the 1876 Act.

Lord Shaftesbury spoke eloquently in the House of Lords in favour of strict controls in 1876, suggesting, inter alia, as we have noted,[106] that animals not only have immortal souls, but are far more deserving of them than some humans. Other species are as much God's creatures as we are, he opined, adding that "no physical pain can possibly equal the injury caused by the moral degradations of the feelings which such barbarous experiments must naturally induce."[107] The fourth earl of Carnarvon, who was in charge of the "strong" 1876 Act, was called away from Parliament by his wife's sudden illness, and subsequent death, immediately before the bill was to be presented. It included a prohibition on experimentation on horses, dogs, and cats, and as we have seen, would have required complete anaesthesia throughout all invasive experiments on animals. In the weeks before the bill could reappear, a concerted effort by the medical profession, supported by Darwin, persuaded the home secretary to weaken it. The compromise bill as practised did not meet even the limited expectations held by Lord Shaftesbury. Consequently, in 1879 he joined with Lord Truro to introduce a bill for the total abolition of vivisection. Even though it stood very little chance of enactment, it was given far more serious consideration than any similar act would be given today. Even though she played no role in the Victoria Street Society, Florence Bramwell Booth, the Salvation Army leader, was a fellow contemporary abolitionist who coupled her anti-vivisection views with vegetarianism, abstinence from alcohol, and the campaign against cruel sports, along with her ministries to the poor of the East End of London. Christian opposition to vivisection stretched from the representatives of the most traditional churches to those of the most evangelical. Christianity, especially but far from solely evangelical Christianity, stood far more on the side of the animals than did Darwin and his allies.

The idea of the inevitability of progress has become so much a part of the Western ethos that we take it for granted that the culture, laws, and values of contemporary society are in some manner ahead of the past. This is certainly not necessarily so. An instructive example may be found in the words of the 1979 Advisory Committee Report to the British Home Office, treated at the time as a significant step in positive attitudes toward the restriction of animal experimentation: "Infliction of pain on animals, then, amounts to cruelty when the pain is not compensated by the consequential good. The human good envisaged must be a serious and necessary good, not a frivolous or dispensable one, if the infliction of pain on animals is to be ethically acceptable."[108] The Reverend William Kirby, dubbed "the father of entomology," had cut to the quick of the matter already in 1815: "Every degree of unnecessary pain becomes cruelty."[109] And so felt Truro, Shaftesbury, Booth, and Lord Chief Justice Coleridge. On the basis of their words, we can see that they were far in advance of the 1979 report.

The celebrated social reformer and author of *Modern Painters* (1843-47), *The Seven Lamps of Architecture* (1849), and *The Stones of Venice* (1851-53), John Ruskin, followed the bishop of Oxford as speaker at a meeting of the Oxford branch of the Victoria Street Society in 1884, declaring that "it is not the question whether animals have a right to this or that in the inferiority they are placed in to mankind, it is a question of what relation they have to God, and what is the true sense of feeling as taught to them by Christ the Physician ... These scientific pursuits are now, defiantly, provokingly, insultingly separated from the science of religion; they are all carried on in defiance of what has hitherto been held to be compassion and pity, and to the great link which binds together the whole creation from its Maker to the lowest creatures."[110] Ruskin was appalled when funds were voted by the Oxford Senate in March 1885 to allow for a physiology laboratory in which vivisection would be practised. After a few days of soul searching, he resigned his Chair, as he wrote to the *Pall Mall Gazette* (24 April 1885), "on the Monday following the vote endowing vivisection in the University, solely as a consequence of that vote."

The devout Robert Browning, like Tennyson a vice president of the Victoria Street Society, wrote two anti-vivisection poems, "Tray" (1879) and "Arcades Ambo" (1889).[111] In the first, a dog saves a child's life and is rewarded with the prospect of being vivisected so that his extraordinary behaviour may be understood. In the second, Browning pronounces vivisection both cowardly and haphazard. In 1875 he wrote of vivisection that he "would rather submit to the worst of deaths, than have a single dog or cat tortured on the pretence of saving me a twinge or two."[112] Christina Rossetti, an author of primarily religious verse, distributed pamphlets for the Victoria Street Society and provided a dozen autographed copies of a specially composed poem for an anti-vivisection fundraising bazaar.[113] In "To what purpose is this waste?" (1872), she expressed her egalitarian animal sensibilities: "The tiniest living thing / That soars on feathered wing, / Or crawls among the long grass out of sight / Has just as good a right / To its appointed portion of delight / As any King."[114] Less poetically, but no less pointedly, Queen Victoria "advised" and "warned" her ministers several times – the height of her constitutional rights – on the topic of vivisection. For example, she had her secretary write to the new evangelical prime minister, William Ewart Gladstone, in April 1881 that "the Queen has seen with pleasure that Mr. Gladstone takes an interest in the dreadful subject of vivisection, in which she has done all she could, and she earnestly hopes that Mr. Gladstone will take an opportunity of speaking strongly against a practice which is a disgrace to humanity and Christianity."[115]

Cardinal Henry Edward Manning said in 1882 that "vivisection is a detestable practice ... Nothing can justify, no claim of science, no conjectural result, no hope for discovery, such horrors as these. Also, it must be remembered that whereas these torments, refined and indescribable, are certain, the result is altogether conjectural – everything about the result is uncertain, but the certain infraction of the first laws of mercy and humanity."[116] The head of the Roman Catholic Church in England opposed precisely the grounds of Darwin's support for vivisection on the basis of their "infraction of the first laws of mercy and humanity." In a speech to the Victoria Street Society in 1887, he added: "Our moral obligation and moral duty is to Him who made [the animals], and, if we wish to know the limit and broad outline of our obligation, I say at once it is His nature and His perfections, and among those perfections, one is most profoundly that of eternal mercy ... [animals] should be used in conformity to His own perfections, which is His own law, and therefore, our law. It would seem to me that the practice of vivisection, as it is now known and now exists, is at variance with those moral perfections."[117] The context of "now exists" is the 1876 Act to control vivisection and its consequences – an act that was too strong for Darwin and too weak for the head of the Roman Catholic Church in England.

Writing around the same time in the United States, the celebrated theologian, author, and pastor of Tabernacle Church in Brooklyn, Thomas de Witt Talmage, questioned:

> Have you ever thought that Christ came among other things to alleviate the sufferings of the animal creation? ... Not a kennel in all the centuries, not a robbed bird's nest, not a worn-out horse on the tow-path, not a herd freezing in the poorly-built cow pen, not a freight car bringing the beeves to market without water through a thousand miles of agony, not a surgeon's room witnessing the struggles of the fox or rabbit or pigeon or dog in the horrors of vivisection, but has an interest in the fact that Christ was born in

a stable surrounded by animals. He remembers that night, and the prayer He heard in their pitiful moan He will answer in the punishment of those who maltreat them.[118]

Talmage was certainly not known primarily for his animal sympathies. But there can be no doubt that he had them and that his readers and congregation appreciated them. A few years later, the devout physician Elijah Buckner remarked of the vivisectors that they "have caused every conceivable manner of torment that the devilish nature of a man could invent ... Can [the vivisector] locate, in that innocent, bloody piece of flesh before him, in the form of a dog, where that true and everlasting love it has for its master is to be found?"[119] Writing of those who, like Darwin, opposed strenuous restrictions on vivisection, he claimed that they become "so hardened by selfishness that if the life of a lower animal ... stands in his supposed way to notoriety he regards them as mere things to be destroyed ... It seems a great misfortune ... that a few inhumane physiologists should insist on shocking, not only the Christian world, but the heathen world, by using the most sickening tortures ... on the living bodies of sensitive beings."[120] In fact, Buckner wanted to prohibit vivisection in its entirety and called upon his fellow Christians to bring the matter to an end:

> The efforts which are being made to-day by many of the best people in the civilized world to prohibit vivisection should have the support of all Christian people of all nations, and I believe the cause does have their support. Right here is where the line should be drawn; if a man or woman refuses to help support this great humane cause, he or she cannot possibly be a Christian according to the teachings of the Bible, which says, "Open thy mouth for the dumb ... be merciful ... do justly, love mercy, and walk humbly ... a righteous man regardeth the life of his beast ... be ye harmless as doves ... he shall be judged without mercy who shows no mercy."[121]

Basil Wilberforce, Anglican archdeacon of Westminster Abbey, who refused to speak on behalf of the RSPCA because its members were too weak on the vivisection question, observed in a 1909 sermon at the abbey: "I believe that no greater cruelty is perpetuated on this earth than that which is committed in the name of science in some physiological laboratories ... The cause which we are championing is no fanatical protest based on ignorant sentimentality, but a claim of simple justice not only on the transcendent truths of the immanence of the divine truth in all that lives, but also upon the irrefutable logic of ascertained fact."[122] Contrary to Darwin, it was not "the tenderness of their hearts ... and their profound

ignorance" that, for Wilberforce, lay behind the claims of the abolitionists
but rather, the archdeacon insisted, "a claim of simple justice" based "upon
the irrefutable logic of ascertained fact." I can find no evidence of anyone
joining the "cause which we are championing" on the basis of their commitment to evolutionary principles, although Thomas Hardy came quite
close to it in the same year that Wilberforce was giving it a Christian as
well as a secular basis.[123] But there are very significant numbers who found
the basis for their opposition to vivisection to lie in firmly held Christian
conviction.

Not only does the evidence offered here suggest the depth of revulsion
in which vivisection was held by prominent Christians, but it also indicates that the animal sensibilities of such authorities, whether right or
wrong on the practical issue of vivisection, were far stronger than those of
the most celebrated proponents of evolution by natural selection – at least
on this issue, the most prominent animal-ethics issue of the time. None
of this should indicate to us that all those who conceived of themselves
first and foremost as Darwinians were of one mind nor that those for whom
adherence to Christianity was a primary self-identification were of another.
Indeed, the codiscoverer of evolution by natural selection, Alfred Russel
Wallace, was as adamant an anti-vivisectionist as Lord Chief Justice Coleridge or Lord Shaftesbury, although he played no role in the great vivisection debate itself. Still, he wrote a correspondent, "nothing but total
abolition will meet the case of vivisection."[124] In his 1911 book, *The World
of Life,* Wallace waxed eloquent on the subject. But he opposed animal
experimentation not for the sake of the animals but because of the depravity involved in the act of experimentation itself. It was the researchers who
were degraded, not the animals.

The customary tale of how Darwin's theory of evolution occasioned the
most fundamental revolution in animal ethics needs to be rethought and
retold. From a moral perspective, Darwinism added nothing that had not
long been proclaimed. From a practical perspective, a number of the more
prominent Darwinians – Darwin himself, Huxley, and Wallace, for a start
– were far less sympathetic to the animal cause than many of the more
prominent Christians. Indeed, in the later nineteenth century, Christianity was far more significant in bringing about a revolution in attitudes to
animals than was evolution.

The reality is that the history of human attitudes is far more complex,
indeed convoluted, than we are customarily led to believe. Much of our
contemporary analysis reads more like prejudicial ideology than history.
The prevailing premises of the history of animal ethics require a thorough
reinvestigation – and one in the scientific spirit if not according to the

scientific method. Thucydides claimed that history is philosophy leading by examples. But if we have our history wrong, we will learn from the wrong examples. If we have our history wrong – and we do – our collective social consciousness will err about who we are to the extent that who we are depends on knowing aright who we were.

9

The Moral Status of Animals

PRACTICAL JUDGMENT, REASONABLE
PARTIALITY, AND SPECIES NEEDS

I

THE THEME OF THIS BOOK HAS BEEN that there is no orthodoxy in the history of animal ethics, or even more emphatically, that
those ideas treated as orthodox by a large proportion of contemporary commentators are decidedly not so. They have played a prominent
role but also have been continually questioned and opposed. The history
of animal ethics is, in fact, no less complex than the history of human culture. Many of those who have discussed the history of animal ethics have
approached it prejudicially, and many more who have lacked the prejudice
against our forbears have been persuaded by, or at least have followed
uncritically, those whose minds were bent on scoring points rather than
on uncovering the complex dimensions of a historical reality.

We have seen in this book that, contrary to the conventional view, many
historical figures of commonly acknowledged repute who have ventured
into print on the subject of animal souls have declared them immortal –
and that several more who have declared them otherwise have deemed this
very fact a reason for treating animals with greater moral consideration in
the here and now. Moreover, some who have been severely chastised for
proclaiming the virtue of the human domination of nature have considered it a prerequisite for the benevolent treatment of animals rather than
a means to their oppression. Others have declared loudly that animals are
ends in themselves and have denounced those who would regard animals
as solely intended for human use. Further, the much vaunted claim that
increased sensibility to animals was stimulated by Charles Darwin's theory
of evolution does not stand up to careful scrutiny. The sensibility has
existed in perpetuity, and to the extent that it became more pronounced
in the Darwinian age, its consequence was anything but Darwinian. In
fact, those whose source of inspiration was quite other than Darwinian

displayed a far greater sensibility to animals, at least on the issue of animal experimentation. Moreover, some throughout history were convinced that there was something superior about the nature of animals and that, at least in some respects, their lives and morals were decidedly preferable. Perhaps most who employed the paradigm used it primarily to ridicule humanity's pretensions. However, there was also an element of sincerity in the message. In short, the reality of the history of animal ethics is astoundingly different from the customary tales we are told. The dominant version of the history of attitudes to animals has been written as ideological myth rather than as history.

II

How, then, does the *history* of animal ethics relate to the *philosophy* of animal ethics? Many have treated the whole history as the tragedy that current ethical theory and consequent practice must overcome, and I would certainly concur that much needs to be done to raise the status and treatment of animals. But, I would suggest, very often contemporary philosophizing has not helped us very far in approaching the solutions because it has, by and large, remained in the realm of the ahistorical and nonethological abstract.[1] History, cognitive ethology, and animal-welfare science are often ignored or, when used, often misinterpreted in the customary debate. There is in fact no Platonic Form of the Just waiting to be plucked from the skies and applied by the philosopher king. Or, at least, none has been found that satisfies the task. Ethics, as Aristotle wisely understood, is one of the practical sciences, which we best approach by first discovering, as a guide to our investigation, what it is that people believe rather than by dismissing popular opinion, à la Plato, as beneath the province of the lovers of wisdom. Common opinion of justice is not justice itself but contains the seeds of justice, which logical investigation will help us to clarify. Despite Plato's dismissal of public opinion, he too realized that teaching ethics meant reminding people of what was already implicit in their unexpressed views. And, as Kant recognized aright, a good part of this ethic is consistency – applying the same rules in the same way to all relevant beings (even if, in the long run, he failed to recognize which beings were relevant). To effect telling change one must build on what is already there. The elusive, but not fictional, "common man" – and woman – is less likely to be controlled by untenable but seemingly convincing abstract theories. Hence their views offer an appropriate starting point. An analysis of the history of attitudes to animals is, then, a prerequisite for understanding animal ethics well. If we are to determine the ideal, our path must be through the historically real as witnessed through the development of character and culture – the attitudes, emotions, and beliefs pertaining to animals.

Moreover, in determining what is due to animals, we need to know the particular natures of particular animals. If we have to know the needs and wants of humans in order to decide what it is to which they are entitled, then so too must we know the needs and wants of other species to know what it is to which they are entitled. If, as Bernard Rollin argues, we decide the importance of freedom of speech and of assembly by looking to human nature, so must we acknowledge that "animals, too, have natures – the pigness of the pig, the cowness of the cow ... which are as essential to their well-being as speech and assembly are to us."[2] Indeed, historically, the prevailing cultural view has been that animals need to be distinguished according to their species characteristics in terms both of their needs and of their capabilities. As Aristotle argued in the *Nicomachean Ethics:* "Now if what is healthy or good is different for people and for fish, but what is white or straight is always the same, everyone would say that what is [philosophically] wise is the same, while what is practically wise is different ... That is why people say that certain animals are also practically wise, namely, those that appear capable of forethought about their own lives."[3]

There are different goods for different species, and those that display practical wisdom have different needs from those that lack this capacity. Each species must be understood on its own terms. Indeed, it follows from such a principle that it is speciesist to talk of animal rights – or, indeed, of animal wellbeing – in general and in the abstract. When we talk of human rights, we refer to those rights appropriate to the human species. To talk, then, of animal rights is to treat humans according to their species characteristics while lumping all other species together in a generic category. To discuss rights in such terms is already to treat the human as a special case. On the other hand, to talk separately of the needs or rights or wellbeing of the human, the dolphin, the giraffe, and the assassin bug is to diminish the effect of this easy error. It is the specific character of their species being that is at issue, not just their animal being in general. Even then, one has to wonder just how helpful the customary language of abstract philosophy is in addressing the thorny issues of animal ethics. The philosophical approach may be of great value for understanding the moral relations among human beings, for which it was developed. But when it is extended beyond the human realm, it serves to confuse as much as to enlighten. We expect far more from it than it is capable of delivering. Or at least much of what has been offered raises more questions than it ever settles.

Much of the study of animal ethics that is devoted to the interest of animals begs the question of the relation of rights to responsibilities or mistakenly assumes that one cannot have any rights unless one has some responsibilities. Thus it also raises the question of whether the tiger, who might be said to have a right not to be harmed by a human, has a corresponding obligation to the human – and, if not, whether this special responsibility of the human being ascribes a separate moral standing to the human. With regard to humans, we deem the intent or character of the actor relevant to the moral approbation or disapprobation of his or her conduct. Did he harm his neighbour wilfully or inadvertently? Yet we would find such distinctions only rarely of significance in discussing animal behaviour – and then only in discussing that of domesticated species or perhaps of the great apes, in which case, at least with respect to the former, then perhaps quite inappropriately. We do not normally regard other species, with the possible exception of companion animals and apes, maybe whales and dolphins, as possessing a moral sense extending beyond that sense associated with their primordial evolutionary nature and at most toward their own immediate relatives. If we owe a moral responsibility to other species, and they none to us, then it is inappropriate to discuss the human-animal moral relationship in the language of rights, as such language requires a certain equality in principle and – in the case of animal rights, I would suggest, if it is to be logical – a certain species reciprocity. At least in some respects, it is inappropriate to include animals in the moral community, for community appears to involve at least the potential for reciprocity. But excluding animals from the moral community *in this sense* does not for a moment deny that we have a moral commitment to the maintenance of many of their interests. Nor does it deny that we belong to a moral community with them in the sense of sharing much, though far from all, of our very natures in common.

Humans are expected to resist some of their urges if they are to be moral. We neither require nor expect this of other species – again perhaps companion animals excluded as well as, perhaps somewhat more justifiably, the great apes and maybe the megafauna of the sea. But even here we do not expect it in anything like the manner or degree to which we expect it of humans. To the extent that morality is concerned with restrictions on appetites, admittedly only a part of its province, it does not, in general, pertain to nonhuman animals, for we rarely consider them capable of restricting their appetites. As Aristotle explains:

> Since one can commit injustice without yet being an unjust person, what
> sort of unjust acts make the person who commits them, such as a thief, an

adulterer, or a pirate, unjust in each type of injustice? Or is the quality of the act irrelevant? For a person might have sex with a woman knowing who she was, but through feeling rather than the first principle of rational choice. So he commits injustice, but he is not unjust; a person is not a thief, for example, though he stole, or an adulterer, though he committed adultery, and so on. But a person who acts like this from rational choice is unjust and wicked ... Similarly, a person is just when he acts justly by rational choice, but acts justly if he merely acts voluntarily.[4]

Now one might wish to suggest that Aristotle's categories of justice are inappropriate. According to Ovid, Cinyras's daughter Myrrha questioned whether extant sexual mores were unnatural and hence unjust, implying that to follow nature would be more just than to rein in her appetites.[5] And in the Enlightenment, Pierre Proudhon proclaimed all property theft, by which standard Aristotle's thief would be merely one who recaptured an original entitlement. Perhaps it would be argued that the adulterers who failed to control their mutual lust were unjust persons but less unjust than those who decided consciously to commit adultery – that is, who acted out of "the first principle of rational choice." To make a clear distinction between unjust persons and persons who commit unjust acts might be thought too fine a distinction – even though it is a customary consideration in court cases, when, in sentencing, past conduct is considered and when descriptions of deleterious social and other conditions are offered as explanations of how an act may have been a consequence of something other than rational choice or otherwise out of character. Yet these are the appropriate minutiae of ethical deliberation in human matters. What should be clear is that such issues have very little, if any, bearing on any discussion of animal ethics.

If they did, we would revert to the medieval practice of the beast trials. When we read of these trials we deem them unjust precisely because it is inappropriate to expect nonhuman animals to have responsibility to humans or to live by human standards of law. While complex animals act voluntarily and from feeling, no one would expect animals to act from the "first principle of rational choice." We *might* say that they act justly or unjustly but not that they are just or unjust beings acting from rational choice. Animal morality, if it exists at all, as some would claim, is clearly not of this order. For animals, moral behaviour, to the extent that it may be so described, is a matter of acting within the requirements of, not against the natural impulses of, their species being. For humans, morality frequently consists in resisting the temptations of their passions in favour of a consideration for others, whose interests our passions are inclined to neglect.

As Aristotle explains, "the self-controlled person, knowing that his appetites are bad, because of reason does not follow them."[6] With the *possible* exception of companion animals, great apes, and the giants of the seas, and then often inappropriately, we do not imagine that other species control their appetites or have responsibilities to members of species other than their own. Indeed, this is precisely why we customarily object to claims that an animal acted in a manner "unjust and wicked" when it was merely obeying its natural proclivities. "We do not call animals temperate or intemperate,"[7] as Aristotle says. When a newly dominant lion kills the infant offspring sired by a previously dominant lion, we explain this behaviour by talking of the lion's instincts, which are seen to instruct it to act from principles of natural selection. No similar behaviour would be tolerated on the part of a human. If the human male is driven by lustful urges detrimental to other individual members of the species in order to act in a manner consistent with what may be regarded as the dictates of natural selection, we consider him moral precisely to the extent that he resists these urges. We make no such requirements of the lion. Human behaviour becomes just and benevolent precisely to the extent that it is prompted by motives different from those seen in other species. One should be reminded of Mark Twain's admonition that the human being is the inferior animal precisely to the extent that, unlike other species, it has a moral sense, whose admonitions it frequently ignores.[8] And, in the same way, Aristotle deems humans the highest of animals, *except* when they act from motives other than those consistent with a morality that differentiates them from the other animals. When they do so act, for Aristotle, humans become the lowest of animals in the sense that nonhuman animals are always amoral, while humans may be moral or immoral.[9] Humans can be the highest or the lowest according to their adherence to the dictates of rational moral choice. The animals have no place in the ladder of morality, for there is no action they can perform that is deemed to break the moral rules.

The question of animal morality raises the issue of whether the wild goat has any rights with respect to the wolf, or the fox any responsibilities to poultry. And if the answer is in the negative, as it presumably must be, then the language of rights is of decidedly limited application in animal ethics, which is not to say that it is entirely useless, unless animal ethics is concerned solely with the human responsibility to animals. At the very least, a negative answer indicates that there is here no human-animal reciprocity. Some philosophers, such as Kant and Whewell,[10] and those who followed them, have argued that, *accordingly*, animals do not possess rights and that therefore we owe them no responsibilities in and for themselves. Yet their claim is only of significance if the animals' *rights* are what are at issue. Just because they may be deemed not to have rights, philosophically speaking, does not mean that they do not possess what are often called rights. The fact that animals have legitimate needs, distinguished according to their species characteristics, and are commonly recognized to possess such needs, is *prima facie* sufficient for their interests to have earned consideration. They possess not the universal rights that are claimed for *and by* humans but interests as the beings that they are with the needs, wants, and purposes that they have.

To deny that the philosophical language of animal rights is entirely persuasive is not to claim that all the considerations that the animal-rights advocates proclaim for animals are necessarily inappropriate. At issue, rather, is that the philosophical language that they use to justify these claims is not very helpful and, indeed, allows the opponents of animal interests to use this philosophical inadequacy to deny animals some of the moral considerations that they deserve. Indeed, the opponents of animal rights are able to go further in claiming that the possession of rights depends upon the ability to claim them, which animals do not, and cannot, possess. But this raises the thorny issue of "marginal cases," an issue that is philosophically important – potentially leading us toward answering the question of how one may apply rights to animals without restricting these rights to those who can claim them for themselves – but an issue that may nevertheless lead us away from our understanding of the relationship of history to ethics. Our task is to consider what alternatives may be offered if the *abstract language* of rights is not very helpful in animal ethics and if the concern for pain and suffering alone are insufficient.

III

I have intimated that the quest might be best undertaken by looking to the specific needs of different species. Unfortunately, some of the most persuasive arguments in the history of ethics dissuade us from taking the "naturalistic" course of inquiring into the specific needs of individual species or, indeed, into the needs of animals in general – and no less so the needs of humans – in determining ethical questions. Famously, the Scottish utilitarian David Hume observed in *A Treatise of Human Nature* (1739-40) that, in discussions of ethics, authors commonly make

> observations concerning human affairs; when of a sudden I am surpriz'd to find, that instead of the usual copulations of propositions, *is,* and *is not,* I meet with no proposition that is not connected with an *ought,* or an *ought*

not. The change is imperceptible; but is however of the last consequence. For as the *ought,* or *ought not,* expresses some new relation or affirmation, 'tis necessary that it should be observ'd and explain'd; and at the same time that a reason should be given, for what seems altogether inconceivable, how this new relation can be a deduction from others which are entirely different from it.[11]

In Hume's view, you cannot get there from here, logically speaking. No *ought* can be derived from an *is.*

Of almost equal fame is G.E. Moore's pronouncement in *Principia Ethica* (1903) of what he called "the naturalistic fallacy" – although, surprisingly, without mentioning David Hume – whereby he argued that the good could not be *defined* in natural terms, that it was, in essence, like yellow, which may be recognized and may be explained in terms of light vibrations but cannot itself be defined. "Good," then, he argued, "if we mean by it that quality which we assert to belong to a thing, when we say that thing is good, is incapable of any definition, in the most important sense of that word. The most important sense of 'definition' is that in which a definition states what are the parts which invariably compose a certain whole; and in this sense 'good' has no definition because it is simple and has no parts."[12] The good is simple, real, and knowable but not definable. It cannot be understood in and of itself in terms of any attributes that it may be said to possess. In short, from the perspective of animal ethics, an understanding of the "good" for the human, the giraffe, the gorilla, or the guinea pig cannot be derived from an understanding of the human's, the giraffe's, the gorilla's, or the guinea pig's species characteristics.

Despite the *philosophical* wisdom of such subtleties, commentators on animal ethics over the centuries – and on human ethics, too, for that matter – have often ignored these niceties. For instance, in his remarkable study of the history of attitudes to animals to the end of the eighteenth century, *Love for Animals and How It Developed in Great Britain* (1928), Dix Harwood observed that "as human opinion wavered from century to century, two distinct attitudes toward beasts have stood out prominently. Sometimes men have held the anthropomorphic view that animals and men are very much alike with the same emotions and similar mental powers ... At other times men have held stubbornly to the anthropocentric opinion that it is a man's world and that an unbridgeable chasm yawns between the human race and other species."[13] Those who were persuaded of the "anthropomorphic" view proclaimed our obligations to other animals, whereas the "anthropocentrists" were more inclined to deny them. For centuries, people have *in fact* decided that ethics is subject to being understood in naturalistic terms and that the relationship between human and animal ethics may be understood in large part by the degree to which humans and animals are alike or are different – and although it is usually subliminal, also by the degree to which the pig shares the characteristics of the dog, the bee the characteristics of the polar bear. Whether humans and animals have "the same emotions and similar mental powers," including the capacity for pain and suffering, has been a customary criterion of whether or not humans owe other species a moral responsibility.

On the basis of the pronouncements of Hume and Moore, one might question the legitimacy of the customary interpretations, whether from the "anthropomorphic" or "anthropocentric" perspective. Yet when we read Hume's own analysis of the human-animal relationship, we find him taking an essentially similar, if apparently incongruous, path – one decidedly in opposition to his philosophical pronouncements about the impossibility of deriving an *ought* from an *is*. In *A Treatise of Human Nature,* he has chapters on "Of the Reason of Animals," "Of the Pride of Animals," and "Of the Love and Hatred of Animals," in which he likens the attributes of other animals to those of humans. No truth, he tells us,

> appears to be more evident, than that beasts are endow'd with thought and reason as well as men ... 'Tis from the resemblance of the external action of animals to those we ourselves perform, that we judge their internal likewise to resemble ours ...
>
> Every thing is conducted by springs and principles, which are not peculiar to man, or any one species of animals ... Love in animals, has not only for its object animals of the same species, but extends itself farther ... 'Tis evident that *sympathy*, or the communication of passions takes place among the animals, no less than among men.[14]

And there is much more in this vein. From such evidence of what Hume sees as the relevant human-animal similarities, he draws the relevant – but, from Hume's own perspective, logically inappropriate – moral conclusion that "we should be bound by the laws of humanity to give gentle usage to these creatures."[15] The attributes shared alike by human and nonhuman animals are seen to have an immediate bearing on the human-animal ethical relationship. Whatever the soundness of Hume's philosophical insight that moral propositions may not be deduced logically from factual information, when the pudding comes to the cooking, *perceived facts* are its ingredients.

How can this be reconciled with the view that there is a necessary disjuncture between the *is* and the *ought?* Hume never provides a convincing solution to the dilemma despite a few sensible comments about experience and probabilities. To be sure, he tells us that moral judgments are a matter of feeling rather than of rational conviction,[16] but this observation evades, rather than solves, the problem, for what are feelings if not facts of nature? The solution was left to Charles Taylor – some might suggest that Aristotle arrived there earlier, with his concept of *eudaemonia* – writing over two centuries later than Hume. Taylor concurs that "we can never say that 'good' *means* 'conducive to human happiness', as Moore saw." Nonetheless, he adds convincingly, "that something is conducive to human happiness, or in general to the fulfilment of human needs, wants, and purposes, is a *prima facie* reason for calling it good, which stands unless countered." The use of the term, he remarks, "is unintelligible outside of any relationship to wants, needs, and purposes."[17] Determining what is good for humans thus involves knowing what it is that fulfills humans as the beings that they are with the needs, wants, and purposes that they have.

In light of the historically successful demonstration of human-animal continuity, not least through the researches of Borelli, Goethe, and the Darwins, it may be said to follow prima facie that to understand what is good for other species is also to understand what fulfills them as the beings that they are with the needs, wants, and purposes that they have. Thus that ethical and factual claims have historically gone hand in hand is both understandable and appropriate. The similarities and/or differences between humans and other species are relevant to the human-animal ethical relationship. Nonetheless, the comparisons have often been marred in that, on the whole, the similarities and/or differences between humans and animals have been stressed, when, in fact, the similarities between humans and some species are significantly greater than the similarities between other species – a fact that renders ethical decision making rather more complex. As Aristotle wrote, with his customary cautious wisdom,

> if we are to say that the science that concerns our own particular advantage is wisdom, there will be many wisdoms; for there is not a single science concerned with the good of all creatures, but each kind of good has a different science – just as there is not a single science of medicine for all beings.
>
> It makes no difference if it is claimed that a human being is superior to all the other animals. For there are other things far more divine in nature than human beings, such as – to take the most obvious example – the things constituting the cosmos.[18]

A further hindrance to our understanding of the human-animal relationship and its ethical implications lies in the fact that the perceived relevant criteria have not remained constant over the centuries. Indeed, they have varied from reason to kinship, a moral sense, sociality, a common soul, self-consciousness, sentience, and will. Nonetheless, whether humans and other species are seen to possess similar relevant attributes – usually some form of what are seen as mental attributes – has historically been the determining factor of whether other species are entitled to moral consideration as the beings that they are with the needs, wants, and purposes that they have.

As we noted in the first chapter, however, there are instances in which no appropriate comparisons can be made, such as when, for example, in the capacities for echolocation and flight, the animal possesses abilities beyond the reach of humans. The trouble with the introduction of ideas such as this is that it makes discussion of an animal ethic far more difficult. There are no meaningful terms in which the discussion may take place insofar as there are no possibilities, or at least no reasonably easy possibilities, of appropriate and meaningful human-animal comparison with regard to the fulfillment of needs. Still, it can be said that whatever might impede echolocation for bats and flight for birds is a hindrance to their fulfillment as the beings that they are with the needs that they have and that, consequently, whatever may hinder these functions is prima facie unethical – at least until all the factors respecting the fulfillment of the competing needs of other animals are taken into consideration. An equally relevant question is whether species other than *Homo sapiens* have any responsibility to assist other species in reaching their goals.

IV

When Jeremy Bentham declared that the relevant question with regard to whether animals are entitled to moral consideration revolved around whether they could suffer rather than whether they could reason or speak – undoubtedly, the best-known proposition of any philosopher among those interested in animal ethics – he offered no evidence or explanation. He relied on the likelihood of humans searching their ethical sensibilities and recognizing that he was right. And, indeed, when we look inside ourselves, we all seem to concur that there is at least some truth in the proposition. Bentham has raised to the level of consciousness, and given linguistic form to, an important element of the human moral sense. We all seem to understand that suffering is a relevant criterion. However, many will conclude, important as Bentham's statement is, that it is insufficient. Many will say that animate life itself matters. For William Blake, referring to animated nature, "every thing that lives is holy."[19] And as Albert Schweitzer wrote, "this early influence upon me of the commandment not to kill or torture other creatures is the great experience of my childhood and youth. By the side of that all others are insignificant."[20] For Schweitzer, while the infliction of pain and suffering is of great significance, life itself also deserves to be maintained for no other reason than that it is animate life. Indeed, life itself is needed if the animal is to satisfy the needs that it has as the being that it is.

If it is *only* suffering that is at issue – and if we can kill an animal instantaneously without its being aware of its plight – then the killing of an animal, or for that matter a human, is not even a misdemeanour. Those who think that such killing is indeed a moral felony will recognize that Bentham's bald statement can be used to oppose animal interests as much as to support them. In *The Principles of Penal Law* (1810), Bentham observes that "it ought to be lawful to kill animals, but not to torment them."[21] Suffering, but not death, is to be avoided. One wonders if he would have applied the same principle to his doted-upon cat, Sir John Langbourne. Certainly, the principle is not applied by Bentham to humans. Moreover, since it is clear that some animals are capable of suffering considerably more than others, there must be a very distinct hierarchy of animals and animal interests if we are to rely on the issue of animal suffering alone. But Bentham goes beyond even this hierarchy in the *Deontology* (1819), where he tells us that "we deprive animals of life, and this is justifiable – their pains do not equal our enjoyment. There is a balance of good."[22] For Bentham animal suffering is on a lower level than human suffering, unless – and admittedly the meaning is a little opaque here – he means no more than the unverified empirical claim that the pleasure humans derive from eating animals is greater than the victims' pain in being killed. "The quantum of suffering is not commensurate for all species" is perhaps the more generous interpretation. And if it is not simply a matter of counting the amount of suffering, there must be something that indicates that human beings are, for Bentham, somehow of a different order of being. We can "deprive animals of life" by different criteria from that which would be allowed if we were to seek to deprive a human of life. Human suffering counts for more than animal suffering even if the animal suffering is greater. The distinction between humans and other species is far greater than we would have imagined on being told only that the relevant question was "can they *suffer?*" If, however, we follow Blake and Schweitzer in recognizing the sacredness of animal life itself or Tom Regan in talking of the life of a mature mammal as having an inherent value, then the distinctions between the species diminish. They do not, however, disappear, unless we take the value inherent in life per se as the sole relevant criterion – which would mean that the life of the flea were as sacred as that of the bonobo. Yet if we restrict the idea of "inherent value" to mature mammals, we put at risk the relative status of many other species. I cannot see why the boa constrictor, the shark, the owl, or the nightingale should in principle count a great deal less. Nor would it seem prima facie appropriate to count the life of a squirrel as equal in worth to that of a gorilla or a dolphin. If circumstances dictated that the life of either a squirrel or a chimpanzee had to be sacrificed, most would opt for maintaining the life of the chimpanzee without much hesitation – and would consider their choice a moral one – however much they might decry the necessity of the choice. The choice would involve a tragedy, even if it reflected a reality. And this is because there is a common assumption that the mental state of the chimpanzee is of a different order from that of the squirrel and, indeed, from that of many much more complex animals. And such a conclusion appears to be born out by the findings of empirical research.[23] Moreover, we should remember that from the medieval period on, there were many who thought of the great apes as a lost tribe of humans.

In general, traditional utilitarians have found themselves on the horns of a philosophical dilemma in discussing animal ethics. Thus, for example, John Stuart Mill declared famously in *Utilitarianism* (1863) that "it is better to be a human being dissatisfied than a pig satisfied."[24] For Mill, the intellectual pleasures were qualitatively superior to the sensual pleasures. Human pleasures were thus decidedly superior to porcine pleasures. There was something essentially superior in the life of the intellect, possessed by humans alone. Indeed, Mill was resurrecting the views of Plato and Aristotle in new, and rather uncomfortable, raiment. However, years later, in response to the jibe of the Cambridge scientist, Kantian philosopher, and theologian Dr. William Whewell that utilitarianism would make it our duty to increase the pleasure of pigs or geese rather than that of humans, Mill declared: "We are perfectly willing to stake the whole question on this one issue. Granted that any practice causes more pain to animals than it gives pleasures to man; is that practice moral or immoral? And, if exactly in proportion as human beings raise their heads out of the slough of selfishness, they do not with one voice answer 'immoral,' let the morality of the principle of utility for ever be condemned."[25] It would appear that Mill is now abandoning his previous consciously expressed commitment to quality over quantity. It is now simply the quantum of pleasure that is at stake. The issue may thus be solved by Bentham's felicific calculus. Yet, as we have seen, both Bentham and Mill denied on other occasions the commensurability of animal and human pain and suffering. What this suggests is that when we come to judge the relative merit of animal and human interests, there is more than one criterion at stake, and they are not necessarily compatible with each other. Indeed, they raise confusion in the ranks of the traditional utilitarians – perhaps especially when we consider the pain imposed upon the goat by the wolf.

The quality of pleasures and pains, enjoyment and suffering, the quantity of pleasures and pains, enjoyment and suffering, the very estimation of the value of animate life itself, and the needs and wants of different species, all enter our minds, and we juggle them there, hoping that the ideas will somehow cohere but knowing that, if we are to remain logical, they will not, at least not completely, and that we will be compelled to make judgments that will permit us some partial accommodation among the competing principles and interests.

Peter Singer, who brought Bentham's view of animal suffering to prominence, acknowledges that there is a distinct hierarchy among animal interests, at least when it is only the quantum of pleasure that is at stake; and he does not appear to want to exceed this consideration, at least to the extent that the capacity for suffering may be deemed to correspond to rational capacity. In *Animal Liberation,* Singer denounced those who claimed of the animal liberationists, in the words of Dr. Irving Weissman, that "'some of these people believe that every insect, every mouse, has as much right to life as a human' ... It would be interesting to see Dr. Weissman name some prominent Animal Liberationists who hold this view. Certainly (assuming only that he was referring to the right to life of a human being with mental capacities very different from those of the insect and the mouse) the position described is not mine. I doubt that it is held by many – if any – in the animal liberation movement."[26] The differing mental capacities, with differing capacities for pain and suffering, ascribe higher and lower status, and corresponding treatment, to different species. Verily, simply another version of the Great Chain of Being! Yet on other occasions, Singer takes what some may consider a quite different approach, arguing that, for example, if it is wrong to inflict pain on a human infant for no good reason, then it is "equally wrong to inflict the same amount of pain on a horse for no good reason."[27] Here the equine and human species appear to be on a par.

Many would accept the proposition that if one could demonstrate different characteristics among different species, this would entitle animals to very different treatment. Peter Singer seems to accept the principle behind this view when he states that: "The basic principle of equality does not require equal or identical *treatment;* it requires equal consideration. Equal consideration for different beings may lead to different treatment and different rights."[28] This is reminiscent of Aristotle's principle of equality, but for the Stagirite, in some matters numerical equality is appropriate and in others proportionate equality – clearly differing conceptions of equality. We might initially wonder what is meant, for Singer, by "basic" equality, remembering, for example, that equality of condition and equality of opportunity, like numerical and proportionate, are competing concepts in human ethics. Which is more "basic"? We might wonder to what the *treatment* should be proportionate – suffering, real or potential, different animal needs, or different animal capacities? We might wish to conclude that Singer's proposition requires that equal consideration of animals who suffered differently would mean that those who suffered more pain would be entitled to more benefits. On this basis, cruelly treated rats would be entitled to more benefits than pampered pets. Or would it mean that those who had the *capacity* for greater suffering would be entitled to more benefits? On this basis, pampered pets would be entitled to more benefits than cruelly treated rats. What must be clear is that Singer's proposition does not always help us to answer *practical* ethical questions, other than in a rather simple manner. Certainly, it is not twisting the Singer argument too far to declare it a potential boon to those who support extensive animal experimentation. If there are considerable differences between the capacity for pain and suffering of those animals that are experimented upon and the capacity for pain and suffering of the human animal, in whose interests the experiments are conducted, all in line with the human propensity for mental development so prized by Singer, then animal experimentation could be declared quite consistent with at least some versions of the utilitarianism Singer represents, even though Singer himself would appear to oppose such invasion. While philosophical argument raises to the level of consciousness the inner truths of our minds (or souls, to borrow Buckner's practice of equal substitution),[29] it does not, for a moment, allow us an escape from making practical judgments among competing principles in deciding on any given occasion what constitutes the right thing to do. And it is the relative weights given, usually subconsciously, to the competing principles that form the basis of our decisions. Nowhere does Singer's philosophy provide a rationale for where the greater weight should be given, other than in unidimensional terms.

Of course, other philosophers have suggested that there are other criteria involving a far greater measure of practical equality. For example, Tom Regan talks of complex animals as having "inherent value," as being "subjects of a life," a concept that appears to approximate that of self-consciousness, and all beings that are such subjects of a life are said to share their rights in common, which of course seems to contradict Singer's principle that "equal consideration" may lead to "different rights." Animals that are subjects of a life – generally "mentally normal mammals of a year or more" – will have "beliefs and desires; perception, memory, and a sense of the future, including their own future; an emotional life together with feelings of pleasure and pain; preference – and welfare – interests; the ability to initiate action in pursuit of their desires and goals; a psychophysical identity over time; and an individual welfare in the sense that their experiential life fares well or ill for them, logically independently of their utility for others and logically independently of their being the object of anyone else's interests."[30]

Regan then claims that complex animals possess these attributes. Unfortunately for the creation of a compelling ethic, many ethologists disagree with him. Some claim that only humans, dolphins, chimpanzees, and gorillas possess self-consciousness – that is, a conscious sense of the self as an autonomous individual, as an "I." Others claim that while animals certainly have desires, to claim that they also possess "beliefs" is misleading. If they do possess beliefs, they do so neither at a self-conscious level nor in the form of what humans call their "beliefs" – and the possession of such beliefs is something that may be far more significant for some species than for others. Some doubt that animals have any sense of their own future, others that if a few species possess such a sense, it is not of a future of more than a few minutes' duration. One of the difficulties involved in the debate is that there is adamant disagreement about the needs, wants, desires, and capacities of each species. If we look at the claims of those who conduct empirical research on animal behaviour, we will find that their conclusions about the capacities of, say, pigs, poultry, gorillas, orangutans, and cattle differ fundamentally from the conclusions of those who speculate about the nature of these animals' being; and not just of the difference between, say, humans and pigs but between pigs and poultry. The relevant question is whether, if it were demonstrated empirically that many mammals do not possess the characteristics Regan ascribes to them, he would be willing to declare that these animals do not possess the rights he has ascribed to them. At least for the present, there is no consensus on the nature of species. Nonetheless, it would appear reasonable to suggest that, until we possess such information, generalizations about animal wellbeing – or, if one prefers, animal rights – cannot be discussed in a manner that will allow competing parties to settle disputes about the legitimate treatment of animals. And, if this is so, then a great deal of our energy should be put into acquiring this kind of information. If we are to develop a compelling, or even generally persuasive, animal ethic, then we must expect more from empirical studies than from abstract formulae.[31] Indeed, the developers of abstract formulae should readily submit to empirical verification the claims about the nature of animals that they use to support their ethic. The study of ethics and the study of animal-welfare science and ethology are each essential to a sound understanding of our responsibilities to other species. Separately, they offer a measure of understanding. Taken together, they offer a great deal more.

V

It has always been one of the most inconvenient aspects of the study of ethics that it has, by and large, failed to provide compelling arguments about what is good, philosophically speaking, either for humans or for animals. Yet without descending into the despondent slough of moral relativism, Aristotle demonstrates the nature of ethics in a manner that is customarily forgotten today. He tells us that the study of the moral sense is a part of the practical rather than the theoretical sciences. The moral sense is a part of our calculative reason, which is concerned not with knowledge for its own sake but with knowledge for the sake of action. It is the concern of the city rather than of the school; that is, it belongs to our practical rather than to our theoretical concerns: "in practical studies the end consists not in contemplation and knowing about each point" in a philosophical manner "but rather in acting upon them."[32] Ethics involves not *sophia* (philosophical wisdom) but *phronesis* (practical wisdom), which is that form of reason that enables people to decide correctly the requirements of moral virtue by choosing among the competing options, many of which have an element of merit in them. Indeed, both ethics and politics are so much a part of practical rather than theoretical knowledge that, for Aristotle, it is the *Spoudaios,* the person of mature practical wisdom, rather than the intellectual, who is to be the political leader and who makes the most suitable role model. In the *Nicomachean Ethics* (1141b), Aristotle tells us that Anaxagoras and Thales as well as similar scholars are philosophically wise but not practically so. Their extraordinary abstract knowledge is godlike but practically useless. You cannot have precision, cannot have rules in ethics that may simply be obeyed. Ethics, like politics, is a matter of making practical judgments in the light of experience with goodwill and good socialization. "We must be satisfied," as Aristotle observes, "to indicate the truth with a rough and general sketch: when the subject and basis of a discussion consist of matters that hold good only as a general rule, but not always, the conclusions reached must be of the same order."[33] And while the objective knowledge provided by rational intuition (*nous*)[34] constitutes the unexpressed – and, indeed, ineffable – universal principles, the decisions themselves are the product of our practical judgments in practical circumstances. And, as is quite clear to Aristotle, they involve the satisfaction of human needs.[35] Of course, some will object that intuitionism oversimplifies ethics and that if the intuitionist's view were correct, then we would all share the same judgments, which we patently do not. This easy denunciation of the intuitionist's view, at least of its unusual Aristotelian variety, fails on the ground that there is no good reason why different persons should not intuit different things. If the intuitions relate to our needs, if they reflect primal memories of our needs, if our differing needs have been developed at different stages of evolutionary history, and if our intuitions are thus vestiges in our minds of past needs, just as the appendix is a vestige of our body's past, then we may intuit competing ideas. These competitions will be both among persons and within persons. We reflect on these competing promptings and act on our deliberations thereof. Thus ethics and politics require not that we follow abstract rules but that we make concrete judgments, selecting from the alternatives before us, none of which is likely to be unequivocally appropriate and a number of which will have commendable elements.

Aquinas followed Aristotle self-consciously in this form of reasoning, stating that "all other animals are able to discern by inborn skill what is useful and what is injurious; just as the sheep naturally regards the wolf as his enemy ... Man, however, has a natural knowledge of the things which are essential for his life only in a general fashion, inasmuch as he has power of attaining knowledge of the particular thing necessary for life by reasoning from universal principles."[36] Although these universal principles are rational, they are not themselves derived by the use of our reason but exist as rational intuitions, or *nous,* within our being. They are, if you like, instincts, just as the animals have instincts, but with less immediate practical relevance, although they may involve a recognition of the kinds of needs that are to be fulfilled. In short, practical reason determines the course to be followed by the intuitions. No doubt, too, this was what Lord Chief Justice Mansfield (1705-93), head of the English judiciary, had in mind when he advised his judges to "consider what justice requires and decide accordingly. But never give your reasons; for your judgement will probably be right, but your reasons will certainly be wrong." Practical knowledge will help the person of mature ethical wisdom to make the right judgments; such knowledge, Aristotle says, is concerned "not with what is eternal and unchanging, nor with what comes into being, but with what someone might puzzle about. For this reason it is concerned with the same thing as practical reason."[37] Confounding this practical knowledge with abstruse argumentation is more likely to interfere with sound ethical judgment than to improve it. Once we have raised to the level of consciousness the intuitions of our soul, the purpose of philosophical analysis is to give more elaborate linguistic form to these intuitions and to investigate the implications of their application in particular circumstances. The *idea* of the good – the product of the intuitions – is objective, simple, and knowable. What is conducive to the good is variable, complex, and debatable largely because human and animal needs are multiple; it is relative to the conditions and circumstance in which the intuitions are to be applied. In ethics we deliberate, above all else, about this relative complexity. We are thus compelled to make judgments by weighing the strengths and weaknesses of the alternatives before us.

While the truths of philosophy are, for Aristotle, humanity's ultimate quest, in practical matters of ethics and politics, the task is better left in less abstract, abstruse, and ethereal form. Ethics and politics are the province of the practical person. Thus, whereas Plato derides popular opinion, demeaning it to the level of sophistry, for Aristotle popular opinion is the starting point of ethics. What the people believe is a prima facie indication of the just, and when one has philosophized from this starting point, if one deviates too much in one's conclusions from this starting point, then one should rethink the matter. Popular opinion often has the foundations of sounder wisdom than the opinions of the greatest minds. Certainly, for Aristotle, popular opinion is not a determinant of what is right, but it is both a guide and a check. The philosopher is capable of both the greatest wisdom and the greatest blunders. Plato is the greatest of philosophers, Aristotle opines, but one of whom one should be wary in practical matters of politics and ethics. His brilliance may serve to divorce him from practical truths.

Nous, or rational intuition, is the source of our ethical knowledge, but it is not an easy matter to know whether we have interpreted it and its implications correctly, and here popular opinion serves as an indicator to us of whether we have interpreted well. To be sure, popular opinion is inconsistent. Thus "everyone agrees that justice in distribution must be in accordance with some merit, but not everyone means the same by merit ... the just is a sort of proportion."[38] And this is precisely what is always at issue. What are the proportions? What is practically meritorious? Logical arguments are necessary to arrive at appropriate answers, but they should always start from the underlying opinions of the prevailing culture, which are the summation of its historical wisdom. A knowledge of the history of attitudes allows us to make better practical judgments. We come to understand both the possibilities that history allows us and the limitations that history imposes upon us. This does not mean that we cannot move toward radical conclusions but only that they are safer if we move toward them piecemeal, with an earnest endeavour to understand why people hold the views that they do and with a respect and consideration for the culture of society as the accumulated wisdom of a society's collective mind. This does not mean for a moment that we should be satisfied with the status quo but simply that we can wisely advocate change only once we have understood the circumstances and beliefs that have come to constitute the status quo. To act otherwise is more likely to disrupt than to advance the social and moral order. After all, as James Joyce wisely recognized, abstract speculation tends to take control of our minds and to lead us in other directions than those encouraged by common sense and intuition. If we recognize the limitations to human reason and philosophical capacity, we will always check our conclusions by reference to the lessons learned from our more mundane thoughts.

One of the most important aspects of popular culture with regard to animals is that they are not seen under the general concept of animality but according to the differentiating characteristics of particular species and even of individuals. They are pigs, rhinoceroses, rabbits, and hamsters (and humans) before they are animals. We will understand the generality better if we look at the specificity more closely. We should thus proceed from the specific to the general, starting not with ideas of *animal* rights and *animal* wellbeing but with the wellbeing and rights of cobras, racoons, and zebras. Moreover, since the wellbeing and the rights of the bird and the cat, the rat and the weasel, the zebra and the lion cannot be squared, we should never expect full and final answers from any philosophical ethic, however prima facie persuasive, however philosophically brilliant.

VI

If Aristotle is right, a careful nonideological analysis of the history of animal ethics will not only provide an appropriate understanding of the development of Western culture, but also be a guide and a check in our attempts to develop an appropriate animal ethic. The reality is a guide to the morality – although, of course, not at all an unerring one. Prevailing beliefs will inevitably need refinement and refraction, but they are the appropriate starting point because the elusive "common man and woman" possesses an instructive instinct. This instinct is one that philosophical wisdom may improve upon but it may only refute the belief with overwhelming evidence of the idea's inadequacy.

Indeed, while in philosophizing we usually ascribe interests to individuals, in common parlance we ascribe them to levels of community – or belongingness. The idea is expressed most fully in Mencius's Confucian claim, encountered in the first chapter, that "the superior man feels concern for creatures, but he is not benevolent to them. He is benevolent to the people but he does not love them. He loves his parents, is benevolent to the people and feels concern for creatures."[39] The words may not satisfy the Western mode of expression, but the meaning is clear enough and one that pervades the Western ethic as much as the Chinese: we owe an obligation to all creatures, but our obligation is above all to our immediate kin, then to people in general, and finally to the realm of the animals. To be sure, it is a conception that requires considerable refinement, but it is nonetheless a valuable point of departure. Relationships are at least as important as individuals.

We have a sense of community with all, but a greater sense with close kin than with humanity at large, and close kin in turn have a greater sense compared to animals at large. Moreover, for Aristotle, a community "is kept together by proportionate reciprocation. For some people seek to return evil for evil – otherwise they feel like slaves – or good for good – otherwise no exchange takes place, and it is exchange that holds them together."[40] Indeed, in the most complete community, we share our identities with others. If it happens to them it happens also to us. If there is no belongingness, there is no possibility of doing good, and belongingness must of necessity be accorded differently to different beings. It is thus in the very nature of the dispensation of good that it must be accorded in different degrees. The proportion due to a spouse is different from that due to a stranger, and that due to a stranger is different from that due to a mouse, *at least generally speaking* – a rider necessary for all Aristotelian moral philosophizing.

Today, the overriding principle that we find expressed in the study of ethics is that of universalizability. That is, each is to count as one and no more than one. Yet this idea is itself not universal. Few cultures would subscribe to it. That one has a greater responsibility to kin than to stranger is commonly accepted not only as an aspect of human psychology, but also as the primary dictate of morality itself. And, indeed, the idea of universalizability is a relatively late introduction to Western culture, a historical product of the Enlightenment. It is given its firmest expression in Jeremy Bentham's famous statement that the "community is a fictitious *body*, composed of the individual persons who are considered as constituting as it were its *members*. The interest then of the community is what? – The sum of the interests of the several members who compose it."[41] The statement is prima facie persuasive in a culture entirely convinced that individuals are the only ends in themselves, as Kant postulated. Yet, of course, the principle confers confusion on animal ethics, for those who counted as one and no more than one were humans. The principle itself does not provide an answer as to whether animals at all, in general, or only according to species characteristics are included.

Recognizing that *in fact* many continue to think of collectivities as in some manner ends, that *in fact* we identify ourselves not merely as individuals but in our belongingness with others, some philosophers have begun to use the concepts of "reasonable partiality" and "unreasonable impartiality"[42] as qualifications of the liberal egalitarian ethic that reigns in Western culture – or more correctly, in the Western philosophical expression of this culture. It is, so the argument goes, at least sometimes reasonable to prefer the interests of those to whom we are more closely related, biologically or otherwise, over the interests of humanity as a whole. Indeed, historically, alongside the notion of individuals as ends in themselves – an ever-present conception, but one given far greater emphasis in contemporary Western culture – such collectivities as the family, the clan, the tribe, the church, the nation, and even the school and the club were also seen as ends in themselves with their own organic lives, their own goals, their own rights, and their own purposes. The relationship to other individuals is seen to be as important as, if not more important than, the individuals who comprise the relationship – hence the traditional view that divorce is unacceptable because the act of marriage has created a new entity, a new organism, that is to all intents and purposes a new living entity: an end in itself. The same is true of a nation. Locke believed that governments, but not national communities, could be dissolved by majority vote because, even though for Locke national communities arise initially out of a compact among individuals, they become thereby the permanent embodiment of this compact. National communities, but not governments, are organisms. If both individuals and communities are ends in themselves, then there will be an inevitable conflict in ethical deliberation in the attempt to find some accommodation between the competing legitimate principles.

Naturally, this has led to a recognition of the tension between the citizen and the individual as an integral part of the contradictions that pervade society. This tension received particular consideration in the writings of Plato, Aristotle, and Rousseau. Indeed, society cannot be understood in the absence of a recognition that the human mind is inhabited by contradictions between, for example, love and honour, altruism and egotism, kindness and honesty (and its concomitant loyalty and honesty), individual and community, public and private, justice and prudence, the brute and the angel, ideology and philosophy, the sacred and the profane, order and liberty, reason and feeling (although perhaps not as frequently and in the manner commonly surmised),[43] and convention and nature, including

conventional and natural communities, back to nature and progressivist
ideals, and the yearning for simplicity and concomitant desire for complexity – as we witnessed so clearly in Chapter 4, exemplified most fully
in the writings and life of Thoreau. We are individuals and members of
several communities at the same time, and thus life is what Plato called "an
existence-in-tension." While there may be an absolute good, simple and
knowable – that is, known in its Form – what is conducive to the achievement of this good is complex and contradictory, involving the balancing
of incompatibilities to the degree that they may be balanced at all.

In the *Nicomachean Ethics* Aristotle discussed the matter in considerable
detail with regard to the idea of *philia,* customarily rendered as "friendship."[44] Yet it is more than we mean by friendship. Aristotle is writing of
a form of love in which we seek the good of the friend without reference to
our own good or to that of our other compatriots. Roger Crisp has observed
that "relationship" and "personal relationship" are alternative translations.
He continues, the "verb *philein* is translated as 'to love.' A broader notion
than friendship as we understand it, *philia* includes not only familial relations of non-humans as well as humans, but business partnerships and the
natural kinship felt by one human being with another."[45] This relationship
allows for what is now called "reasonable partiality" in ethical matters. With
not dissimilar effect, David Hume states:

> Now it appears, that in the original frame of our mind, our strongest atten-
> tion is confin'd to ourselves; our next is extended to our relations and acquain-
> tance; and 'tis only the weakest which reaches to strangers and indifferent
> persons. This partiality, then, and unequal affection, must not only have an
> influence on our behaviour and conduct in society, but even on our ideas of
> vice and virtue; so as to make us regard any remarkable transgression of such
> a degree of partiality, either by too great an enlargement, or contraction of
> the affections as vicious and immoral. This we may observe in our common
> judgments concerning actions, where we blame a person, who either cen-
> ters all his affections in his family, or is so regardless of them, as, in any
> opposition of interest, to give the preference to a stranger, or mere chance
> acquaintance. From all which it follows, that our natural uncultivated ideas
> of morality, instead of providing a remedy for the partiality of our affections,
> do rather conform themselves to that partiality, and give it an additional force
> and influence.[46]

It is not merely our interests but our "uncultivated ideas of morality," the
impulses of our soul, that require partiality in favour of community.

In 1869, as we noted in the first chapter, W.E.H. Lecky extended the argument to animals explicitly: "At one time, the benevolent affections merely embrace the family, soon the circle expanding includes first a class, then a nation, then a coalition of nations, then all humanity and finally its influence is felt in dealings of man with the animal world ... there is such a thing as a natural history of morals, a defined and regulated order, in which our feelings are unfolded."[47] The imperatives of human psychology are embedded most firmly first in familial associations and then in associations with loved ones along the lines of Aristotelian *philia*. What this should indicate to us is that just as our respect for humans in general develops out of our filial associations, so too does respect for animal life in general have its source in our love for our companion animals, which is embedded in our thinking of them first as part members and then as full members of our family. We come to respect animals in general as an extension of our love and affection for our companion animals, just as we come to respect human beings in general as an extension of our love and affection for our own kith, clan, and kin.

Any ethic that ignores the "belongingness" imperatives of psychology may be intellectually sophisticated but of less effect than if it were to acknowledge the human-animal bond not merely as a fact but as the foundation of the human-animal ethic. It must equally recognize the fact, even if it is not ultimately justifiable in the prevalent terms of contemporary academic ethics, that we will prefer some humans over others and some animals over others on the basis of similarity to ourselves, perceived potential harm to ourselves, their beauty, accomplishments, helpfulness to us, and their relationship to us. And, of course, we might wish to suggest some of these factors as morally irrelevant. But if we try to replace them on the grounds that they interfere with notions of each counting as one or of the equal "inherent value" of each complex animal, we do so at the peril of diminishing the affections, attractions, and respect that made us alive to the entitlements of the animal realm in the first place.

It is in Aristotle's *Nicomachean Ethics* that the idea of "reasonable partiality" is most fully developed, although it is not there called by this name. There we read that "good people will be friends for each other's sake, because they are friends in so far as they are good."[48] Clearly, since we cannot be friends with all – certainly not equal friends with all – act for the sake of all others, and engage in a *philia* relationship with all, we must give preference to some over others, even beyond the preference that we normally acknowledge for immediate kin – i.e., parents, siblings, and cousins – as well as for others with whom we have a sense of belonging derived from some other source than choice. As Aristotle says, "one cannot be a friend – in the sense of complete friendship – to many people, just as one cannot be in love with many people at the same time."[49] Nonetheless, for others, one may have "goodwill," "which seems to be a characteristic of friendship, but still it is not friendship,"[50] and "affection," which, of course, ensures them consideration, if not equal consideration, with those who are complete friends. Aristotle does recognize this as a *kind* of friendship but regards it as a friendship between a superior and an inferior (or, perhaps, between one who imagines himself a superior and one who relates as an inferior): "In all friendships involving superiority, the affection must be proportional as well. The better, that is to say, must be loved more than he loves, and so must the more useful, and each of the others likewise; when the affection is in accordance with merit, then a kind of equality results, which is of course thought to be a mark of friendship."[51] Although it is not Aristotle's concern, we can see that this kind of friendship relates to that of the companion human with a companion animal, which, as has been suggested, is a primary source of respect for animalkind in general and without which no general concern for the interests of nonhuman animals would have been derived. Indeed, it explains why many of us are quite willing to acknowledge our obligations toward animals without requiring a concomitant responsibility on their part toward us. Of course, if "reasonable partiality" is an idea that helps us to understand the moral basis of the human-animal ethic, it should not dissuade us from a recognition of the fact that what we more commonly encounter is "unreasonable partiality," where the interests of the animals receive far from adequate consideration.

VII

Why, one might ask, are animals entitled to ethical consideration at all? What attributes do they possess that entitle them to consideration? That historically, by and large, they have been accorded a moral status, even if lower than what many of us would have liked, is a prima facie sound reason for considering them so entitled, which stands unless countered. The fact that they possess needs, purposes, and wants, those characteristics that appear relevant to ascribing moral status to humans, is a prima facie reason for ascribing moral status to animals as well. That scientists and scholars have demonstrated an unbroken human-animal continuity, both historically and anatomically, is a sufficient reason for ascribing moral status to animals until and to the degree that solid arguments to the contrary may be offered. And the degree to which they are entitled to consideration will relate in some measure to the place they are seen to occupy on the scale of being. Of course, not all have acknowledged our continuity with the animals. For instance, in the *Ethica* (1677) Baruch Spinoza writes: "I do not deny that beasts feel; what I deny is that we may not consult our own advantage and use them as we please, treating them in the way which best suits us; for their nature is not like ours, and their emotions are naturally different from human emotions."[52] Spinoza, however, has not offered us any reason other than self-interest for his position. In light of human-animal continuity, Spinoza is logically required in making his case to demonstrate how their nature differs from ours, how their emotions are naturally different from our own, and in what manner these differences may be relevant. And this he singularly fails to do. As it stands, his account amounts to saying no more than that even though animals may experience pain and suffering, I refuse to countenance this as a reason for treating them as anything other than means. Again, this is not to say that there are no relevant differences between humans and other animals. It is to say that the presumption must be against such relevant differences until demonstrated.

Earlier we noted that, among the moderns, in 1680 Giovanni Borelli demonstrated the unity of life, showing how the same laws governed the wings of birds, the fins of fishes, and the legs of insects. Goethe proved in around 1780 the existence of the *os maxillare* in humans, thus showing that there was no distinctive human structure. In the 1790s Erasmus Darwin evinced all the principles of evolution in embryo, save for natural selection. In 1801, and more extensively and elaborately in 1809, Jean-Baptiste Lamarck developed the first full-fledged evolutionary theory. Around 1828 Etienne Geoffroy Saint-Hilaire declared from evolutionary ideas that "there is, philosophically speaking, only a single animal." In 1836 Ralph Waldo Emerson announced in *Nature* that "each creature is only a modification of the other; the likeness in them is more than the difference, and their radical law is one and the same." And, of course, in 1858 Charles Darwin – and, at the same time, Alfred Russel Wallace – first accounted publicly for the manner in which evolution takes place: natural selection. Whether of an evolutionary bent or not, all concurred: human and animal were essentially one. But they were not only "one" anatomically. Most concurred that this anatomical similarity was positive evidence for ascribing them moral status. To emphasize again, it followed that if we and they were similar in so many respects, then, unless it could be demonstrated otherwise, the supposition must be that they are alike, too, in possessing the right to ethical treatment – at least other than in ways in which the differences may be demonstrated. Logically, there must be a prima facie presumption in favour of animals being entitled to moral consideration in the same manner as are humans. Any deviance therefrom requires solid evidence, both philosophical and scientific. It is up to those who deny that animals are entitled to ethical consideration to demonstrate the validity of their contention. None of this is to deny in principle that relevant differences may be adumbrated. It is to indicate that, if we are to accord animals different rights and treatment, then we must be offered both sophisticated empirical as well as philosophical arguments.

Any notion of a human-animal ethical equality is doomed to failure in practice both because, even if animals do possess the relevant attributes, they are deemed by the vast majority of humans to hold them in lower degree and because in practice the sense of belongingness that lies at the root of ethical consideration must, at least in the minds of a majority, always put the human interest ahead of the animal interest. But then one could argue that if the idea of reasonable (or unreasonable) partiality hinders a notion of human-animal ethical equality, it may be said to be equally injurious to equality among humans. Whether appropriate or not, the reality is that, despite occasional exceptions, humans consider animals to have less of the relevant attributes for ethical consideration, less potential for suffering (although not necessarily less potential for pain), a lower capacity for reasoning, and, if any at all, a lower capacity for moral behaviour. Moreover, except with companion animals, which are often felt in Western society to be full family members, the animal is usually seen to be more distant in kinship than other human kin. Even companion animals will be deemed, generally speaking, lower on the scale than human siblings and offspring. To be sure, the animal is largely regarded more positively today than in the historical past because, in part and in general, humans no longer have to fear "wild" animals as predators. Certainly, the human-animal interest continues – indeed, increases – in questions of animal experimentation and intensive farming, but the fact remains that, generally speaking, the human is not at major risk from poisonous snakes, tigers, lions, or grizzly bears. Even the serious illnesses, including the plague, historically associated with black rats, have evaporated – in part because the black rats were exterminated by the fiercer but less infectious brown rats and in part because brick and stone dwellings were more resistant to infestation.[53] Nor, if intensive farming were eliminated, would we have to fear starvation. The historical circumstance is relevant to our sensibilities. Still, we should not forget that antipathy to once legitimately feared animals sometimes remains once the danger is long past. The fear of the wolf in Britain did not disappear once the animal had been extirpated, and the animal-loving Charles Dickens, because of his youthful experiences working in a Thames blacking factory, retained his horror of rats throughout his life.[54]

VIII

However much the Great Chain of Being has been replaced by evolution by natural selection as the appropriate *biological* model for understanding the human-animal relationship, the reality is that the chain is the *moral* model that pervades most minds, including those of the Darwinians. The chain, with the medieval elements of angels and cherubim either discarded or at the very least downplayed, remains the model that Darwinians, Christians, Hindus, Jaina, Muslims, and Jews, along with many others, employ as the ideational concept by which to determine the human-animal ethical relationship. Whatever the science or the religion may indicate, when it comes in practice to deciding how animals should be treated, it is the implicit idea of the chain of being that is operative. Without it, Darwinians would not be able to justify animal experimentation, and Hindus would not be able to determine which animal the human soul may enter into in accord with its merit, and Buddhists would not be able to assign particular animals to the level of the higher and lower orders of existence. Nor would anyone be able to justify the use of buffalo to pull the plough or of donkeys to carry the burden.

However much we are impressed by the animals, and however much we liken them to humans, in the end there are, with few exceptions, status distinctions that we constantly make. For example, writing in her epistolary *Instinct Displayed* (1821), the Quaker philanthropist Priscilla Wakefield, whom we met in the first chapter, tells her correspondent: "It will give you pleasure to see how far the animal feelings can approach to the moral virtue peculiar to rational and responsible beings; and what a union of the most affectionate and the most hostile qualities can exist in the same creature, both springing from a noble, generous disposition."[55] Despite the approximation of animal to human feelings, and even though the animal is of a "noble, generous disposition," it is only the human who is "rational and responsible." Wakefield finds "a charm ... throughout all the tribes of animated nature," but it cannot erase the fact that the human is superior in status:

> There are, indeed, innumerable gradations of intelligence, as of the other qualities with which the animal kingdom is endowed; in like manner as the different orders of beings approach each other so closely, and are so curiously united by links, partaking of the nature of those above and those below, that it requires a discerning eye to know what rank to assign them. Thus quadrupeds and birds are assimilated to each other by the bat; the inhabitants of the waters to those of the land, by amphibious animals; animals to vegetables,

by the leaf insect, and by plants that appear to have sensation; and animate to inanimate by the oyster, the molluscae, and the sea anemones.

The ranks are discernible, even if with a degree of difficulty. Even with reason – "the proud prerogative of man" – human superiority, while not to be doubted, is not readily discerned by Wakefield. Thus "reason and instinct have obvious differences; yet the most intelligent animals, in some of their actions, approach so near to reason, that it is really surprising how small the distinction is."[56] Nonetheless, the humans are always at the apex, and each of the others occupies its appropriate niche.

The gradations of status are perhaps not quite as clear as Wakefield suggests. If the choice between saving the life of a fly and the life of a mature oak were at issue, most would opt for the oak not merely for environmental reasons, but because in most people's minds, there is something of a higher order about a grand tree – which, of course, gives the lie to the universal preference for animal over vegetal interests. Moreover, snakes, which have developed from lizards, losing their ears and their legs in the process, may still be determined "lower" than their ancestors. The model of relative status may be more complex than Wakefield imagines.

Nonetheless, Wakefield has used the chain of being to elevate the status of animals in the public mind in a manner generally consistent with the common conception of the "scale of creatures," as the model is alternatively called. What is equally certain is that others used the chain to emphasize human superiority, employing the same criteria. What is of the greatest significance, however, is that the idea of a chain of being is pervasive in human conceptualization – and as much today as in earlier centuries. Thus, given the nature of human psychology, ethical theories insisting that each is to count as one and no more than one, especially with regard to animals, are relevant in reality only in the classroom – and can have little sway in the real world of ethics and politics. If there is an inevitable conflict between abstract ethical theory and empirical reality, one has to wonder how valuable the ethical theory can really be. If the ideas of the chain and of community are inconsistent with prevailing ethical theory, it is prevailing ethical theory that must be looked at askance.

IX

If we cannot rely on prevailing ethical analyses in the field of animal ethics with any great degree of confidence, despite its occasional successful illuminations, what alternatives do we have? It might be worth first looking in an unlikely place: the writings of Immanuel Kant, after Descartes the philosopher most commonly derided by those who declare themselves the friends of animal interests. Having engaged in philosophical argument decidedly detrimental to animal interests, Kant concludes, as we noted earlier:

> If a dog has served his master long and faithfully, his service, on the analogy of human service, deserves reward, and when the dog has grown too old to serve, his master ought to keep him till he dies ... the more we come in contact with animals and observe their behavior, the more we love them, for we see how great is their care for their young ... a master who turns out his ass or his dog because the animal can no longer earn its keep manifests a small mind ... Leibniz used a tiny worm for the purposes of observation, and then carefully replaced it with its leaf on the tree so that it should not come to any harm through any act of his. He would have been sorry – a natural feeling for a humane man – to destroy such a creature for no reason ... We can judge the heart of a man by his treatment of animals.[57]

In the case of the dog too old to serve, Kant is repeating almost the exact words of Plutarch in his "Life of Marcus Cato," one of the very favourite philosophers of animal advocates. In the case of animals caring for their young, Kant acknowledges the justifiability of love for animals, an idea surely incompatible with the view that we have responsibilities only to humans. It is difficult to conceive how one may love without a concern for the object of the love in and for itself. In the cases of the ass, the dog, and the caterpillar, he acknowledges *practical* human responsibility toward animals, even in trying to avoid *philosophical* acknowledgment of it. And his claim that treatment of animals is a primary basis for judgment of the worth of a human indicates again that we should behave compassionately toward nonhuman animals. Even if we reject his philosophical claim that animals are not ends in themselves, we can see that in practice his required behaviour toward animals is very similar to that of many persons who would claim that animals are ends in themselves. If we have a complaint against Kant, it should be more for the inadequacies of his philosophy than for the inadequacies of his humanity. If we are ultimately dissatisfied with Kant – and, ultimately, animal advocates will be – what this tells us is that in animal ethics it is not merely the recognition of the need for just treatment that is at issue. There is also the issue of respect for the being as the animal that it is, with a corresponding recognition of its right to consider itself an "I" that is to be acknowledged *by us* as an end in him- or herself, even if the lion cub cannot claim the same right from the new animal master of the pride, or the sheep from the wolf, or the fly from the spider.

If there is anything to be learned from Kant, it is that magnanimity, goodwill, affection, and compassion are owed to animals, even if we are philosophically obscurantist enough to require that the route be taken via our obligation to fellow humans. In less obfuscating reality, we owe the animals compassion because, as Rousseau recognized above all others in *Discourse on Inequality* (1754), it is a part of the human natural law that we should do so. In explaining the natural law, he tells us:

> As long as [man] does not resist the inner compulsion of compassion, he will never do harm to another man, or even to another sentient being, except in those legitimate cases where, since his own preservation is involved, he is obliged to give preference to himself. By this means, the old debate concerning the applicability to animals of natural law can be put to an end, for it is clear that, bereft of understanding and liberty, they cannot recognize this law, but since they share to some extent in our nature by virtue of the sensibility with which they are endowed, it will be thought they must also participate in natural right, and that man is bound by some kind of duty towards them. It seems, in fact, that if I am obligated to do no harm to my fellow man, it is less because he is a rational being than because he is a sensitive being since sensitivity is a quality which is common to man and beast. This should at least give the beast the right not to be needlessly mistreated by the man.[58]

The natural law and natural right are of limited applicability to animals because the animals have no understanding of it. They participate in it only in part because of their intellectual limitations. But this does not hinder the fact of our duties toward animals as required by the natural law. Years before Bentham came to a similar conclusion, Rousseau recognized the importance of sentience and the lesser importance of rationality to questions of morality. In the end, however, it is the dictates of the "inner compulsion of compassion" that require our consideration for animals. Because "they share to some extent in our nature," we have obligations toward them, at least to the extent that they do share our nature. And, for Rousseau, this extent requires that we not consume the flesh of our fellow beings, just as we would not consume the flesh of our fellow humans.[59] For Rousseau, it is precisely because the primary characteristic that requires our not harming humans also extends to animals that we must have a similar kind of obligation toward them. If we should have a prior consideration for humankind, it is not by a great degree. In practice, the debate has almost never been about whether animals are entitled to be treated well but about the degree to which the interests of humans should be preferred when human and animal interests are in conflict. And to the extent that we are more humane, we will always give the animal the benefit of any doubt, although we will not do so for all animals equally.

In *Emile* (1762) Rousseau described this intuited compassion that is the source of our morality in some detail, indicating that it enjoins empathy for our fellow creatures:

> Emile ... will begin to have gut reactions at the sounds of complaints and cries, the sight of blood flowing will cause him an ineffable distress before he knows whence comes this new movement within him ...
>
> Thus is born pity, the first sentiment that touches the human heart according to the order of nature. To become sensitive and pitying, the child must know that there are beings like him who suffer what he has suffered, who feel the pains he has felt, and that there are others whom we ought to conceive of as being able to feel them too. In fact, how do we let ourselves be moved by pity if not by transporting ourselves outside of ourselves and identifying with the suffering animal, by leaving, as it were, our own being to take on its being? It is not in ourselves, it is in him that we suffer. Thus, no one becomes sensitive until his imagination is animated and begins to transport him out of himself.[60]

Compassion for our fellow beings derives from the very psychology of what it is to be truly human. Why, then, if compassion for other animals is a part of the very constitution of our souls, do we practise it so little? In part, it is a matter of the relative weight that we place on human and animal interests. There is no magic formula for deciding the appropriate weight. It is a matter of *phronesis,* judgments derived from practical reason undertaken with as much impartiality and benevolent human character as we can muster. But, more important, in part it is a matter of power. Often we do not act as we know justice requires precisely because there is no penalty for ignoring the dictates of our intuition, which in turn persuades our intuitions to be less enterprising in their appearances. We fail to act justly because we have the power to act unjustly without incurring an appropriate punishment. Moreover, this power is instrumental in our managing to justify those actions that arise from self-interest and relegate the animals to a lowly place. If Hobbes is decidedly wrong in his view that humans are everywhere and always self-interested, he is decidedly right that we cannot rely on human altruism. The state is necessary to enforce the better part of humanity's compassionate inner constitution. And education is necessary to arouse people to an awareness of all the iniquities practised against our fellow creatures, of which they may be unaware or of which, against the existing dictates of their self-interest, they must be persuaded. If no final philosophical answers can be given to questions of the degree of responsibility we owe our brethren, at least we can have absolutely no doubt that there are myriad problems to be solved in issues of animal experimentation, intensive farming, hunting, and the like, where far too little consideration is given to those who share so much in common with us.

The virtue ethic, associated initially with Aristotle, asks us not what obligations we have or what rules we should follow but what particular personal characteristics we should cultivate in order to live a good life. Angus Taylor has enumerated these virtues as, for example: "courage, honesty, loyalty, benevolence, industriousness, compassion, and more generally, all the traits that help us to get along with others in society and to face the hardship of life. Virtues stand in contrast to vices, such as cruelty, cowardice, and dishonesty."[61] This ethic is often thought inimical to the consideration of the interests of animals. Yet it certainly need not be so. Even if it tends to lay emphasis on human-to-human concerns, one need only ask to whom we should be loyal and to whom be compassionate in order to recognize that animals might have a meaningful place in an ethic primarily concerned with making humans the best they can be as humans. As we have seen, Rousseau has a ready understanding of our natural compassion and empathy. Yet we can ask the question from a different perspective. Are animals the kinds of beings for whom our affection, love, respect, awe, and/or admiration are warranted? Does acting compassionately toward other species allow us to be the best we can be as human beings? Surely, a few moments' impartial introspection will produce the answer "yes," although in different respects and according to the characteristics of particular animals. Of course, if our interests interfere with our objectivity, "no" will be the result. A further few moments' consideration would produce an acknowledgment that we should be compassionate and considerate to our fellow beings. After all, animals, human animals included, are a cluster of evolved needs and powers that must be met and exercised if these needs and powers are not to lead to frustration, boredom, and alienation from one's species nature. And nowhere in our investigations have we found a justification for treating the needs, wants, and purposes of nonhuman animals in a manner entirely different from the appropriate treatment for human beings. While I would not limit myself to utilitarian premises as the sole criterion of the good – although I would certainly acknowledge sentience a matter of prime importance – I would still employ a variant of John Stuart Mill's response to the Kantianism of William Whewell: "If exactly in proportion as human beings raise their heads out of the slough of selfishness they do not with a preponderant

voice acclaim nonhuman animals entitled to ethical consideration in proportion to their individual and species needs, wants, and purposes, so will
the need for consideration of animal interests for ever be condemned."[62]
What matters is not *proving* that animals deserve consideration. Almost all
people do accept – and, as a serious history of animal ethics teaches us,
almost all people always have accepted – that animals are entitled to some
consideration. Unfortunately, many have ignored the promptings of their
conscience, have managed to convince themselves that while animals matter, whenever the animal and human interests are in conflict humans should
always win. The task, then, is to strengthen the sensibilities through education, ensure far stronger legislation, and promote a lessening of the distinctions customarily made between humans and nonhuman animals, all
in the light of the lessons learned from history that indicate both the practical opportunities that prevailing culture allows us and the practical limitations that it imposes upon us. As Aristotle understood with such clarity,
getting one's ethics right involves, in important part, making the right
choices about one's role models, which involves understanding history
aright. If we have history wrong, and if our role models have in turn mis-
represented historical reality, then our ethical appreciations will be tar-
nished. In the final analysis, there is nothing more conducive to allowing
our heightened sensibilities to exercise sound and effective influence on
our practical judgments and political practice than coming to know the
relative needs, purposes, and wants of different species through empirical
investigation of their natures.

Notes

INTRODUCTION

1 Mary Shelley, *The Last Man,* vol. 3 (1826; reprint, New York: Bantam Books, 1994), ch. 2, 353-54.

2 Elijah Buckner, *The Immortality of Animals, and the Relation of Man as Guardian, from a Biblical and Philosophical Hypothesis* (Philadelphia: George W. Jacobs, 1903), 161.

3 George Boas, *The Happy Beast in French Thought of the Seventeenth Century* (1933; reprint, New York: Octagon Books, 1966). See also Arthur O. Lovejoy and George Boas, "The Superiority of the Animals," Chapter 13 in *Primitivism and Related Ideas in Antiquity* (1935; reprint, Baltimore: Johns Hopkins University Press, 1997), 389-420. The chapter concentrates on the thought of Democritus, the Cynics, the New Comedians, Diodorus Siculus, Ovid, Seneca, Pliny the Elder, and Plutarch. The term "animalitarianisni" replaces that of "theriophily" in *Primitivism and Related Ideas in Antiquity,* but when the idea is referred to in recent literature, "theriophily" is almost always the term used.

4 For an extract of the argument, see Rod Preece, *Awe for the Tiger, Love for the Lamb: A Chronicle of Sensibility to Animals* (Vancouver: UBC Press; London: Routledge, 2002), 46-48.

5 Jean-Jacques Rousseau, Preface to *Discourse on the Foundations of Inequality among Men,* in *Rousseau's Political Writings,* ed. Alan Ritter and Julia Conaway Bondanella, trans. Bondanella (New York: W.W. Norton, 1988), 28.

6 Ibid., 29.

7 Ibid., 13.

8 Ibid., 19-20.

9 See William Godwin, *Life of Geoffrey Chaucer,* 2 vols. (London: Richard Phillips, 1803), vol. 1, 120.

10 See William St. Clair, *The Godwins and the Shelleys* (New York: W.W. Norton, 1989), 128.

11 Charles Kingsley, *Yeast* (1851; reprint, London: Macmillan, 1890), ch. 10, 42.

12 Mary Midgley, *Beast and Man: The Roots of Human Nature* (1979; reprint, London: Routledge, 1995), 17 n. 19.

13 Howard Williams, *The Ethics of Diet: A Catena of Authorities Deprecatory of the Practice of Flesh-Eating* (London: F. Pitman, 1883).

14 Dix Harwood, *Love for Animals and How It Developed in Great Britain* (New York: privately printed, 1928; reprint edited, introduced, and annotated by Rod Preece and David Fraser, Lampeter, Wales: Mellen Animal Rights Library, Historical List, vol. 10, 2002).

15 Rod Preece, *Animals and Nature: Cultural Myths, Cultural Realities* (Vancouver: UBC Press, 1999).

16 Quoted in John McManners, ed., *The Oxford History of Christianity* (Oxford: Oxford University Press, 1993), 11-12.

17 See Keith Akers, *The Lost Religion of Jesus: Simple Living and Nonviolence in Early Christianity* (New York: Lantern, 2000), 24ff., 233.

18 Stephen H. Webb, *Good Eating* (Grand Rapids, MI: Brazos Press, in the series Christian Practice of Everyday Life, 2001), 119.

19 See ibid., chs. 5-6.

20 Augustine, *Of the Morals of the Catholic Church, Nicene and Post-Nicene Fathers of the Christian Church,* 1st series, vol. 4 (1872; reprint, Grand Rapids: E.B. Erdmann, [1995?]), 33, 73.

21 Benedicta Ward, ed. and trans., *The Sayings of the Desert Fathers* (Kalamazoo: Cistercian, 1975), xxiv.

22 Ibid., 81.

23 David Winston, ed. and trans., *Philo of Alexandria: The Contemplative Life, Giants and Selections* (New York: Paulist Press, 1981), 249, 54.

24 Saint Athanasius, *The Life of St. Anthony the Great* (Willits, CA: Eastern Orthodox Books), 11, 72, 74, 96.

25 See Helen Waddell, ed. and trans., *The Desert Fathers* (New York: Vintage, 1998), 19, citing *Historia Lausiaca,* as edited by Hervetus in the sixteenth century.

26 Waddell, Foreword to Ward, ed. and trans., *The Sayings of the Desert Fathers,* xix.

27 "History of the Monks of Egypt," in Waddell, ed. and trans., *The Desert Fathers,* 51.

28 "Cicero to Marcus Marius, late August 55 BC," in *Cicero's Letters to His Friends [ad familiares],* trans. D. Shackleton Bailey (London: Penguin, 1967), 81.

29 "The Sayings of the Fathers," in Waddell, ed. and trans., *The Desert Fathers,* 160.

30 "The *Pratum Spirituale* of John Moschus," in Waddell, ed. and trans., *The Desert Fathers,* 173-74.

31 Helen Waddell, *Beasts and Saints* (1934; reprint, London: Constable, 1970), xi. "Louted" means "bowed." Robert Mannyng of Brunne (1269-1340) was the author of *Handlyng Synne* – perhaps most appropriately rendered as a "Treatise on Sins" – which was a free translation of William of Wadington's *Manuel des Peschiez.* See Harwood, *Love for Animals,* 56, 59.

32 Anna Bonus Kingsford, *Perfect Way in Diet: A Treatise Advocating a Return to the Natural and Ancient Food of Our Race* (Kila, MT: Kessinger, [1906?]), 37. This book is a translation by Kingsford, with supplements, of her doctoral thesis *Thèse pour le doctoral: De l'alimentation végétale chez l'homme* (1880).

33 Henry Chadwick, "The Early Christian Community," in *The Oxford History of Christianity,* ed. John McManners, 38.

34 James Turner, *Reckoning with the Beast* (Baltimore: Johns Hopkins University Press, 1980), 2.

35 From Geoffrey Grigson, ed., *The Oxford Book of Satirical Verse* (Oxford: Oxford University Press, 1983), 352-53.

36 By contrast, since the sixteenth century, "beef" has found occasional usage in reference to ox or to any animal of the ox kind and, in its plural form as "beeves," has received slightly more extensive usage since the fourteenth century. I have encountered one instance where the word "venison" was used for "deer" – in a 1633 translation of Rabelais's *Gargantua and Pantagruel* (prt. 1, ch. 55) – perhaps precisely because it was a translation and the French usage would not have been informed by a history similar to that of the English usage.

37 James Serpell, *In the Company of Animals* (1986; reprint, Oxford: Basil Blackwell, 1988), 158.

38 Walter Scott, *Waverley Novels*, vol. 9, *Ivanhoe* (Edinburgh: Robert Caddell, 1848), 28-29. See St. Clair, *The Godwins and the Shelleys*, 10.

39 Jim Mason states that "the flesh from chickens, ducks or geese needs no euphemism because the animals are small and, as birds, more remote in degree of kinship." *An Unnatural Order: Uncovering the Roots of Our Domination of Nature and Each Other* (New York: Simon and Schuster, 1993), 174.

40 The Cheyenne creation myth is reproduced in Preece, *Awe for the Tiger*, 9-10.

41 Quoted in Paul Johnson, *The Birth of the Modern: World Society 1815-1830* (London: Phoenix, 1991), 408.

42 Mason, *An Unnatural Order*, 175-76.

43 Quoted in Keith Thomas, "The First Vegetarians," in *The Animal Movement*, ed. Kelly Wand (San Diego: Greenhaven Press, 2003), 32. The quotation is from Hazlitt's *The Plain Speaker*.

44 For the record, I am a vegetarian.

45 Carol J. Adams, *The Sexual Politics of Meat: A Feminist-Vegetarian Critical Theory* (1990; reprint, New York: Continuum, 2002), 115.

46 *Dictionary of National Biography* (London: Smith, Elder and Co., 1887), vol. 5, 218. Sir John Elliott's medical knowledge is derided, but his vegetarianism is ignored (vol. 17, 270-71). Two authors one might have expected to be treated ill on account of their radical reputations escaped almost unscathed. For example, the self-proclaimed Jacobin John Oswald, author of *The Cry of Nature* (1791) is described thus: "From intercourse with the Brahmins he imbibed certain curious beliefs. Although not accepting all their doctrines – for he was professedly an atheist – he shared their repugnances to flesh, from which he abstained on the professed ground of humanity, but was accustomed to drink wine plentifully." Percy Bysshe Shelley's vegetarianism passed by almost unnoticed: "About this time [i.e., 1812] he adopted the vegetarian system of diet, to which he adhered with more or less constancy when in England, but seems to have generally discarded when abroad."

47 Ibid., vol. 2, 285.

48 Adams, *The Sexual Politics of Meat*, 112.

49 Ibid.

50 Notably, *The Contrast* (1792), designed by Lord George Murray and engraved by Thomas Rowlandson, James Gillivray's *French Liberty and British Slavery* (1792), and Isaac Cruikshank's *French Happiness/English Misery* (1793), all designed to discredit supporters of the French Revolution. See Deborah Kennedy, *Helen Maria Williams and the Age of Revolution* (Lewisburg: Bucknell University Press, 2002), 105.

51 Adams, *The Sexual Politics of Meat*, 115.

52 Keith Thomas, *Man and the Natural World: Changing Attitudes in England 1500-1800* (London: Penguin, 1984), 295.

53 Ibid., 296.

54 *The Gentleman's Magazine and Historical Chronicle* 95, prt. 2 (July-December 1825): 642.

55 April 1789. Quoted in George Nicholson, *On the Primeval Diet of Man: Vegetarianism and Human Conduct toward Animals* (1801; reprint edited, introduced, and annotated by Rod Preece, Lampeter: Mellen Animal Rights Library, 1999), 198. The same extract can also be found in William H. Drummond, *The Rights of Animals and Man's Obligation to Treat Them with Humanity* (London: John Mardon, 1838), 97-98. The two accounts vary slightly, mainly with regard to punctuation, but the differences in no manner affect the message.

56 William Thackeray, *Vanity Fair: A Novel without a Hero* (1848; reprint, New York: Oxford University Press, 1984), 219.

57 Carol J. Adams, Introduction to Howard Williams, *The Ethics of Diet: A Catena of Authorities Deprecatory of the Practice of Flesh-Eating* (Urbana: University of Illinois Press, 2003), xiv.

58 Steven G. Kellerman, "Fish, Flesh, and Foul: The Anti-Vegetarian Animus," *American Scholar* 69, 4 (Autumn 2000): 85-97.

CHAPTER 1: IN QUEST OF THE SOUL

1 Donald Tannenbaum and David Schultz, *Inventors of Ideas: An Introduction to Western Political Philosophy,* 2nd ed. (Belmont, CA: Wadsworth/Thomson, 2004), 2. They suggest that perhaps Socrates introduced the idea. In fact, while Socrates may have elaborated the philosophical concept of the soul, he certainly did not introduce it as a philosophical image. It was common currency by at least the time of Pythagoras. And, of course, the idea of the soul as such (i.e., in its nonphilosophical dimensions) goes back to at least the first burials, perhaps some eighty thousand years ago.

2 H.V. Morton, *Through Lands of the Bible* (London: Methuen, 1938), 8, 12.

3 J. Howard Moore, *The Universal Kinship,* ed. Charles Magel (1906; reprint, Fontwell: Centaur, 1992), 5.

4 Ibid., xxxv.

5 See Richard Sorabji, *Animal Minds and Human Morals: The Origins of the Western Debate* (Ithaca: Cornell, 1993), 215.

6 Edward Clodd, Preface to 2nd revised edition of *Pioneers of Evolution: From Thales to Huxley, with an Intermediate Chapter on the Causes of Arrest of the Movement* (Kila, MT: Kessinger Publishing LLC, [1907?]), v-vi. It is interesting that Clodd sees evolution as a "movement" rather than as only a scientific theory. Moreover, Clodd's "intermediate chapter" is a useful reminder that in the early years of the twentieth century, evolutionary theory was falling out of favour. Its ultimate victory would have come as a surprise to many in the first quarter of the century.

7 Peter Singer, *Animal Liberation,* 2nd ed. (New York: New York Review of Books, 1990), 171.

8 See Adrian Desmond and James Moore, *The Life of a Tormented Evolutionist* (London: Michael Joseph, 1992), 427.

9 See pp. 350-57.

10 Elijah Buckner, *The Immortality of Animals, and the Relation of Man as Guardian, from a Biblical and Philosophical Hypothesis* (Philadelphia: George W. Jacobs, 1903), 31, 177, 204.

11 John Steinbeck, *The Log from the Sea of Cortez* (1941; reprint, London: Penguin, 1995), ch. 21, 78.

12 Quoted in Arthur O. Lovejoy, *The Great Chain of Being: A Study in the History of an Idea* (New York: Harper, 1960), 193, citing *Allgemeine Naturgeschichte und Theorie des Himmels* (1755), 133.

13 "On seeing a lost greyhound in winter, lying upon the snow in the fields," in *The Early Poems of John Clare 1804-1822*, vol. 1, ed. Eric Robinson and David Powell (Oxford: Clarendon Press, 1989), 203.

14 William H. Drummond, *The Rights of Animals and Man's Obligation to Treat Them with Humanity* (London: John Mardon, 1838), 22, 175.

15 Jean-Baptiste Lamarck, *Histoire naturelle des animaux sans vertèbres*, vol. 1 (Paris: J.B. Baillière, 1835), 335: "Étant dépourvus du sentiment, n'ayant pas même de leur existence."

16 Quoted in William F. Karkeek, "On the Future Existence of the Brute Creation, Part II," *The Veterinarian* 12, 142 (1839): 749.

17 Joseph Hamilton, *Animal Futurity: A Plea for the Immortality of the Brutes* (Belfast: C. Aitchison, 1877), 28.

18 Lovejoy, *The Great Chain of Being*, viii.

19 Peter J. Bowler, *The Norton History of the Environmental Sciences* (New York: W.W. Norton, 1993), 337.

20 Charles Darwin, *The Descent of Man, and Selection in Relation to Sex*, 2nd ed. (New York: A.L. Burt, 1874), 74. Darwin goes a little further than Clodd and writes of "very different degrees."

21 Bowler, *The Norton History of the Environmental Sciences*, 339.

22 "It'll Grow Back," *Globe and Mail*, 26 July 2003, F9.

23 Quoted in Hamilton, *Animal Futurity*, 66.

24 Ibid., 78.

25 W. Lauder Lindsay, *Mind in the Lower Animals*, vol. 1 (London: C. Kegan Paul, Tench, 1879), vii.

26 Ibid., viii.

27 Ibid., xi.

28 Marjorie Spiegel, *The Dreaded Comparison: Human and Animal Slavery* (New York: Mirror Books, 1996), 22.

29 See Henry Fairfield Osborn, *From the Greeks to Darwin: The Development of the Evolution Idea through Twenty-Four Centuries* (1929; reprint, New York: Arno Press, 1975).

30 Christine Kenyon-Jones, *Kindred Brutes: Animals in Romantic-Period Writing* (Aldershot: Ashgate, 2001), 205.

31 Michael W. Fox, *Inhumane Society* (New York: St. Martin's, 1990), 232.

32 Spiegel, *The Dreaded Comparison*, 21.

33 For example, in the 2nd edition of 1874 (see note 20), instances may be found on pages 24, 25, 40, 61, 62, 73, 81, 87, 98, 105, 125, 133, 135, 138, 153, and elsewhere.

34 Darwin, *The Descent of Man*, 191.

35 Ibid., 98.

36 Ibid., 40.

37 Ibid., 133.

38 Ibid., 105.

39 Ibid., 194.

40 Ibid., 199.

41 Ibid., 193.

42 Spiegel, *The Dreaded Comparison,* 34.

43 See, for example, an 1876 letter to one of his daughters, quoted in James Rachels, *Created from Animals: The Moral Implications of Darwinism* (Oxford: Oxford University Press, 1991), 213, citing Francis Darwin, ed., *The Life and Letters of Charles Darwin,* vol. 3 (1888; reprint, New York: Basic Books, 1959), 202-3.

44 Alfred Russel Wallace, *World of Life: A Manifestation of Creative Power, Directive Mind and Ultimate Purpose* (1911; reprint, New York: Moffat, Yard, 1916), 411.

45 Darwin, *The Descent of Man,* 167.

46 Ibid., 135.

47 Drummond, *The Rights of Animals,* 101.

48 Richard Milner, *The Encyclopedia of Evolution: Humanity's Search for Its Origins* (New York: Henry Holt, 1990), 201.

49 Rachels, *Created from Animals,* 64.

50 Spiegel, *The Dreaded Comparison,* 20.

51 Darwin, *The Descent of Man,* 143.

52 Mary Midgley, *Beast and Man: The Roots of Human Nature* (1979; reprint, London: Routledge, 1995), 208.

53 Mary Midgley, *Animals and Why They Matter* (Athens: University of Georgia Press, 1983), 90.

54 Ibid., 141. Darwin is, however, circumspect about the matter, noting in *The Descent of Man,* for example, that "without the accumulation of capital the arts could not progress" (153) and that "As Mr. [Walter] Bagehot has remarked [in *Fortnightly Review,* 1 April 1868, 452], we are apt to look at progress as normal in human society; but history refutes this. The ancients did not even entertain the idea, nor do the Oriental nations at the present day. According to another high authority, Sir Henry Maine [in *Ancient Law,* 1861, 22], 'the greatest part of mankind has never shown a particle of desire that its civil institutions should be improved'" (150).

55 Quoted in Milner, *The Encyclopedia of Evolution,* 32.

56 Darwin, *The Descent of Man,* 78.

57 See Erik Trinkaus and Pat Shipman, *The Neandertals: Changing the Image of Mankind* (New York: Alfred A. Knopf, 1993), 10.

58 T.H. Huxley, *Man's Place in Nature, and Other Collected Essays* (1863; reprint, New York: D. Appleton, 1900), 152.

59 For a graphic example, see Buckner, *The Immortality of Animals,* 119.

60 David N. Livingstone, *Darwin's Forgotten Defenders: The Encounter between Evangelical Theology and Evolutionary Thought* (Edinburgh: Scottish Academic Press, 1987), 108.

61 Darwin, *The Descent of Man,* 119ff. The opening words of Chapter 4 are: "I fully subscribe to the judgment of those writers who maintain that of all the differences between man and the lower animals, the moral sense or conscience is by far the most important."

62 Ibid., 83.

63 Ibid., 84.

64 Ibid., 54.

65 Ernest Barker, *Greek Political Theory: Plato and His Predecessors* (1918; reprint, London: Methuen, 1957), 73.

66 On Baldwin, Smuts, and Koestler, see Bowler, *The Norton History of the Environmental Sciences,* 454-55.

67 James L. Wiser, *Political Philosophy: A History of the Search for Order* (Englewood Cliffs, NJ: Prentice-Hall, 1983), 260.

68 See note 6 above.

69 John Locke, *Essay concerning Human Understanding,* vol. 2 (1690; reprint, London: H. Hills, 1710), 49.

70 Ernest Thompson Seton, *Wild Animals I Have Known* (1898 edition; facsimile reprint, London: Penguin, 1987), 12.

71 Moore, *The Universal Kinship,* 5.

72 John Rodman, "The Liberation of Nature?" *Inquiry* 20 (1972): 94.

73 *A Dictionary of the English Language in which the Words are Deduced from their Originals Explained in their Different Meanings, and Authorized by the Names of the Writers in Whose Works they are Found. By Samuel Johnson, Ll. D. Abridged from the Reverend H.J. Todd's Corrected and Enlarged Quarto Edition by Alexander Chalmers, F.S.A.* (London: C. and J. Rivington, 1824; facsimile reprint, New York: Barnes and Noble, 1994), 683.

74 Ibid., 689.

75 Ibid., 313.

76 Buckner, *The Immortality of Animals,* 29.

77 W.C. Spooner, "On the Non-Immortality of Animals," *The Veterinarian* 12, 150 (June 1840): 376.

78 *Cassell's Latin Dictionary* (London: Cassell and Company, 1948), 40.

79 Ibid., 868.

80 Juvenal, *Satire XV,* line 150.

81 William Gifford, ed. and trans., *Juvenal's Satires with The Satires of Persius* (1854; reprint, London: Dent, 1954), 15.

82 Quoted in Latin in Drummond, *The Rights of Animals,* 205: Quicquid est illud quod sentit, quod sapit, quod viget, caeleste et divinum est, ideoque aeternum.

83 Plotinus, *Enneads,* IV, 7, 6.

84 Benoit Patar, ed., *Le traité de l'âme de Jean Bouridan* [*de prima lectura*] (c. 1330; reprint, Louvain: Éditions de l'I.S.P., 1991). Despite the title, the work was, of course, written in Latin.

85 *Funk and Wagnalls New Practical Standard Dictionary of the English Language,* vol. 2 (1946; reprint, New York: Funk and Wagnalls, 1956), 1247.

86 Ibid., vol. 2, 848. Buckner, *The Immortality of Animals,* 30, undertakes a similar, if far less extensive analysis, via Webster's dictionary.

87 Hamilton, *Animal Futurity,* 50.

88 It is a matter of some minor interest, although I have been unable to determine of what significance, that there is no adjective corresponding to soul in the same manner that "spiritual" corresponds to spirit.

89 Buckner, *The Immortality of Animals,* 23-24.

90 Quoted in Jonathan Barnes, ed. and trans., *Early Greek Philosophy* (London: Penguin, 1987), 89.

91 See Daniel A. Dombrowski, *The Philosophy of Vegetarianism* (Amherst: University of Massachusetts Press, 1984), 50.

92 Martin West, "Early Greek Philosophy," in John Boardman, Jasper Griffin, and Oswyn Murray, eds., *The Oxford History of the Classical World* (1986; reprint, Oxford: Oxford University Press, 1993), 113.

93 Ibid.

94 Ibid., 114.

95 See Midgley, *Beast and Man,* especially the Introduction to the revised edition.

96 Quoted in Lewis G. Regenstein, *Replenish the Earth: A History of Organized Religion's Treatment of Animals and Nature, Including the Bible's Message of Conservation and Kindness toward Animals* (New York: Crossroad, 1991), 184, citing Rabbi Solomon Ganzfried, *Code of Jewish Law* (New York: Hebrew Publishing Company, 1962), bk. 4, ch. 19, 184.

97 Quoted in Richard H. Schwartz, "Tsa'ar Ba'alei Chayim: Judaism and Compassion for Animals," in *Judaism and Animal Rights: Classical and Contemporary Responses,* ed. Roberta Kalechofsky (Marblehead, MA: Micah Publications, 1992), 61. The Hebrew phrase is the mandate not to cause harm to animals.

98 W.E.H. Lecky, *The History of European Morals from Augustus to Charlemagne,* vol. 1 (New York: D. Appleton, 1869), 103.

99 Wollstonecraft's was the most emphatic and unequivocal: "In what does man's pre-eminence over the brute creation consist? The answer is as clear as that a half is less than the whole; in Reason." *Mary Wollstonecraft: Political Writings* (1792; reprint, ed. Janet Todd, Oxford: Oxford University Press, 1994), 50.

100 Mill was not, however, consistent on the point, sometimes choosing – as, for example, in his famous debate with Dr. William Whewell – a more egalitarian criterion of sentience. See John Stuart Mill, "Three Essays on Religion," in *Essays on Ethics, Religion and Society,* in *Works of John Stuart Mill,* vol. 10 (Toronto: University of Toronto Press, 1965), 185-87.

101 Quoted in Philip J. Ivanhoe, *Ethics in the Confucian Tradition: The Thoughts of Mencius and Yang-Ming* (Atlanta: Scholar's Press, 1990), 13, citing *Mencius,* Harvard-Yenching Institute Sinological Index Series, no. 17:54/7A/45.

102 Peter Singer, *Animal Liberation,* 2nd ed. (New York: New York Review of Books, 1990), 188, 189.

103 Jim Mason, *An Unnatural Order: Uncovering the Roots of Our Domination of Nature and Each Other* (New York: Simon and Schuster, 1993), 33.

104 Rachels, *Created from Animals,* 87.

105 See, for example, Plutarch, 224-27.

106 Aristotle, *On Man in the Universe: Metaphysics, Parts of Animals, Ethics, Politics, Poetics,* edited with an Introduction by Louise Ropes Loomis (1943; reprint, New York: Grammercy, 1971), 5.

107 Thomas Hobbes, *Leviathan,* ed. C.B. MacPherson (London: Penguin, 1981), prt. 1, ch. 6, 124.

108 Ibid., prt. 1, ch. 4, 100.

109 Aristotle, *The Politics of Aristotle,* ed. and trans. Ernest Barker 1.8.11-12 (1946; reprint, London: Oxford University Press, 1952), 21.

110 Thomas Hobbes, *The English Works of Thomas Hobbes of Malmesbury,* vol. 5, ed. Sir William Molesworth (London: G. Bohn, 1839), 187-88.

111 Nonetheless, these are abuses rather than uses of speech. In *Leviathan,* prt. 1, ch. 4, Hobbes tells us: "To these Uses [of speech] there are also four correspondent Abuses ... Fourthly, when they use them to grieve one another; for seeing nature hath armed living creatures, some with teeth, some with horns, and some with hands, to grieve an enemy, it is but an Abuse of speech, to grieve him with the tongue, unless it be one whom we are obliged to govern; and then it is not to grieve, but to correct and amend."

112 Hobbes, *Leviathan,* prt. 1, ch. 4, 109.

113 Ibid., prt. 1, ch. 3, 98.

114 Thomas Hobbes, *The Elements of Law: Natural and Politic,* ed. Ferdinand Tönnies (New York: Barnes and Noble, 1969), xv.

115 Hobbes, *Leviathan,* prt. 1, ch. 11, 161.

116 Ibid., ch. 12.

117 Aristotle, *Parts of Animals,* 2.10, in *On Man in the Universe,* 62.

118 Ibid., 1.3.46-47.

119 Ibid., 1.4.49, 50.

120 Ibid., 2.2.52.

121 Aristotle, *Nicomachean Ethics,* trans. Martin Oswald (Indianapolis: Bobbs-Merrill, 1962), 17.

122 Aristotle, *The Politics of Aristotle,* 1.2.10.5-6; 1.2; 1.5.8.13.

123 Ibid., 7.13.12.314.

124 "The slave is entirely without the faculty of deliberation; the female indeed possesses it, but in a form which remains inconclusive; and if children also possess it, it is only in an immature form." Ibid., 1.13.7.

125 Ibid., 7.1.4.280.

126 Aristotle, *Parts of Animals,* 1.3.47.

127 See chs. 4 and 5.

128 Buckner, *The Immortality of Animals.* Graphic examples are offered on pages 13, 23, 28, 29, 39, and elsewhere.

129 Ibid., 170, 184, 189.

130 Ibid., 115, 189, 115.

131 Whether Plutarch was being strictly fair to the Stoics is a moot point. What is of significance, however, is how the treatment illuminates the question of rationality and the possession of an immortal soul.

132 Plutarch, *Plutarch's Moralia,* vol. 12, trans. Harold Cherniss and William C. Helmblod (1927; reprint, London: William Heinemann, 1957), 327.

133 See James E. Gill, "Theriophily in Antiquity: A Supplementary Account," *Journal of the History of Ideas* 30 (July-September 1969): 409.

134 Pierre Bayle, *Historical and Critical Dictionary: Selections,* trans. Richard H. Popkin (Indianapolis: Bobbs-Merrill, 1965), 215 n. B.

135 All three examples are to be found in the 1961 *Compact Edition of the Oxford English Dictionary,* vol. 2, 2621, 1840 edition, citing consecutively, p. 189.

136 Priscilla Wakefield, *Instinct Displayed, in a Collection of Well-Authenticated Facts, Exemplifying the Extraordinary Sagacity of Various Species of the Animal Creation,* 4th ed. (London: Harvey and Darton, 1821), vii-ix.

137 Karkeek, "On the Future Existence of the Brute Creation," 660 (italics in the original).

138 S.T. Coleridge, "To a Young Ass: Its Mother Being Tethered Near It," in *The Poetical Works of Samuel Taylor Coleridge,* ed. Ernest Hartley Coleridge (London: Henry Fowde, 1912), 74-76.

139 S.T. Coleridge, "The Rime of the Ancient Mariner," in ibid., 209.

140 Fyodor Dostoevsky, *The Brothers Karamazov,* trans. Richard Pavear and Larissa Volokhonsky (1879-80; reprint, New York: Alfred A. Knopf, 1992), 238.

141 Ibid., 241-42.

142 Of course, Darwin wrote of the *Descent of Man,* referring to the derivation of humankind, rather than the "Ascent of Man," as Twain implies. Yet, in the sense of the *progression* of animalkind, Darwin writes in terms of humanity having "ascended" to the apex.

143 M. Twain, "Man's Place in the Animal World" (1896), in Mark Twain, *Collected Tales: Sketches, Speeches, & Essays, 1891-1910*, ed. Louis J. Budd (New York: Library of America, 1992), 207-10, 212, 216.

144 Buckner, *The Immortality of Animals*, 56.

145 Christopher Smart, *Jubilate Agno*, ed. W.H. Bond (1760; reprint, London: Rupert Hart Davis, 1954), Fragment B2, line 719.

146 Jane Goodall, *Through a Window: My Thirty Years with the Chimpanzees of Gombé* (Boston: Houghton Mifflin, 1990).

147 Thus James L. Wiser says: "humanity has both reason and speech. Whereas such thinkers as Heraclitus and Aristotle understood these faculties to be indications of humankind's appropriateness for political life, Hobbes sees them as sources of social disruption. For example, reason produces innovation and innovation threatens the social order. Speech, in turn, is the means by which individuals can propagate false opinions, and false opinions, according to Hobbes, are one of the causes of warfare. Finally, human beings' extreme vainglory encourages them to take offense even in those situations where no damage was intended. Thus human conflict is a continual possibility even among well-intentioned citizens." *Political Philosophy: A History of the Search for Order* (Englewood Cliffs, NJ: Prentice Hall, 1983), 202.

148 "Hinder" but not prevent. Many birds are especially adept at "lying" to predators about the location of their nests, eggs, and chicks, leading them a merry dance away from those they are protecting.

149 Thus Declan Kibberd in his Introduction to *Ulysses* (London: Penguin, 1992) writes: "those who know how to feel often have no capacity to express themselves, and by the time they have acquired the expressive capacity, they have all but forgotten how to feel ... languages have a far tighter hold on mankind than mankind ever gains on them" (xli, xliii).

150 Adapted from Charles Dickens, *A Tale of Two Cities* (1859; reprint, London: Penguin, 1970), 35. I have substituted "animals," "animal," "it," and "they," for "times," "age," "epoch," "season," "spring," "winter," and "we."

151 James Turner, *Reckoning with the Beast: Animals, Pain and Humanity in the Victorian Mind* (Baltimore: Johns Hopkins University Press, 1980), 23.

152 Hamilton, *Animal Futurity*, x, xi.

153 Ibid., 18.

154 William F. Karkeek, *The Veterinarian* 13, 142 (1839): 748. In the same journal for August 1840, Karkeek adumbrates their contributions. See pages 514-16. Crousaz is perhaps Jean-Pierre de Crousaz (1663-1730), Swiss professor of logic.

155 Not a very profitable example; see pp. 83-84

156 Buckner, *The Immortality of Animals*, 73-89.

157 Quoted in ibid., 37-38.

158 Buckner, *The Immortality of Animals*, 18.

159 Ibid., 19.

160 Ibid., 87.

161 Hieronymus Rorarius, *Quod animalia bruta saepe Ratione utantur melius Homine. Libro duo: Qua recensuit Dissertatione Historica-Philosophica de Anima Brutorum*. Annotated by Geor. Hein. Ribovius (Helemstadtii: Chri. Frider. Weygandi, 1728).

162 David-Renaud Boullier, *Essai philosophique sur l'Âme des Bêtes, où l'on traite de son existence & de sa nature. Et où l'on mêle par occasion Diverses Reflexions sur la nature de*

la Liberté, sur celle de nos Sensations, sur l'Union de l'Âme & du Corps, sur l'immortalité de l'Âme &c. Et où l'on réfute diverses Objections de M. Bayle (Amsterdam: F. Changuion, 1728).

163 (A critical history of works on the souls of animals, the views of the philosophers of antiquity, and those of modern philosophers on this topic.)

164 Drummond, *The Rights of Animals*, 197.

165 Ezra Abbot, Appendix 2 of "Nature, Origin and Destiny of the Souls of Brutes," in William Rounseville Alger, *A Critical History of the Doctrine of a Future Life* (1862; reprint, Philadelphia: George W. Childs, 1864), 868.

166 Ibid., 632.

CHAPTER 2: PERIPATETIC SOULS

1 Wendy Doniger, ed., *The Laws of Manu* (Harmondsworth: Penguin, 1991), 12, 57, 283-84.

2 Elijah Buckner, *The Immortality of Animals* (Philadelphia: George W. Jacobs, 1903), 88.

3 Burton Watson, ed., *Lotus Sutra* (New York: Columbia, 1993), 74, 77.

4 Nicholas J. Saunders, *Animal Spirits* (Boston: Little, Brown, 1995), 22.

5 Quoted in Christopher Key Chapple, *Nonviolence to Animals, Earth and Self in Asian Traditions* (Albany: State University of New York Press, 1993), 27.

6 Aquinas, *Summa contra Gentiles,* vol. 4, trans. Charles J. O'Neil (Grand Bend: Notre Dame University Press, 1975), 42, sec. 3.

7 The belief in transmigration did occur in ancient Egypt but not with any great frequency. Occasionally, human souls united with those of gods or entered an animal for a time, either willingly or unwillingly. I can find no persuasive evidence for any direct Egyptian influence. However, one of Plato's arguments allows for human and animal souls to choose their bodily destination, which appears closer to Egyptian than to Indian doctrine.

8 See, for example, William H. Drummond, *The Rights of Animals and Man's Obligation to Treat Them with Humanity* (London: John Mardon, 1838), 29.

9 It has become customary to specify dates as being either CE (common era) or BCE (before common era). However, the commonality is restricted to the Abrahamic tradition. The terms could thus appear demeaning to all outside that tradition. While BC and AD are also agenda-laden, it is now well known that they do not coincide with the birth of Christ and are thus artificial, almost arbitrary, and for this reason preferable.

10 Charles H. Kahn, *Pythagoras and the Pythagoreans: A Brief History* (Indianapolis: Hackett Publishing, 2001), 3-4. I am heavily indebted to Kahn's groundbreaking study for much of my understanding of the development of Pythagoreanism. Also of value has been Daniel A. Dombrowski, *The Philosophy of Vegetarianism* (Amherst: University of Massachusetts Press, 1984), ch. 3.

11 Tacitus, *Annals* ([AD 100?]; reprint, Franklin Center, PA.: Franklin, 1982), 101.

12 Plato, *Statesman*, 271d-274c, in *Plato,* trans. Harold N. Fowler and W.R.M. Lamb (London: W. Heinemann, 1925).

13 See page 72.

14 Quoted in C. Northcote Parkinson, *The Evolution of Political Thought* (London: University of London Press, 1958), 20.

15 Quoted in John Passmore, *Man's Responsibility for Nature* (London: Duckworth, 1974), 7, citing Chuang Tsu.

16 *The New Jerusalem Bible: Standard Edition* (New York: Doubleday, 1998), 3. *The New Jerusalem Bible* is a modern translation of the original scriptures undertaken by l'Ecole Biblique in Jerusalem in the 1950s. All further Biblical quotations are taken from this translation unless otherwise noted.

17 Geoffrey Chaucer, "The Nun's Priest's Tale," in *The Canterbury Tales,* trans. Nevill Coghill (London: Cresset Press, 1986), 136.

18 Richard Erdoes and Alfonso Ortiz, eds., *American Indian Myths and Legends* (1984; reprint, London: Pimlico, 1997), 111-14. The story is based on a rendering by George A. Dorsey in 1905, which is entirely consistent with that reported by the American Museum of Natural History in 1899. The story is repeated in Rod Preece, *Awe for the Tiger, Love for the Lamb: A Chronicle of Sensibility to Animals* (Vancouver: UBC Press; London and New York: Routledge, 2002), 8-9.

19 Hesiod, *Works and Days,* in *Theogony and Works and Days,* trans. M.L. West (Oxford: Oxford University Press, 1988), 42.

20 See "How the Buffalo Hunt Began," in Margot Edmonds and Ella E. Clark, eds., *Voices of the Winds: Native American Legends* (New York: Facts on File, 1989), 184-85. The story is repeated in Preece, *Awe for the Tiger, Love for the Lamb,* 12-13.

21 Immanuel Kant, "Duties toward Animals and Spirits," in *Lectures on Ethics* ([1780?]; modern edition 1930, trans. Louis Infield; reprint, New York: Harper and Row, 1963), 240.

22 Joseph Conrad, *Nostromo* (1904; reprint, London: J.M. Dent, 1995), prt. 1, ch. 8, 81.

23 Ibid., prt. 2, ch. 6, 156.

24 See Mathias Guenther, *Tricksters and Trancers: Bushman Religion and Society* (Bloomington: Indiana University Press, 1999), 126-45. A short extract is to be found in Preece, *Awe for the Tiger, Love for the Lamb,* 15.

25 Watson, ed., *Lotus Sutra,* 75.

26 Samuel Richardson, *Pamela* (1741; reprint, New York: W.W. Norton, 1993), Letter 32, 142.

27 See Rod Preece, *Awe for the Tiger, Love for the Lamb,* 116.

28 Porphyry, *Vita Pythagorae,* as quoted in Kahn, *Pythagoras and the Pythagoreans,* 105.

29 Plato, *Laws,* trans. Trevor J. Saunders (Baltimore: Penguin, 1970), 395.

30 "Epistle 7," in *Plato's Epistles,* trans. Glenn R. Morrow (Indianapolis: Bobbs-Merrill, 1962), 230.

31 Plato, *Statesman,* trans. J.B. Skemp (Indianapolis: Bobbs-Merrill, 1957), 97.

32 See, for example, the issue for June 1840, 511.

33 Quoted in Simon Heffer, *Moral Desperado: A Life of Thomas Carlyle* (London: Phoenix, 1995), 293.

34 See Peter Levi, *Tennyson* (New York: Charles Scribner's Sons, 1993), 152.

35 See Richard Sorabji, *Animal Minds and Human Morals: The Origins of the Western Debate* (Ithaca: Cornell University Press, 1993), 192-93.

36 Ibid., 65-68, 72, 97-99, 101, 190-93.

37 "Hindu Perspectives on the Use of Animals," in Tom Regan, ed., *Animal Sacrifices: Religious Perspectives on the Use of Animals in Science* (Philadelphia: Temple University Press, 1986), 200-5.

38 Plato, *Republic,* trans. Richard W. Sterling and William C. Scott (New York: W.W. Norton, 1985), 290 (600b1-5). For the "realities and illusions," see 289 (599a9).

39 Ibid., 225, bk. 7 (531c6-7).

40 See Kahn, *Pythagoras and the Pythagoreans,* 3-4.

41 Ibid., 51.

42 Plato, *Phaedo,* in *The Trial and Death of Socrates: Four Dialogues,* trans. Benjamin Jowett ([1875?]; reprint, New York: Dover, 1992), 63.

43 See Kahn, *Pythagoras and the Pythagoreans,* 50; and Plato, *Gorgias: A Revised Text with Introduction and Commentary by E.R. Dodds* (Oxford: Clarendon Press, 1959), 303, 381.

44 Plato, *Phaedo,* in *The Trial and Death of Socrates,* 74, 75.

45 Ibid., 80-81.

46 In Brahminic Hinduism there are four *varna* (literally "colour"), or castes: the *Brahmin,* whose tasks are religious, intellectual, and instructional, corresponding to Plato's philosopher kings; the *Kshatriya,* the warriors, whose tasks are protective and military, corresponding to Plato's auxiliaries; the *Vaishya,* who perform agricultural and business functions, corresponding to Plato's artisans; and the *Shudra,* who are involved in menial labour, tasks undertaken by noncitizens in Plato's republic. There are also outcastes who correspond to slaves.

47 Plato, *Republic,* trans. H. Spens (Glasgow: Robert and Andrew Foulis, 1763), 427-28.

48 Quoted in Arthur O. Lovejoy and George Boas, *Primitivism and Related Ideas in Antiquity* (1935; reprint, Baltimore: Johns Hopkins University Press, 1997), 396, citing Th. Kock, *Com. att. fragm.* 3, 63-64.

49 See Florence M. Firth, ed., with an introduction by Annie Besant, *The Golden Verses of Pythagoras and Other Pythagorean Fragments* (1905; reprint, Kila, MT: Kessinger, 1997). The Preface indicates that what has been compiled are "the best and most reliable of the sets of Ethical Verses attributed to the Pythagoreans." Reliance is primarily on "Rowe's translation from the French of André Dacier (1707)" and, to a lesser degree, on "Hall's translation from the Greek," undertaken in 1657. Other parts are from "Bridgman's translation ... (1804)" of the "Golden Sentences of Democrates, the Similitudes of Demophilus, and the Pythagorean Symbols." It should be clear that it is not the "Golden Verses" of Pythagoras himself that are purportedly being cited but verses of Pythagoreans who lived many hundreds of years after Pythagoras and who possessed very little reliable evidence as to what Pythagoras had ever claimed. As a reflection of the historical confusion, it is worth noting that, according to some sources, Pythagoras believed beans to be the most valuable food and that, according to others, Pythagoras forbad the consumption of beans in his community. Edouard Schure, a late-nineteenth-century French Gnostic mystic, wrote a very readable account of *Pythagoras and the Delphic Mysteries* (English-language edition [1907?]; reprint, Kila, MT: Kessinger, n.d.), which is still popular in some circles. Unfortunately, Schure repeats the old barb of Moderatus that "Plato, at great trouble and cost, obtained through [the Pythagorean] Archytas a manuscript of the Master" (9) – hence Plato was a plagiarist and his writings were no more than a repetition of Pythagoras. "The essence of the system" of Pythagoras, Schure indicates, "comes down to us in the *Golden Verses* of Lysis, the commentary of Hierocles, fragments of Philolaus and in the Timaeus of Plato, which contains the cosmogony of Pythagoras" (9). What follows is an enjoyable, but quite inaccurate, journey through Greek thought, reaching the conclusion that "In truth this [Pythagorean] edifice was never destroyed. Plato, who took from Pythagoras the whole of his metaphysics, had a complete idea thereof, though he unfolded it with less clearness and precision" (179). Suffice it to say that Schure's conclusions, although not without an occasional insight, are rejected by such modern

Pythagorean scholars as Walter Burkert, Carl Huffman, Charles H. Kahn, and Dominic O'Meara – who calls the *Golden Verses* "pseudo-Pythagorean" in *Pythagoras Revived: Mathematics and Philosophy in Late Antiquity* (Oxford: Clarendon Press, 1989), III. The verses are, however, given rather more credence by Martin West, "Early Greek Philosophy," in *Oxford History of the Classical World*, ed. John Boardman, Jasper Griffin, and Oswyn Murray (Oxford: Oxford University Press, 1993), which tells us that Pythagoras "bequeathed to his disciples in south Italy a quantity of brief maxims, catechisms, and enigmatic sayings, some expressing old religious taboos, others cosmological or eschatological dogmas" (114).

50 Dombrowski, *The Philosophy of Vegetarianism,* 36.

51 "Vegetarian," it should be pointed out, is said to derive not from vegetable but from the Latin "vegetus," meaning "lively, vigorous, active" (*Cassell's Latin Dictionary,* 1948, 602). However, although I have read the claim in several places, I have not been able to verify the origins independently. The reason I am skeptical is that other information found at the same sources is less than convincing. Thus, for example, in one of them, a list of vegetarians includes Socrates (I have been able to discover no evidence of his vegetarianism), Plato (definitely not, although it is possible that he intended his republic to be a vegetarian city on ascetic grounds), Sir Isaac Newton (decidedly not so, although Voltaire tells us that he had meats prepared in his kitchen without the customary cruelty), Voltaire (an advocate but not a practitioner), Rousseau (likewise), Benjamin Franklin (just for a short time in his youth), Lamartine (in private, but not in public, on his own testimony), Thoreau (no, although he thought that there was something to be said for it), Rabindranath Tagore (certainly not; indeed, he was that rarity, a Brahmin who not only hunted turtles, killing them cruelly, but even ate beef). Vegetarian societies, or at least a large proportion of them, are as careless in their evidence as some of the scholars who write on animal issues. One of the reasons that I am skeptical about the origins of the usage of "vegetarian" referring to "lively, vigorous, active" is that already in the early nineteenth century, we find Percy Bysshe Shelley writing "On the Vegetable System of Diet" with decided reference to the kind of food consumed. Even earlier, John Frank Newton, apparently Shelley's vegetarian mentor, wrote *Return to Nature, or A Defence of the Vegetable Regimen* (London: T. Cadell and W. Davies, 1811). It is probable that this kind of usage was persuasive in the 1840s, when the term "vegetarian" appears to have been coined. The Oxford English Dictionary gives the first usage as 1842, but its general application is said to follow the founding of the Vegetarian Society at Ramsgate, Kent, in 1847.

52 All three biographies may be found in Kenneth Sylvan Guthrie, comp. and trans., *The Pythagorean Sourcebook and Library* (Grand Rapids: Phanes Press, 1988), 57-122 (Iamblichus); 123-35 (Porphyry); 141-56 (Diogenes Laertius). There is also a short anonymous account that was preserved by Photius (137-40).

53 Ovid, *Metamorphoses,* trans. Mary M. Innes (London: Penguin, 1955), bk. 15, 337.

54 Ibid., 339.

55 Quoted in Howard Williams, *The Ethics of Diet: A Catena of Authorities Deprecatory of the Practice of Flesh-Eating* (London: F. Pitman, 1883), 7.

56 See the biographies in Guthrie, comp. and trans., *The Pythagorean Sourcebook and Library,* 137, 130.

57 *Statesman,* 269-74; *Statesman,* 288e; *Laws,* 847e; *Republic,* 332c; *Laws,* 667b; *Republic,* 404c. See Dombrowski, *The Philosophy of Vegetarianism,* 23, 58.

58 Ibid., 62-63.
59 Williams, *The Ethics of Diet*, vi.
60 *Republic*, 373b, in *The Great Dialogues of Plato*, trans. W.H.D. Rouse (New York: Mentor, 1956), 169.
61 Ibid., 371a.
62 Williams, *The Ethics of Diet*, 8.
63 For example, in the *Republic* Plato allows for freedom of movement between the classes of society, based upon intellectual potential. Yet the age at which intellectual training of the guardian class is to begin effectively bars children among the ranks of the artisans from taking their rightful place.
64 From Jonathan Barnes, ed. and trans., *Early Greek Philosophy* (London: Penguin, 1987), 84.
65 From ibid., 87.
66 Seneca, *Ad Lucilium Epistolae Morales,* vol. 3, trans. Richard M. Gummere (New York: G.P. Putnam's Sons, 1925), 241-43.
67 From Barnes, ed. and trans., *Early Greek Philosophy,* 82.
68 From ibid., 200.
69 From ibid., 196.
70 From ibid., 199-200.
71 See Kahn, *Pythagoras and the Pythagoreans*, 149.
72 Ibid., 150.
73 Ibid.
74 See Sorabji, *Animal Minds and Human Morals,* 189.
75 Ibid.
76 Arnobius, *The Seven Books of Arnobius Adversus Gentes*, trans. Hamilton Bruce and Hugh Campbell, in *Ante-Nicene Christian Library: Translations of the Fathers down to AD 325,* vol. 19 (Edinburgh: T.T. Clark, 1871), 82-84.
77 See Sorabji, *Animal Minds and Human Morals,* 99, 188, 202.
78 Karl Popper, *The Open Society and Its Enemies,* vol. 2 (London: Routledge and Kegan Paul, 1957), 272.
79 Keith Thomas, *Man and the Natural World: Changing Attitudes in England, 1500-1800* (London: Penguin, 1984), 138.
80 Thomas Young, *An Essay on Humanity to Animals,* ed. Rod Preece (1798; reprint, Lampeter: Mellen Animal Rights Library, 2001), 140.
81 Quoted in Peter Singer, *Animal Liberation,* 2nd ed. (New York: New York Review of Books, 1990), 201.
82 *Compact Edition of the Oxford English Dictionary,* vol. 2 (Oxford: Oxford University Press, 1971), 1675.
83 Quoted in Thomas, *Man and the Natural World,* 138, citing Laurence Clarkson, *Look About You* (1659), 98.
84 John Wilmot, "Satyr," in *The Poems of John Wilmot, Earl of Rochester,* ed. Keith Walker (Oxford: Basil Blackwell, 1984), 91, line 4.
85 John Gay, "On Walking the Streets by Day," book 2 of *Trivia*, in *John Gay: Poetry and Prose,* vol. 1, ed. Victor A. Dearing with the assistance of Charles E. Beckworth (Oxford: Clarendon Press, 1974), 150 (italics in the original).
86 Henry Fielding, *The Champion: Containing a Series of Papers, Humorous, Moral, Political, and Critical Edited by Henry Fielding,* no. 56, 22 March 1739 (1940). Microform.

87 Soame Jenyns, *A Free Inquiry into the Nature and Origin of Evil,* 2nd ed. (London: R. and J. Dodsley, 1757; facsimile reprint, New York: Garland, 1976), 76-77.

88 See Dix Harwood, *Love for Animals and How It Developed in Great Britain* (1928; reprint, ed. Rod Preece and David Fraser, Lampeter: Mellen Animal Rights Library, 2002), 331.

89 Guthrie, comp. and trans., *The Pythagorean Sourcebook and Library,* 13.

90 William Wordsworth, *The Works of William Wordsworth* (Ware: Wordsworth, 1994), 588, stanza 5, lines 58-63; stanza 4, lines 36-41.

91 Quoted in James King, *William Blake: His Life* (London: Weidenfeld and Nicholson, 1991), 145.

92 Charles Baudelaire, *Selected Poems from "Flowers of Evil,"* trans. Wallace Fowlie (1857; reprint, New York: Dover Publications, 1992), 8. One may, however, choose to interpret the poem as referring to ahistorically imagined days earlier in his historical life.

93 Henry Fielding, *Joseph Andrews,* in *Joseph Andrews and Shamela* (1742 and 1749; reprint, Oxford: Oxford University Press, 1980), bk. 3, ch. 1.

94 Richard Ellmann, *Oscar Wilde* (London: Penguin, 1988), 9.

95 Isaac Bashevis Singer, *Law and Exile: An Autobiographical Trilogy* (New York: Farrar, Strauss and Giroux, 1997), 19.

96 West, "Early Greek Philosophy," 116.

97 Juan Mascaró, trans., *The Upanishads* (London: Penguin, 1965), 124.

98 Giordano Bruno, *Cause, Principle, and Unity: Five Dialogues,* trans. Jack Lindsay (New York: International Publishers, 1962), 81.

99 Quoted in John Ray, *The Wisdom of God Manifested in the Works of Creation* (London: Samuel Smith, 1691; facsimile reprint, New York: Garland, 1979), 127-29.

100 Johann Gottfried Herder, "Ideas for a Philosophy of the History of Mankind," bk. 3, sec. 6, in F.M. Barnard, ed. and trans., *Herder on Social and Political Culture* (Cambridge: Cambridge University Press, 1969), 255.

101 (Ideas toward a philosophy of nature.)

102 Ralph Waldo Emerson, "The Over-Soul," in *Essays: First Series,* in *The Selected Writings of Ralph Waldo Emerson,* ed. Brooks Atkinson (New York: The Modern Library, 1992), 237.

103 Ralph Waldo Emerson, *Nature,* in ibid., 5, 12, 21, 23.

104 Hegel too has something to say of the world-soul, or "world-spirit" (Weltgeist), as he calls it. Yet Hegel means something substantially different, and I have not included his words in the body of the text. For Hegel, the world-soul is the driving force of history: "I saw the Emperor [Napoleon at Jena in 1806] – this world soul – riding out of the city on reconnaissance ... I adhere to the view that the world-spirit has given the age marching orders. These orders are being obeyed. The world-spirit, this essential power, proceeds irresistibly like a closely-drawn armoured phalanx." Quoted in Paul Johnson, *The Birth of the Modern: World Society, 1815-1830* (London: Phoenix, 1991), 813-14.

105 See Rod Preece, *Animals and Nature: Cultural Myths, Cultural Realities* (Vancouver: UBC Press, 1999), 259-62.

106 Williams, *The Ethics of Diet,* 10.

107 Plato, *Euthyphro,* in *The Trial and Death of Socrates: Four Dialogues* (New York: Dover, 1992), 12.

108 See Preece, *Awe for the Tiger, Love for the Lamb,* 233-35.

109 John Steinbeck, *Cannery Row* (1945; reprint, London: Penguin, 1994), ch. 2, 18.

110 Ibid., ch. 26, 133.

111 Quoted in Seamus Deane's notes to James Joyce, *A Portrait of the Artist as a Young Man* (1916; reprint, London: Penguin, 1993), 324.

112 See Ben Lazare Mijuskovic, *The Achilles of Rationalist Arguments: The Simplicity, Unity and Identity of Thought and Soul from the Cambridge Platonists to Kant: A Study in the History of an Argument* (The Hague: Martinus Nijhoff, 1974).

113 *The Veterinarian* 12, 142 (1839): 653-67; 12, 143 (1839): 748-58; 12, 144: 773-85; 13, 150: 245-347; and 13, 152: 513-22.

114 Christian legend has it that, during the persecutions of Decian (c. 250), seven martyrs were entombed in a grotto on the northern slopes of Mount Pion, near Ephesus. In one of the versions of the legend, a faithful dog saves the seven men. The story was made popular by Goethe.

115 William Rounseville Alger, *The Destiny of the Soul: A Critical History of the Doctrine of a Future Life,* 10th ed., vol. 1 (Philadelphia: George W. Childs, 1878), 36.

CHAPTER 3: A NATURAL HISTORY OF ANIMAL SOULS

1 For a lengthy treatment of this topic, see Rod Preece, *Animals and Nature: Cultural Myths, Cultural Realities* (Vancouver: UBC Press, 1999), especially ch. 7.

2 Christopher Key Chapple, *Nonviolence to Animals, Earth, and Self in Asian Traditions* (Albany: State University of New York Press, 1993), 42.

3 Peter Knudtson and David Suzuki, *Wisdom of the Elders* (Toronto: Stoddart, 1992), 13, 15.

4 For a learned discussion of the concept and its relation to justice, see Robert Murray, *The Cosmic Covenant: Biblical Themes of Justice, Peace and the Integrity of Creation* (London: Sheed and Ward, 1992).

5 Edward Payson Evans, "Ethical Relations between Man and Beast," *Popular Science Monthly* (September 1894), included in Evans's *Evolutional Ethics and Animal Psychology* (New York: W.D. Appleton, 1897), 82-104; Lynn White Jr., "The Historical Roots of Our Ecologic Crisis," *Science* 155 (1967): 1203-7; William Leiss, *The Domination of Nature* (1972; reprint, Montreal and Kingston: McGill-Queen's University Press, 1984), 48; John Passmore, *Man's Responsibility for Nature: Ecological Problems and Western Traditions* (London: Duckworth, 1974), 6; Peter Singer, *Animal Liberation,* 2nd ed. (New York: New York Review of Books, 1990), 187; Keith Thomas, *Man and the Natural World: Changing Attitudes in England, 1500-1800* (London: Penguin, 1984), 17-18; Roderick Frazier Nash, *The Rights of Nature: A History of Environmental Ethics* (Madison: University of Wisconsin Press, 1984), 50-52; Jim Mason, *An Unnatural Order: Uncovering the Roots of Our Domination of Nature and Each Other* (New York: Simon and Schuster, 1993), 26-28.

6 Theodore Hiebert, *The Yahwist's Landscape: Nature and Religion in Early Israel* (New York: Oxford University Press, 1996), 157. Hiebert explains that the Pentateuch – the first five books of the Old Testament, or Hebrew Bible – is thought to consist of "a combination of four different sources or documents, authored by four different authors living at different times in Israelite history" (24). The oldest have been attributed to the Yahwist and are designated J (from Jahwist in German). These older narratives, it is hypothesized, "were later incorporated into a new edition of Israel's beginnings prepared by Priestly Writer(s)/Editor(s) (abbreviated P)" (24).

7 Elijah Buckner, *The Immortality of Animals* (Philadelphia: George W. Jacobs, 1903), 17-18, especially the latter.

8 More recent renderings present the same message. Thus, according to *The New Jerusalem Bible* translation, "God said ... let them be masters of the fish of the sea, the birds of heaven, the cattle, all the wild animals and all the creatures that creep along the ground."

9 Andrew Linzey, *Christianity and the Rights of Animals* (New York: Crossroad, 1991), 25-28.

10 Elijah Judah Schochet, *Animal Life in Jewish Tradition: Attitudes and Relationships* (New York: KTAV Publishing House, 1984), 46-79. The phrase "the delicate tool" is at page 63.

11 Matthew Hale, *The Primitive Organization of Mankind* (London: n.p., 1677), sec. 4, ch. 8.

12 Quoted in William Youatt, *The Obligation and Extent of Humanity to Brutes* (1839; reprint, ed. Rod Preece, Lampeter: Mellen Animal Rights Library, 2003), 40.

13 James Thomson, "Spring," in *The Seasons and The Castle of Indolence,* ed. James Sambrook (New York: Oxford University Press, 1972), 9, line 241.

14 John Brown, *Self-Interpreting Bible* (1776; reprint, Glasgow: Blackie, 1834), 2.

15 Thomas Young, *An Essay on Humanity to Animals* (1798; reprint, ed. Rod Preece, Lampeter: Mellen Animal Rights Library, 2001), 59.

16 George Nicholson, *On the Primeval Diet of Man: Vegetarianism and Human Conduct toward Animals* (1801; reprint, ed. Rod Preece, Lampeter: Mellen Animal Rights Library, 1999), 12.

17 Joseph Ritson, *An Essay on Abstinence from Animal Food as a Moral Duty* (London: Richard Phillips, 1802), 164.

18 Quoted in Youatt, *The Obligation and Extent of Humanity to Brutes,* 41.

19 William H. Drummond, *The Rights of Animals and Man's Obligation to Treat Them with Humanity* (London: John Mardon, 1838), 2, 21, 47, 48, 167, 50.

20 William F. Karkeek, "An Essay on the Future Existence of the Brute Creation," *The Veterinarian* 12, 144 (December 1839): 750, 751. Karkeek also quotes Drummond at considerable length on the matter at pages 751-52.

21 Ibid., 751.

22 Ibid.

23 Anne Brontë, *Agnes Grey* (1847; reprint, London: Penguin, 1988), 105-6.

24 Youatt, *The Obligation and Extent of Humanity to Brutes,* 12.

25 Quoted in H.E. Carter, "The Veterinary Profession and the RSPCA: The First Fifty Years," *Veterinary History* n.s. 6, 2 (1989/90): 65.

26 *Politics,* 1.8.11-12; *Historia Animalium,* 588a8; *On the Parts of Animals,* 1.5; 2.2; 2.10; *Poetics,* ch. 7.

27 Quoted in Drummond, *The Rights of Animals,* 188, citing Plutarch's *Lives,* vol. 1, 218 n.

28 The quotations from both Oppian and Alpers are to be found in Mary Midgley, *Beast and Man: The Roots of Human Nature* (1979; reprint, London: Routledge, 1995), 211, citing Antony Alpers, *Dolphins* (London: Routledge, 1965), 43, 158-59.

29 Claudius Aelianus, *On the Characteristics of Animals,* 2, 6, excerpted in Rod Preece, *Awe for the Tiger, Love for the Lamb: A Chronicle of Sensibility to Animals* (Vancouver: UBC Press; London and New York: Routledge, 2002), 55-56.

30 Drummond, *The Rights of Animals,* 198.

31 Midgley, *Beast and Man,* prt. 4, 201ff.

32 Quoted in Jeffrey Myers, *D.H. Lawrence: A Biography* (New York: Vintage Books, 1992), 42.

33 *Queen Mab,* stanza 8, in *The Works of P.B. Shelley* (Ware: Wordsworth, 1994), 32.

34 *Bhagavatam* 6, quoted in G. Naganathan, *Animal Welfare and Nature: Hindu Scriptural Perspectives* (Washington, DC: Center for the Respect of Life and Environment, 1989), 5.

35 Quoted in Jon Wynne-Tyson, *The Extended Circle: An Anthology of Humane Thought* (Fontwell: Centaur, 1990), 140.

36 *Acaranga Sutra,* 1.1.3 quoted in Chapple, *Nonviolence to Animals, Earth and Self in Asian Traditions,* 11.

37 Juan Mascaró, ed. and trans., *Dhammapada* (London: Penguin, 1973), 74.

38 Quoted in Ronald Isaacs, *Animals in Jewish Thought and Tradition* (Northvale, NJ: Jason Aaronson, 2000), 85, citing Even HaEzer 5:14.

39 Quoted in ibid., 85, citing Orach Chayyim 223:6.

40 Quoted in ibid., 86, citing Mechilta to Exodus 23:12.

41 Quoted in ibid., 87, citing Kitzur Schulchan Aruch 186:1.

42 *Shabbat* 77b, quoted in Lewis G. Regenstein, *Replenish the Earth: A History of Organized Religion's Treatment of Animals and Nature, Including the Bible's Message of Conservation and Kindness toward Animals* (New York: Crossroad, 1991), 189, citing Barry Freundel, "The Earth Is the Lord's: How Jewish Tradition Views Our Relationship to the Environment," *Jewish Action* (Summer 1990): 24-25.

43 Quoted in Myers, *D.H. Lawrence: A Biography,* 271.

44 See Rod Preece, *Animals and Nature,* 18-20.

45 Big Bill Neidjie, *Speaking for the Earth: Nature's Law and the Aboriginal Way* (Washington, DC: Center for Respect of Life and Environment, 1991), 35, reprinted from Big Bill Neidjie, Stephen Davis, and Allan Fox, *Kakadu Man* (Northryde, NSW: Angus and Robertson, n.d.).

46 Margot Edmonds and Ella E. Clark, eds., *Voices of the Winds: Native American Legends* (New York: Facts on File, 1989), 184-85.

47 See Mathias Guenther, *Tricksters and Trancers: Bushman Religion and Society* (Bloomington: Indiana University Press, 1999), 127.

48 Burton Watson, ed. and trans., *Lotus Sutra* (New York: Columbia, 1993), 74, 77.

49 *The Tibetan Book of the Dead* (Boston: Shambhala, 1992), 223.

50 Burton Watson, Introduction to *Lotus Sutra,* xiv.

51 Quoted in Daniel J. Boorstin, *The Creators: A History of Heroes of the Imagination* (New York: Vintage Books, 1993), 21-22, translation by Edward Conze, in *Buddhist Scriptures* (London: Penguin, 1973).

52 Epiphanius, *Panarion,* 30.16, cited in Keith Akers, *Lost Religion of Jesus: Simple Living and Nonviolence in Early Christianity* (New York: Lantern Books, 2000), 26-27.

53 There is one instance in which Christ seems to counter this. In Mark 1:40-45, he tells the leper whom he has healed to go to the temple and make a proper sacrifice. For an explanation of this event, see Stephen H. Webb, *Good Eating* (Grand Rapids, MI: Brazos Press, 2001), 94.

54 See ibid., 96-97.

55 See ibid., 98.

56 Stephen H. Webb indicates that "some scholars infer from Acts 2:46 that the early Christians sacrificed at the temple" (ibid., 94). The verse reads: "Each day, with one heart, they regularly went to the Temple, but met in their houses for the breaking of bread; they shared their food gladly and generously" (*New Jerusalem Bible*). In the King

James version, "meat" is used instead of "food," and since flesh was usually acquired at the Temple alone, a ready inference is that they might have sacrificed there. However, if nonflesh food is all that is implied, such an inference is unnecessary.

57 The Biblical account is to be found in 2 Kings 22-23.

58 Hope MacLean, "Species Reasoning," *Canadian Forum* 72 (1993): 45.

59 See page 78.

60 Mohandas K. Gandhi, *How to Serve the Cow* (1934). Quoted in Marvin Harris, *Cows, Pigs, Wars and Witches: The Riddles of Culture* (1974; reprint, New York: Vintage, 1989), 26.

61 *Young India*, 18 November 1926.

62 Göran Burenhult, "Life and Death among the Toraja," in *Traditional Peoples Today: Continuity and Change in the Modern World*, ed. Göran Burenhult (San Francisco: Harper, 1994), 62.

63 D.H. Lawrence, *Apocalypse* (1906), quoted in Myers, *D.H. Lawrence: A Biography*, 353.

64 Singer, *Animal Liberation*, 191.

65 John Steinbeck, *The Log from the Sea of Cortez* (London: Penguin, 1995), 41.

66 Ibid., 53.

67 Simon Davis, *The Archeology of Animals* (New Haven: Yale University Press, 1987), 113.

68 See page 116.

69 Davis, *The Archeology of Animals*, 102.

70 Jonathan Kingdon, *Self-Made Man: Human Evolution from Eden to Extinction?* (New York: John Wiley, 1993), 29.

71 Gary Kowalski, *The Souls of Animals* (Walpole, NH: Stillpoint, 1991), 104.

72 Quoted in Drummond, *The Rights of Animals*, 203, citing William Gifford, ed. and trans., *Juvenal's Satires with The Satires of Persius* (1854; reprint, London: Dent, 1954). The Latin, it may be said, is not quite so decisive as the English, although not far from it:

> Sensum a cœlesti demissum traximus arce,
> Cujus egent prona et terram spectantia. Mundi
> Principio indulsit communis conditor illis
> Tantum animas, nobis animum quoque.

73 See page 65.

74 See Michael Allen Fox, *Deep Vegetarianism* (Philadelphia: Temple University Press, 1999), 14.

75 Alexander Pope, "On Cruelty to the Brute Creation" (sometimes called "Against Barbarity to Animals"), *Guardian* 61 (21 May 1713), in *The Tatler and Guardian* (New York: Bangs Brother, 1852). Such articles were untitled, and the titles now ascribed to them are additions of later commentators.

76 William F. Karkeek, "On the Furture Existence of the Brute Creation," *The Veterinarian* 12, 142 (1839): 656.

77 W.C. Spooner, "On the Non-Immortality of Animals," *The Veterinarian* 13 (March 1840): 247.

78 Quoted in Marian Scholtmeijer, *Animal Victims in Modern Fiction: From Sanctity to Sacrifice* (Toronto: University of Toronto Press, 1993), 17.

79 Humphry Primatt, *The Duty of Mercy and the Sin of Cruelty to Brute Animals* (1776; reprint, ed. Richard D. Ryder, Fontwell: Centaur, 1992), 21.

80 Quoted in Nicholson, *On the Primeval Diet of Man*, 71-72, citing Richard Dean, *Essay*

on the Future Life of Brutes, introduced with Observations upon Evil, its Nature and Origins (Manchester: n.p., 1767).

81 Quoted in Thomas, *Man and the Natural World,* citing Norman Callan, ed., *The Collected Poems of Christopher Smart,* vol. 1, 290.

82 Quoted in Drummond, *The Rights of Animals,* 45.

83 Jeremy Bentham, *An Introduction to the Principles of Morals and Legislation* (1789; reprint, ed. J.H. Burns and H.L.A. Hart, London: Methuen, 1982), 17, 4b, 282 (italics in the original).

84 Hilda Kean, *Animal Rights: Political and Social Change in Britain since 1880* (London: Reaktion Books, 1998), 70.

85 Richard Sorabji, *Animal Minds and Human Morals* (Ithaca: Cornell, 1993), 210.

86 Primatt, *The Duty of Mercy and the Sin of Cruelty to Brute Animals,* 23. Provocatively, Primatt adds a footnote to the word "soul": "It is of no consequence as to the case now before us, whether the soul is, as some think, only a power, which cannot exist without the body; or as is generally supposed, a spiritual substance, that can exist, distinct and separate from the body."

87 Ibid., 32.

88 Ibid., 33.

89 Ibid., 25, 126-27 (italics in the original).

90 Drummond, *The Rights of Animals,* 199.

91 Youatt, *Obligation and Extent of Humanity to Brutes,* 253, 254 (italics in the original).

92 Ibid., 6.

93 Singer, *Animal Liberation,* 191.

94 Midgley, *Beast and Man,* 34.

95 Drummond, *The Rights of Animals,* 15.

96 Ibid., 22.

97 See, for example, Henry Chadwick, "The Early Christian Community," in John McManners, ed., *The Oxford History of Christianity* (1990; reprint, Oxford: Oxford University Press, 2002), 52. Writing of Justin Martyr, an important early-second-century figure, Chadwick informs us that "Justin was convinced that in the highminded Stoic ethics of human brotherhood, and especially in the other-worldly Platonic metaphysics, there was much for a Christian to welcome." We are frequently reminded today of Christianity's neo-Platonism but less often than might be appropriate of its Stoicism.

98 Singer, *Animal Liberation,* 200.

99 Angus Taylor, *Magpies, Monkeys and Morals: What Philosophers Say about Animal Liberation* (Peterborough: Broadview, 1999), 23; Barbara Noske, *Beyond Boundaries: Humans and Animals* (Montreal: Black Rose Books, 1997), 46; Randy Malamud, "Poetic Animals and Animal Souls," *Society and Animals* 6, 3 (1998): 263-67. To be fair to Malamud, he is explicit only about Cartesianism and, a little more opaquely, cultural anthropologists (267). Nonetheless, the Western Christian tradition as a whole is implied (263); William Leiss, *The Domination of Nature* (1972; reprint, Montreal and Kingston: McGill-Queen's University Press, 1984); Mason, *An Unnatural Order: Uncovering the Roots of Our Domination of Nature and Each Other,* especially 21-49 and 210-99.

100 Dom Ambrose Agius, *God's Animals* (London: Catholic Study Circle for Animal Welfare, 1973).

101 Quoted in Jonathan Barnes, *Early Greek Philosophy* (London: Penguin, 1987), 291, citing Simplicius, *Commentary on the Physics,* 152.

102 See especially "On the Use of Reason by 'Irrational' Animals," Plutarch, *Essays* (London: Penguin, 1972), 385-98.

103 Against conventional wisdom, Howard Williams argues that Seneca remained a Pythagorean in private if not in public. See *The Ethics of Diet: A Catena of Authorities Deprecatory of the Practice of Flesh Eating* (London: F. Pitman, 1883), 27ff.

104 Seneca, *Ad Lucilium Epistulae Morales*, Letter 58.

105 Plotinus, *Enneads*, 3, in *Plotinus*, vol. 3, trans. A.H. Armstrong (London: William Heinemann, 1967), 67.

106 Porphyry, *On Abstinence from Animal Food*, trans. Thomas Taylor ([1793]; reprint, Fontwell: Centaur, 1965), 48.

107 Quoted in Buckner, *The Immortality of Animals*, 89.

108 Ibid.

109 "Dumb animals" refers to creatures that do not possess the capacity for complex speech. The notion of "dumb" as "stupid" was a twentieth-century innovation.

110 Arnobius, *The Seven Books of Arnobius Adversus Gentes*, vol. 19, 82-84.

111 Quoted in Buckner, *The Immortality of Animals*, 37-38.

112 William F. Karkeek, "On the Future Existence of the Brute Creation," *The Veterinarian* 12, 144 (December 1839): 781.

113 Ibid., 35-36.

114 Genesis 6:12-13, 9:11, 9:17.

115 Regenstein, *Replenish the Earth*, 43.

116 Karkeek, "An Essay on the Future Existence of the Brute Creation," 785.

117 Lactantius, *The Father of the Church, Lactantius, the Divine Institutes, Books I-VII*, trans. Francis McDonald (Washington: Catholic University of America Press, 1964), 185-86.

118 From *De Incarnatione*, quoted in Andrew Linzey and Tom Regan, eds., *Animals and Christianity: A Book of Readings* (London: SPCK, 1989), 98-99, citing Athanasius, *Contra Gentes and De Incarnatione*, trans. Robert W. Thomson (Oxford: Clarendon Press, 1971).

119 Quoted in Daniel A. Dombrowski, *The Philosophy of Vegetarianism* (Amherst: University of Massachusetts Press, 1984), 142.

120 Quoted by C.W. Hume, *Universities Federation for Animal Welfare Theological Bulletin*, no. 2 (1962): 3.

121 Richard Sorabji, *Animal Minds and Human Morals: The Origins of the Western Debate* (Ithaca: Cornell University Press, 1993), 202.

122 Bk. 2, ch. 17.

123 Augustine, *Of the Morals of the Catholic Church, Nicene and Post-Nicene Fathers of the Christian Church*, 1st series, vol. 4 (1872; reprint, Grand Rapids: E.B. Erdmann, [1995?]), 33, 73.

124 See Webb, *Good Eating*, 153, citing Blake Leyerle, "Clement of Alexandria on the Importance of Table Etiquette," *Journal of Early Christian Studies* 3, 2 (Summer 1995): 125-41.

125 It is worth noting that Augustine, who relies primarily on Plato among the classics, refers to animals as "meant for our use," a doctrine customarily attributed to Aristotle. Yet Aristotle's works had been lost by this time – to be rediscovered from Arab sources in the twelfth century. Either the doctrine was far more pervasive than sometimes thought or Aristotle's influence was being felt via other writers, perhaps Xenophon, who had argued earlier in *Memorabilia* – in fact, before Aristotle – that animals are intended for human use.

126 St. Augustine, *The Confessions of St. Augustine* (London: Longman's Green, 1897), bk. 4, ch. 3, 72.

127 Ibid., bk. 7, ch. 12, 179-80.

128 St. Augustine, *Concerning the City of God against the Pagans,* 4th ed., trans. Henry Bettenson (Harmondsworth: Penguin, 1980), bk. 19, ch. 14, 872-73.

129 William Langland, *Piers the Ploughman,* translated into modern English with an introduction by J.F. Goodridge (Harmondsworth: Penguin, 1959), prt. 2, bk. 11, 138.

130 Andrew Linzey, "Christianity and the Rights of Animals," *Animals' Voice* (Los Angeles), (August 1989): 45.

131 Webb, *Good Eating,* 29.

132 Quoted in Andrew Linzey, *Animal Theology* (Urbana: University of Illinois Press, 1995), 56, citing Nietzsche, *Thoughts Out of Season* (1873; reprint, Edinburgh: T.N. Foulis, 1909), 154.

133 Quoted in ibid., 56, citing Vladimir Lossky, *The Mystical Theology of the Eastern Church* (1973), 111.

134 Saint Francis of Assisi, *Admonitions,* in *Francis and Clare: The Complete Works,* trans. Regis J. Armstrong and Ignatius C. Brady (New York: Paulist Press, 1982), 29.

135 Thomas of Celano, *First Life of St. Francis,* in Marion A. Habig, ed., *St. Francis of Assisi: Writings and Early Biographies* (Chicago: Franciscan Herald Press, 1983), bk. 1, ch. 21, 228; bk. 1, ch. 29, 297.

136 St. Bonaventure, *Life of St. Francis,* in Ewart Cousins, trans., *Bonaventure: The Soul's Journey into God, The Tree of Life and The Life of St. Francis* (Mahwah, NJ: Paulist Press, 1978), 254.

137 John Passmore, "The Treatment of Animals," *Journal of the History of Ideas* 36 (1975): 215-16.

138 W.E.H. Lecky, *A History of European Morals from Augustus to Charlemagne,* vol. 2 (New York: D. Appleton, 1869), 182-83.

139 *The Canterbury Tales,* trans. Nevill Coghill (London: Cresset Press, 1992), 279.

140 See Kean, *Animal Rights,* 37, 113.

141 Lecky, *A History of European Morals?,* 183.

142 Quoted in Dom Ambrose Agius, *God's Animals* (1970; reprint, London: Catholic Study Circle for Animal Welfare, 1973), 45, citing *Ark* 18: 6.

143 Quoted in Joyce E. Salisbury, *The Beast Within: Animals in the Middle Ages* (New York: Routledge, 1994), 19.

144 Quoted in Agius, *God's Animals,* 12, citing Alban Butler, *Lives of the Saints,* vol. 2, revised and supplemented by Rev. Hurbert Thurston and Donald Attwater (New York: P.J. Kennedy and Sons, n.d.), 362.

145 Edward Maitland, *Anna Kingsford: Her Life, Letters, Diary and Work,* vol. 2, 3rd ed. (London: G. Redway, 1913), 312.

146 Quoted in Michael W. Fox, *Inhumane Society* (New York: St. Martin's Press, 1990), 91.

147 Webb, *Good Eating,* 27.

148 Aquinas, *Summa Theologica,* 1.92. Aristotle's comment is in *The Generation of Animals,* vol. 2.3.

149 See, for example, pages 104-7.

150 *Summa Theologica* 1.96.2, Reply to Objection 3.

151 Ibid. (italics in the original).

152 Ibid., 1.96.1, Reply to Objection 4.

153 Quoted in Agius, *God's Animals,* citing Catherine of Siena's *Letters,* vol. 1, 237.

154 Quoted in Arthur Helps, *Animals and Their Masters* (1872; reprint, London: Chatto and Windus), 124.

155 Henry Chadwick, "The Early Christian Community," in *The Oxford History of Christianity*, ed. John McManners (1990; reprint, Oxford: Oxford University Press, 2002), 42.

156 Quoted in Dix Harwood, *Love for Animals and How It Developed in Great Britain* (1928; reprint, ed. Rod Preece and David Fraser, Lampeter: Mellen Animal Rights Library, 2002), 37. The full text may be found in Priscilla Barnum, ed., *Dives et Pauper* (Early English Text Society, 1976; reprint, New York: Kraus, 1973), 1, 2, 36. "Sap. V" is ch. 5 of the Book of Wisdom.

157 Ron Baxter, *Bestiaries and Their Users in the Middle Ages* (London: Sutton Publishing/ Courtauld Institute, 1998), 209.

158 Kenneth Siam, ed., *Fourteenth Century Verse and Prose* (1921; reprint, Oxford: Oxford University Press, 1985), 41-42.

159 Irma A. Richter, ed., *Selections from the Notebooks of Leonardo da Vinci* (Oxford: Oxford University Press, 1977), 61.

160 Ibid., 245.

161 Ibid.

162 *The Literary Works of Leonardo da Vinci*, 2nd ed., vol. 2, enlarged and revised by Jean Paul Richter and Irma A. Richter (London: Phaidon, 1970), 293. The first edition was edited by Jean Richter alone in 1883.

163 Giorgio Vasari, *The Great Masters* (1550; reprint, ed. Michael Sonino, trans. Gaston Du C. De Vere, New York: Park Lane, 1988), 93-94.

164 From original document in the library of Windsor Castle.

165 From *Quaderni d'Anatomia II*, 14, housed in the Royal Library at Windsor. This was kindly brought to my attention by David Hurwitz. The word "vegetarian" was interpolated by the translator Jean Paul Richter, who was the first to decipher Leonardo's notebooks. It is clear both from Italian usage and from the context that it is vegetarian food that is meant. It is worth recalling that there was no explicit term for vegetarianism before the mid-nineteenth century, either in English or in any other European language. And, of course, vegetarianism is itself a misnomer, since vegetarians eat a great deal more than vegetables. Terms like "simple food" or "natural food" were thus quite common, although most referred to vegetarians as Pythagoreans, in Italian, too, as is clear from the title of Antonia Celestina Cocchi's *Del vitto pitagorico per usa della medicina* (Florence, 1743) (The Pythagorean diet in medical practice). It should be noted, too, that when Percy Bysshe Shelley wrote his vegetarian pamphlets in the early nineteenth century, he entitled his first piece *A Vindication of Natural Diet*, his second *On the Vegetable System of Diet*.

166 *The Literary Works of Leonardo da Vinci*, 298.

167 Quoted in Nicholson, *On the Primeval Diet of Man*, 80, citing Marianne Stark, *Letters from Italy, between 1792 and 1798* (1800), 2 vols., Letter 14.

168 Alphonse de Lamartine, *Les Confidences* (1848; reprint, trans. Rod Preece, Paris: Hachette, 1893), 78.

169 The report is given with more details of the actual diet in Anna Bonus Kingsford, *The Perfect Way of Diet: A Treatise Advocating a Return to the Natural and Ancient Food of Our Race* (Kila, MT: Kessinger, [1906?]), 37-38.

170 Vasari, *The Great Masters*, 96.

171 Quoted in Harwood, *Love for Animals*, 160-61, citing *Notes and Queries*, ser. 8, 2, 233.

172 Quoted in Webb, *Good Eating*, 176, citing Colleen McDannell and Bernhard Lang, *Heaven: A History* (New York: Vintage Books, 1990), 153.

173 *The Complete Works of St. John of the Cross*, vol. 2, trans. E. Allison Peers (London: Burns, Oates and Washburn, 1935), stanza 5, 50-51.

174 In Lucretius, *On the Nature of the Universe* (1951; reprint, trans. R.A. Latham, London: Penguin, 1994), the phrase is nicely rendered as "each species develops according to its own kind, and they all guard their specific character in obedience to the laws of nature" (bk. 5, lines 922-24, 152).

175 Michel de Montaigne, "Apology for Remond Seybond," in *Selected Essays* (New York: Oxford University Press, 1983), 326.

176 Denis Saurat, *Milton: Man and Thinker* (1925; reprint, New York: Haskell, 1970), 141-42, quoting from *Treatise of Christian Doctrine*, vol. 4, 195.

177 Ibid., 143.

178 Quoted in ibid., 293.

179 See ibid., 301ff.

180 Quoted in the original in Hester Hastings, *Man and Beast in French Thought of the Eighteenth Century* (1936; reprint, New York: Johnson Reprint Corporation, 1973), 22, citing *Le Comte de Gabelis*, vol. 2 (1670; reprint, Londres: Frères Vaillant, 1742), 229-30, my translation.

181 Richard Overton, *Man's Mortalitie* (1643; reprint, ed. Harold Fisch, Liverpool: University of Liverpool Press, 1968), 126.

182 See, for example, A. Denney, *Descartes' Philosophical Letters* (Oxford: Clarendon, 1970); John Cottingham, *A Descartes Dictionary* (Oxford: Blackwell, 1993); and Gary Steiner, "Descartes on the Moral Status of Animals," *Archiv für Geschichte der Philosophie* 80, 3 (1998): 268-91.

183 Hastings, *Man and Beast*, 21-22.

184 Quoted in Harwood, *Love for Animals*, 98, citing *De la Recherche de la Vérité* (Paris, 1678), vi, 2, vii.

185 Pierre Gassendi, *Exercises in the Form of Paradoxes in Refutation of the Aristoteleans*, in *The Selected Works of Pierre Gassendi*, trans. Craig B. Brush (New York: Johnson Reprint Corporation, 1972), 86.

186 Pierre Gassendi, *Metaphysical Colloquy, or Doubts and Rebuttals concerning the Metaphysics of René Descartes*, Rebuttal to Meditation 2, Doubt 7, in ibid., 197-98.

187 Quoted in Williams, *The Ethics of Diet*, 104, citing Gassendi, "De Virtutibus," in *Physics*, bk. 2 (italics in the original).

188 Quoted in Harwood, *Love for Animals*, 102, citing "Epistola Prima H. Mori ad Renatum Cartesium" (First letter of Henry More to René Descartes), in Henry More, *Collection of Several Philosophical Works* (London, 1712). The Latin original is to be found in René Descartes, *Oeuvres de Descartes*, vol. 5 (Paris: Charles et Paul Tannery, 1897), 243.

189 Thomas, *Man and the Natural World*, 139.

190 Quoted in Ben Lazare Mijuskovic, *The Achilles of Rationalist Arguments: The Simplicity, Unity and Identity of Thought and Soul from the Cambridge Platonists to Kant: A Study in the History of an Argument* (The Hague: Martinus Nijhoff, 1974), 26, from John Smith, "A Discourse Demonstrating the Immortality of the Soul," in E.T. Campagnac, *The Cambridge Platonists* (Oxford, 1901), 107 (italics in the original).

191 Ralph Cudworth, *True Intellectual System of the Universe*, vol. 4 (London: Richard Royston, 1678), 151.

192 Quoted in Harwood, *Love for Animals,* 104, citing John Norris, *An Essay towards the Theory of the Ideal or Intelligible World,* vol. 2 (London, 1701), 44.

193 See, especially, pages 156-58.

194 For the context of this comment, see the remarks of Pierre Bayle reported on page 157.

195 Quoted in the original in Hastings, *Man and Beast,* 57, citing Charles Bonnet, "Essai de Psychologie," in *Oeuvres d'histoire naturelle et de philosophie,* vol. 8 (Neuchatel: Fauche, 1783), 106-7, my translation.

196 See Hastings, ibid., 63.

197 Quoted in the original in ibid., 38, citing *Le Nouveau Gulliver, ou Voyage de Jean Gulliver, fils du Capitaine Lemuel Gulliver, par l'abbé Desfontaines (Voyages Imaginaires, XV),* 1st ed. (Paris: Quérard, 1730), 266, my translation.

198 See ibid., 39 n. 1.

199 Quoted in the original in ibid., 81 n. 1, my translation.

200 See, for example, Johann Friedrich Blumenbach, "Über die Liebe der Thiere," *Göttingische Magazin der Wissenschaften und Literatur* (1781).

201 Voltaire, *Traité sur la Tolérance* (1763; reprint, trans. Rod Preece, Paris: Flammarion, 1989), 170-71.

202 Julien Offray De La Mettrie, *Man a Machine* (La Salle: Open Court Publishing, 1953), 87, 93, 97, 98.

203 Ibid., 157.

204 Quoted in part in the original in Hastings, *Man and Beast,* 44, citing Julien Offray De La Mettrie, *Oeuvres philosophiques, nouvelle édition corrigée ...,* vol. 1 (Amsterdam, 1774), i and 67, my translation.

205 Quoted in Buckner, *The Immortality of Animals,* 88-89.

206 Quoted in the original in Hastings, *Man and Beast,* 30, citing Bernard Nieuwentyt, *L'Existence de Dieu démontrée par les merveilles de la nature* (Amsterdam, 1710), 354, my translation.

207 *The Diaries of John Evelyn,* vol. 4, ed. E.S. de Beer (Oxford: Clarendon Press, 1955), 106.

208 George Fox, *To All Sorts of People in Christendom* [1673?], in *Gospel Truth Demonstrated in a Collection of Doctrinal Books Given Forth by that Faithful Minister of Jesus Christ, George Fox; Containing Principles Essential to Christianity and Salvation, Held among People Called Quakers,* vol. 1 (Philadelphia: Marcus T.C. Gould, 1831), 320.

209 Many modern Quakers deny that Fox was a vegetarian and claim that he did not espouse animals' spiritual immortality.

210 Quoted in Thomas, *Man and the Natural World,* 139.

211 Richard Overton, *Man's Mortalitie* (1643; reprint, ed. Harold Fisch, Liverpool: University of Liverpool Press, 1968), 26.

212 Thomas Babington Macaulay, *History of England,* vol. 1 (London: Longman, 1854), 161.

213 Pierre Bayle, *Historical and Critical Dictionary: Selections,* trans. Richard H. Popkin (Indianapolis: Bobbs-Merrill, 1965), 214.

214 Ibid., 221, 222, 223, 224.

215 Ibid., 235.

216 Ibid., 236.

217 Ibid., 238, 239. The words within quotation marks are from Leibniz, *Histoire des Ouvrages des Savants,* February 1696 (italics in the original).

218 Quoted in Harwood, *Love for Animals,* 166, citing David Hartley, *Observations of Man* (1749), bk. 2, 233.

219 Thomas, *Man and the Natural World*, 139-40 (italics in the original).

220 Joseph Butler, *The Analogy of Religion, Natural and Revealed, to the Constitution and Course of Nature* (1736; reprint, London: Longman, 1834), 17 (italics in the original).

221 Quoted in Nicholson, *On the Primeval Diet of Man*, 113.

222 Quoted in Harwood, *Love for Animals*, 165, citing John Hildrop, *Free Thoughts on the Brute Creation*, in *Works*, vol. 1 (London, 1754), 214.

223 Quoted in Thomas, *Man and the Natural World*, 140, citing Richard Dean, *Essay on the Future Life of Brutes*, vol. 2 (1767), 49.

224 See Harwood, *Love for Animals*, 170.

225 Quoted in Nicholson, *On the Primeval Diet of Man*, 71-72.

226 Quoted in Arthur Helps, *Animals and Their Masters* (1872; reprint, London: Chatto and Windus, 1883), 87, citing Abraham Tucker, *The Light of Nature Pursued*, vol. 5 (1754; reprint, London, 1777), prt. 3, ch. 19.

227 Quoted in Thomas, *Man and the Natural World*, 140, citing George Cheyne, *An Essay on Regimen* (1740), 86-87.

228 Ibid.

229 *The Poetical Works of Anna Seward, with Extracts from her Literary Correspondence*, vol. 2, ed. Walter Scott (Edinburgh: Ballantyne, 1810), 60 (italics in the original).

230 See page 91.

231 Soame Jenyns, *A Free Inquiry into the Nature and Origin of Evil*, 2nd ed. (London: R. and J. Dodsley, 1757; facsimile reprint, New York: Garland, 1976), 36.

232 Quoted in Drummond, *The Rights of Animals*, 199.

233 Quoted in ibid., 204.

234 Quoted in the original in Hastings, *Man and Beast*, 75-76, citing *Suite des erreurs et de la vérité* (Salomonopolis, 1784), 338, my translation.

235 John Wesley, "The General Deliverance" (1788), *Sermons on Several Occasions*, 2nd series, Sermon 60, in *The Works of John Wesley*, vol. 6 (Grand Rapids, MI: Zondervan, n.d.; reprint of the Wesleyan Conference Office edition, London, 1872), 244, 248.

236 Quoted in Buckner, *The Immortality of Animals*, 81-82.

237 Ibid., 78-79.

238 Ibid., 84.

239 Ibid., 79.

240 "On the Death of a Favourite Old Spaniel," in *Poems* (Bristol: Joseph Cottle, 1797), 133.

241 *A Lay Sermon* (1817), in *The Collected Works of Samuel Taylor Coleridge*, vol. 6, ed. R.J. White (London: Routledge and Kegan Paul, 1972), 183 n. 6 (italics in the original). The Socinian creed is named for Faustus (1539-1604) and Lailius Socinius (1525-62), who denied the doctrine of the trinity, the natural depravity of humankind, vicarious atonement, and eternal punishment.

242 Jules Michelet, *The Bible of Humanity*, trans. Vincenzo Calfia (New York: J.W. Bouton, 1877), 37.

243 Jules Michelet, *The Bird* (1879; reprint, trans. W.H. Davenport Adams, London: Wildwood House, 1981), 271.

244 Quoted in Paul Johnson, *Modern Times* (1991; reprint, New York: Perennia, 2001), 70.

245 George Eliot, *Middlemarch* (1871-72; reprint, New York: Alfred A. Knopf, 1991), bk. 1, ch. 3, 26.

246 David Winston, ed. and trans., *Philo of Alexandria: The Contemplative Life, Giants, and Selections* (New York: Paulist Press, 1981), 295.

247 See pages 125, 277-78.

248 Edward Carpenter, *Towards Democracy* (1883; reprint, London: Unwin Brothers, 1926), 174.

249 *Parliamentary Debate* (London: HMSO), 26 May 1876.

250 See Drummond, *The Rights of Animals,* 209.

251 Henry S. Salt, *Animals' Rights Considered in Relation to Social Progress* (Clarks Summit, PA: Society for Animal Rights, 1980), 162.

252 Quoted in Buckner, *The Immortality of Animals,* 83.

253 Ibid., 82-83.

254 George Macdonald, *Hope of the Gospel,* in *Hope of the Universe* (London: Ward, Lock, Bowden, 1892), converted to e-text by Joahnneson Printing and Publishing, <http://www.johannesen.com/HopeoftheGospelComplete.htm>.

255 Quoted in Buckner, *The Immortality of Animals,* 79-80.

256 Louis Agassiz, *Contributions to the Natural History of the United States,* vol. 1 (Boston: Little, Brown, 1857), 64-66. Agassiz included the following in a lengthy footnote to page 65: "A close study of the dog might satisfy every one of the similarity of his impulses with those of man, and those impulses are regulated in a manner which discloses psychical faculties in every respect of the same kind as those of man; moreover, he expresses by his voice his emotions and his feelings, with a precision which may be as intelligible to man as the articulated speech of his fellow men. His memory is so retentive that it frequently baffles that of man. And though all these faculties do not make a philosopher of him, they certainly place him in that respect upon a level with a considerable proportion of poor humanity."

257 Quoted in Buckner, *The Immortality of Animals,* 86-87.

258 Ibid., 84.

259 See page 65.

260 Buckner, *The Immortality of Animals,* 237.

261 Quoted in Frederick Brown, *Zola: A Life* (New York: Farrar, Strauss, Giroux, 1995), 542.

262 Quoted in Martin Seymour-Smith, *Hardy* (London: Bloomsbury, 1995), 240.

263 P.N. Furbank, *E.M. Forster: A Life,* vol. 2 (London: Martin Secker and Warburg, 1978), 216 n. 1.

264 "Letter to William Sotheby," in *Collected Letters of Samuel Taylor Coleridge,* vol. 2, ed. Earl Leslie Griggs (Oxford: Oxford University Press, 1956), 459 (italics in the original).

CHAPTER 4: RETURN TO NATURE

1 Quoted in Robert D. Richardson, *Emerson: The Mind on Fire* (Berkeley: University of California Press, 1995), 101.

2 Quoted in Frederick Brown, *Zola: A Life* (New York: Farrar, Strauss, Giroux, 1995), 88.

3 Samuel Butler, *Erewhon,* 2nd ed. (London: Page and Company, 1907), 277.

4 Ibid., 49.

5 Ibid., 51.

6 Ibid., 54.

7 Quoted in Arthur O. Lovejoy and George Boas, *Primitivism and Related Ideas in Antiquity* (1935; reprint, Baltimore: Johns Hopkins University Press, 1977), 46-47.

8 Thomas Hobbes, *Leviathan* (1651; reprint, ed. Michael Oakeshott, New York: Collier Books, 1962), 100.

9 John Locke, *Two Treatises of Government* (1690; reprint, ed. Peter Laslett, New York: New American Library, 1965), 321 (italics in the original).

10 It has, however, occasionally been given different applications. Thus, for example, in Kenneth Grahame's *The Golden Age* (1895), the concept refers to the days of childhood.

11 H.A. Guerber, *Myths and Legends of the Middle Ages: Their Origins and Influence in Literature and Art* (1909; reprint, London: G.G. Harrap, 1926), 241.

12 Mircea Eliade, *The Myth of the Eternal Return, or Cosmos and History* (Princeton: Princeton University Press, 1974).

13 Quoted in Leo Tolstoy, *A Confession and Other Religious Writings,* Chapter 16, "The Law of Love and the Law of Violence" (London: Penguin, 1987), 211.

14 For a discussion of Max Weber's concept of the "ideal type," see Lewis A. Coser, *Masters of Sociological Thought: Ideas in Historical and Social Context* (Albany, NY: International Thomson Publishing, 1977), 223-24.

15 In William Makepeace Thackeray's *Vanity Fair* (1848), "Arcadian Simplicity," the title of Chapter 11, refers to the bumblingly and uncouthly bucolic. This use is relatively rare but reflects the increasingly widely held Victorian belief – which we can already witness in embryo in Elizabethan and Stuart times – that the city is superior in sophistication, education, rationality, civility, and worldliness.

16 The unicorn is a fabled, pure white, equine animal with a horn in the middle of its forehead. It was once thought to inhabit parts of the Middle East and India, and "sightings" were reported frequently throughout the known world in the medieval era and even later. Nicholas J. Saunders explains: "Different kinds of unicorn occur in legends worldwide ... The Greek physician, Ctesias, in a fourth century BC history of Persia and Assyria, was the first to describe a unicorn as being equine, 'a kind of wild ass, white with a dark red head and an 18-inch [45 cm] horn on its forehead, which, when ground to powder, yields a certain remedy against epilepsy and the most potent poison.'" Nicholas J. Saunders, *Animal Spirits* (Boston: Little, Brown, 1995), 151. Ctesias's unicorn is a synthesis of three animals: the onager, an Asian wild donkey; the blackbuck, an antelope in which one horn of the male sometimes atrophies; and the Indian rhinoceros. Saunders indicates that "a link between virginity and the unicorn goes back to a Graeco-Roman tradition associating the creatures with Artemis and Diana." In Greek mythology Artemis was Mistress of the Animals and an accomplished hunter – that is, both the protector and the slayer of animals. Diana was her Roman counterpart. "Christianity adapted this tradition by modifying the image of the fierce unicorn until it became a symbol of the Immaculate Conception, and its horn a form of spiritual penetration" (ibid.).

17 Quoted from Lovejoy and Boas, *Primitivism and Related Ideas in Antiquity,* 27.

18 Hesiod, *Theogony and Works and Days* (Oxford: Oxford University Press, 1988), 40, xix. There is in fact an earlier description of a "golden age" society in Book 9 of Homer's *Odyssey,* although the term itself is not used, nor is their any mention of historical stages: "And we came to the land of the Cyclôpes, a froward and a lawless folk, who trusting to the deathless Gods plant not aught with their hands, neither plough: but, behold, all these things spring for them in plenty, unsown and untilled, wheat and barley, and vines, which bear great clusters of the juice of the grape, and the rain of Zeus gives them increase. They have neither gatherings for council nor oracles of law, but they dwell in hollow caves on the crests of the high hills, and each one utters the law to his children and his wives, and they reck not one of another." *The Odyssey of Homer,* trans. S.H. Butcher and A. Land (1909; reprint, New York: P.F. Collier, 1969), 118-19. Also,

although much later, in Plutarch's "On the Use of Reason by 'Irrational' Animals," Ulysses acknowledges the life of ease of the land of the Cyclopes to be superior to that which may be lived in his own "harsh Ithaca." No doubt Plutarch was relying on Homer's *Odyssey* for his information. Plutarch, *Essays*, trans. Robin Waterfield (London: Penguin, 1972), 385.

19 Quoted in Lovejoy and Boas, *Primitivism and Related Ideas in Antiquity*, 33, citing Diels, *Fragmente der Vorsokratiker*, vol. 1, 4th ed. (1922), 271-72, 273.

20 See Stephen H. Webb, *Good Eating* (Grand Rapids, MI: Brazos Press, 2001), 154.

21 Letter of 6 May 1898. Quoted in Henri Troyat, *Tolstoy* (New York: Doubleday, 1967), 580. Like Thoreau, although without his self-conscious awareness of the paradox, Tolstoy also finds sympathy with modernity. Thus, in *What Is Religion, and of What Does Its Essence Consist?* (1902), Tolstoy expresses his admiration for the conquest of nature if undertaken by free men: "It is easy to conquer nature and to construct railways, steamships, museums and so forth if you do not spare human lives. The Egyptian pharaohs were proud of their pyramids and we too admire them, forgetting about the lives of millions of slaves sacrificed in constructing them. In the same way we admire our exhibition palaces, iron-clads and transoceanic cables, forgetting with what we pay for it all. We should only feel proud of it all when it is done freely, by free men and not by slaves." *A Confession and Other Religious Writings* (London: Penguin, 1987), 102.

22 Elijah Buckner, *The Immortality of Animals* (1903; reprint, ed. Rod Preece, Lampeter: Mellen Animal Rights Library, 2003), 23, 24, 43.

23 Ibid., ch. 5.

24 John Wesley, "The General Deliverance" (1788), *Sermons on Several Occasions*, 2nd series, Sermon 60, in *The Works of John Wesley*, vol. 6 (Grand Rapids, MI: Zondervan, n.d.; reprint of the Wesleyan Conference Office edition, London, 1872), 240-48.

25 Emily Brontë, *Wuthering Heights* (1847; reprint, Franklin Center: Franklin, 1979), ch. 24, 294. Emily is, obviously, trying to reverse traditional sexual stereotypes. In sister Charlotte's *Jane Eyre* (ch. 8, also 1847), the eponymous heroine offers an image very similar to that of Linton, although with just a hint of Arcadia in the rocks and ruins: "I feasted on the spectacle of ideal drawings, which I saw in the dark; all the work of my own hands: freely pencilled houses and trees, picturesque rocks and ruins, Cuyp-like groups of cattle, sweet paintings of butterflies hovering over unblown roses, of birds picking at ripe cherries, of wrens' nests enclosing pearl-like eggs, wreathed about with young ivy sprays." There were three famous seventeenth-century Dutch Cuyp painters. Charlotte Brontë is almost certainly referring to Aelbert Cuyp (1620-91), today one of the most celebrated of landscape painters, whose cows are indeed evocative of rustic health.

26 George Sand, *Indiana* (1831; reprint, Oxford: Oxford University Press, 1994), prt. 4, ch. 29, 241.

27 Jeffrey Myers, *D.H. Lawrence: A Biography* (New York: Vintage Books, 1992), 69.

28 From Geoffrey Grigson, ed., *The Oxford Book of Satirical Verse* (Oxford: Oxford University Press, 1983), 70-71.

29 Bill Moyers, *Genesis: A Living Conversation* (New York: Doubleday, 1996), 56-57.

30 George Eliot, *Mill on the Floss* (Franklin Center: Franklin, 1981), bk. 4, ch. 1, 279.

31 Peter J. Bowler, *The Norton History of the Environmental Sciences* (New York: W.W. Norton, 1993), 168.

32 Keith Thomas, *Man and the Natural World: Changing Attitudes in England, 1500-1800* (London: Penguin, 1984), 299.

33 Gilbert White, *The Natural History of Selborne* (1789; reprint, ed. W.S. Scott, London: Folio Press, 1962), Letter 13, to the Hon. Daines Barrington, 109.

34 Ibid., Letter 9, to Thomas Pennant, Esq., 19.

35 Ibid., Letter 13, to the Hon. Daines Barrington, 109.

36 Ibid., Letter 24, to the Hon. Daines Barringon, 139.

37 Cyparissus unwittingly killed a stag whom he loved and spent the remainder of his life in mourning. The story is told in Ovid's *Metamorphoses,* bk. 10, and is repeated in Rod Preece, *Awe for the Tiger, Love for the Lamb: A Chronicle of Sensibility to Animals* (Vancouver: UBC Press; London and New York: Routledge, 2002), 53-54.

38 For an analysis of the conflicts, see Arthur Guy Lee's Introduction to *The Eclogues* (London: Penguin, 1984), 19-22. He notes, for example, that "*Eclogue* VI does indeed transport us to an imaginary land where Silenus may be found asleep in a cave where Fauns and wild beasts will dance to his singing, but some of the things he sings about can hardly be called Arcadian: Chaos, the Flood, theft, rape, bestiality, infanticide and cannibalism" (21). On the basis of the "ideal types" that I have proposed, some of these aspects may be deemed inimical to the Arcadian idea, but they are far more incompatible with that of Eden. Despite the conceptual conflicts, after Virgil's *Eclogues,* Arcadia came to be regarded as the symbol of rural joy. Geographically, Arcadia is the mountainous district of central Greece, where, in ancient times, Pan was worshipped. See James Thomson, *The Seasons and the Castle of Indolence,* ed. James Sambrook (1972; reprint, Oxford: Clarendon Press, 1991), 72, note to line 1301.

39 Virgil, *The Eclogues,* 65.

40 Ibid., 71. The Latin is "Faunosque ferasque uideres / ludere," suggesting, as with the English "feral," not merely wild as the antithesis to domesticated, but wild in the sense of savage and cruel.

41 Ibid., 41.

42 Ibid., 105.

43 Quoted in Edwin Haviland Miller, *Salem Is My Dwelling Place: A Life of Nathaniel Hawthorne* (Iowa City: Iowa University Press, 1991), 216.

44 *The Annotated Walden: Walden, or Life in the Woods by Henry D. Thoreau, Together with "Civil Disobedience,"* ed. Philip Van Doren Stern (New York: Barnes and Noble, 1970), 246-47.

45 Ibid., 192.

46 Ibid., 295.

47 "In communist society, where nobody has one exclusive sphere of activity but each can become accomplished in any branch he wishes, society regulates the general production and thus makes it possible for me to do one thing today and another tomorrow, to hunt in the morning, fish in the afternoon, rear cattle in the evening, criticize [i.e., philosophize] after dinner, just as I have in mind, without ever becoming hunter, fisherman, cowherd, or critic [i.e., philosopher]." David McLellan, ed., *Karl Marx: Selected Writings* (Oxford: Oxford University Press, 1977), 169.

48 Herbert Marcuse, *One Dimensional Man: Studies in Ideology of Advanced Industrial Society* (Boston: Beacon Press, 1964).

49 *The Annotated Walden,* 222.

50 Ibid., 223.

51 Ibid., 250-51.

52 Ibid., 279.

53 Ibid., 339.

54 Ibid., 344.

55 Ibid., 346.

56 Despite the name, the Jardin des Plantes was also the Paris zoological garden.

57 Quoted in Richardson, *Emerson: The Mind on Fire*, 141-42.

58 See pages 81-82.

59 Victor Hugo, *Les Misérables* (1862; reprint, trans. Norman Denny, London: Penguin, 1976), prt. 1, bk. 5, ch. 5, 164.

60 Others accord the honour to his son Enoch. See George Boas, *Essays on Primitivism and Related Ideas in the Middle Ages* (1948; reprint, Baltimore: Johns Hopkins University Press, 1997), 193.

61 Quoted in Mason Wade, *Margaret Fuller: Whetstone of Genius* (New York: Viking, 1940), 4-5.

62 Aldous Huxley, *Island* (1962; reprint, London: Flamingo, 1994), ch. 8, 134.

63 John Steinbeck, *The Log from the Sea of Cortez* (London: Penguin, 1995), 80.

64 Ann Radcliffe, *The Romance of the Forest* (1791; reprint, Oxford: Oxford University Press, 1986), prt. 3, ch. 16, 243.

65 Quoted in William H. Drummond, *The Rights of Animals and Man's Obligation to Treat Them with Humanity* (London: John Mardon, 1838), 57.

66 See McLellan, ed., *Karl Marx*, 121-22.

67 See Helen Waddell, ed. and trans., *The Desert Fathers* (New York: Vintage, 1998), 56-57.

68 See Michael Seidlmayer, *Currents of Medieval Thought with Special Reference to Germany* (Oxford: Basil Blackwell, 1960), 9.

69 McLellan, ed., *Karl Marx*, 179.

70 Angus Taylor has pointed out to me wisely that the subject of Marx's view of history is the human being with "feet firmly on the solid ground ... exhaling and inhaling all the forces of nature" (*Economic and Philosophical Manuscripts*) – i.e., the goal is not to transcend nature per se. Also in a well-known passage in *Capital*, Marx mentions animal labour, indicating that "a spider conducts operations which resemble those of the weaver, and a bee would put many a human architect to shame by the construction of its honeycomb cells. But what distinguishes the worst architect from the best of bees is that the architect builds the cell in his mind before he constructs it in wax. At the end of every labour process, a result emerges which had already been conceived by the worker at the beginning, hence already existed ideally. A man not only effects a change of form in the materials of nature; he also realizes (*Verwirklichkeit*) his own purpose in those materials." In other words, for Marx, there is a fundamental difference between human labour and animal labour, despite the resemblances. Yet I must respond that this begs the question of the value of the intellect if its employment by humans at least in certain instances results in poorer consequences than would actions undertaken without it.

71 Karl Marx and Friedrich Engels, *Manifest der Kommunistischen Partei* (1848; reprint, Berlin: Dietz Verlag, 1958), 7: "Die Geschichte aller bisherigen Gesellschaft ist die Geschichte von Klassenkämpfen."

72 See note 2 above.

73 See ch. 1, note 143 above (italics in the original).

74 See note 47 above.

75 Quoted in John Passmore, *Man's Responsibility for Nature: Ecological Problems and Western Tradition* (London: Duckworth, 1974), 24, citing David McLellan, ed., *Marx's Grundrisse* (London, 1971), 94.

76 "On the Vegetable System of Diet," in *The Complete Works of Percy Bysshe Shelley*, vol. 6, ed. Roger Ingpen and Walter E. Peck (New York: Gordian Press, 1965), 344 (italics in the original).

77 Howard Williams, *The Ethics of Diet: A Catena of Authorities Deprecatory of the Practice of Flesh-Eating* (London: F. Pitman, 1883), x-xi.

78 Quoted in Hilda Kean, *Animal Rights: Political and Social Change in Britain since 1880* (London: Reaktion Books, 1998), 133, citing Isabella Ford, *Women and Socialism* (Independent Labour Party, 1907), 7, 11.

79 Bowler, *The Norton History of the Environmental Sciences*, 144.

80 The designation is that of Friedrich Engels in his pamphlet *Socialism: Scientific and Utopian* (1880). Along with Fourier, Engels numbers Henri de Saint-Simon and Robert Owen as the prime examples, and as subsidiaries he lists Thomas Muentzer during the sixteenth-century German peasant wars, the Leveller Movement in the seventeenth-century English Civil War, and Gracchus Babeuf and the Conspiracy of Equals during the French Revolutionary period. Prototypical, yet immature, socialist theories he said could be found in the sixteenth- and seventeenth-century literary utopias of Thomas More and Tomasso Campanella and in the eighteenth-century communist musings of Morelly and the abbé de Mably. For Engels, Marx is, of course, the representative of scientific socialism. As should be clear, I regard Marx not as *anti*-utopian but as *differently* utopian from those to whom Engels assigns the category.

81 Quoted in James L. Wiser, *Political Philosophy: A History of the Search for Order* (Englewood Cliffs: Prentice Hall, 1983), 329.

82 See William St. Clair, *The Godwins and the Shelleys* (New York: W.W. Norton, 1989), 97.

83 Quoted in Jeffrey Myers, *D.H. Lawrence: A Biography* (New York: Vintage Books, 1992), 172.

84 Lois Whitney, *Primitivism and the Idea of Progress in English Popular Literature of the Eighteenth Century* (1934; reprint, New York: Octagon Press, 1973), 1.

85 Paul Johnson, *Modern Times: The World from the Twenties to the Nineties* (New York: Harper Collins, 2001), 203.

86 *Paradise Lost*, IX, 782-84.

87 Ibid., IX, 1000-1.

88 Quoted in Bill Moyers, *Genesis: A Living Conversation*, 4.

89 Eusebius, *Ecclesiastical History*, 5.1.25-26, in Philip Schaff and Henry Wace, eds., *The Church History of Eusebius* (Grand Rapids, MI: William B. Eerdman, 1961), 214.

90 See page 72.

91 Quoted in Richard Erdoes and Alfonso Ortiz, eds., *American Indian Myths and Legends* (1984; reprint, London: Pimlico, 1997), 111.

92 Joseph Campbell, *Historical Atlas of World Mythology*, vol. 1, *The Way of the Animal Powers*, prt. 1, *Mythologies of the Primitive Hunters and Gatherers* (New York: Harper and Row, 1988), 14.

93 Joseph Campbell, *Historical Atlas of World Mythology*, vol. 2, *The Way of the Seeded Earth*, prt. 3, *Mythologies of the Primitive Planters: The Middle and Southern Americas* (New York: Harper and Row, 1989), 331. This bears some resemblance to Genesis 6:12, where "God looked upon the earth, and, behold, it was corrupt; for all flesh had corrupted his way upon the earth." It is certainly consistent with this verse that humans might have followed animals in their breach of the vegetarian law.

94 Thomas Young, *An Essay on Humanity to Animals* (1798; reprint, ed. Rod Preece, Lampeter: Mellen Animal Rights Library, 2001), 75.

95 William Paley, *Moral and Political Philosophy* (1785), bk. 2, ch. 11, in *The Works of William Paley, D.D., Archdeacon of Carlisle* (Philadelphia: Crissy Markley, 1850), 44.

96 A variant of this view was still being maintained toward the close of the nineteenth century. Thus the anthropologist duke of Argyll in *Primeval Man* (1869) advanced the view, in line with that of Richard Whately, Archbishop of Dublin, that humans were in origin civilized beings but then regressed to "a savage state."

97 Ann-Robert Jacques Turgot, "Philosophical Tableau on the Successive Advancements of the Human Mind" (1750) and *Reflections on the Production and Distribution of Wealth* (1758); Marquis de Condorcet, *Sketch for a Historical Picture of the Progress of the Human Mind* (1795).

98 Quoted in St. Clair, *The Godwins and the Shelleys*, 106.

99 For an examination of Pope's ideas about the Golden Age, see Maynard Mack, *Alexander Pope: A Life* (New York: W.W. Norton, 1988), 512, 522-50, 588, 898.

100 See "The Supposed Primitivism of Rousseau's *Discourse on Inequality*" (1923), in Arthur O. Lovejoy, *Essays in the History of Ideas* (New York: George Braziller, 1955), 14-37.

101 John Oswald, *The Cry of Nature, or An Appeal to Mercy and to Justice on Behalf of the Persecuted Animals* (1791; reprint, ed. Jason Hribal, Lampeter: Mellen Animal Rights Library, 2000).

102 Joseph Ritson, *An Essay on Abstinence from Animal Food as a Moral Duty* (London: Richard Phillips, 1802).

103 John Frank Newton, *The Return to Nature, or A Defence of the Vegetable Regimen* (London: T. Cadell and W. Davies, 1811).

104 *Madame Bovary: A Story of Provincial Life* (1865; reprint, New York: Oxford University Press, 1985), 179.

105 Nikolai Gogol, *Dead Souls* (1842; reprint, London: Penguin, 1961), prt. 2, ch. 3, 324-25.

106 See Troyat, *Tolstoy*, 482.

107 Ibid., 305.

108 See Peter Ackroyd, *Dickens* (Toronto: Stewart House, 1991), 551.

109 Ibid.

110 Mary Wollstonecraft has in mind William Blackstone, author of *Commentaries on the Laws of England* (1765-68); Edmund Burke, author of *Reflections on the Revolution in France* (1790); and those who had initially supported the French Revolution but did a volte-face in light of the accuracy of Burke's predictions of the course of the revolution, its cruelties, and its failures. By 1796 Wollstonecraft herself could be numbered among them in that her *Letters Written During a Short Residence in Sweden, Norway and Denmark* showed due deference to all the Burkean anti-rationalist categories.

111 *Vindication of the Rights of Woman*, in *Mary Wollstonecraft: Political Writings* (1792; reprint, ed. Janet Todd, Oxford: Oxford University Press, 1994), 78, 79.

112 Henry James, *The Bostonians* (1886; reprint, London: Penguin, 1986), bk. 2, ch. 28, 268.

113 This view is not held by all feminists. For example, in *The Sexual Politics of Meat: A Feminist-Vegetarian Critical Theory* (1991; reprint, New York: Continuum, 2002), Carol Adams says: "I do not believe women are innately more caring than men, or have an essential pacifist quality. But many of my feminist-vegetarian sources did believe this" (22). Of course, Adams understands that if she had taken what she deems "an essentialist position of women," she would not have been able to blame almost everything on men because of their chosen behaviour rather than blame almost everything on men because of their innate characteristics. There are, of course, legitimate differences

of position to be taken between the extremes of the nature-nurture controversy. Thus one may argue that men have a greater innate disposition to aggression and to a corresponding certain type of courage without believing women incapable of the most heinous and destructive of behaviour and also argue that as a more gender-egalitarian society is approached, such behaviour will increase in women without it ever becoming as predominant as in males.

114 See George Boas, *The Happy Beast in French Thought of the Seventeenth Century* (1933; reprint, New York: Octagon Books, 1966), 11-13.

115 Voltaire, *Candide and Other Stories,* trans. Roger Pearson (Oxford: Oxford University Press, 1990), 61-62.

116 Ibid., 239.

117 Voltaire, *Traité sur la tolérance* (1763; reprint, trans. Rod Preece, Paris: Flammarion, 1989), 171. See also Voltaire, "Bêtes," in *Dictionnaire philosophique* (1764; reprint, Paris: Garnier, 1961), 33.

118 Voltaire, *Candide and Other Stories,* 106.

119 Ibid., 125.

120 Mary Wollstonecraft, *Original Stories from Real Life, with Conversations Calculated to Regulate the Affections and Form the Mind to Truth and Goodness* (London: J. Johnson, 1788).

121 See Chloe Chard's Introduction (xvi) and Notes (382) to Radcliffe, *The Romance of the Forest.*

122 Arthur O. Lovejoy, "Monboddo and Rousseau," in *Essays in the History of Ideas,* 38. First published in *Modern Philology* 30 (1933): 275-96.

123 James Boswell, *The Life of Samuel Johnson LL.D.,* vol. 1 (1791; reprint, London: J.M. Dent and Sons, 1906), 358 (italics in the original).

124 Otto Gierke, *Natural Law and the Theory of Society, 1500 to 1800* (1934; reprint, Boston: Beacon Press, 1957), 96.

125 James Burnet (Lord Monboddo), *Origin and Progress of Language,* vol. 1, 2nd ed. (1774), 147, as cited in Lovejoy, "Monboddo and Rousseau," 42.

126 Knight, *Lord Monboddo and Some of His Contemporaries* (1900), 73, as cited in Lovejoy, "Monboddo and Rousseau," 43.

127 Jean-Jacques Rousseau, *Rousseau's Political Writings,* ed. Allan Ritter and Julia Conaway Bondanella, trans. Bondanella (New York: W.W. Norton, 1988), 13.

128 *Spectator* 2, 62 (11 May 1711). Reprinted in Alex Chalmers, ed., *The Spectator: A New Edition Corrected from the Originals,* vol. 7 (New York: E. Sargent and M.A. Ward, 1810), 14-15.

129 Quoted in John Vyvyan, *In Pity and in Anger* (1969; reprint, Marblehead, MA: Micah, 1988), 26, citing *The Plays of William Shakespeare,* vol. 7, ed. Samuel Johnson (London, 1765), 279. Johnson was commenting on the rebuke against the Queen in *Cymbeline* when she proposed to experiment with poison on animals. With justice, Johnson assumed opposition to animal experimentation to be Shakespeare's own view, as later did Bernard Shaw. See Preece, *Awe for the Tiger, Love for the Lamb,* 346.

130 *Rambler* 33 (1750); *Adventurer* 67 (1753).

131 Quoted in Jackson Bate, *Samuel Johnson* (San Diego: Harcourt, Brace, 1979), 139.

132 Lewis has "man" instead of "beast." I take this to be a typographical error, since Lewis's commentary implies that he understands "beast" at this point.

133 *Essay on Man,* epist. 3.5.146.

134 Lewis appears to misunderstand Pope's concept of "self-love," which corresponds more closely to a sense of self-worth or sense of dignity rather than to anything approaching

narcissism. This is a matter of some moment because it would appear to be the basis for Lewis's claim of "self-contradictions."

135 Lewis here has a footnote that reads: "This supposition implies a change in the nature of beasts, as well as of man: for beasts avoid man, and prey upon him, which have never been subject to his attacks."

136 George Cornewall Lewis, *Remarks on the Use and Abuse of Some Political Terms* (London: B. Fellowes, 1832; facsimile reprint, Columbia: University of Missouri Press, 1970), 185-87.

137 Daniel A. Dombrowski, *The Philosophy of Vegetarianism* (Amherst: University of Massachusetts Press, 1984), 19.

138 Quoted in Newton, *Return to Nature,* 89. "Shambles" is the traditional term for an abattoir. The "sacred records" that Evelyn had in mind probably included the Hindu *Santi-parva,* the Buddhist *Buhaddharma Purana,* and the Persian *Zend-Avesta.*

139 For Pope, see *Guardian* 61 (21 May 1713), and for Locke, see *Some Thoughts concerning Education* (1693), ed. John Yolton and Jean Yolton (Oxford: Oxford University Press, 1998).

140 Job 12:7-9:

> You have only to ask the cattle, for them to instruct you, and the birds of the sky for them to inform you.
>   The creeping things of the earth will give you lessons, and the fish of the sea provide you with an explanation: there is not one such creature but will know that the hand of God has arranged things like this.

141 Ivan Turgenev, *Fathers and Sons* (1861; reprint, Franklin Center: Franklin, 1985), ch. 26, 221-22.

142 See George Boas, *Essays on Primitivism and Related Ideas in the Middle Ages* (1948; reprint, Baltimore: Johns Hopkins University Press, 1997), 13.

143 Huxley, *Island,* ch. 3, 17 (italics in the original).

144 Alice Chandler, *A Dream of Order: The Medieval Ideal in Nineteenth Century Literature* (London: Routledge and Kegan Paul, 1971).

145 Eliot, *Mill on the Floss,* ch. 12, 117.

146 Chandler, *A Dream of Order,* 231-32.

147 See Preece, *Awe for the Tiger, Love for the Lamb,* 302.

148 Charles Kingsley, *Yeast,* in *Westward Ho! Hypatia and Yeast* (1851; reprint, London: Macmillan and Co., 1890), ch. 5, 24.

149 Quoted in Richardson, *Emerson: The Mind on Fire,* 491.

150 Newton, *Return to Nature,* 66.

151 See page 3

152 Jean-Jacques Rousseau, *Discourse on the Origin and Foundations of Inequality among Men,* in *Rousseau's Political Writings,* 28.

153 Ibid., 29.

154 Ibid., 7.

155 Jean-Jacques Rousseau, *Emile, or On Education* (1762; reprint, trans. Allan Bloom, New York: Basic Books, 1979), 55.

156 Ritson, *An Essay on Abstinence from Animal Food as a Moral Duty,* 235.

157 George Nicholson, *On the Primeval Diet of Man: Vegetarianism and Human Conduct toward Animals* (1801; reprint, ed. Rod Preece, Lampeter: Mellen Animal Rights Library, 1999), 14.

158 Ibid., 13.

159 Ibid., 18.

160 Ibid.

161 Ibid., 115-54.

162 Quoted in Deborah Kennedy, *Helen Maria Williams and the Age of Revolution* (Lewisburg: Bucknell University Press, 2002), 26.

163 *The Devils*, vol. 2, 21, quoted in Joseph Frank, *Dostoevsky: The Miraculous Years, 1865-71* (Princeton: Princeton University Press, 1995), 490.

164 Fyodor Dostoevsky, *The Brothers Karamazov* (1880; reprint, trans. Richard Pevear and Larissa Volokhonsky, New York: Alfred A. Knopf, 1992), 294, 319.

165 June Dwyer, *John Masefield* (New York: Ungar, 1987), 101. The quotation from Masefield's *The Everlasting Mercy* can be found at 31.

166 Mikhail Sholokhov, *And Quiet Flows the Don* (1929; reprint, trans. Stephen Garry, London: Penguin, 1967), prt. 2, ch. 4, 255.

167 Victor Hugo, *Les Misérables,* prt. 3, bk. 4, ch. 4, 574.

168 Quoted in Lovejoy and Boas, *Primitivism and Related Ideas in Antiquity,* 389. I am heavily indebted to Chapter 13 on "The Superiority of the Animals," 389-420, for many of the classical quotations used in the remainder of this chapter, as acknowledged in the respective notes.

169 Charles Darwin, *The Descent of Man, and Selection in Relation to Sex,* 2nd ed. (New York: A.L. Burt, 1874), 54.

170 Letter dated 17 April 1872. Quoted in Troyat, *Tolstoy,* 347.

171 Quoted in Marjorie Spiegel, *The Dreaded Comparison: Human and Animal Slavery* (New York: Mirror Books, 1996), 23-24.

172 J.R. Ackerley, *My Dog Tulip* (1956; reprint, New York: Poseidon, 1990), ch. 2, 43.

173 George Sand, *Indiana* (1831; reprint, trans. Sylvia Raphael, Oxford: Oxford University Press, 1994), prt. 4, ch. 29, 239.

174 Quoted in Lovejoy and Boas, *Primitivism and Related Ideas in Antiquity,* 391.

175 Plutarch, "Whether Land or Sea Animals are Cleverer," 20.974A, quoted in Jonathan Barnes, ed. and trans., *Early Greek Philosophy* (London: Penguin, 1987), 262.

176 Quoted in Barnes, ed. and trans., *Early Greek Philosophy,* 266. Barnes warns us that many of the ascriptions he cites from Democritus "are at best dubious" (265).

177 Quoted in Sir Ernest Barker, *Greek Political Theory: Plato and His Predecessors* (1918; reprint, London: Methuen, 1957), 73.

178 Ibid., 209.

179 *Aristophanes: Four Comedies,* trans. Dudley Fitts ([1965?]; reprint, New York: Harvest Books, 2003).

180 Quoted in Lovejoy and Boas, *Primitivism and Related Ideas in Antiquity,* 392-93, citing Dio Chrysostom, *Disc. VI,* 21-23, 26-28, vol. 1, ed. J. von Arnim (1893), 88-89.

181 Barker, *Greek Political Theory: Plato and His Predecessors,* 106-7.

182 "La propriété c'est le vol," from Pierre-Joseph Proudhon, *Qu'est-ce que la propriété?* (1840), ch. 1.

183 Quoted in Lovejoy and Boas, *Primitivism and Related Ideas in Antiquity,* 393, citing Dio Chrysostom, *Disc. X,* 16.

184 Quoted in McManners, ed., *The Oxford History of Christianity,* 16.

185 Quoted in James E. Gill, "Theriophily in Antiquity: A Supplementary Account," *Journal of the History of Ideas* 30 (July-September 1969): 404. See the whole of Gill's article for a lengthy treatment of Seneca on animals.

186 Ibid., 405, citing Seneca, *De ira,* vol. 2, 397.

187 Roger Masters, ed., *Jean-Jacques Rousseau: The First and Second Discourses,* trans. Roger D. Masters and Judith R. Masters (New York: St. Martin's Press, 1964), 141.

188 Quoted in Lovejoy and Boas, *Primitivism and Related Ideas in Antiquity,* 394, citing Th. Kock, *Com. att. fragm.,* vol. 2 (Leipzig, 1884), prt. 1, 503-4.

189 Quoted in ibid., citing ibid., 507.

190 Quoted in ibid., 395, citing ibid., 504.

191 For Henry Fielding, see *Covent-Garden Journal,* 22 February 1751-52; for comte de Buffon, see *Histoire naturelle,* as cited in Drummond, *The Rights of Animals,* 101; for Coleridge, see "To a Young Ass: Its Mother Being Tethered Near It" (1794); for Laurence Sterne, see *A Sentimental Journey through France and Italy* (1768); and for Robert Louis Stevenson, see *Travels with a Donkey in the Cévennes* (1879).

192 Quoted in Lovejoy and Boas, *Primitivism and Related Ideas in Antiquity,* 395-96, citing Th. Kock, *Com. att. fragm.,* vol. 3, 158.

193 James E. Gill, "Theriophily in Antiquity: A Supplementary Account," *Journal of the History of Ideas* 30 (July-September 1969): 403.

194 See, for example, pages 44-45 and 77.

195 Plutarch, "Whether Land or Sea Animals Are Cleverer," in *Plutarch's Moralia,* vol. 12, trans. Harold Cherniss and William C. Helmblod (1927; reprint, London: William Heinemann, 1957), 369.

196 R.G. Bury, trans., *Sextus Empiricus,* vol. 1 (London: William Heinemann, 1958), 95.

197 Quoted in Lovejoy and Boas, *Primitivism and Related Ideas in Antiquity,* 397, citing *De posteritate Caini,* vol. 2, ed. P. Wendland (Berlin, 1897), 160-62, and *De finibus,* vol. 2, xiii, 40; S.O. Dickerman, *De argumentis quibusdam e structura hominis et animalium petitis* (Halle, 1909).

198 D. Shackleton Bailey, trans., *Cicero's Letters to His Friends* (London: Penguin, 1967), 81.

199 *Historia animalium,* 612b, 614b, 612a, 588:18. See Lovejoy and Boas, *Primitivism and Related Ideas in Antiquity,* 391.

200 See Lovejoy and Boas, *Primitivism and Related Ideas in Antiquity,* 399.

201 Quoted in ibid., 400, citing Seneca, *De ira,* vol. 2, ed. E. Hermes (Leipzig, 1917), 16.

202 Ovid, *Metamorphoses,* trans. Mary M. Innes (London: Penguin, 1955), bk. 10, 233.

203 Quoted in A. Passerin d'Entrèves, *Natural Law: An Introduction to Legal Philosophy* (1951; reprint, London: Hutchinson, 1970), 29, citing Ulpian, *Digest,* I.i.i.

204 Quoted in Wiser, *Political Philosophy,* 80.

205 Paul Sigmund, *Natural Law in Political Thought* (Cambridge: Winthrop Publishers, 1971), 38.

206 Pliny the Elder, *Natural History: A Selection,* trans. John F. Healy (London: Penguin, 1991), bk. 7, 74.

207 See Buckner, *The Immortality of Animals* (1903), 141.

208 Kingsley, *Yeast,* ch. 2, 9.

209 It is notable that the view of both Homer and Plutarch seems to be the very antithesis of that of Karl Marx. Whereas Homer and Plutarch appear to prefer a life of leisure over a life of labour, Marx has the wisdom to understand that it is through labour – although only labour of a nondemeaning kind – that we are fulfilled. Of course, the view held by Homer and Plutarch is more "realistic" in the sense that theirs is the view of how most of us imagine that we are fulfilled.

210 Plutarch, *Essays,* trans. Robin Waterfield, introduced and annotated by Ian Kidd (London: Penguin, 1912), 385, 386, 387, 389, 392, 397, 398.

211 John Stuart Mill, *Utilitarianism,* in *Utilitarianism, Liberty, Representative Government* (1863; reprint, London: J.M. Dent, 1910), 9.

212 Mary Warnick, ed., *John Stuart Mill: Utilitarianism, On Liberty, Essay on Bentham* (New York: New American Library, n.d.), 259.

213 Tolstoy, *A Confession and Other Religious Writings,* 159-60.

214 Ibid., 86, 87.

215 See page 60.

216 See for example, ch. 3 of *What Is Religion, and of What Does Its Essence Consist?* (London: Penguin, 1997).

217 Leo Tolstoy, *Resurrection* (London: Penguin, 1966), ch. 1, 19.

218 Boas, *Primitivism and Related Ideas in the Middle Ages,* 193.

219 Quoted in Lovejoy and Boas, *Primitivism and Related Ideas in Antiquity,* 402. The rendering by John F. Healy is rather less certain. His translation offers us: "We cannot confidently say whether she is a good parent to mankind or a harsh stepmother" (*Natural History,* 74).

220 Irma A. Richter, ed., *Selections from the Notebooks of Leonardo da Vinci* (Oxford: Oxford University Press, 1977), 245.

221 Quoted in ibid., 194, citing *Contra Celsum,* vol. 4, ed. Paul Koetschau (Leipzig, 1899), lxxvi, in *Opera,* vol. 1, 346ff.

222 *The Wars of Alexander: An Alliterative Romance, translated chiefly from the Historia Alexandri Magni de preliis,* Early English Text Society, extra series 47 (1886; reprint, New York: Kraus, 1973), 5582ff.

223 See Preece, *Awe for the Tiger, Love for the Lamb,* 68-73.

224 See ibid., 73-74.

225 William Langland, *Piers the Ploughman,* translated into modern English with an introduction by J.F. Goodridge (Harmondsworth: Penguin, 1959), prt. 2, bk. 11, 136-38.

226 See pages 125-26.

227 Arnobius, *The Seven Books of Arnobius Adversus Gentes,* in *Ante-Nicene Christian Library: Translations of the Fathers down to AD 325,* vol. 19, trans. Hamilton Bruce and Hugh Campbell (Edinburgh: T.T. Clark, 1871), 84.

228 Lactantius, *The Father of the Church, Lactantius, the Divine Institutes, Books I-VII,* trans. Francis McDonald (Washington: The Catholic University of America Press, 1964), 186.

229 Quoted in Michael W. Fox, *St. Francis of Assisi, Animals and Nature* (Washington, DC: Center for Respect of Life and Environment, 1989), 3.

230 Walter Map, *De Nugis Curialium: Courtiers' Trifles,* ed. and trans. M.R. James (Oxford: Clarendon Press, 1983), vol. 1, i; sec. 5, 7.

CHAPTER 5: THERIOPHILY REDIVIVUS

1 Peter Singer, *Animal Liberation,* 2nd ed. (New York: New York Review of Books, 1990), 198.

2 Richard D. Ryder, *Animal Revolution: Changing Attitudes towards Speciesism,* rev. ed. (Oxford: Berg, 2000), 39.

3 Mary Midgley, *Animals and Why They Matter* (Athens: University of Georgia Press, 1983), 11. Midgley makes the remark in the context of a discussion of Auguste Comte's *Religion of Humanity.* In fact, Comte made a great deal of the human "link with other species" in the *Religion of Humanity.* See Rod Preece, *Awe for the Tiger, Love for the*

Lamb: A Chronicle of Sensibility to Animals (Vancouver: UBC Press; London and New York: Routledge, 2002), 251-53.

4 Quoted in Michael Seidlmayer, *Currents of Medieval Thought with Special Reference to Germany* (Oxford: Basil Blackwell, 1960), 157. For a general discussion of the topic, see Jacob Burkhardt, *The Civilization of the Renaissance in Italy* (1860; reprint, London: Phaidon, 1960): In the dark ages "man was conscious of himself only as a member of a race, people, party, family, or corporation – only through some general category." With the onset of the Renaissance, however, "man became a spiritual *individual* and recognized himself as such" (81, italics in the original).

5 Seidlmayer, *Currents of Medieval Thought*, 108.

6 See page 108.

7 Jim Mason, *An Unnatural Order: Uncovering the Roots of Our Domination of Nature and Each Other* (New York: Simon and Schuster, 1993), 37.

8 Roderick Frazier Nash, *The Rights of Nature: A History of Environmental Ethics* (Madison: University of Wisconsin Press, 1989), 17-18.

9 Ibid., 18.

10 See Ryder, *Animal Revolution*, rev. ed., 49.

11 Quoted in Keith Thomas, *Man and the Natural World: Changing Attitudes in England, 1500-1800* (London: Penguin, 1984), 157, citing William Hinde, *A Faithful Remonstrance of the Happy Life and Holy Death of John Bruen* (1625; reprint, London, 1641), 31-32.

12 James Serpell, *In the Company of Animals: A Study of Human-Animal Relationships* (Oxford: Oxford University Press, 1983), 137.

13 Thomas, *Man and the Natural World*, 35. The words quoted by Serpell are on page 41.

14 *The Literary Works of Leonardo da Vinci*, vol. 2, 2nd ed., enlarged and revised by Jean Paul Richter and Irma A. Richter (1st ed. 1883, ed. Jean Paul Richter; 2nd ed. 1939; reprint, London: Phaidon, 1970), 302.

15 Quoted in Singer, *Animal Liberation*, 199. Singer has "Picola" instead of "Pico."

16 In his *Examen vanitationis doctrinae gentium*.

17 Thomas à Kempis, *The Imitation of Christ* (Bombay: St. Paul Society, 1986), 194. I am indebted to Saba Alemayehu for bringing this passage, and the precise source for the following note, to my attention.

18 Singer, *Animal Liberation*, 101.

19 "St. Philip Romolo Neri," in *Catholic Encyclopedia*, n.p.

20 Ibid.

21 George Boas, *The Happy Beast in French Thought of the Seventeenth Century* (1933; reprint, New York: Octagon Books, 1966).

22 See Boas, *The Happy Beast*, ch. 4, 191 and 188.

23 Pierre Bayle, *Historical and Critical Dictionary: Selections,* trans. Richard H. Popkin (Indianapolis: Bobbs-Merrill, 1965), 213. The title of the essay is quite misleading. Only two out of forty-two pages, plus a half-dozen or so subsidiary comments, are devoted to Rorarius. The remainder concerns the question of animal souls and the ideas of Leibniz.

24 Ibid.

25 Ibid., 214.

26 Ibid., 216-17.

27 Ibid., 230-31.

28 Boas, *The Happy Beast*, 38.

29 Extracts of the Latin text of Rorarius's work can be found in ibid., 38, 39, 40.

30 Pliny the Elder, *Natural History: A Selection,* trans. John F. Healy (London: Penguin, 1991), bk. 8.

31 Quoted in Richard Overton, *Man's Mortalitie* (1643; reprint, ed. Harold Fisch, Liverpool: University of Liverpool Press, 1968), 68.

32 At least I have not encountered reported instances in the early medieval period. In part, however, this may be because there are few records from this period; indeed, court records of the later medieval and even Renaissance periods are scarcely abundant. In *The Beast Within: Animals in the Middle Ages* (New York: Routledge, 1994), Joyce E. Salisbury says that we "know of 93 cases of criminal proceedings against animals, and the earliest in 1266" (39). This contrasts with the 189 reported by Harwood (see note 33 below), but I assume that Salisbury's "93 cases" refers to the Middle Ages alone. The later thirteenth century seems to be about the right time, which is also the time when we hear the first rumblings that would become the roar of the Renaissance.

33 This figure is given in Dix Harwood, *Love for Animals and How It Developed in Great Britain* (1928; reprint, ed. Rod Preece and David Fraser, Lampeter: Mellen Animal Rights Library, 2002), 8. In *Notre-Dame of Paris* (1831), bk. 8, Victor Hugo describes such a trial in an ecclesiastical court at some length, based on solid research of court records.

34 Harwood, *Love for Animals and How It Developed in Great Britain,* 8-9. Harwood's sources were Robert W. Chambers, *The Book of Days* (Philadelphia, 1899), i, 126-29; Edward Payson Evans, *The Criminal Prosecution and Capital Punishment of Animals* (London, 1906); and, especially for the details of this case, Walter Woodman Hyde, "Prosecution and Punishment of Animals and Lifeless Things in the Middle Ages and Modern Times," *University of Pennsylvania Law Review* 64: 696-730.

35 Boas, *The Happy Beast,* 44.

36 Ibid., 47.

37 It was published at Augustanus Ambergae. I have been unable to determine where this is.

38 Boas, *The Happy Beast,* 48.

39 See page 81.

40 See page 3.

41 Quoted in Boas, *The Happy Beast,* 11: "Qu'il vaut mieux estre ignorant, que sçavant; Que le pauvre villageois est plus à son aise que n'est le citoyen."

42 Quoted in ibid.: "[L]'estime grandement l'ordonnance des Lucquois, que nul faisant profession de lettres, ou en qualité de docteur, puisse obtenir aucun office, ou magistrat en leur parlement: Car ils craignent que ces gens de lettres, par leur grand sçavoir, dont ils presument tant de leurs personnes, ne perturbent la tranquilité et le bon ordre de leur Republique."

43 Ibid., 24-25.

44 See page 192.

45 Boas managed to find a rare French-language copy, from which he quotes at some length. The relevant words here are "les uns des plumes, les autres du poil, autres du cuir, autres des escailles et toisons" (*The Happy Beast,* 16).

46 Quoted in ibid.: "Physiciens lesquels sous ombre d'un recipé muet rend, et font decipé, et faut acheter bien cher le travail de celui qui bien souvent nous cause la mort, car la plupart de leurs médicines laxatives ne sont autres choses que vrais marteaux pour assomer les hommes."

47 Ibid.

48 Victor Hugo, *Les Misérables* (1862; reprint, trans. Norman Denny, London: Penguin, 1976), prt. 1, bk. 4, ch. 5, 164.

49 Boas, *The Happy Beast,* 21-22.

50 See page 225-26.

51 Boas, *The Happy Beast,* 29.

52 Ibid., 35.

53 Quoted in Troyat, *Tolstoy,* 396.

54 Quoted in Elijah Buckner, *The Immortality of Animals* (Philadelphia: George W. Jacobs, 1903), 243-44.

55 See page 248, and Stelio Cro, *The Noble Savage: Allegory of Freedom* (Waterloo: Wilfrid Laurier University Press, 1990).

56 Michel de Montaigne, *Selected Essays* (1943; reprint, New York: Oxford University Press, 1982), 213, 214, 213.

57 Ibid., 215.

58 Ibid., 216.

59 Ibid., 326. The quotation from Virgil is from *Georgics,* bk. 4, lines 263-65.

60 Ibid., 326.

61 Ibid.

62 Quoted in Thomas, *Man and the Natural World,* 159, citing *Essays of Montaigne,* vol. 2, trans. Florio, 126.

63 Montaigne, *Selected Essays,* 323.

64 The quotation is from Lucretius, *De rerum natura,* bk. 5. In one translation, *On the Nature of the Universe* (1951; reprint, trans. R.A. Latham, London: Penguin, 1994), 152, the phrase is nicely rendered as "each species develops according to its own kind, and they all guard their specific characters in obedience to the laws of nature."

65 Montaigne, *Selected Essays,* 342-43.

66 Ibid., 342.

67 Ibid., 343.

68 Ibid., 344.

69 Ibid., 348.

70 Quoted in Boas, *The Happy Beast,* 52-53: "N'estimez pas que je me moque: Car quant à moy je suis du nombre de ceux qui pensent que nature ait estez trop indulgente envers les autres animaux, au regard de nous."

71 Henry Fielding, *Joseph Andrews,* in *Joseph Andrews and Shamela* (1742; reprint, Oxford: Oxford University Press, 1980), bk. 1, ch. 11, 44.

72 Quoted in Boas, *The Happy Beast,* 54: "nous sommes contraints de confesser que c'est une très-bonne et salutaire médicine de n'user point de médicine."

73 Quoted in ibid., 55: "Qui entretient les hérésies, qui nourrit les procez, qui rend un homme adultère de la femme de son voisin, sinon la mesure parole?"

74 John Frank Newton, *The Return to Nature, or A Defence of the Vegetable Regimen* (London: T. Cadell and W. Davies, 1811), 63-64.

75 See page 220.

76 Mark Twain, "Dick Baker's Cat," in *Short Stories* (London: Penguin, 1993), 105, 107 (italics in the original).

77 (Minor tracts or letters.)

78 Quoted in Boas, *The Happy Beast,* 63-64: "les Chats se persuadent peut-être que les Rats et les Souris ne sont que pour engraisser."

79 Quoted in ibid., 119: "ne se trouveroit pas différente ou esloignée de la Raison."

80 See ibid.

81 Charles Darwin, *The Descent of Man, and Selection in Relation to Sex*, 2nd ed. (New York: A.L. Burt, 1874), 104.

82 Quoted in Boas, *The Happy Beast*, 132: "une legère ombre de nos lumières at une foible imitation de nostre raisonnememnt."

83 See F. Bourdy, "La saignée chez le cheval dans l'Antiquité tardive," *Revue de médecine vétérinaire* 139, 12 (1988): 1181-84; Adam Grasmück, *Die Blutentziehung als Heilmittel bei den ältesten Völkern mit besonderer Besichtigung der Tierheilkunde*, dissertation, Ludwig Maxmilian University, Munich, 1922; "L'art vétérinaire antique: Considérations sur les saignées pratiquées par les hippiatres grecs," *Recueil de médecine vétérinaire* 98 (1922): 209-34.

84 W. Rieck, "126 verschiedene Venaesectiones um 1550," *Veterinärische Mitteilungen* 9, 3 (1929): 9-15.

85 Giovanni Battista Ferraro, *Libri quattro: de' quali si tratta delle razze, delle discipline del cavalcare, e di molte altre cose appertinenti a si fatto essercitio* (Campania: Gio. Domeninico Nibio and Francesco Scaglione, 1560). The third book is devoted primarily to veterinary medicine, and the fourth to equine surgery.

86 A. Kapp, "Das Aussputzen des Halses und der Zähnen, sowie das Aderlassen bei den Pferden, ein Privileg der Lepizieger Schmiedegesellen," *Deutsche Schmiedezeitung* 52 (1936): 1056, as cited by R. Froehner, *Veterinärhistorische Mitteilungen* 17, 4 (1937): 31-32.

87 Quoted in Boas, *The Happy Beast*, 137: "Les oiseaux qui prennent l'essor, nè fondent jamais sur la proye quand elle est trop eloignée."

88 Quoted in ibid.: "L'image, par example, que la brebis a du loup, n'est ny l'image de ce loup-cy, ni celle de celuy-là ... parce qu'il y a en tous les loups quelque chose de semblable ... que l'Auteur de la nature a gravé en sa mémoire l'image de cette ressemblance."

89 Quoted in ibid., 139: "Le bon-heur supernaturel est un bien très vaste, et l'entendement des bestes est fort borné; La félicité est un bien purement spirituel, et l'entendement des bestes est une faculté comme corporelle."

90 Quoted in ibid.: "Le bien purement spirituelle est incorruptible, et les facultés comme corporelles sont perissables."

91 Marie de Ratin Chantal, marquise de Sévigné.

92 Quoted in Boas, *The Happy Beast*, 141: "Des Machines qui aiment, des machines qui ont une élection pour quelqu'un, des machines qui sont jalouses, des machines qui craignent! Allez, allez, vous vous moquez de nous; jamais Descartes n'a prétendu nous le faire croire."

93 Lord Bolingbroke, *The Works of Henry St. John, Lord Viscount Bolingbroke*, vol. 5 (1754; reprint, London: D. Mallet, 1809), 344.

94 Quoted in Boas, *The Happy Beast*, 141: "Vous dites que les Bêtes sont des Machines aussi-bien que des Montres? Mais mettez une Machine de Chien et une Machine de Chienne l'une aupres de l'autre, il en pourra résulter une troisième petite Machine; au lieu que deux Montres seront l'une auprès de l'autre toute leur vie, sans faire jamais une troisième Montre. Or nous trouvons par notre Philosophie ... que toutes les choses qui étant deux ont la vertu de se faire trois, sont d'une noblesse bien élevée au-dessus de la Machine."

95 Quoted in ibid., 143: "Elle seroit bien ridicule de croire qu'un animal tout nud, que la nature même en mettant au jour ne s'étoit pas souciée de fournir des choses nécessaires à le conserver, fût comme eux capable de raisonner. Encore, ajoutoient ils, si c'étoit un animal qui approchât un peu davantage de notre figure; mais justement le plus dissemblable, et le plus affreux; enfin une bête chauve, un oiseau plumé, une chimère amassée de toutes sortes de natures, et qui fait peur à toutes: l'homme, dis-je, si sot et vain, qu'il se persuade que nous n'avons des faits que pour lui: l'homme qui soutient

qu'on ne raisonne que par le rapport des sens, et qui cependant a les sens les plus foibles, les plus tardifs, et les plus faux d'entre toutes les créatures: l'homme enfin que la nature, pour faire de tout, a créé comme les monstres mais en qui pourtant elle a infus l'ambition de commander à tous les animaux, à l'exterminer."

96 Quoted in ibid., 143-44: "Premièrement, puisqu'il est si effronté de mentir, en soutenant qu'il ne l'est pas; secondement, en ce qu'il rit comme un fou; trosièmement, en ce qu'il pleure comme un sot; quatrièmement, en ce qu'il se mouche comme un vilain; cinquièmement, en ce qu'il est plumé comme un galeux; sixièmement, en ce qu'il porte la queue devant; septièmement, en ce qu'il a toujours une quantité de petis grès quarrés dans la bouche, qu'il n'a pas l'esprit de cacher ni d'avaler; huitièmement, et pour conclusion, en ce qu'il lève en haut tous les matins, ses yeux, son nez at son large bec, colle ses mains ouvertes la pointe au ciel, plat contre plat, et n'en faut qu'une attachée, comme s'il s'ennuyoit d'en avoir deux libres, se casse les jambes par moitié, ensorte qu'il tombe sur ses gigots; puis avec des paroles magiques qu'il bourdonne, j'ai pris garde que ses jambes rompues se r'attachent, et qu'il se relève aussi gai qu'auparavant."

97 Molière, *The Miser*, Act 3, Scene 1, in Molière, *Comedies*, trans. Donald M. Frame (New York: Oxford University Press, 1985), 299.

98 Quoted in Boas, *The Happy Beast*, 147:

De tous côtés, docteur, voyant les hommes fous,
Qu'il diroit le bon coeur, sans en être jalous ...
Ma foi, non plus que nous, l'homme n'est qu'une bête.

99 Quoted in ibid., 147-48:

Cependant nous avons la raison pour partage;
Et vous en ignorez l'usage.
Innocens animaux, n'en soyez point jaloux.

100 Quoted in ibid., 153, from *Discours à Madame de la Sablière,* written before 1679: "Qu'en elle tout se fait sans choix et par ressorts: / Nul sentiment, point d'âme; en elle tout est corps."

101 See ibid., 152. The reference is to Bobaci, burrowing squirrels found in Poland and adjoining countries. They are sometimes called Polish marmots and sometimes Polish foxes.

102 See pages 231-32.

103 Margaret Cavendish, Marchioness of Newcastle, *Philosophical Letters* (London, 1664), 40-41. Spelling modernized.

104 Margaret Cavendish, *Poems and Fancies* (London: J. Martin and J. Allestrye, 1653), 102 (italics in the original). Spelling modernized.

105 Ibid., 104 (italics in the original). Spelling modernized. Lacedaemonians are Spartans.

106 Quoted in Thomas, *Man and the Natural World,* 166, citing Thomas Edwards, *Gangraena,* vol. 1 (London, 1646), 20.

107 Quoted in Colin Spencer, *The Heretic's Feast: A History of Vegetarianism* (London: Fourth Estate, 1993), 205, citing Norman Cohn, *The Pursuit of the Millennium* (London: Paladin, 1957).

108 See Preece, *Awe for the Tiger: Love for the Lamb,* 120.

109 "Satyr," in *The Poems of John Wilmot, Earl of Rochester,* ed. Keith Walker (Oxford: Basil Blackwell, 1984), 91, 94, 95, lines 1-11, 114-18, 125-32 (italics in the original).

110 *"Bipes et implumis"* (two-legged and featherless) refers, one presumes, to Plato's definition of man as a two-legged animal without feathers. Diogenes, it is said, plucked a cock, took it to the Academy, and announced: "This is Plato's man."

111 Lines 1-6, 197-200, of *The Beasts' Confession to the Priest* in *Swift: Poetical Works,* ed. Herbert Davis (London: Oxford University Press, 1967), 538, 543-44 (italics in the original).

112 Soame Jenyns, *A Free Inquiry into the Nature and Origin of Evil* (London: R. and J. Dodsley, 1757), 21.

113 Alexander Pope, *An Essay on Man,* 3.4.25-52, in *The Works of Alexander Pope* (Ware: Wordsworth, 1995), 213.

114 Quoted in George Nicholson, *On the Primeval Diet of Man: Vegetarianism and Human Conduct toward Animals* (1801; reprint, ed. Rod Preece, Lampeter: Mellen Animal Rights Library, 1999), 113-14 (italics in the original).

115 From Geoffrey Grigson, ed., *The Oxford Book of Satirical Verse* (Oxford: Oxford University Press, 1983), 252.

116 *The Complete Works of St. John of the Cross,* vol. 2, trans. E. Allison Peers (London: Burns, Oates, and Washburn, 1935), stanza 5, 50-51.

117 Granville Sharp, *A Tract on the Law of Nature and Principles of Action in Man* (London: B. White and E. and C. Dilly, 1777), 177 (italics in the original).

118 See Hilda Kean, *Animal Rights: Political and Social Change in Britain since 1800* (London: Reaktion Books, 1998), 59.

119 William H. Drummond, *The Rights of Animals and Man's Obligation to Treat Them with Humanity* (London: John Mardon, 1838), 132-33. The lines quoted are from John Milton's *Paradise Lost.*

120 Auguste Comte, *System of Positive Philosophy, or Treatise on Sociology,* vol. 1, trans. Henry Dix Hutton (New York: Burt Franklin, [1877?]), 495.

121 Quoted in Jon Wynne-Tyson, *The Extended Circle: An Anthology of Humane Thought* (Fontwell: Centaur, 1990), 345.

122 Buckner, *The Immortality of Animals* (2003), 157.

123 Ibid., 159.

124 Ibid., 269.

125 Ibid., 152.

Chapter 6: Symbiosis

1 William Leiss, *The Domination of Nature* (1972; reprint, Montreal and Kingston: McGill-Queen's University Press, 1994), 48.

2 Jim Mason, *An Unnatural Order: Uncovering the Roots of Our Domination of Nature and Each Other* (New York: Simon and Schuster, 1993), 36.

3 Peter Singer, *Animal Liberation,* 2nd ed. (New York: New York Review of Books, 1990), 199.

4 Quoted in Irma A. Richter, *Selections from the Notebooks of Leonardo da Vinci* (Oxford: Oxford University Press, 1977), 245.

5 Quoted in ibid., 61.

6 Francis Bacon, "Of Goodness and Goodness of Nature," in *Essays, or Councils, Civil and Moral of Sir Francis Bacon* (1597; reprint, London: R. Chiswell et al., 1706), 30-31. Spelling modernized.

7 *De Augmentis Scientarium* (1627), in Francis Bacon, *The Philosophical Works of Francis Bacon,* ed. John M. Robertson (1905; reprint, Freeport, NY: Books for Libraries Press, 1970), 586.

8 Victor Hugo, *Les Misérables* (1862; reprint, trans. Norman Denny, London: Penguin, 1976), 995.

9 Quoted in Adrian Desmond and James Moore, *Darwin: The Life of a Tormented Evolutionist* (New York: Warner, 1992), 449.

10 Quoted in Michael Holroyd, *Bernard Shaw,* vol. 1 (London: Penguin, 1990), 87, 218.

11 Auguste Comte, *Theory of the Future of Man,* in *Système de politique positive, ou traité de sociologie, instituant la religion de l'humanité,* vol. 4 (Paris: Carilian: Goeuvry, 1854), 359, 225. I have borrowed Henry Dix Hutton's translation.

12 Thomas Hardy, *Jude the Obscure* (1895; reprint, Ware: Wordsworth, 1995), 8, 10, 284, 294.

13 Quoted in Joanna Cullen Brown, *Let Me Enjoy the Earth: Thomas Hardy and Nature* (London: Allison and Busby, 1990), 278-79, citing *Academy and Literature* (17 May 1902).

14 October 1906, quoted in ibid., 281.

15 May 1910, quoted in ibid., 279.

16 John Steinbeck, *The Log from the Sea of Cortez* (1951; reprint, London: Pelican, 1995), 170.

17 Michael Shelden, *Orwell: The Authorized Biography* (London: Minerva, 1991), 217.

18 Howard Williams, *The Ethics of Diet: A Catena of Authorities Deprecatory of the Practice of Flesh-Eating* (London: F. Pitman, 1883), 8.

19 Johanna Angermeyer, *My Father's Island: A Galapagos Quest* (London: Viking, 1989), 81ff.

20 Auguste Comte, *Theory of the Great Being,* in *System of Positive Philosophy, or Treatise on Sociology,* vol. 4, trans. Henry Dix Hutton (New York: Burt Franklin, [1877?]), 33, 43.

21 Bradford Torrey, *A Rambler's Leave* (Boston: Houghton Mifflin, 1889), 103-5.

22 Ralph H. Lutts, *The Nature Fakers: Wildlife, Science and Sentiment* (Golden, CO: Fulcrum, 1990), 194-96. Cited in Rod Preece and Lorna Chamberlain, *Animal Welfare and Human Values* (Waterloo: Wilfrid Laurier University Press, 1993), 243-46, wherein I provide a general critique of Lutts's hunting-sympathetic philosophy.

23 Ibid., 195.

24 Desmond Morris, *The Animal Contract* (London: Virgin, 1990), 84.

25 Ibid., 85.

26 Ibid., 86.

27 Steinbeck, *Log from the Sea of Cortez,* 178-79.

28 Quoted in Jay Parini, *John Steinbeck: A Biography* (London: Minerva, 1994), 582.

29 John Steinbeck, "Appendix: About Ed Ricketts," in *Log from the Sea of Cortez,* 264.

30 Steinbeck, *Log from the Sea of Cortez,* 218.

31 James Lovelock, *Gaia: A New Look at Life on Earth* (London: Oxford University Press, 1979).

32 Lynn White Jr., "The Historical Roots of Our Ecologic Crisis," *Science* 155 (1967): 1203-7.

33 James Serpell, *In the Company of Animals: A Study of Human-Animal Relationships* (Oxford: Oxford University Press, 1983), 122.

34 Singer, *Animal Liberation,* 188, 189.

35 See page 104-7.

36 Porphyry, *On Abstinence from Animal Food,* trans. Thomas Taylor ([1793]; reprint, Fontwell: Centaur, 1965), 140.

37 See page 51.

38 Quoted in Arthur O. Lovejoy, *The Great Chain of Being: A Study in the History of an Idea* (New York: Harper, 1960), 100, citing *The Guide to the Perplexed,* bk. 3, ch. 14.

39 Margaret Cavendish, *Poems and Fancies: Written by the Right Honourable the Lady Newcastle* (London: Martin and Allystrye, 1653), 113. Spelling modernized.

40 Quoted in William H. Drummond, *The Rights of Animals and Man's Obligation to Treat Them with Humanity* (London: John Mardon, 1838), 122.

41 John Ray, *The Wisdom of God Manifested in the Works of Creation* (London: Samuel Smith, 1691; facsimile reprint, New York: Garland, 1979), 127-29.

42 Quoted in Lovejoy, *The Great Chain of Being,* 124, citing *Principia,* vol. 3, 3.

43 Quoted in ibid., citing *Oeuvres de Descartes,* vol. 4 (Paris: Charles et Paul Tannery, 1897), 292.

44 Quoted in ibid., 127, citing *Pensées* 72 (1, 70).

45 See page 233.

46 Lovejoy, *The Great Chain of Being,* 186.

47 Quoted in ibid., 187, citing *Traité de l'existence de Dieu,* vol. 1, 2.

48 Quoted in ibid., citing *Traité de l'existence de Dieu,* vol. 1, 2: "On ne trouverait plus d'animaux féroces que dans les forêts reculées, et on les réserverait pour exerciser la hardiesse, la force et l'adresse du genre humain, par un jeu qui représenterait la guerre, sans qu'on eût jamais besoin de guerre véritable entre les nations."

49 Quoted in ibid., 162, citing *Traité de l'existence de Dieu,* vol. 2, v: "il arrive même souvent que je sois plus parfait de me taire que de parler."

50 See Rod Preece, *Awe for the Tiger, Love for the Lamb: A Chronicle of Sensibility to Animals* (Vancouver: UBC Press; London and New York: Routledge, 2002), 173-74.

51 Arthur Schopenhauer, *On the Basis of Morality,* 2nd ed., trans. E.F.J. Payne (1st ed. 1841; 2nd ed. 1860; reprint, Indianapolis: Bobbs-Merrill, 1965), 96.

52 "Metamorphose der Tiere," in *Goethe: Sämtliche Werke, nach Epochen seines Schaffens* (München: Carl Hanser Verlag, 1992), vol. 13, prt. 1, 153. My translation.

53 Letter to William Sotheby, 1802, *Collected Letters of Samuel Taylor Coleridge,* vol. 2, ed. Earl Leslie Griggs (London: Oxford University Press, 1956), 459.

54 Lines 8-10, *Visions of the Daughters of Albion,* in *Blake: Complete Writings,* ed. Geoffrey Keynes (London: Oxford University Press, 1966), plate 8, 195.

55 Lines, 26-27, *The Book of Thel,* prt. 2, in ibid., 174.

56 Soame Jenyns, *A Free Inquiry into the Nature and Origin of Evil,* 2nd ed. (London: R. and J. Dodsley, 1757; facsimile reprint, New York: Garland, 1976), 21-22.

57 Joseph Ritson, *An Essay on Abstinence from Animal Food as a Moral Duty* (London: Richard Phillips, 1802), 231-33 (italics in the original).

58 This and the following poem quoted in Christine Kenyon-Jones, *Kindred Brutes: Animals in Romantic-Period Writing* (Aldershot: Ashgate, 2001), 43.

59 See Preece, *Awe for the Tiger, Love for the Lamb,* 222.

60 This and the following poem quoted in Jon Wynne-Tyson, *The Extended Circle: A Commonplace Book of Animal Rights* (1985; reprint, New York: Paragon, 1989), 78.

61 William Kirby and William Spence, *An Introduction to Entomology, or Elements of Natural History of Insects,* vol. 1 (London: Longman, Hurst, Rees, Orme, and Brown, 1822), 387.

62 Anne Brontë, *Agnes Grey* (1847; reprint, London: Penguin, 1988), 79.

63 George Sand, *Indiana* (1831; reprint, trans. Sylvia Raphael, Oxford: Oxford University Press, 1994), 190.

64 Drummond, *The Rights of Animals,* 27.

65 Ibid., 48.

66 Ibid.

67 William F. Karkeek, "On the Future Existence of the Brute Creation," *The Veterinarian* 12, 142 (1839): 750-51.

68 W.C. Spooner, "On the Non-Immortality of Animals," *The Veterinarian* 13 (June 1840): 376-77.

69 Jules Michelet, *The Bird* (1879; reprint, trans. W.H. Davenport Adams, London: Wildwood House, 1981), 67 (italics in the original).

70 Quoted in Roderick Frazier Nash, *The Rights of Nature: A History of Environmental Ethics* (Madison: University of Wisconsin Press, 1989), 39.

71 J. Howard Moore, *The Universal Kinship* (1906; reprint, ed. Charles Magel, Fontwell: Centaur, 1992), 324 (italics in the original).

CHAPTER 7: EVOLUTION, CHAIN, AND CATEGORICAL IMPERATIVE

1 Rod Preece, *Animals and Nature: Cultural Myths, Cultural Realities* (Vancouver: UBC Press, 1999) constitutes something of an exception, but even there my comments are peripheral and scattered.

   Goethe's animal orientations are also mentioned in my article with David Fraser: "The Status of Animals in Biblical and Christian Thought: A Study in Colliding Values," *Society and Animals* 8, 3 (2000): 245-63, but not at any length. Finally, I translated a few pieces from Goethe for my *Awe for the Tiger, Love for the Lamb: A Chronicle of Sensibility to Animals* (Vancouver: UBC Press; London and New York: Routledge, 2002). The arguments of this chapter were first sketched in a peremptory fashion in a talk to the Centre for Applied Ethics at the University of British Columbia in November 1999 but lay dormant while I completed other tasks. Nonetheless, they have benefitted from the critiques I received on that occasion. For the convenience of an anglophone audience, I have, wherever possible, used English-language secondary works to illustrate my points. Nonetheless, similar points could be made using German-language illustrations.

2 Charles Darwin, *The Origin of Species by Means of Natural Selection, or The Preservation of Favoured Races in the Struggle for Life* (1859; reprint, ed. J.W. Burrow, London: Penguin, 1983), 54-55.

3 Marian Scholtmeijer, *Animal Victims in Modern Fiction: From Sanctity to Sacrifice* (Toronto: University of Toronto Press, 1993), 25.

4 The significant contributions of all four are acknowledged in, to take but two examples, Jon Wynne-Tyson, *The Extended Circle: A Commonplace Book of Animal Rights* (1985; reprint, New York: Paragon, 1989), and Richard D. Ryder, *Animal Revolution: Changing Attitudes towards Speciesism* (Oxford: Basil Blackwell, 1989). Goethe goes unmentioned.

5 James Rachels, *Created from Animals: The Moral Implications of Darwinism* (Oxford: Oxford University Press, 1991).

6 Elijah Buckner, *The Immortality of Animals* (Philadelphia: George W. Jacobs, 1903), 176.

7 Rachels, *Created from Animals,* 14-16.

8 See Darwin, *The Origin of Species.*

9 Scholtmeijer, *Animal Victims in Modern Fiction,* 25.

10 See Preece, *Awe for the Tiger, Love for the Lamb,* 193-94, 195, 183-84, 185, 225, 190, 187-88, 189.

11 Stuart Atkins, "On Goethe's Classicism," in *Goethe Proceedings: Essays Commemorating the Goethe Sesquicentennial at the University of California, Davis,* ed. Clifford A. Bernd et al. (Columbia, CA: Camden House, 1984), 12.

12 Ute Lischke, *Lily Braun, 1865-1916: German Writer, Feminist, Socialist* (Rochester, NY: Camden House, 2000), 15-16.

13 Ibid., 68.

14 J.G. Robertson, *The Life and Work of Goethe* (1932; reprint, New York: Gaskell House, 1973).

15 Ibid., 307.

16 Certainly, however, Albert Bielschowsky's classic *The Life of Goethe* (3 vols., 1908) sinks to the level of hagiography in its elevation of Goethe's evolutionary originality.

17 Arthur O. Lovejoy, *The Great Chain of Being: A Study of the History of an Idea* (1936; reprint, New York: Harper and Row, 1965), 367 n. 76.

18 Nicholas Boyle, *Goethe: The Poet and the Age*, vol. 1, *The Poetry of Desire, 1749-90* (1991; reprint, Oxford: Oxford University Press, 1992), 399.

19 William Rounseville Alger, *A Critical History of the Doctrine of a Future Life* (1860; reprint, Philadelphia: George W. Childs, 1864), 36.

20 Epigraph to Arthur Helps, *Animals and Their Masters* (1872; reprint, London: Chatto and Windus, 1883), ii.

21 See, for example, H.B. Nisbet, *Goethe and the Scientific Tradition* (London: University of London Institute of Germanic Studies, 1972), and G.A. Wells, *Goethe and the Development of Science* (Alphen: Sijthoff and Noordhoff, 1978).

22 Charles Darwin, "Historical Sketch," in *The Origin of Species,* 55.

23 See page 27-38.

24 Henry Fairfield Osborn, *From the Greeks to Darwin: The Development of the Evolution Idea through Twenty-Four Centuries,* 2nd ed. (New York: Charles Scribner's Sons, 1929).

25 For the quotations from Hippolytus and Plutarch (the latter both directly and via Eusebius), see Jonathan Barnes, *Early Greek Philosophy* (London: Penguin, 1987), 72-74.

26 Samuel Pufendorf, *The Law of Nature and Nations* (1688; reprint, trans. C.H. Oldfather and W.A. Oldfather, New York: Oceana, 1931), 531.

27 Some claim our moral obligation to other species precisely on the basis of animal inferiority – in the same manner that infants, the infirm, and the handicapped are entitled to greater moral consideration than the strong. In this vein, see Matthew Scully, *Dominion: The Power of Man, the Suffering of Animals, and the Call to Mercy* (New York: St. Martin's Press, 2002). Also in this vein is Andrew Linzey, "The Moral Priority of the Weak," Chapter 2 of *Animal Theology* (Urbana: University of Illinois Press, 1995). "Generosity," he argues, should be the criterion of our behaviour.

28 For Empedocles via Aristotle and Aelian, see Barnes, *Early Greek Philosophy,* 177, 181.

29 Osborn, *From the Greeks to Darwin,* 52.

30 Darwin, "Historical Sketch," in *The Origin of Species,* 53.

31 Osborn, *From the Greeks to Darwin,* 78.

32 For Plutarch, see, in particular, "The Life of Marcus Cato," "On the Use of Reason by 'Irrational' Animals," and "Whether Land or Sea Animals Are Cleverer," in *Essays,* trans. Robin Waterfield (London: Penguin, 1972). For Lucretius, see *On the Nature of the Universe* (1951; reprint, trans. R.A. Latham, London: Penguin, 1994), passim, but pages 2 and 3552-66 are most illuminating. For Porphyry, see *On Abstinence from Animal Food,* trans. Thomas Taylor ([1793]; reprint, Fontwell: Centaur, 1965). For Montaigne's use of Lucretius, see Michel de Montaigne, "Apology for Raymond Sebond," in *Selected Essays* (New York: Oxford University Press, 1982), 324, 325, 330.

33 Quoted in Philip Wayne, Introduction to Johann Wolfgang Goethe, *Faust, Part Two* (1832; reprint, trans. Philip Wayne, London: Penguin, 1959), 8.

34 Johann Wolfgang Goethe, "Rocky Inlets of the Aegean," Act 2 in *Faust, Part Two* (1832; reprint, trans. Philip Wayne, London: Penguin, 1959), 150.

35 See Aristotle, *Nicomachean Ethics,* trans. Martin Oswald (Indianapolis: Bobbs-Merrill, 1960), ch. 8, especially 215-20.

36 Osborn, *From the Greeks to Darwin,* 63.

37 Hegel tells us that Kant's positing of a separation between knower and known in this manner ensures his failure to grasp reality.

38 Atkins, "On Goethe's Classicism," in *Goethe Proceedings,* 7.

39 Quoted in Albert Bielschowsky, *The Life of Goethe,* vol. 3 (1908; reprint, New York: AMS Press, 1970), 377 n. 16.

40 See Boyle, *Goethe: The Poet and the Age,* vol. 1, 500-1.

41 See Daniel J. Boorstin, *The Discoverers: A History of Man's Search to Know His World and Himself* (1983; reprint, New York: Vintage Books, 1985), 472.

42 *Summa Theologica* 1a.96.4. Quoted in F.C. Copleston, *Aquinas* (1955; reprint, London: Penguin, 1975), 237.

43 See Michael Seidlmayer, *Currents of Medieval Thought with Special Reference to Germany* (Oxford: Basil Blackwell, 1960), 66; Jacob Burkhardt, *Civilization of the Renaissance in Italy* (1860; reprint, London: Phaidon, 1960), 81; Michael Oakeshott, *On Human Conduct* (Oxford: Clarendon Press, 1975), 223.

44 Seidlmayer, *Currents of Medieval Thought,* 66.

45 Aquinas, *Summa contra Gentiles,* vol. 3, trans. Charles J. O'Neil (Grand Bend: Notre Dame University Press, 1975), ch. 80.

46 The quotations from both Buffon and Goldsmith can found in Harriet Ritvo, *The Platypus and the Mermaid and Other Figments of the Classifying Imagination* (1997; reprint, Cambridge: Harvard University Press, 1998), 23.

47 Quoted in Bielschowsky, *The Life of Goethe,* vol. 3, 106.

48 Darwin, *The Origin of Species,* 104, 108.

49 Letter to J.G. Gmelin, 14 February 1747. Quoted in Carl Sagan and Ann Druyen, *Shadows of Forgotten Ancestors: A Search for Who We Are* (New York: Random House, 1992), 274.

50 Darwin, *The Descent of Man,* 170.

51 D. Hart and M.P. Karmel, "Self-Awareness and Self-Knowledge in Humans, Apes and Monkeys," in *Reaching into Thought: The Minds of the Great Apes,* ed. A.E. Russon, K.A. Bard, and S.T. Parker (Cambridge: Cambridge University Press, 1997), 325-47.

52 Quoted in Richard Milner, *The Encyclopedia of Evolution: Humanity's Search for Its Origins* (New York: Henry Holt, 1990), 59.

53 Darwin, *The Origin of Species,* 53-54.

54 Comte de Buffon, *Histoire naturelle,* vol. 5 (1755).

55 See Boorstin, *The Discoverers,* 472; Lovejoy, *The Great Chain of Being,* primarily 268.

56 See Boorstin, *The Discoverers,* 472.

57 Jean-Baptiste Robinet, *De la Nature,* 5. 148, (1765) quoted in Lovejoy, *The Great Chain of Being,* 275.

58 Baron d'Holbach, *Système de la Nature* (1770), prt. 1, ch. 6, quoted in Lovejoy, *The Great Chain of Being,* 269.

59 Ibid., 283.

60 See Boorstin, *The Discoverers,* 472.

61 Arthur O. Lovejoy, "Some Eighteenth-Century Evolutionists," *Popular Science Monthly* (March 1904): 283ff., 323ff. See Hester Hastings, *Man and Beast in French Thought of the Eighteenth Century* (1936; reprint, New York: Johnson Reprint Corporation, 1973), 109.

62 Immanuel Kant, "Duties towards Animals and Spirits," in *Lectures on Ethics* ([1780?]; modern edition 1930, trans. Louis Infield; reprint, New York: Harper and Row, 1963), 241.

63 See Eric Trinkaus and Pat Shipman, *The Neandertals: Changing the Image of Mankind* (New York: Alfred A. Knopf, 1993), 21-22.

64 To take but one example, see the section on "Naturphilosophie" in Milner, *The Encyclopedia of Evolution*, 320.

65 Honoré de Balzac, *Le Père Goriot* (London: Daily Telegraph, [1888?]), vi (my italics). Balzac's comment on "the last words which came from [Goethe's] pen" may refer, exaggeratedly, to his direction of the French translation of the *Metamorphose der Pflanzen*. K.W. Müller wrote *Goethes letzte literarische Tätigkeit* (Goethe's last literary activity) to describe this event and the reception of the translation by Geoffroy Saint-Hilaire at the Académie Française, who was delighted by the prescient recognition of his own evolutionary views. In his report to the Académie, he remarked that "when Goethe came out with his work in 1790 it was little noticed; indeed, scientists came near considering it an aberration. To be sure, there was an error at the bottom of it, but such a one as only a genius can commit. Goethe's only error consisted in allowing his treatise to be published almost half a century too soon, before there were any botanists who were able to study and understand it." Quoted via Müller in Bielschowsky, *The Life of Goethe*, vol. 3, 95.

66 Darwin, *The Origin of Species*, 68.

67 Quoted in Desmond King-Hele, *Erasmus Darwin: A Life of Unequalled Achievement* (London: Giles de la Mare, 1999), 349.

68 Quoted in ibid., 283.

69 Quoted in Bielschowsky, *The Life of Goethe*, vol. 3, 88-89.

70 Johann Wolfgang Goethe, "Das Göttliche," lines 3-6, in *Sämmtliche Werke, nach Epochen seines Schaffens*, vol. 2, trans. Rod Preece (München: Carl Hanser Verlag, 1986-92), sec. 1, 90.

71 Quoted in Robertson, *The Life and Work of Goethe*, 307.

72 Jean-Baptiste Robinet, *De la Nature*, 4 vols., 3rd ed. (Amsterdam: van Harrevelt, 1763-66), 4.1.i.4.

73 Johann Wolfgang Goethe, "Metamorphose der Tiere," in *Sämmtliche Werke, nach Epochen seines Schaffens*, vol. 13, prt. 1, sec. 17, 153-55.

74 Johann Gottfried Herder, "Ideas for a Philosophy of the History of Mankind," bk. 3, sec. 6, in F.M. Barnard, ed. and trans., *Herder on Social and Political Culture* (Cambridge: Cambridge University Press, 1969), 257.

75 Quoted in Bielschowsky, *The Life of Goethe*, vol. 3, 107, citing Eckermann, *Gespräche*, vol. 2, 191.

76 Quoted in ibid., citing n.s., vol. 6, 263.

77 My translation.

> Und umzuschätzen das Geschaffne,
> Damit sich's nicht zum Starren waffne,
> Wirkt ewiges, lebend'ges Tun ...
> Es soll sich regen, schaffend handeln,
> Erst sich gestalten, dann verwandeln;
> Nur scheinbar steht's Momente still.

78 Quoted in Bielschowsky, *The Life of Goethe*, vol. 3, 110.

79 Ibid.

80 My translation.

> Denn solches Los dem Menschen wie den Tieren ward,
> Nach deren Urbild ich mir Besseres bildete,

Dass eins dem anderen, einzeln oder auch geschart,
Sich wiedersteht, sich hassend aneinder drängt,
Bis eins dem andern Übermacht bestätigt.

81 Quoted in Bielschowsky, *The Life of Goethe,* vol. 3, 108.
82 Johann Wolfgang Goethe, *Faust, Part One,* in *Sämmtliche Werke, nach Epochen seines Schaffens,* vol. 6, sec. 1, 629.
83 See page 309.
84 Johann Wolfgang Goethe, *Goethe: Poetical Works,* trans. John Storer Cobb (Boston: Francis A. Niccolls, 1902).
85 H.B. Nisbet, "Lucretius in Eighteenth-Century Germany, with a Commentary on Goethe's 'Metamorphose der Tiere,'" *The Modern Language Review* 81 (1986): 97-115.
86 Lucretius, *On the Nature of the Universe,* 46-47.
87 Nicholas Boyle, *Goethe: The Poet and the Age,* vol. 2, *Revolution and Renunciation, 1790-1803* (Oxford: Clarendon Press, 2000), 418.
88 Quoted in ibid., 35.
89 Quoted in ibid., 40-41.
90 See page 306.
91 My translation. "Handle nur nach derjenigen Maxime, durch die du zugleich wollen kannst, dass sie ein allgemeines Gesetz werde!"; "Handle so, als ob die Maxime deiner Handlung durch deinen Willen zum allgemeinen Naturgesetze werden sollte!"; "Handle so, dass du die Menschheit, sowohl in deiner Person als auch in der Person eines jeden anderen, jederzeit zugleich als Zweck, niemals bloß als Mittel brauchst!" Immanuel Kant, *Grundlegung zur Metaphysik der Sitten* (1785; reprint, Hamburg: Meiner, 1994), 42-43, 52.
92 My translation. "Der Mensch und überhaupt jedes vernünftige Wesen existiert als Zweck an sich selbst nicht bloß als Mittel zum beliebigen Gebrauche für diesen oder jenen Willen." Ibid., 2.
93 Bielschowsky, *The Life of Goethe,* vol. 3, 102.
94 Kant, "Duties toward Animals and Spirits," in *Lectures on Ethics,* 239-41.
95 Lovejoy, *The Great Chain of Being,* 144.
96 Quoted in ibid., 144-45, from "a letter of Leibniz's, usually omitted in the editions of his collected writings."
97 Scholtmeijer, *Animal Victims in Modern Fiction,* 318.
98 Milner, *The Encyclopedia of Evolution,* 201.
99 David Maybury-Lewis, *Millennium: Tribal Wisdom of the Modern World* (New York: Viking, 1992), 36-37.
100 Plotinus, *Enneads,* in *Plotinus,* vol. 3, trans. A.H. Armstrong (London: William Heinemann, 1967), 83-85, 133.
101 Macrobius, *Commentary on Cicero's "Dream of Scipio."* For a more extensive extract, see Preece, *Animals and Nature,* 119.
102 St. Robert Bellarmine, *De ascensione mentis in Deum per scalas creatorum.* See Preece, *Animals and Nature,* 120.
103 John of the Cross, "Spiritual Canticle," stanza 5, in *The Complete Works of St. John of the Cross,* vol. 2, trans. E. Allison Peers (London: Burns, Oates and Washburn, 1935), 50.
104 John Locke, *An Essay concerning Human Understanding,* vol. 2 (1690; reprint, London: H. Hills, 1710), bk. 3, 49.
105 Alexander Pope, *An Essay on Man,* 1.8.5-9 and 3.1.22-26, in *The Works of Alexander Pope* (Ware: Wordsworth, 1995), 148, 167.

106 Sarah Trimmer, *The History of the Robins, with Twenty-four Illustrations from Drawings by Harrison Weir* (London: Griffith and Farrar, n.d.), 136. The title *The History of the Robins* gradually replaced *Fabulous Histories* as the work's commonly given title in the nineteenth century.

107 Priscilla Wakefield, *Instinct Displayed in a Collection of Well-Authenticated Facts Exemplifying the Extraordinary Sagacity of Various Species of the Animal Creation*, 4th ed. (London: Harvey and Darton, 1821), viii.

108 William Youatt, *The Obligation and Extent of Humanity to Brutes, Principally Considered with Reference to the Domesticated Animals* (1839; reprint, ed. Rod Preece, Lampeter: Mellen Animal Rights Library, 2003), 5. The quotation is taken from Pope, *An Essay on Man*, 1.9.267.

109 Johann Wolfgang Goethe, "Limits of Humanity," in *The Poetical Works of J.W. von Goethe*, ed. Edwin Hermann Zeydel, vol. 1 (Chapel Hill: University of North Carolina Press, 1957), 212-13.

110 For a complete translation of the poem, see Preece, *Awe for the Tiger, Love for the Lamb*, 181-82.

111 Quoted in Bielschowsky, *The Life of Goethe*, vol. 3, 102.

112 See page 29-30.

113 W.M.S. Russell and R.L. Burch, *The Principles of Humane Experimental Technique* (London: Methuen), 1959.

114 See page 26.

CHAPTER 8: KINSHIP AND EVOLUTION

1 See page 295.

2 Marian Scholtmeijer, *Animal Victims in Modern Fiction: From Sanctity to Sacrifice* (Toronto: University of Toronto Press, 1993), i.

3 Michael Allen Fox, *Deep Vegetarianism* (Philadelphia: Temple University Press, 1999), 18.

4 Mary Midgley, "Practical Solutions," in *The Status of Animals: Ethics, Education and Welfare*, ed. David Paterson and Mary Palmer (Oxford: CAB International, 1989), 18.

5 Rev. Giles Legood, "A Brief History of British Animal Welfare," *Ark* 188 (Summer 2001), 2.

6 Hilda Kean, *Animal Rights: Political and Social Change in Britain since 1800* (London: Reaktion Books, 1998), 71.

7 Christine Kenyon-Jones, *Kindred Brutes: Animals in Romantic-Period Writing* (Aldershot: Ashgate, 2001), 201.

8 Ibid., 27.

9 Peter Singer, *Animal Liberation*, 2nd ed. (New York: New York Review of Books, 1990), 206.

10 Ted Honderich, ed., *Oxford Companion to Philosophy* (Oxford: Oxford University Press, 1995), 35.

11 In his preface to a new edition of Henry Salt, *Animals' Rights Considered in Relation to Social Progress* (1892; reprint, Clarks Summit, PA: Society for Animal Rights, 1980), vii-viii, Singer is a little more generous, allowing a modest role for John Lawrence, Thomas Young, Lewis Gompertz, Edward Nicholson, and Henry Salt.

12 W.C. Spooner, "On the Future Existence of the Brute Creation," *The Veterinarian* 13 (June 1840): 373.

13 Dr. Michael W. Fox, *Inhumane Society: The American Way of Exploiting Animals* (New York: St. Martin's Press, 1990), 24.

14 Both letters and Hardy's claim of Darwinian adherence are to be found in Joanna Cullen Brown, ed., *Let Me Enjoy the Earth: Thomas Hardy and Nature* (London: Allison and Busby, 1990), 290-91.

15 Charles Dickens, "Inhumane Humanity," *All the Year Round* 15 (17 March 1866).

16 William Youatt, *The Obligation and Extent of Humanity to Brutes, Principally Considered with Reference to the Domesticated Animals* (1839; reprint, ed. Rod Preece, Lampeter: Mellen Animal Rights Library, 2003), 212-26.

17 *Spectator* 120 (18 July 1711), reprinted in Alex Chalmers, ed., *The Spectator: A New Edition Corrected from the Originals,* vol. 2 (New York: E. Sargent and M.A. Ward, 1810), 283-84.

18 Quoted in Dix Harwood, *Love for Animals and How It Developed in Great Britain* (1928; reprint, ed. Rod Preece and David Fraser, Lampeter: Mellen Animal Rights Library, 2002), 326-27.

19 Quoted in John Vyvyan, *In Pity and in Anger* (1969; reprint, Marblehead, MA: Micah, 1988), 26, citing Samuel Johnson, ed., *The Plays of William Shakespeare,* vol. 6 (London, 1765), 279.

20 Quoted in George Nicholson, *On the Primeval Diet of Man: Vegetarianism and Human Conduct toward Animals* (1801; reprint, ed. Rod Preece, Lampeter: Mellen Animal Rights Library, 1999), 179, citing *Monthly Review* (September 1770), 179.

21 John Oswald, *The Cry of Nature, or An Appeal to Mercy and to Justice on Behalf of the Persecuted Animals* (1791; reprint, ed. Jason Hribal, Lampeter: Mellen Animal Rights Library, 2000), 31-33.

22 Quoted in the original in Hester Hastings, *Man and Beast in French Thought of the Eighteenth Century* (1936; reprint, New York: Johnson Reprint Corporation, 1973), 270, my translation, citing Voltaire, "Bêtes," in *Dictionnaire philosophique* (1764; reprint, Paris: Garnier, 1961), 566.

23 Quoted in the original in ibid., 270-71, my translation, citing *Oeuvres,* 1818, IV, Etude, x, 118.

24 Quoted in the original in ibid., 271, my translation, citing *Oeuvres,* 1824, X, Les Trois Règnes (1794-1808), ii, 223.

25 William H. Drummond, *The Rights of Animals and Man's Obligation to Treat Them with Humanity* (London: John Mardon, 1838), 145.

26 Jules Michelet, *The Bird* (1879; reprint, trans. W.H. Davenport Adams, London: Wildwood House, 1981), 148.

27 J. Howard Moore, *The Universal Kinship* (1906; reprint, ed. Charles A. Magel, Fontwell: Centaur, 1992), and *The New Ethics* (London: E. Bell; Chicago: S.A. Block, 1907).

28 For this and more of Chrysostom's animal-sympathetic attitudes, see his *Homilies of John Chrysostom on the Epistle of St. Paul to the Romans,* Homily 19, "The Liturgy of St. Basil."

29 C.W. Hume, *Universities Federation for Animal Welfare Theological Bulletin,* no. 2 (1962), citing Saint John Chrysostom, *Homilies of John Chrysostom on the Epistle of St. Paul to the Romans,* Homily 19, "The Liturgy of St. Basil."

30 Letter to William Sotheby, 10 September 1802, *Collected Letters of Samuel Taylor Coleridge,* vol. 2, ed. Earl Leslie Griggs (Oxford: Oxford University Press, 1956), 864.

31 Elijah Buckner, *The Immortality of Animals* (Philadelphia: George W. Jacobs, 1903), 175.

32 W. Lauder Lindsay, *Mind in the Lower Animals,* vol. 1 (London: C. Kegan Paul, Tench, 1879), 125.

33 Charles Darwin, *The Descent of Man, and Selection in Relation to Sex,* 2nd ed. (New York: A.L. Burt, 1874), 139.

34 Quoted in James Rachels, *Created from Animals: The Moral Implications of Darwinism* (Oxford: Oxford University Press, 1991), 213.

35 Darwin, *The Descent of Man,* 78.

36 Instances of these for each of the authors mentioned can be found in Rod Preece, *Awe for the Tiger, Love for the Lamb: A Chronicle of Sensibility to Animals* (Vancouver: UBC Press; London and New York: Routledge, 2002).

37 Darwin, *The Descent of Man,* 74.

38 Ibid., 89.

39 Ibid., 111.

40 Ibid., 110.

41 Ibid., 142-43.

42 Ibid., 139.

43 Quoted in Buckner, *The Immortality of Animals,* 56-57.

44 William Rounseville Alger, *A Critical History of the Doctrine of a Future Life* (1860; reprint, Philadelphia: George W. Childs, 1864), 632.

45 Youatt, *The Obligation and Extent of Humanity to Brutes,* 49-101.

46 Kean, *Animal Rights,* 45, citing *The Gardens and Menageries of the Zoological Society Delineated, 1830-31* (London, 1831), 99-100, 92-93, 190.

47 Title page to Youatt, *The Obligation and Extent of Humanity to Brutes.*

48 Ibid., 55.

49 William F. Karkeek, "On the Future Existence of the Brute Creation," *The Veterinarian* 12, 142 (October 1839): 660.

50 Youatt, *The Obligation and Extent of Humanity to Brutes,* 55.

51 Ibid., 56.

52 Ibid., 58. Youatt capitalized "association of ideas."

53 Ibid., 59.

54 Ibid., 60.

55 Ibid., 66.

56 Ibid., 67ff.

57 Voltaire, "Bêtes," in *Dictionnaire philosophique* (1764; reprint, Paris: Garnier, 1961), 33, my translation.

58 Youatt, *The Obligation and Extent of Humanity to Brutes,* 75 (italics in the original).

59 Ibid., 77-78.

60 Ibid., 88-101.

61 Ibid., 50.

62 Ibid., 65 (italics in the original).

63 Ibid., 11-36.

64 Richard Sorabji, *Animal Minds and Human Morals: The Origins of the Western Debate* (Ithaca: Cornell University Press, 1993), 210.

65 Quoted in Richard Overton, *Man's Mortalitie* (1643; reprint, ed. Harold Fisch, Liverpool: University of Liverpool Press, 1968), 26.

66 Joseph Butler, *Analogy of Religion, Natural and Revealed, to the Constitution and Course of Nature* (1736; reprint, London: Longman, 1834), 28.

67 Drummond, *The Rights of Animals,* 5-6, 204.

68 Henry Salt, *Animals' Rights Considered in Relation to Social Progress* (1892; reprint, Clarks Summit, PA: Society for Animal Rights, 1980), 162, citing J.G. Wood, *Man and Beast, Here and Hereafter* (1874).

69 Porphyry, *On Abstinence from Animal Food,* trans. Thomas Taylor ([1793]; reprint, Fontwell: Centaur, 1965).

70 For an extract of the argument, see Preece, *Awe for the Tiger, Love for the Lamb,* 46-48.

71 Lactantius, *Divinae Instituiones,* 3.10, in *The Father of the Church, Lactantius, the Divine Institutes, Books I-VII,* trans. Francis McDonald (Washington: Catholic University of America Press, 1964), 185-86.

72 Buckner, *The Immortality of Animals,* 163.

73 Ralph Waldo Emerson, *Nature* (1836; reprint, Harmondsworth: Penguin, 1995), 31.

74 See Introduction, page 18.

75 Quoted in Desmond King-Hele, *Erasmus Darwin: A Life of Unqualified Achievement* (London: Giles de la Mare, 1999), 291.

76 Ibid., 301.

77 Coleridge, *Collected Letters of Samuel Taylor Coleridge,* vol. 4, 574-75.

78 Honoré de Balzac, *Le Père Goriot* (London: Daily Telegraph, [1888?]), vi.

79 See David N. Livingstone, *Darwin's Forgotten Defenders: The Encounter between Evangelical Theology and Evolutionary Thought* (Edinburgh: Scottish Academic Press, 1987).

80 In a letter from Asipovo, 1 November 1910. Quoted in William L. Shirer, *Love and Hatred: The Stormy Marriage of Leo and Sofya Tolstoy* (New York: Simon and Schuster, 1994), 350-51.

81 George Bernard Shaw, Preface to *The Doctor's Dilemma, Getting Married, and the Shewing-Up of Blanco Posnet by Bernard Shaw* (London: Constable, 1911), 99-100.

82 Quoted in Michael Holroyd, *Bernard Shaw,* vol. 2 (Harmondsworth: Penguin, 1990), 58.

83 Peter Kropotkin, *Mutual Aid: A Factor of Evolution* (1st ed. 1902; 2nd ed. 1914; reprint, Boston: Extending Horizon Books, [1955?]), especially 5-6, 57-59, and 74-75. In 1890 Kropotkin wrote articles on "Mutual Aid among Animals" in the journal *Nineteenth Century.*

84 Buckner, *The Immortality of Animals,* 176.

85 Stephen Coleridge, *Memories* (London: John Lane, 1913), 234-35.

86 Stephen Coleridge, *Vivisection: A Heartless Science* (London: John Lane, 1916), 15.

87 Quoted in Adrian Desmond and James Moore, *Darwin: The Life of a Tormented Evolutionist* (New York: Warner, 1991), 427.

88 Quoted in Rachels, *Created from Animals,* 214.

89 Quoted in ibid., 215.

90 Quoted in ibid., 216.

91 Desmond and Moore, *Darwin: The Life of a Tormented Evolutionist,* 621.

92 Thomas Young, *An Essay on Humanity to Animals* (1798; reprint, ed. Rod Preece, Lampeter: Mellen Animal Rights Library, 2001), 133.

93 Quoted in Kean, *Animal Rights,* 36, citing Arthur Broome, *SPCA Founding Statement* (London, 1824), 2.

94 Quoted, at much greater length, in Drummond, *The Rights of Animals,* 164.

95 Lewis G. Regenstein, *Replenish the Earth* (New York: Crossroad, 1991), 95; C.W. Hume, *The Status of Animals in the Christian Religion* (1957; reprint, Potter's Bar: UFAW, 1980), 34. Regenstein also errs in his belief that the 1858 Linnean Society papers announcing evolution by natural selection were "a co-authored ... joint paper."

96 Youatt, *The Obligation and Extent of Humanity to Brutes,* 205.
97 Quoted in H.E. Carter, "The Veterinary Profession and the RSPCA: The First Fifty Years," *Veterinary History* n.s. 6, 2 (1989/90): 69.
98 Quoted in ibid., 70.
99 See Peter Ackroyd, *Dickens* (Toronto: Stewart House, 1991), 464-65, 541-42, 698-70.
100 Dickens, "Inhumane Humanity," *All the Year Round* 15 (17 March 1866): 240.
101 See Morton N. Cohen, *Lewis Carroll: A Biography* (New York: Alfred A. Knopf, 1995), 205, 350-52, 367.
102 Charles Dodgson, "Some Popular Fallacies about Vivisection," *Fortnightly Review* (June 1875): 854.
103 Wilkie Collins, *Heart and Science: A Story of the Present Time* (Toronto: Rose, 1883), preface and ch. 23.
104 Quoted in Kean, *Animal Rights,* 103, citing Frances Power Cobbe, *The Right of Tormenting: A Meeting of the Scottish Society for the Total Suppression of Vivisection* (London, 1881), 8.
105 Lord Chief Justice Coleridge, "The Nineteenth Century Defenders of Vivisection," *Fortnightly Review* 38 (February 1882): 236.
106 See page 167.
107 *Parliamentary Debates* (London: HMSO), 26 May 1876.
108 See Rod Preece and Lorna Chamberlain, *Animal Welfare and Human Values* (Waterloo: Wilfrid Laurier University Press, 1993), 59.
109 Quoted in Drummond, *The Rights of Animals,* 169.
110 Quoted in Jon Wynne-Tyson, *The Extended Circle: A Commonplace Book of Animal Rights* (1985; reprint, New York: Paragon, 1989), 289, citing John Ruskin, "Speech on Vivisection," Oxford, 9 December 1884. I have changed the quotation as it appears in Wynne-Tyson's text from the past tense to the present tense. It was common in the nineteenth century to report both parliamentary and public proceedings in the past tense.
111 Robert Browning, *The Poetical Works of Robert Browning,* vol. 2 (London: Smith, Elder, 1896), 596, 753.
112 Quoted in Donald Thomas, *Robert Browning: A Life within Life* (London: Weidenfeld and Nicholson, 1982), 265.
113 See Jan March, *Christina Rossetti: A Literary Biography* (London: Pimlico, 1994), 433ff.
114 Christina Rossetti, *The Complete Poems of Christina Rossetti,* vol. 2, ed. R.W. Crump (Baton Rouge: Louisiana State University Press, 1986), 43-44.
115 Quoted in Vyvyan, *In Pity and in Anger,* 162, citing Gladstone Papers in British Museum.
116 Quoted in Wynne-Tyson, *The Extended Circle,* 196, citing Henry Edward Manning, Speech of 21 June 1882.
117 Quoted in Andrew Linzey and Tom Regan, eds., *Animals and Christianity: A Book of Readings* (London: SPCK, 1989), 165-66, citing Henry Edward Manning, Speech of 9 March 1887, contained in the National Anti-Vivisection Society and Catholic Study Circle for Animal Welfare, *Speeches against Vivisection,* undated leaflet.
118 Quoted in Buckner, *The Immortality of Animals,* 62-63.
119 Ibid., 191, 192.
120 Ibid., 280-81.
121 Ibid., 282.
122 Quoted in Wynne-Tyson, *The Extended Circle,* 400-1.
123 See pages 333-34.
124 Quoted in Wynne-Tyson, *The Extended Circle,* 392.

CHAPTER 9: THE MORAL STATUS OF ANIMALS

1 As exceptions, among the philosophers, I would mention Bernard E. Rollin, David Degrazia, Mary Midgley, and James Rachels. Among the scientists, David Fraser and Ian J.H. Duncan deserve mention.

2 Bernard E. Rollin, "Animal Production and the New Social Ethic for Animals," in *Food Animal Well-Being* (West Lafayette: Purdue University Office of Agricultural Research Programs, 1993), 3-13 at 6. See also Rollin's *Farm Animal Welfare: Social, Bioethical and Research Issues* (Ames: Iowa State University Press, 1995), 17-18.

3 Aristotle, *Nicomachean Ethics,* ed. and trans. Roger Crisp (Cambridge: Cambridge University Press, 2000), (1141a), bk. 6, ch. 7, 109.

4 Ibid., (1134a, 1136a), bk. 5, ch. 8, 95-96.

5 See pages 222-23.

6 Aristotle, *Nicomachean Ethics,* ed. and trans. Crisp, (1145b), bk. 7, ch. 1, 120.

7 Ibid., (1149b), bk. 7, ch. 6, 130.

8 See page 60.

9 See page 58.

10 See pages 388, 391-92.

11 David Hume, *A Treatise of Human Nature* (1739-40; reprint, ed. L.A. Selby-Bigge and P.H. Nidditch, Oxford: Clarendon Press, 1978), bk. 3, prt. 1, sec. 1, 469.

12 G.E. Moore, *Principia Ethica* (1903; reprint, Cambridge: Cambridge University Press, 1959), 9.

13 Dix Harwood, *Love for Animals and How It Developed in Great Britain* (1928; reprint, ed. Rod Preece and David Fraser, Lampeter: Mellen Animal Rights Library, 2002), 6. Harwood's use of the terms "anthropomorphic" and "anthropocentric" may jar the expectations of the modern reader. In behavioural biology "anthropomorphic" has come to mean attributing human traits to other species, and "anthropocentric" is used by ethicists in reference to moral thinking that considers only human interests; in both cases a degree of error is often implied.

14 Hume, *A Treatise of Human Nature,* bk. 1, prt. 3, sec. 16, 167-68; bk. 2, prt. 2, sec. 12, 397-98 (italics in the original).

15 David Hume, *An Enquiry concerning the Principles of Morals* (1775), prt. 1, sec. 3, 152, in *Enquiries concerning the Human Understanding and concerning the Principles of Morals,* ed. L.A. Selby-Bigge (Oxford: Clarendon Press, 1902), 190-91. Strictly speaking, Hume is not at this point referring to animals per se but to an imaginary "species of creatures intermingled with men." However, in the next paragraph, he adds: "This is plainly the situation of men, with regard to beasts."

16 See A.H. Basson, *David Hume* (Harmondsworth: Penguin, 1958), 86-112.

17 Charles Taylor, "Neutrality in Political Science," in *Philosophy, Politics and Society,* 3rd series, ed. Peter Laslett and W.G. Runciman (Oxford: Basil Blackwell, 1967), 55, 54.

18 Aristotle, *Nicomachean Ethics,* ed. and trans. Crisp, (1141a/b), bk. 6, ch. 7, 109.

19 William Blake, *Visions of the Daughters of Albion,* line 10, in *Blake: Complete Writings,* ed. Geoffrey Keynes (London: Oxford University Press, 1966), plate 8, 195.

20 Albert Schweitzer, "Feeling for Animal Life," in *A Treasury of Albert Schweitzer,* ed. Thomas Kiernan (1965; reprint, New York: Grammercy Books, 1994), 17.

21 Jeremy Bentham, *The Principles of Penal Law,* in *The Works of Jeremy Bentham,* vol. 10 (1843; reprint, New York: Russell, 1962), 549.

22 Quoted in W.E.H. Lecky, *History of European Morals from Augustus to Charlemagne,* vol. 1 (New York: D. Appleton, 1875), 47.

23 See, for example, D. Hart and M.P. Karmel, "Self-Awareness and Self-Knowledge in Humans, Apes and Monkeys," in *Reaching into Thought: The Minds of the Great Apes*, ed. A.E. Russon, K.A. Bard, and S.T. Parker (Cambridge: Cambridge University Press, 1997), 325-47.

24 John Stuart Mill, *Utilitarianism*, in *Utilitarianism, Liberty and Representative Government* (London: J.M. Dent, 1909), 10.

25 John Stuart Mill, "Three Essays on Religion," in *Essays on Ethics, Religion and Society*, in *Works of John Stuart Mill*, vol. 10 (Toronto: University of Toronto Press, 1965), 187.

26 Peter Singer, *Animal Liberation*, 2nd ed. (New York: New York Review of Books, 1990), 171.

27 Ibid., 15.

28 Singer, *Animal Liberation*, 2.

29 See page 54-55.

30 Tom Regan, *The Case for Animal Rights* (Berkeley: University of California Press, 1983), 278, 243.

31 There is much I would like to add to this discussion on the conflict between animal and environmental interests as well as on such issues as culling. I have, however, dealt with them before in Rod Preece and Lorna Chamberlain, *Animal Welfare and Human Values* (Waterloo: Wilfrid Laurier University Press, 1993). And while I would take a much stronger stance on these issues today than I did in 1993, the criteria I employed there are essentially those that I would employ today.

32 Aristotle, *Nicomachean Ethics*, ed. and trans. Crisp, (1179a/b), bk. 10, ch. 9, 199.

33 Aristotle, *Nicomachean Ethics*, trans. Martin Oswald (Indianapolis: Bobbs-Merrill, 1962), 5.

34 In *Beast and Man: The Roots of Human Nature* (1979; reprint, London: Routledge, 1995), Mary Midgley has claimed that "the intuitionists' idea that *no* facts are relevant, that there is always a 'naturalistic fallacy' involved in reasoning from them, is quite incoherent" (149 n. 11). This may be a sound critique of G.E. Moore and his followers, but I am not convinced that it applies to Aristotle's "intuitionism" – to the extent that this is the appropriate term. I imagine that Aristotle regarded *nous* as capable of intuiting what natural facts are relevant.

35 See ibid., 250 n. 11: Aristotle "understands morality as the expression of natural human needs."

36 Aquinas, *On the Governance of Rulers*, trans. Gerald Phelan (New York: Sheed and Ward, 1938), 34-35.

37 Aristotle, *Nicomachean Ethics*, ed. and trans. Crisp, (1142a), bk. 6, ch. 8, 112.

38 Ibid., (1131a), bk. 5, ch. 3, 86.

39 Quoted by Philip J. Ivanhoe, *Ethics in the Confucian Tradition: The Thoughts of Mencius and Yang-Ming* (Atlanta: Scholar's Press, 1990), 13, citing Mencius, *Analects*, Harvard-Yenching Institutes Sinological Index Series (HYSIS) no. 17: 54/7A/45.

40 Aristotle, *Nicomachean Ethics*, ed. and trans. Crisp, (1132b/1133a), bk. 5, ch. 5, 89.

41 Jeremy Bentham, *Fragment on Government and an Introduction to the Principles of Morals and Legislation* (1789; reprint, ed. Wilfred Harrison, Oxford: Basil Blackwell, 1967), 126.

42 See, for example, Thomas Nagel, *Equality and Partiality* (New York: Oxford University Press, 1991).

43 See Joseph Butler, *Sermons* (1726; reprint, ed. W.R. Matthews, London: Bell, 1969), especially Sermon 5, sec. 3, as quoted in Midgley, *Beast and Man*, 261.

44 As I indicated in the Acknowledgments, I am indebted to Bernard Rollin for drawing

my attention to the significance of Aristotle and Hume for the question of "reasonable partiality."

45 Roger Crisp, Glossary to Aristotle, *Nicomachean Ethics,* ed. and trans. Crisp, 207.

46 Hume, *A Treatise of Human Nature,* bk. 3, prt. 2, sec. 2, 488-89.

47 W.E.H. Lecky, *The History of European Morals from Augustus to Charlemagne,* vol. 2 (New York: W.D. Appleton, 1869), 185.

48 Aristotle, *Nicomachean Ethics,* ed. and trans. Crisp, (1157b), bk. 8, ch. 4, 149.

49 Ibid., (1158a), bk. 8, ch. 6, 150.

50 Ibid., (1166b), bk. 9, ch. 5, 171.

51 Ibid., (1158b), bk. 8, ch. 7, 152.

52 Quoted in Midgley, *Beast and Man,* 339, citing Baruch Spinoza, *Ethica,* prt. 4, proposition 37, no. 1.

53 See S.T. Bindoff, *Tudor England* (Harmondsworth: Penguin, 1950), 15.

54 See Peter Ackroyd, *Dickens* (Toronto: Stewart House, 1990), 72.

55 Priscilla Wakefield, *Instinct Displayed in a Collection of Well-Authenticated Facts Exemplifying the Extraordinary Sagacity of Various Species of the Animal Creation,* 4th ed. (London: Harvey and Darton, 1821), 192.

56 Ibid., 171, vii-ix.

57 Immanuel Kant, "Duties towards Animals and Spirits," in *Lectures on Ethics* ([1780?]; modern edition 1930, trans. Louis Infield; reprint, New York: Harper and Row, 1963), 239-41.

58 Jean-Jacques Rousseau, Preface to *Discourse on the Origins and Foundations of Inequality among Men,* in *Rousseau's Political Writings,* ed. Alan Ritter and Julia Conaway Bondanella, trans. Bondanella (New York: W.W. Norton, 1988), 7.

59 A commendation he managed to ignore in his personal life.

60 Jean-Jacques Rousseau, *Emile, or On Education* (1762; reprint, trans. Allan Bloom, New York: Basic Books, 1979), 55.

61 Angus Taylor, *Animals and Ethics: An Overview of the Philosophical Debate* (Peterborough: Broadview Press, 2003), 29-30.

62 Compare John Stuart Mill, on page 371.

Select Bibliography

Ackerley, J.R. *My Dog Tulip*. 1956. Reprint, New York: Poseidon, 1990.

Ackroyd, Peter. *Dickens*. Toronto: Stewart House, 1991.

Adams, Carol. *The Sexual Politics of Meat: A Feminist-Vegetarian Critical Theory*. 1990. Reprint, New York: Continuum, 2002.

Agassiz, Louis. *Contributions to the Natural History of the United States*. Boston: Little, Brown, 1857.

Agius, Dom Ambrose. *God's Animals*. London: Catholic Study Circle for Animal Welfare, 1973.

Akers, Keith. *The Lost Religion of Jesus: Simple Living and Nonviolence in Early Christianity*. New York: Lantern, 2000.

Alger, William Rounseville. *A Critical History of the Doctrine of a Future Life*. 1860. Reprint, Philadelphia: George W. Childs, 1864.

–. *The Destiny of the Soul. A Critical History of the Doctrine of a Future Life*. 10th ed. Philadelphia: George W. Childs, 1878.

Angermeyer, Johanna. *My Father's Island: A Galapagos Quest*. London: Viking, 1989.

Aquinas. *On the Governance of Rulers*. Translated by Gerald Phelan. New York: Sheed and Ward, 1938.

–. *Summa contra Gentiles*. Translated by Charles J. O'Neil. Grand Bend: Notre Dame University Press, 1975.

Aristophanes. *Aristophanes: Four Comedies*. Translated by Dudley Fitts. [1965?] Reprint, New York: Harvest Books, 2003.

Aristotle. *On Man in the Universe: Metaphysics, Parts of Animals, Ethics, Politics, Poetics*. Edited by Louis Ropes Loomis. 1943. Reprint, New York: Grammercy, 1971.

–. *Nicomachean Ethics*. Translated by Martin Oswald. Indianapolis: Bobbs-Merrill, 1962.

–. *Nicomachean Ethics*. Edited and translated by Roger Crisp. Cambridge: Cambridge University Press, 2000.

–. *The Politics of Aristotle*. Edited and translated by Ernest Barker. London: Oxford University Press, 1952.

Arnobius. *The Seven Books of Arnobius Adversus Gentes*. Translated by Hamilton Bruce and Hugh Campbell. In *Ante-Nicene Christian Library: Translations of the Fathers down to AD 325*. Edinburgh: T.T. Clark, 1871.

Athanasius. *Contra Gentes and De Incarnatione.* Translated by Robert W. Thomson. Oxford: Clarendon Press, 1971.

–. *The Life of St. Anthony the Great.* Willis, CA: Eastern Orthodox Books, n.d.

Atkins, Stuart. "On Goethe's Classicism." In *Goethe Proceedings: Essays Commemorating the Goethe Sesquicentennial at the University of California, Davis.* Columbia, CA: Camden House, 1984.

Augustine. *Concerning the City of God against the Pagans.* 4th ed. Translated by Henry Bettenson. Harmondsworth: Penguin, 1980.

–. *The Confessions of St. Augustine.* London: Longman's Green, 1897.

–. *Of the Morals of the Catholic Church, Nicene and Post Nicene Fathers of the Christian Church.* Grand Rapids, MI: E.B. Erdeman, [1995?].

Bacon, Francis. *Essays, or Councils, Civil and Moral of Sir Francis Bacon.* 1597. Reprint, London: R. Chiswell, 1706.

–. *The Philosophical Works of Francis Bacon.* Edited by John M. Robertson. 1905. Reprint, Freeport, NY: Books for Libraries Press, 1970.

Bailey, D. Shackleton, trans. *Cicero's Letters to His Friends.* London: Penguin, 1967.

Balzac, Honoré de. *Le Père Goriot.* London: Daily Telegraph, [1888?].

Barker, Ernest. *Greek Political Theory: Plato and His Predecessors.* 1918. Reprint, London: Methuen, 1957.

Barnard, Frederick M., ed. and trans. *Herder on Social and Political Culture.* Cambridge: Cambridge University Press, 1969.

Barnes, Jonathan. *The Presocratic Philosophers.* London: Routledge and Kegan Paul, 1979.

–, ed. and trans. *Early Greek Philosophy.* London: Penguin, 1987.

Basson, A.H. *David Hume.* Harmondsworth: Penguin, 1958.

Bate, Jackson. *Samuel Johnson.* San Diego: Harcourt, Brace, 1979.

Baudelaire, Charles. *Selected Poems from "Flowers of Evil."* Translated by Wallace Fowlie. 1857. Reprint, New York: Dover Publications, 1992.

Baxter, Ron. *Bestiaries and Their Users in the Middle Ages.* London: Sutton Publishing/Courtauld Institute, 1998.

Bayle, Pierre. *Historical and Critical Dictionary: Selections.* Translated by Richard H. Popkin. Indianapolis: Bobbs-Merrill, 1965.

Bellow, Saul. *The Bellarosa Connection.* New York: Penguin, 1989.

–. *Collected Stories.* New York: Viking, 2000.

Bentham, Jeremy. *An Introduction to the Principles of Morals and Legislation.* 1789. Reprint, edited by J.H. Burns and H.L.A. Hart, London: Methuen, 1982.

–. *The Works of Jeremy Bentham.* Vol. 10. 1843. Reprint, New York: Russell, 1962.

Bielschowsky, Albert. *The Life of Goethe.* 1908. Reprint, New York: AMS Press, 1970.

Bindoff, S.T. *Tudor England.* Harmondsworth: Penguin, 1950.

Blake, William. *Blake: Complete Writings.* Edited by Geoffrey Keynes. London: Oxford University Press, 1966.

Blumenbach, Johann Friedrich. "Über die Liebe der Thiere." *Göttingische Magazin der Wissenschaften und Literatur,* 1781.

Boas, George. *Essays on Primitivism and Related Ideas in the Middle Ages.* 1948. Reprint, Baltimore, MD: Johns Hopkins University Press, 1997.

–. *The Happy Beast in French Thought of the Seventeenth Century.* 1933. Reprint, New York: Octagon Books, 1966.

Bolingbroke, Lord. *The Works of Henry St. John, Lord Viscount Bolingbroke.* Vol. 5. 1754. Reprint, London: D. Mallett, 1809.

Bonaventure, St. *Bonaventure: The Soul's Journey into God, the Tree of Life and the Life of St. Francis.* Mahwah, NJ: Paulist Press, 1978.

Boorstin, Daniel J. *The Creators: A History of Heroes of the Imagination.* New York: Vintage Books, 1993.

–. *The Discoverers: A History of Man's Search to Know His World and Himself.* New York: Vintage Books, 1985.

Boswell, James. *The Life of Samuel Johnson LL.D.* 1791. Reprint, London: J.M. Dent, 1906.

Boullier, David-Renaud. *Essai philosophique sur l'Âme des Bêtes, où l'on a traité de son existence & de sa nature. Et où l'on mêle par occasion Diverses Reflexions sur la Liberté, sur celle de nos sensations, sur l'Union de l'Âme et du Corps, sur l'immortalité de l'Âme &c. Et où l'on réfute diverses Objections de M. Bayle.* Amsterdam: F. Changuion, 1728.

Bourdy, F. "La saignée chez le cheval dans l'Antiquité tardive." *Revue de médecine vétérinaire* 139, 12 (1988): 1181-84.

Bowler, Peter J. *The Norton History of the Environmental Sciences.* New York: W.W. Norton, 1993.

Boyle, Nicholas. *Goethe: The Poet and the Age.* Vol. 1, *The Poetry of Desire, 1749-90.* Oxford: Oxford University Press, 1992.

–. *Goethe: The Poet and the Age.* Vol. 2, *Revolution and Renunciation, 1790-1803.* Oxford: Clarendon Press, 2000.

Brontë, Anne. *Agnes Grey.* 1847. Reprint, London: Penguin, 1988.

Brontë, Emily. *Wuthering Heights.* 1847. Reprint, Franklin Center: Franklin, 1979.

Brown, Frederick. *Zola: A Life.* New York: Farrar, Strauss, Giroux, 1995.

Brown, Joanna Cullen. *Let Me Enjoy the Earth: Thomas Hardy and Nature.* London: Allison and Busby, 1990.

Brown, John. *Self-Interpreting Bible.* 1776. Reprint, Glasgow: Blackie, 1834.

Browning, Robert. *The Poetical Works of Robert Browning.* London: Smith, Elder, 1896.

Bruno, Giordano. *Cause, Principle, and Unity: Five Dialogues.* Translated by Jack Lindsay. New York: International Publishers, 1962.

Buckner, Elijah. *The Immortality of Animals, and the Relation of Man as Guardian, from a Biblical and Philosophical Hypothesis.* Philadelphia: George W. Jacobs, 1903.

–. *The Immortality of Animals.* 1903. Edited, introduced, annotated by Rod Preece. Lampeter: Mellen Animal Rights Library, 2003.

Burenhult, Göran, ed. *Traditional Peoples Today: Continuity and Change in the Modern World.* San Francisco: Harper, 1994.

Burkhardt, Jacob. *Civilization of the Renaissance in Italy.* 1860. Reprint, London: Phaidon, 1960.

Butler, Joseph. *The Analogy of Religion, Natural and Revealed, to the Constitution and Course of Nature.* 1736. Reprint, London: Longman, 1834.

Butler, Samuel. *Erewhon.* 2nd ed. London: Page and Company, 1907.

Campbell, Joseph. *Historical Atlas of World Mythology.* Vol. 1, *The Way of the Animal Powers.* New York: Harper and Row, 1988.

–. *Historical Atlas of World Mythology.* Vol. 2, *The Way of the Seeded Earth.* New York: Harper and Row, 1989.

Carpenter, Edward. *Towards Democracy.* 1883. Reprint, London: Unwin Brothers, 1926.

Carter, H.E. "The Veterinary Profession and the RSPCA: The First Fifty Years." *Veterinary History* n.s. 6, 2 (1989/90).

Cavendish, Margaret. *Poems and Fancies: Written by the Right Honourable the Lady Newcastle*. London: J. Martin and J. Allestrye, 1653.

Celano, Thomas of. *First Life of St. Francis*. In *St. Francis of Assisi: Writings and Early Biographies*. Edited by Marion A. Habig. Chicago: Franciscan Herald Press, 1983.

Chalmers, Alex, ed. *The Spectator: A New Edition Corrected from the Originals*. New York: Sargent and Ward, 1810.

Chandler, Alice. *A Dream of Order: The Medieval Ideal in Nineteenth Century Literature*. London: Routledge and Kegan Paul, 1971.

Chapple, Christopher Key. *Nonviolence to Animals, Earth and Self in Asian Traditions*. Albany: State University of New York Press, 1993.

Chaucer, Geoffrey. *The Canterbury Tales*. Translated by Nevill Coghill. London: Cresset Press, 1986.

Clare, John. *The Early Poems of John Clare*. Edited by Eric Robinson and David Powell. Oxford: Clarendon Press, 1989.

Clodd, Edward. *Pioneers of Evolution: From Thales to Huxley, with an Intermediate Chapter on the Causes of Arrest of the Movement*. Kila, MT: Kessinger, [1907?].

Coleridge, Lord Chief Justice John. "The Nineteenth Century Defenders of Vivisection." *Fortnightly Review* 38 (February 1882).

Coleridge, Samuel Taylor. *Collected Letters of Samuel Taylor Coleridge*. Edited by Earl Leslie Griggs. London: Oxford University Press, 1956.

–. *The Collected Works of Samuel Taylor Coleridge*. Vol. 6. Edited by R.J. White. London: Routledge and Kegan Paul, 1972.

–. *The Poetical Works of Samuel Taylor Coleridge*. Edited by Ernest Hartley Coleridge. London: Henry Frowde, 1912.

Coleridge, Stephen. *Memories*. London: John Lane, 1913.

–. *Vivisection: A Heartless Science*. London: John Lane, 1916.

Collins, Wilkie. *Heart and Science: A Story of the Present Time*. Toronto: Rose, 1883.

Comte, Auguste. *System of Positive Philosophy, or Treatise on Sociology*. Translated by Henry Dix Hutton. New York: Burt Franklin, [1877?].

Conrad, Joseph. *Nostromo*. 1904. Reprint, London: J.M. Dent, 1995.

Copleston, F.C. *Aquinas*. 1955. Reprint, London: Penguin, 1975.

Coser, Lewis A. *Masters of Sociological Thought: Ideas in Historical and Social Context*. Albany, NY: International Thomson Publishing, 1977.

Cottingham, John. *A Descartes Dictionary*. Oxford: Blackwell, 1993.

Cro, Stelio. *The Noble Savage: Allegory of Freedom*. Waterloo: Wilfrid Laurier University Press, 1990.

Cudworth, Ralph. *True Intellectural System of the Universe*. Vol. 4. London: Richard Royston, 1701 [1678].

Darwin, Charles. *The Descent of Man, and Selection in Relation to Sex*. 2nd ed. New York: A.L. Burt, 1874.

–. *The Origin of Species by Means of Natural Selection, or The Preservation of Favoured Races in the Struggle for Life*. 1859. Reprint, edited by J.W. Burrow, London: Penguin, 1983.

Davis, Simon. *The Archeology of Animals*. New Haven: Yale University Press, 1987.

Denney, A. *Descartes' Philosophical Letters*. Oxford: Clarendon, 1970.

D'Entrèves, A. Passerin. *Natural Law: An Introduction to Legal Philosophy*. 1951. Reprint, London: Hutchinson, 1970.

Descartes, René. *Oeuvres de Descartes*. Vols. 4 and 5. Paris: Charles et Paul Tannery, 1897.

Desmond, Adrian, and James Moore. *Darwin: The Life of a Tormented Evolutionist.* London: Michael Joseph, 1992.

Dickens, Charles. "Inhumane Humanity." *All the Year Round* 15 (17 March 1866).

–. *A Tale of Two Cities.* 1859. Reprint, London: Penguin, 1970.

Dodgson, Charles. "Some Popular Fallacies about Vivisection." *Fortnightly Review* (June 1875).

Dombrowski, Daniel A. *The Philosophy of Vegetarianism.* Amherst: University of Massachusetts Press, 1984.

Doniger, Wendy, ed. *The Laws of Manu.* Harmondsworth: Penguin, 1991.

Dostoevsky, Fyodor. *The Brothers Karamazov.* Translated by Richard Pavear and Larissa Volokhonsky. 1879-80. Reprint, New York: Alfred A. Knopf, 1992.

Drummond, William H. *The Rights of Animals and Man's Obligation to Treat Them with Humanity.* London: John Mardon, 1838.

Dwyer, June. *John Masefield.* New York: Ungar, 1987.

Edmonds, Margot, and Ella E. Clark, eds. *Voices of the Winds: Native American Legends.* New York: Facts on File, 1989.

Eliade, Mircea. *The Myth of the Eternal Return, or Cosmos and History.* Princeton: Princeton University Press, 1974.

Eliot, George. *Middlemarch.* 1871-72. Reprint, New York: Alfred A. Knopf, 1991.

–. *Mill on the Floss.* 1860. Reprint, Franklin Center: Franklin, 1981.

Ellmann, Richard. *Oscar Wilde.* London: Penguin, 1998.

Emerson, Ralph Waldo. *The Selected Writings of Ralph Waldo Emerson.* Edited by Brooks Atkinson. New York: Modern Library, 1992.

Erdoes, Richard, and Alfonso Ortiz, eds. *American Indian Myths and Legends.* 1984. Reprint, London: Pimlico, 1987.

Evans, Edward Payson. *Evolutional Ethics and Animal Psychology.* New York: D. Appleton, 1897.

Evelyn, John. *The Diaries of John Evelyn.* Edited by E.S. de Beer. Oxford: Clarendon Press, 1955.

Ferraro, Giovanni Battista. *Libri quattro: de' quali si tratta delle raze, delle discipline del cavalcare, e di molti altre cose appertinenti a si fatto essercitio.* Campania: Gio. Domeninico Nibio and Francesco Scaglione, 1560.

Fielding, Henry. *The Champion: Containing a Series of Papers, Humorous, Moral, Political, and Critical Edited by Henry Fielding.* Microform.

–. *Joseph Andrews and Shamela.* 1742 and 1749. Reprint, Oxford: Oxford University Press, 1980.

Firth, Florence M., ed. 1905. *The Golden Verses of Pythagoras and Other Pythagorean Fragments.* Reprint, Kila, MT: Kessinger, 1997.

Flaubert, Gustave. *Madame Bovary: A Story of Provincial Life.* 1865. Reprint, New York: Oxford University Press, 1985.

Fox, George. *Gospel Truth Demonstrated in a Collection of Doctrinal Books Given Forth by that Faithful Minister of Jesus Christ, George Fox; Containing Principles Essential to Christianity and Salvation, Held among People Called Quakers.* Philadelphia: Marcus T.C. Gould, 1831.

Fox, Michael Allan. *Deep Vegetarianism.* Philadelphia: Temple University Press, 1999.

Fox, Michael W. *Inhumane Society: The American Way of Exploiting Animals.* New York: St. Martin's, 1990.

Francis of Assisi, Saint. *Francis and Clare: The Complete Works.* Translated by Regis J. Armstrong and Ignatius C. Brady. New York: Paulist Press, 1982.

Frank, Joseph. *Dostoevsky: The Miraculous Years, 1865-71.* Princeton: Princeton University Press, 1995.

Furbank, P.N. *E.M. Forster: A Life.* London: Martin Secker and Warburg, 1978.

Gassendi, Pierre. *The Selected Works of Pierre Gassendi.* Translated by Craig B. Bush. New York: Johnson Reprint Corporation, 1972.

Gay, John. *John Gay: Poetry and Prose.* Edited by Victor A. Dearing. Oxford: Clarendon Press, 1974.

Gierke, Otto. *Natural Law and the Theory of Society, 1500-1800.* 1934. Reprint, Boston: Beacon Press, 1957.

Gifford, William, ed. and trans. *Juvenal's Satires with the Satires of Persius.* 1854. Reprint, London: Dent, 1954.

Gill, James E. "Theriophily in Antiquity: A Supplementary Account." *Journal of the History of Ideas* 30 (July 1969).

Godwin, William. *Life of Geoffrey Chaucer.* 2 vols. London: Richard Phillips, 1803.

Goethe, Johann Wolfgang. *Faust, Part Two.* 1832. Reprint, translated by Philip Wayne, London: Penguin, 1959.

–. *Goethe: Poetical Works.* Translated by John Storer Cobb. Boston: Francis A. Niccolls, 1902.

–. *Goethe: Sämmtliche Werke, nach Epochen seines Schaffens.* Translated by Rod Preece. München: Carl Hanser Verlag, 1986-92.

–. *The Poetical Works of J.W. von Goethe.* Edited by Edwin Hermann Zeydel. Chapel Hill: University of North Carolina Press, 1957.

Gogol, Nikolai. *Dead Souls.* 1842. Reprint, London: Penguin, 1961.

Grigson, Geoffrey, ed. *The Oxford Book of Satirical Verse.* Oxford: Oxford University Press, 1983.

Guenther, Matthias. *Tricksters and Trancers: Bushman Religion and Society.* Bloomington: Indiana University Press, 1999.

Guerber, H.A. *Myths and Legends of the Middle Ages: Their Origins and Influence in Literature and Art.* 1909. Reprint, London: G.G. Harrap, 1926.

Guthrie, Kenneth Sylvan, comp. and trans. *The Pythagorean Sourcebook and Library.* Grand Rapids, MI: Phanes Press, 1988.

Hale, Matthew. *The Primitive Origination of Mankind.* London: n.p., 1677.

Hamilton, Joseph. *Animal Futurity.* Belfast: C. Aitchison, 1877.

Hardy, Thomas. *Jude the Obscure.* 1885. Reprint, Ware: Wordsworth, 1995.

Harris, Marvin. *Cows, Pigs, Wars and Witches: The Riddles of Culture.* 1974. New York: Vintage Books, 1989.

Harwood, Dix. *Love for Animals and How It Developed in Great Britain.* 1928. Reprint, edited by Rod Preece and David Fraser, Lampeter: Mellen Animal Rights Library, 2002.

Hart, D., and M.P. Karmel. "Self-Awareness and Self-Knowledge in Humans, Apes and Monkeys." In *Reaching into Thought: The Minds of the Great Apes.* Edited by A.F. Russon, K.A. Bard, and S.T. Barker. Cambridge: Cambridge University Press, 1997.

Hastings, Hester. *Man and Beast in French Thought of the Eighteenth Century.* 1936. Reprint, New York: Johnson Reprint Corporation, 1973.

Heffer, Simon. *Moral Desperado: A Life of Thomas Carlyle.* London: Phoenix, 1995.

Helps, Arthur. *Animals and Their Masters*. 1872. Reprint, London: Chatto and Windus, 1883.

Hesiod. *Theogony and Works and Days*. Translated by M.L. West. Oxford: Oxford University Press, 1988.

Hiebert, Theodore. *The Yahwist's Landscape: Nature and Religion in Early Israel*. New York: Oxford University Press, 1996.

Hobbes, Thomas. *The Elements of Law – Natural and Politic*. Edited by Ferdinand Tönnies. New York: Barnes and Noble, 1969.

–. *The English Works of Thomas Hobbes of Malmesbury*. Vol. 5. Edited by Sir William Molesworth. London: G. Bohn, 1839.

–. *Leviathan*. Edited by C.B. MacPherson. London: Penguin, 1981.

Holroyd, Michael. *Bernard Shaw*. London: Penguin, 1990.

Homer. *The Odyssey of Homer*. Translated by S.H. Butcher and A. Land. 1909. Reprint, New York: P.F. Collier, 1969.

Honderich, Ted, ed. *Oxford Companion to Philosophy*. Oxford: Oxford University Press, 1995.

Hugo, Victor. *Les Misérables*. 1862. Reprint, translated by Norman Denney, London: Penguin, 1976.

–. *Notre-Dame of Paris*. 1831. Translated by John Sturrock. London: Penguin, 1986 [1978].

Hume, C.W. *The Status of Animals in the Christian Religion*. 1957. Reprint, Potter's Bar: UFAW, 1980.

Hume, David. *Enquiries concerning the Human Understanding and concerning the Principles of Morals*. Edited by L.A. Selby-Bigge. Oxford: Clarendon Press, 1902.

–. *A Treatise of Human Nature*. 1739-40. Reprint, edited by L.A. Selby-Bigge and P.H. Nidditch, Oxford: Clarendon Press, 1978.

Huxley, Aldous. *Island*. 1962. Reprint, London: Flamingo, 1994.

Huxley, Thomas Henry. *Man's Place in Nature and Other Collected Essays*. 1863. Reprint, New York: D. Appleton, 1900.

Isaacs, Ronald. *Animals in Jewish Thought and Tradition*. Northvale, NJ: Jason Aaronson, 2000.

Ivanhoe, Philip J. *Ethics in the Confucian Tradition: The Thoughts of Mencius and Yang-Ming*. Atlanta: Scholar's Press, 1990.

James, Henry. *The Bostonians*. 1886. Reprint, London: Penguin, 1986.

Jenyns, Soames. *A Free Inquiry into the Nature and Origin of Evil*. 2nd ed. 1757. Reprint, New York: Garland, 1976.

John of the Cross, Saint. *The Complete Works of St. John of the Cross*. Translated by E. Allison Peers. 2 vols. London: Burns, Oates, and Washburn, 1935.

Johnson, Paul. *The Birth of the Modern: World Society, 1815-1830*. London: Phoenix, 1991.

–. *Modern Times: The World from the Twenties to the Nineties*. 1991. Reprint, New York: Perennia, 2001.

Joyce, James. *Finnegans Wake*. London: Faber and Faber, 1975.

–. *A Portrait of the Artist as a Young Man*. 1916. Reprint, London: Penguin, 1993.

–. *Ulysses*. 1922. Reprint, London: Penguin, 1992.

Kahn, Charles H. *Pythagoras and the Pythagoreans: A Brief History*. Indianapolis: Hackett, 2001.

Kalechofsky, Roberta, ed. *Judaism and Animal Rights: Classical and Contemporary Responses*. Marblehead, MA: Micah, 1992.

Kant, Immanuel. *Grundlegung zur Metaphysik der Sitten.* 1785. Reprint, Hamburg: Meiner, 1994.

–. *Lectures on Ethics.* [1780?]. Modern edition 1930, translated by Louis Infield. Reprint, New York: Harper and Row, 1963.

Kapp, A. "Das Aussputzen des Halses und der Zähnen, sowie das Aderlassen bei den Pferden, ein Privileg der Leipziger Schmiedegesellen." *Deutsche Schmiedezeitung* 52, 1056 (1936).

Karkeek, William F. "On the Future Existence of the Brute Creation. *The Veterinarian* (1839-40).

Kean, Hilda. *Animal Rights: Political and Social Change in Britain since 1800.* London: Reaktion Books, 1998.

Kellerman, Steven G. "Fish, Flesh, and Foul: The Anti-Vegetarian Animus." *American Scholar* 69, 4 (Autumn 2000): 85-97.

Kennedy, Deborah. *Helen Maria Williams and the Age of Revolution.* Lewisburg: Bucknell University Press, 2002.

Kenyon-Jones, Christine. *Kindred Brutes: Animals in Romantic-Period Writing.* Aldershot: Ashgate, 2001.

King, James. *William Blake: His Life.* London: Weidenfeld and Nicholson, 1991.

Kingdon, Jonathan. *Self-Made Man: Human Evolution from Eden to Extinction?* New York: John Wiley, 1993.

King-Hele, Desmond. *Erasmus Darwin: A Life of Unqualified Achievement.* London: Giles de la Mare, 1999.

Kingsford, Anna Bonus. *Perfect Way: A Treatise Advocating a Return to the Natural and Ancient Food of Our Race.* Kila, MT: Kessinger, [1906?].

Kingsley, Charles. *Yeast.* In *Westward Ho! Hypatia and Yeast.* 1851. Reprint, London: Macmillan, 1890.

Kirby, William, and William Spence. *An Introduction to Entomology, or Elements of Natural History of Insects.* London: Longman, Hurst, Rees, Orme, and Brown, 1822.

Knudtson, Peter, and David Suzuki. *Wisdom of the Elders.* Toronto: Stoddart, 1992.

Kowalski, Gary. *The Souls of Animals.* Walpole, NH: Stillpoint, 1991.

Kropotkin, Peter. *Mutual Aid: A Factor of Evolution.* 1st edition 1902. Second edition 1914. Reprint, Boston: Extending Horizon Books, [1955?].

Lactantius. *The Father of the Church, Lactantius, the Divine Institutes, Books I-VII.* Translated by Francis McDonald. Washington: Catholic University of America Press, 1964.

Lamarck, Jean-Baptiste. *Histoire naturelle des animaux sans vertèbres.* Paris: J.B. Baillière, 1835.

Lamartine, Alphonse de. *Les Confidences.* 1848. Reprint, Paris: Hachette, 1893.

Langland, William. *Piers the Ploughman.* Translated into modern English with an introducion by J.F. Goodridge. Harmondsworth: Penguin, 1959.

Lecky, W.E.H. *A History of European Morals from Augustus to Charlemagne.* New York: D. Appleton, 1869.

Legood, Gilles. "A Brief History of British Animal Welfare." *Ark* 188 (Summer 2001).

Leiss, William. *The Domination of Nature.* 1972. Reprint, Montreal and Kingston: McGill-Queen's University Press, 1984.

Levi, Peter. *Tennyson.* New York: Charles Scribner's Sons, 1993.

Lewis, George Cornewall. *Remarks on the Use and Abuse of Some Political Terms.* 1832. Reprint, Columbia: University of Missouri Press, 1984.

Lindsay, William Lauder. *Mind in the Lower Animals.* 2 vols. London: C. Kegan, Paul, Tench, 1879.

Linzey, Andrew. *Animal Theology.* Urbana: University of Illinois Press, 1995.

–. *Christianity and the Rights of Animals.* New York: Crossroad, 1991.

–, and Tom Regan, eds. *Animals and Christianity: A Book of Readings.* London: SPCK, 1989.

Lischke, Ute. *Lily Braun, 1865-1916: German Writer, Feminist, Socialist.* Rochester, NY: Camden House, 2000.

Livingstone, David N. *Darwin's Forgotten Defenders: The Encounter between Evangelical Theory and Evolutionary Thought.* Edinburgh: Scottish Academic Press, 1987.

Locke, John. *Essay concerning Human Understanding.* 1690. Reprint, London: H. Hills, 1710.

–. *Thoughts concerning Education.* Edited by John Yolton and Jean Yolton. Oxford: Clarendon Press, 1998.

–. *Two Treatises of Government.* 1690. Reprint, edited by Peter Laslett, New York: New American Library, 1965.

Lovejoy, Arthur O. *Essays in the History of Ideas.* New York: George Braziller, 1955.

–. *The Great Chain of Being: A Study in the History of an Idea.* 1933. Reprint, New York: Harper, 1960.

–. "Some Eighteenth-Century Evolutionists." *Popular Science Monthly* (March 1904).

–, and George Boas. *Primitivism and Related Ideas in Antiquity.* 1935. Reprint, Baltimore: Johns Hopkins University Press, 1997.

Lovelock, James. *Gaia: A New Look at Life on Earth.* London: Oxford University Press, 1979.

Lucretius. *On the Nature of the Universe* [De rerum natura]. Translated by R.A. Latham. 1951. Reprint, London: Penguin, 1994.

Lutts, Ralph H. *The Nature Fakers: Wildlife, Science, and Sentiment.* Golden, CO: Fulcrum, 1990.

Macaulay, Thomas Babington. *History of England.* London: Longman, 1854.

Macdonald, George. *Hope of the Universe.* London: Ward, Lock, Bowden, 1892.

Mack, Maynard. *Alexander Pope: A Life.* New York: W.W. Norton, 1988.

Maclean, Hope. "Species Reasoning." *Canadian Forum* 72 (1993).

McLellan, David, ed. *Karl Marx: Selected Writings.* Oxford: Oxford University Press, 1977.

McManners, John, ed. *The Oxford History of Christianity.* Oxford: Oxford University Press, 1993.

Maitland, Edward. *Anna Kingsford: Her Life, Letters, Diary and Work.* 3rd ed. London: G. Redway, 1913.

Malamud, Randy. "Poetic Animals and Animal Souls." *Society and Animals* 6, 3 (1998): 263-67.

Map, Walter. *De Nugis Curialium: Courtiers' Trifles.* Edited and translated by M.R. James. Oxford: Clarendon Press, 1983.

March, Jan. *Christina Rossetti: A Literary Biography.* London: Pimlico, 1994.

Marcuse, Herbert. *One Dimensional Man: Studies in the Ideology of Advanced Industrial Society.* Boston: Beacon Press, 1964.

Marx, Karl, and Friedrich Engels. *Manifest der Kommunistischen Partei.* 1848. Reprint, Berlin: Dietz Verlag, 1958.

Mascaró, Juan, ed. and trans. *Dhammapada.* London: Penguin, 1973.

–, *The Upanishads*. London: Penguin, 1965.

Mason, Jim. *An Unnatural Order: Uncovering the Roots of Our Dominion of Nature and Each Other*. New York: Simon and Schuster, 1993.

Maybury-Lewis, David. *Millennium: Tribal Wisdom of the Modern World*. New York: Viking, 1992.

Mettrie, Julien Offray De La. *Man a Machine*. La Salle: Open Court Publishing, 1953.

Michelet, Jules. *The Bible of Humanity*. Translated by Vincenzo Calfia. New York: J.W. Bouton, 1877.

–. *The Bird*. 1879. Reprint, translated by W.H. Davenport Adams, London: Wildwood House, 1981.

Midgley, Mary. *Animals and Why They Matter*. Athens: University of Georgia Press, 1983.

–. *Beast and Man: The Roots of Human Nature*. 1979. Reprint, London: Routledge, 1995.

–. "Practical Solutions." In *The Status of Animals: Ethics, Education and Welfare*. Edited by David Paterson and Mary Palmer. Oxford: CAB International, 1989.

Mijuskovic, Ben Lazare. *The Achilles of Rationalist Arguments: The Simplicity, Unity and Identity of Thought and Soul from the Cambridge Platonists to Kant: A Study in the History of an Argument*. The Hague: Martinus Nijhoff, 1974.

Mill, John Stuart. *Utilitarianism, Liberty, Representative Government*. London: J.M. Dent, 1910.

–. *Works of John Stuart Mill*. Vol. 10. Toronto: University of Toronto Press, 1965.

Miller, Edwin Haviland. *Salem Is My Dwelling Place: A Life of Nathaniel Hawthorne*. Iowa City: Iowa University Press, 1991.

Milner, Richard. *The Encyclopedia of Evolution: Humanity's Search for Its Origins*. New York: Henry Holt, 1990.

Molière. *Comedies*. Translated by Donald M. Frame. New York: Oxford University Press, 1985.

Montaigne, Michel de. *Selected Essays*. New York: Oxford University Press, 1983.

Moore, G.E. *Principia Ethica*. 1903. Reprint, Cambridge: Cambridge University Press, 1959.

Moore, J. Howard. *The New Ethics*. London: E. Bell, 1907.

–. *The Universal Kinship*. 1906. Reprint, edited by Charles Magel, Fontwell: Centaur, 1992.

Morris, Desmond. *The Animal Contract*. London: Virgin, 1990.

Morton, H.V. *Through Lands of the Bible*. London: Methuen, 1938.

Moyers, Bill. *Genesis: A Living Conversation*. New York: Doubleday, 1996.

Murray, Robert. *The Cosmic Covenant: Biblical Themes of Justice, Peace and the Integrity of Creation*. London: Sheed and Ward, 1992.

Myers, Jeffrey. *D.H. Lawrence: A Biography*. New York: Vintage Books, 1992.

Naganathan, G. *Animal Welfare: Hindu Scriptural Perspectives*. Washington, DC: Center for Respect of Life and Environment, 1989.

Nagel, Thomas. *Equality and Partiality*. New York: Oxford University Press, 1991.

Nash, Roderick Frazier. *The Rights of Nature: A History of Environmental Ethics*. Madison: University of Wisconsin Press, 1984.

Neidjie, Big Bill. *Speaking for the Earth: Nature's Law and the Aboriginal Way*. Washington, DC: Center for Respect of Life and Environment, 1991.

New Jerusalem Bible. Standard edition. New York: Doubleday, 1998.

Newton, John Frank. *The Return to Nature, or A Defence of the Vegetable Regimen.* London: T. Cadell and W. Davies, 1811.

Nicholson, George. *On the Primeval Diet of Man: Vegetarianism and Human Conduct toward Animals.* 1801. Reprint, edited by Rod Preece, Lampeter: Mellen Animal Rights Library, 1999.

Nisbet, H.B. *Goethe and the Scientific Tradition.* London: University of London Institute of Germanic Studies, 1972.

–. "Lucretius in Eighteenth-Century Germany, with a Commentary on Goethe's 'Metamorphose der Tiere.'" *Modern Language Review* 81 (1986): 97-115.

Noske, Barbara. *Beyond Boundaries: Humans and Animals.* Montreal: Black Rose Books, 1997.

Oakeshott, Michael. *On Human Conduct.* Oxford: Clarendon Press, 1975.

O'Meara, Dominic. *Pythagoras Revived: Mathematics and Philosophy in Late Antiquity.* Oxford: Clarendon Press, 1989.

Osborn, Henry Fairfield. *From the Greeks to Darwin: The Development of the Evolution Idea through Twenty-Four Centuries.* 1929. Reprint, New York: Arno Press, 1975.

Oswald, John. *The Cry of Nature, or An Appeal to Mercy and to Justice on Behalf of the Persecuted Animals.* 1791. Reprint, edited by Jason Hribal, Lampeter: Mellen Animal Rights Library, 2000.

Overton, Richard. *Man's Mortalitie.* 1643. Reprint, edited by Harold Fisch, Liverpool: University of Liverpool Press, 1968.

Ovid. *Metamorphoses.* Translated by Mary M. Innes. London: Penguin, 1955.

Paley, William. *The Works of William Paley, D.D., Archdeacon of Carlisle.* Philadelphia: Crissy Markley, 1850.

Parini, Jay. *John Steinbeck: A Biography.* London: Minerva, 1994.

Parkinson, C. Northcote. *The Evolution of Political Thought.* London: University of London Press, 1958.

Passmore, John. *Man's Responsibility for Nature: Ecological Problems and Western Traditions.* London: Duckworth, 1974.

–. "The Treatment of Animals." *Journal of the History of Ideas* 36 (1975).

Patar, Benoit, ed. *Le traité de l'âme de Jean Bouridan.* Louvain: Éditions de l'I.S.P., 1991.

Plato. *Gorgias.* Edited by E.R. Dodds. Oxford: Clarendon Press, 1959.

–. *The Laws.* Translated by Trevor J. Saunders. Baltimore: Penguin, 1970.

–. *Plato's Epistles.* Translated by Glenn R. Morrow. Indianapolis: Bobbs-Merrill, 1962.

–. *Republic.* Translated by H. Spens. Glasgow: Robert and Andrew Foulis, 1763.

–. *The Republic.* Translated by Richard W. Sterling and William C. Scott. New York: W.W. Norton, 1985.

–. *The Statesman.* In *Plato.* Translated by Harold N. Fowler and W.R.M. Lamb. London: W. Heinemann, 1925.

–. *The Trial and Death of Socrates: Four Dialogues.* Translated by Benjamin Jowett. [1875?]. Reprint, New York: Dover, 1992.

Pliny the Elder. *Natural History: A Selection.* Translated by John F. Healy. London: Penguin, 1991.

–. *Plotinus.* Translated by A.H. Armstrong. London: William Heinemann, 1967.

Plutarch. *Essays.* Translated by Robert Waterfield. London: Penguin, 1972.

–. *Plutarch's Moralia.* Vol. 12. Translated by Harold Cherniss and William C. Helmblod. 1927. Reprint, London: William Heinemann, 1957.

Pope, Alexander. "On Cruelty to the Brute Creation." *Guardian* 61 (21 May 1713).

–. *The Works of Alexander Pope.* Ware: Wordsworth, 1995.

Porphyry. *On Abstinence from Animal Food.* Translated by Thomas Taylor. [1793] Reprint, Fontwell: Centaur, 1965.

Preece, Rod. *Animals and Nature: Cultural Myths, Cultural Realities.* Vancouver: UBC Press, 1999.

–. *Awe for the Tiger, Love for the Lamb: A Chronicle of Sensibility to Animals.* Vancouver: UBC Press; London and New York: Routledge, 2002.

–. "Darwinism, Christianity and the Great Vivisection Debate." *Journal of the History of Ideas* 64, 3 (2003).

–, and Lorna Chamberlain. *Animal Welfare and Human Values.* Waterloo: Wilfrid Laurier University Press, 1993.

–, and David Fraser. "The Status of Animals in Biblical and Christian Thought: A Study in Colliding Values." *Society and Animals: Journal of Human-Animal Studies* 8, 3 (2000): 245-63.

Primatt, Humphry. *The Duty of Mercy and the Sin of Cruelty to Brute Animals.* 1776. Reprint, edited by Richard D. Ryder, Fontwell: Centaur, 1992.

Pufendorf, Samuel. *The Law of Nature and Nations.* 1688. Reprint, translated by C.H. Oldfather and W.A. Oldfather, New York: Oceana, 1931.

Rabelais, François. *Gargantua and Pantagruel.* New York: Alfred A. Knopf, 1994.

Rachels, James. *Created from Animals: The Moral Implications of Darwinism.* Oxford: Oxford University Press, 1991.

Radcliffe, Ann. *The Romance of the Forest.* 1791. Reprint, Oxford: Oxford University Press, 1987.

Ray, John. *The Wisdom of God Manifested in the Works of Creation.* 1691. Reprint, New York: Garland, 1979.

Regan, Tom. *The Case for Animal Rights.* Berkeley: University of California Press, 1983.

–, ed. *Animal Sacrifices: Religious Perspectives on the Use of Animals in Science.* Philadelphia: Temple University Press, 1986.

Regenstein, Lewis G. *Replenish the Earth: A History of Organized Religion's Treatment of Animals and Nature, Including the Bible's Message of Conservation and Kindness toward Animals.* New York: Crossroad, 1991.

Richardson, Robert D. *Emerson: The Mind on Fire.* Berkeley: University of California Press, 1995.

Richardson, Samuel. *Pamela.* 1741. Reprint, New York: W.W. Norton, 1993.

Rieck, W. "126 verschiedene Venaesectiones um 1550." *Vetereinärische Mitteilungen* 9, 3 and 4 (1929): 9-15.

Ritson, Joseph. *An Essay on Abstinence from Animal Food as a Moral Duty.* London: Richard Phillips, 1802.

Ritvo, Harriet. *The Platypus and the Mermaid and Other Figments of the Classifying Imagination.* 1997. Reprint, Cambridge, MA: Harvard University Press, 1998.

Robinet, Jean-Baptiste. *De la Nature.* 4 vols. Amsterdam: Harrevelt, 1763-66.

Rodman, John. "The Liberation of Nature?" *Inquiry* 20 (1972).

Rollin, Bernard E. "Animal Production and the New Social Ethic for Animals." In *Food Animal Well-Being,* 3-13. West Lafayette: Purdue University Office of Agricultural Research Programs, 1993.

–. *Farm Animal Welfare: Social, Bioethical and Research Issues.* Ames: Iowa State University Press, 1995.

Rorarius, Hieronymus. *Quod animalia bruta saepe Ratione utantur melius Homine. Libro* duo: *Qua recensuit Dissertatione Historica-Philosophica de Anima Brutorum.* Annotated by Geor. Hein. Ribovius. Helemstadtii: Chri. Frider. Weygandi, 1728.

Rossetti, Christina. *The Complete Poems of Christina Rossetti.* Edited by R.W. Crump. Baton Rouge: Louisiana State University Press, 1986.

Rouse, W.H.D., trans. *The Great Dialogues of Plato.* New York: Mentor, 1956.

Rousseau, Jean-Jacques. *Discourse on the Origin and Foundations of Inequality among Men.* In *Rousseau's Political Writings.* Edited by Alan Ritter and Julia Conaway Bondanella. Translated by Julia Conaway Bondanella. New York: W.W. Norton, 1988.

–. *Emile, or On Education.* 1762. Reprint, translated by Allan Bloom, New York: Basic Books, 1979.

Ruskin, John. *Unto This Last.* London: G. Allen, 1890.

Russell, W.M.S., and R.L. Burch. *The Principles of Humane Experiment Technique.* London: Methuen, 1959.

Russon, A.E., K.A. Bard, and S.T. Parker, eds. *Reaching into Thought: The Minds of the Great Apes.* Cambridge: Cambridge University Press, 1997.

Ryder, Richard D. *Animal Revolution: Changing Attitudes towards Speciesism.* Oxford: Basil Blackwell, 1989.

–. *Animal Revolution: Changing Attitudes towards Speciesism.* Revised ed. Oxford: Berg, 2000.

Sagan, Carl, and Ann Druyen. *Shadows of Forgotten Ancestors: A Search for Who We Are.* New York: Random House, 1992.

St. Clair, William. *The Godwins and the Shelleys.* New York: W.W. Norton, 1989.

Salisbury, Joyce E. *The Beast Within: Animals in the Middle Ages.* New York: Routledge, 1994.

Salt, Henry S. *Animals Rights' Considered in Relation to Social Progress.* 1892. Reprint, Clarks Summit, PA: Society for Animal Rights, 1980.

Sand, George. *Indiana.* 1831. Reprint, translated by Sylvia Raphael, Oxford: Oxford University Press, 1994.

Saunders, Nicholas J. *Animal Spirits.* Boston: Little, Brown, 1995.

Saurat, Denis. *Milton: Man and Thinker.* 1925. Reprint, New York: Haskell, 1970.

Schaff, Philip, and Henry Wallace, eds. *The Church History of Eusebius.* Grand Rapids, MI: William B. Eerdman, 1961.

Schochet, Elijah Judah. *Animal Life in Jewish Tradition: Attitudes and Relationships.* New York: KTAV, 1984.

Scholtmeijer, Marian. *Animal Victims in Modern Fiction: From Sanctity to Sacrifice.* Toronto: University of Toronto Press, 1993.

Schopenhauer, Arthur. *On the Basis of Morality.* 2nd ed. Translated by E.F.J. Payne. Indianapolis: Bobbs-Merrill, 1965.

Schure, Edouard. *Pythagoras and the Delphic Mysteries.* [1907?]. Reprint, Kila, MT: Kessinger, n.d.

Schwartz, Richard H. "Tsa'ar Ba'alei Chayim: Judaism and Compassion for Animals." In *Judaism and Animal Rights: Classical and Contemporary Responses.* Edited by Roberta Kalechofsky. Marblehead, MA: Micah Publications, 1992.

Schweitzer, Albert. "Feeling for Animal Life." In *A Treasury of Albert Schweitzer.* Edited by Thomas Kiernan. 1965. Reprint, New York: Grammercy Books, 1994.

Scott, Walter. *Waverley Novels.* Vol. 9, *Ivanhoe.* Edinburgh: Robert Cadell, 1848.

Scully, Matthew. *Dominion: The Power of Man, the Suffering of Animals, and the Call to Mercy.* New York: St. Martin's, 2002.

Seidlmayer, Michael. *Currents of Medieval Thought with Special Reference to Germany.* Oxford: Basil Blackwell, 1960.

Seneca. *Ad Lucilium Epistolae Morales.* Translated by Richard M. Gummere. New York: G.P. Putnam's Sons, 1925.

Serpell, James. *In the Company of Animals: A Study of Human-Animal Relationships.* Oxford: Oxford University Press, 1983.

Seton, Ernest Thompson. *Wild Animals I Have Known.* 1898. Reprint, London: Penguin, 1987.

Seward, Anna. *The Poetical Works of Anna Seward, with Extracts from Her Literary Correspondence.* Edited by Walter Scott. Edinburgh: Ballantyne, 1810.

Sextus Empiricus. *Sextus Empiricus.* Translated by R.G. Bury. London: William Heinemann, 1958.

Seymour-Smith, Martin. *Hardy.* London: Bloomsbury, 1995.

Sharp, Granville. *A Tract on the Law of Nature and Principles of Action in Man.* London: B. White and E. and C. Dilly, 1777.

Shaw, George Bernard. *The Doctor's Dilemma, Getting Married and the Shewing-Up of Blanco Posnet by Bernard Shaw.* London: Constable, 1911.

Shelden, Michael. *Orwell: The Authorized Biography.* London: Minerva, 1991.

Shelley, Mary. *The Last Man.* 1826. Reprint, New York: Bantam Books, 1994.

Shelley, Percy Bysshe. *The Complete Works of Percy Bysshe Shelley.* Vol. 6. Edited by Roger Ingpen and Walter E. Peck. New York: Gordian Press, 1965.

–. *The Works of P.B. Shelley.* Ware: Wordsworth, 1994.

Shirer, William L. *Love and Hatred: The Stormy Marriage of Leo and Sofya Tolstoy.* New York: Simon and Schuster, 1994.

Sholokhov, Mikhail. *And Quietly Flows the Don.* 1929. Reprint, translated by Stephen Garry, London: Penguin, 1967.

Siam, Kenneth. *Fourteenth Century Verse and Prose.* 1921. Reprint, Oxford: Oxford University Press, 1985.

Sigmund, Paul. *Natural Law in Political Thought.* Cambridge, MA: Winthrop Publishers, 1971.

Singer, Isaac Bashevis. *Law and Exile: An Autobiographical Trilogy.* New York: Farrar, Strauss and Giroux, 1997.

Singer, Peter. *Animal Liberation.* 2nd ed. New York: New York Review of Books, 1990.

Smart, Christopher. *Jubilate Agno.* 1760. Reprint, edited by W.H. Bond, London: Rupert Hart Davis, 1954.

Sorabji, Richard. *Animal Minds and Human Morals: The Origins of the Western Debate.* Ithaca: Cornell, 1993.

Southey, Robert. *Poems.* Bristol: Joseph Cottle, 1797.

Spencer, Colin. *The Heretic's Feast: A History of Vegetarianism.* London: Fourth Estate, 1993.

Spiegel, Marjorie. *The Dreaded Comparison: Human and Animal Slavery.* New York: Mirror Books, 1996.

Spooner, W.C. "On the Non-Immortality of Animals." *The Veterinarian* (March 1840).

Steinbeck, John. *Cannery Row.* 1945. Reprint, London: Penguin, 1994.

–. *The Log from the Sea of Cortez.* 1941. Reprint, London: Penguin, 1995.

Steiner, Gary. "Descartes on the Moral Status of Animals." *Archiv für Geschichte der Philosophie* 80, 2 (1998): 268-91.

Swift, Jonathan. *Swift: Poetical Works*. Edited by Herbert Davis. London: Oxford University Press, 1967.

Tacitus. *Annals*. [AD 100?]. Reprint, Franklin Center, PA: Franklin, 1982.

Tannenbaum, Donald, and David Schultz. *Inventors of Ideas: An Introduction to Western Political Philosophy*. Belmont, CA: Wadsworth/Thomson, 2004.

Taylor, Angus. *Magpies, Monkeys and Morals: What Philosophers Say about Animal Liberation*. Peterborough: Broadview Press, 1999.

Taylor, Charles. "Neutrality in Political Science." In *Philosophy, Politics and Society*. 3rd series. Edited by Peter Laslett and W.G. Runciman. Oxford: Basil Blackwell, 1967.

Thackeray, William. *Vanity Fair: A Novel without a Hero*. 1848. Reprint, London: Bradbury and Evans, 1854.

Thomas, Donald. *Robert Browning: A Life within Life*. London: Weidenfeld and Nicholson, 1982.

Thomas, Keith. *Man and the Natural World: Changing Attitudes in England, 1500-1800*. London: Penguin, 1984.

Thomson, James. *The Seasons and The Castle of Indolence*. Edited by James Sambrook. 1972. Reprint, New York: Oxford University Press, 1991.

Thoreau, Henry David. *The Annotated Walden: Walden, or Life in the Woods by Henry D. Thoreau, together with "Civil Disobedience."* Edited by Philip Van Doren Stern. New York: Bantam Books, 1970.

The Tibetan Book of the Dead. Boston: Shambhala, 1992.

Tolstoy, Leo. *A Confession and Other Religious Writings*. London: Penguin, 1987.

–. *Resurrection*. London: Penguin, 1966.

–. *What Is Religion, and of What Does Its Essence Consist?* London: Penguin, 1997.

Torrey, Bradford. *A Rambler's Leave*. Boston: Houghton Mifflin, 1889.

Trimmer, Sarah. *The History of the Robins, with Twenty-Four Illustrations from Drawings by Harrison Weir*. London: Griffith and Farrar, n.d.

Trinkaus, Eric, and Pat Shipman. *The Neandertals: Changing the Image of Mankind*. New York: Alfred A. Knopf, 1993.

Troyat, Henri. *Tolstoy*. New York: Doubleday, 1967.

Turgenev, Ivan. *Fathers and Sons*. 1861. Reprint, Franklin Center: Franklin, 1985.

Turner, James. *Reckoning with the Beast: Animals, Pain and Humanity in the Victorian Mind*. Baltimore: Johns Hopkins University Press, 1980.

Twain, Mark. "Man's Place in the Animal World." In *Collected Tales: Sketches, Speeches, and Essays, 1891-1910*. Edited by Louis J. Budd. New York: Library of America, 1992.

Vasari, Giorgio. *The Great Masters*. 1550. Reprint, edited by Michael Sonino, translated by Gaston Du C. De Vere, New York: Park Lane, 1998.

Vinci, Leonardo da. *The Literary Works of Leonardo da Vinci*. 2nd ed. Edited by Jean Paul Richter and Irma A. Richter. London: Phaidon, 1970.

–. *Selections from the Notebooks of Leonardo da Vinci*. Edited by Irma A. Richter. New York: Oxford University Press, 1972.

Virgil. *Eclogues*. Translated by Arthur Guy Lee. London: Penguin, 1984.

Voltaire. *Candide and Other Stories*. Translated by Roger Pearson. Oxford: Oxford University Press, 1993.

–. *Dictionnaire philosophique*. 1764. Reprint, Paris: Garnier, 1961.

–. *Traité sur la Tolérance*. 1763. Reprint, trans. Rod Preece, Paris: Flammarion, 1989.

Vyvyan, John. *In Pity and in Anger.* 1969. Reprint, Marblehead, MA: Micah, 1988.

Waddell, Helen. *Beasts and Saints.* 1934. Reprint, London: Constable, 1970.

Wade, Mason. *Margaret Fuller: Whetstone of Genius.* New York: Viking, 1940.

–, ed. and trans. *The Desert Fathers.* New York: Vintage, 1998.

Wakefield, Priscilla. *Instinct Displayed in a Collection of Well-Authenticated Facts Exemplifying the Extraordinary Sagacity of Various Species of the Animal Creation.* 4th ed. London: Harvey and Darton, 1821.

Wallace, Alfred Russel. *World of Life: A Manifestation of Creative Power, Directive Mind and Ultimate Purpose.* 1911. Reprint, New York: Moffat, Yard, 1916.

Wand, Kelly, ed. *The Animal Rights Movement.* San Diego: Greenhaven, 2003.

Ward, Benedicta, ed. and trans. *The Sayings of the Desert Fathers.* Kalamazoo: Cistercian Publications, 1975.

Warnick, Mary, ed. *John Stuart Mill: Utilitarianism, On Liberty, Essay on Bentham.* New York: New American Library, n.d.

The Wars of Alexander: An Alliterative Romance, chiefly from the Historia Alexanderi Magni de Prellis. Early English Tract Society, extra series, 47. 1886. Reprint, New York: Kraus, 1973.

Watson, Burton, ed. and trans. *Lotus Sutra.* New York: Columbia, 1993.

Webb, Stephen. *Good Eating.* Grand Rapids, MI: Brazos Press, 2001.

Wells, G.A. *Goethe and the Development of Science.* Alphen: Sijthoff and Noordhoff, 1978.

West, Martin. "Early Greek Philosophy." In *The Oxford History of the Classical World.* Edited by John Boardman, Jasper Griffin, and Oswyn Murray. Oxford: Oxford University Press, 1993.

White, Gilbert. *The Natural History of Selborne.* 1789. Reprint, London: Folio Press, 1962.

White, Lynn, Jr. "The Historical Roots of Our Ecologic Crisis." *Science* 155 (1967): 1203-7.

Whitney, Lois. *Primitivism and the Idea of Progress in English Popular Literature of the Eighteenth Century.* 1934. Reprint, New York: Octagon Press, 1973.

Williams, Howard. *The Ethics of Diet: A Catena of Authorities Deprecatory of the Practice of Flesh-Eating.* London: F. Pitman, 1883; reprint, Urbana: University of Illinois Press, 2003.

Wilmot, John. *The Poems of John Wilmot, Earl of Rochester.* Edited by Keith Walker. Oxford: Basil Blackwell, 1984.

Winston, David, ed. and trans. *Philo of Alexandria: The Contemplative Life, Giants and Selections.* New York: Paulist Press, 1981.

Wiser, James L. *Political Philosophy: A History of the Search for Order.* Englewood Cliffs, NJ: Prentice Hall, 1983.

Wollstonecraft, Mary. *Mary Wollstonecraft: Political Writings.* 1792. Reprint, edited by Janet Todd, Oxford: Oxford University Press, 1994.

–. *Original Stories from Real Life, with Conversations Calculated to Regulate the Affections and Form the Mind to Truth and Goodness.* London: J. Johnson, 1788.

Wordsworth, William. *The Works of William Wordsworth.* Ware: Wordsworth, 1994.

Wynne-Tyson, Jon. *The Extended Circle: An Anthology of Humane Thought.* Fontwell: Centaur, 1990.

Yeats, W.B. *A Vision.* Rev. ed. London: Macmillan, 1937.

Youatt, William. *The Obligation and Extent of Humanity to Brutes, Principally Considered with Reference to the Domesticated Animals.* 1839. Reprint, edited by Rod Preece, Lampeter: Mellen Animal Rights Library, 2003.

Young, Thomas. *An Essay on Humanity to Animals.* 1798. Reprint, edited by Rod Preece, Lampeter: Mellen Animal Rights Library, 2001.3

Index